W9-CWE-793

Microbial Polysaccharides and Polysaccharases

Special Publications of the Society for General Microbiology

PUBLICATIONS OFFICER: A. G. CALLELY

1. Coryneform Bacteria, eds. I. J. Bousfield & A. G. Callely
2. Adhesion of Microorganisms to Surfaces, eds. D. C. Ellwood, J. Melling & P. Rutter
3. Microbial Polysaccharides and Polysaccharases, eds. R. C. W. Berkeley, G. W. Gooday & D. C. Ellwood

Microbial Polysaccharides and Polysaccharases

Edited by

R. C. W. BERKELEY
Department of Bacteriology
University of Bristol
Bristol, U.K.

G. W. GOODAY
Department of Microbiology
University of Aberdeen
Aberdeen, U.K.

D. C. ELLWOOD
Microbiological Research Establishment
Porton Down
Salisbury, U.K.

1979

Published for the
Society for General Microbiology
by
ACADEMIC PRESS
London New York San Francisco
A subsidiary of Harcourt Brace Jovanovich, Publishers

ACADEMIC PRESS INC. (LONDON) LTD.
24/28 Oval Road,
London NW1

United States Edition published by
ACADEMIC PRESS INC.
111 Fifth Avenue
New York, New York 10003

Library of Congress Catalogue Card Number: 79-50534
ISBN: 0-12-091450-6

Typed by Joan Wilkins Associates Ltd., London, and
printed in Great Britain by Whitstable Litho Ltd, Whitstable, Kent

CONTRIBUTORS

ATKINS, E.D.T. *H.H. Wills Physics Laboratory, University of Bristol, Royal Fort, Tyndall Avenue, Bristol BS8 1TL.*

BACON, J.S.D. *Rowett Research Institute, Bucksburn, Aberdeen AB2 9SB.*

BERKELEY, R.C.W. *Department of Bacteriology, The Medical School, University Walk, University of Bristol, Bristol BS8 1TD.*

BROOKER, B.E. *National Institute for Research in Dairying, Shinfield, Reading RG2 9AT.*

BROWN, D.E. *Department of Chemical Engineering, UMIST, P.O. Box 88, Sackville Street, Manchester M60 1QD.*

BYRDE, R.J.W. *Long Ashton Research Station, Long Ashton, Bristol BS18 9AF.*

CATLEY, B.J. *Department of Brewing and Biological Sciences, Heriot-Watt University, Edinburgh EH1 1HX.*

ELLOWAY, H.F. *H.H. Wills Physics Laboratory, University of Bristol, Royal Fort, Tyndall Avenue, Bristol BS8 1TL.*

ELLWOOD, D.C. *Microbiological Research Establishment, Porton Down, Salisbury, Wiltshire, SP4 OJG.*

ERIKSSON, K-E. *Swedish Forest Products Research Laboratory, Box 5604, S-11486, Stockholm, Sweden.*

EVANS, C.G.T. *Microbiological Research Establishment, Porton Down, Salisbury, Wiltshire, SP4 OJG.*

EVELEIGH, D.E. *Department of Biochemistry and Microbiology, Cook College, Rutgers - The State University of New Jersey, New Brunswick, New Jersey, 08903, USA.*

GABRIEL, A. *Shell Centre, London SE1 7PG.*

GOODAY, G.W. *Department of Microbiology, Marischal College, Aberdeen AB9 1AS.*

HOBSON, P.N. *Rowett Research Institute, Bucksburn, Aberdeen AB2 9SB.*

ISAAC, D.H. *H.H. Wills Physics Laboratory, University of Bristol, Royal Fort, Tyndall Avenue, Bristol BS8 1TL.*

JARMAN, T.R. *Tate & Lyle Ltd., Group Research and Development Laboratory, P.O. Box 68, Reading RG6 2BX.*

MONTENECOURT, B.S. *Department of Biochemistry and Microbiology, Cook College, Rutgers - The State University of New Jersey, New Brunswick, New Jersey, 08903, USA.*

NORMAN, B.E. *Novo Industri A/S, Enzymes Division, Novo Alle 2880 Bagsvoerd, Denmark.*

POOLE, N.J. *Plant Protection (ICI), Jealott's Hill, Bracknell, Berks RG12 6EY.*

POWELL, D.A. *Unilever Research, Colworth Laboratory, Unilever Limited, Colworth House, Sharnbrook, Bedford MK44 1LQ.*

ROGERS, H.J. *National Institute for Medical Research, Mill Hill,*

London NW7 1AA.
SCHAMHART, D.H.J. *Department of Biochemistry and Microbiology, Cook College, Rutgers - the State University of New Jersey, New Brunswick, New Jersey, 08903, USA.*
SUTHERLAND, I.W. *Department of Microbiology, University of Edinburgh, West Mains Road, Edinburgh EH9 3JG.*
WILDISH, D.J. *Fisheries and Marine Service, Biological Station, St. Andrews, New Brunswick, Canada.*
YEO, R.G. *Microbiological Research Establishment, Porton Down, Salisbury, Wiltshire, SP4 OJG.*

PREFACE

Polysaccharides are ubiquitous - they are by far the most abundant biopolymers on earth. They show a wide variety of chemical composition, and we can now begin to understand how these give rise to the wide variety of physical and biochemical properties shown by polysaccharides.

Polysaccharides are renewable resources, and present a wide range of potential products of use to man. Some, such as cellulose, have been prepared and used by man for centuries; others, such as the bacterial xanthans, are recent products of microbial technology. It is amongst the microbial polysaccharides that we can look for the greatest potential production of useful polysaccharides. The techniques and equipment for the selection and industrial mass culture of microbes are at our disposal.

Hand in hand with the production of polysaccharides must go an understanding of their degradation. Microbes are of paramount importance in the enzymic breakdown of polysaccharides. For man this is doubled-edged, with properties such as cellulolysis and amylolysis leading to great commercial benefit or great commercial loss according to circumstances.

This book is an outcome of the 83rd Ordinary Meeting of the Society for General Microbiology, at the University of Aberdeen, September 1978. It combines the Society Symposium, 'The Microbial Degradation of Polysaccharides' and the Microbial Cell Surfaces and Membranes Group Symposium 'Microbial Extracellular Polysaccharides'. Our intention has been to provide a timely compilation of the current academic and industrial achievements in these topics, and to point to future developments.

ACKNOWLEDGEMENTS

We wish to thank Georges Ware, Cyd van Hemel and Helen Parker for their considerable assistance in compiling the index by computer, Renate Majumdar for the excellence of her typing and Judy Sibley and Sue Corbett of Academic Press for their unfailing helpfulness.

CONTENTS

Chapter 1

MICROBIAL EXOPOLYSACCHARIDES: CONTROL OF SYNTHESIS AND ACYLATION

I.W. SUTHERLAND

Department of Microbiology, University of Edinburgh, West Mains Road, Edinburgh, U.K.

Introduction

Microbial extracellular polysaccharides (exopolysaccharides) range in the complexity of their chemical structures from homopolysaccharides with one or more type of linkage to heteropolysaccharides containing several different monosaccharides, some of which may be present in more than one molar equivalent. In addition, various acyl groups may be present, the most common being acetate as O-acetyl groups and pyruvate in the form of ketals attached to the 3 and 4 or 4 and 6 positions of one of the neutral sugar residues, or more rarely, to a uronic acid (Figure 1). Although the gross structure of a number of exopolysaccharides has now been determined, only two of these can currently be considered to be of industrial importance - alginates [see Jarman, Chapter 2] and the exopolysaccharide of the plant pathogen *Xanthomonas campestris* [Jansson *et al.*, 1975; Melton *et al.*, 1976; Evans *et al.*, Chapter 3; Gabriel, Chapter 8]. Detailed structures of several other polysaccharides produced industrially have not yet been reported. Alginates can be considered essentially as homopolymers of D-mannuronic acid subsequently modified by an extracellular epimerase converting some of the D-mannuronic acid residues to L-guluronic acid [Deavin *et al.*, 1977]. Most heteropolysaccharides are however, composed of repeating units varying in size from disaccharides to hexasaccharides or even larger oligosaccharides. In this respect, *Xanthomonas campestris* polysaccharide (Figure 2) resembles a number of other exopolysaccharides, the structures of which have been elucidated and examples of which are shown in Table 1. Where it differs from these other polymers is that the backbone is a cellulose molecule substituted on alternating glucose units by an acylated trisaccharide (or possibly occasionally smaller oligosaccharide) which may also be pyruvylated.

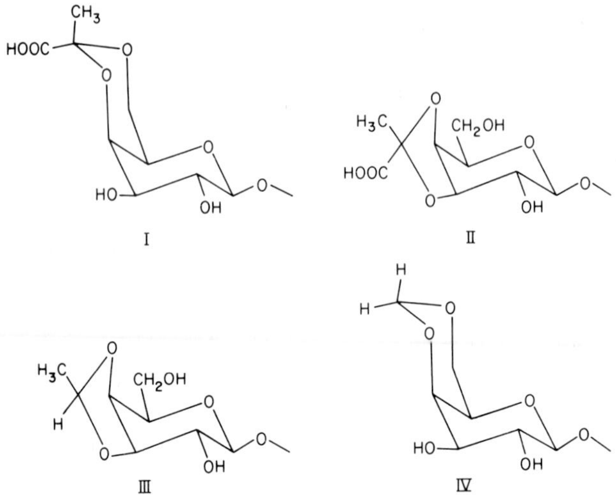

Fig. 1 Ketals attached to the terminal reducing galactose residues of colanic acids from different strains of *Escherichia coli* and *Salmonella typhimurium*. From Garegg *et al*. [1971a, b].

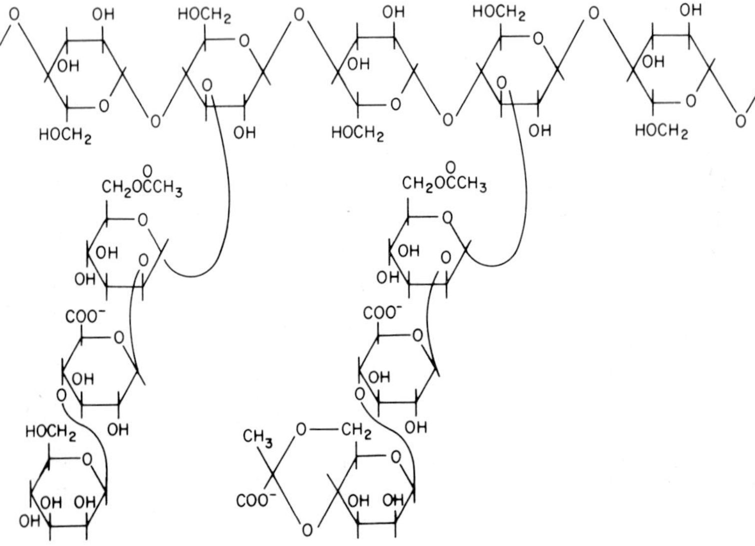

Fig. 2 Structure of extracellular polysaccharide of *Xanthomonas campestris*, according to Jansson *et al*. [1975].

TABLE 1

Repeating unit structures of Klebsiella exopolysaccharides

Type	Components	Structure	Reference
1	Fucose, glucose, glucuronic acid, pyruvate	→3)Glc(1 —β→ 4)GlcA(1 —β→ 4)Fuc(1 —α— GlcA 2,3 — Pyr	Erbing *et al.* [1976]
2	Mannose, glucose, glucuronic acid (various acyl groups)	→4)Glc(1 —α→ 3)Glc(1 —α→ 4)Man(1 ——→ Man 3 ← 1 α-GlcA	Gahan *et al.* [1967] Sutherland [1972]
5	Mannose, glucose, glucuronic acid, pyruvate, acetate	→4)GlcA(1 —β→ 4)Glc(1 —β→ 3)Man(1 —— Glc — Ac; Man 4,6 — Pyr	Dutton and Yang [1973]
7	Mannose, glucose, galactose, glucuronic acid, pyruvate	→3)GlcA(1 —β→ 2)Man(1 → 3)Glc(1 —— GlcA 4 ← 1 Gal; Man 3 ← 1 Gal; Glc 4,6 — Pyr	Dutton *et al.* [1974]
8	Glucose, galactose, glucuronic acid (acetate, pyruvate)	→3)Gal(1 —β→ 3)Glc(1 —α→ 3)Glc(1 —β— Gal ← GlcA	Sutherland [1970]
11	Galactose, glucose, glucuronic acid, pyruvate	→3)Glc(1 —β→ 3)GlcA(1 —β→)Gal(1 —α— GlcA 4 ← 1 Gal Gal 4,6 — Pyr	Thurow *et al.* [1975]
16	Fucose, glucose, glucuronic acid, galactose	→3)-α-D-Glc - (1→4)-β-D-GlcA - (1→4)-α-L-Fuc - (1→ Glc 4 ← 1 β-D- Gal	Chakraborty *et al.* [1977]

TABLE 1 (cont'd)

Type	Components	Structure	Reference
20	Mannose, glucose, galactose, glucuronic acid	$\rightarrow$2)Man(1$\xrightarrow{\alpha}$3)Gal(1$\xrightarrow{\alpha}$ α1 ↑ 3 (on Man) Gal β1 ↑ 3 GlcA	Bebault *et al.* [1973]
21	Mannose, galactose, glucuronic acid, pyruvate	$\rightarrow$3)GlcA(1$\xrightarrow{\alpha}$3)Man(1$\xrightarrow{\alpha}$2)Man(1$\xrightarrow{\alpha}$3)Gal 1$\xrightarrow{\beta}$ α1 ↑ 4 (on GlcA) Gal 4 ‖ 6 Pyr	Choy and Dutton [1973]
24	Mannose, glucose, glucuronic acid	$\rightarrow$4)GlcA(1$\xrightarrow{\alpha}$3)Man(1$\xrightarrow{\alpha}$2)Man(1$\xrightarrow{\alpha}$3)Glc(1$\xrightarrow{\beta}$ 1 ↑ 2 (on GlcA) β-Man	Choy *et al.* [1973]
28	Mannose, glucose, galactose, glucuronic acid	$\rightarrow$2)Gal(1$\xrightarrow{\alpha}$3)Man(1$\xrightarrow{\alpha}$2)Man(1$\xrightarrow{\alpha}$2)Glc(1$\xrightarrow{\beta}$ β1 ↑ 2 (on first Man) GlcA 1 ↑ 3 β-Glc	Curvall *et al.* [1975a]
32	Rhamnose, galactose, pyruvate	$\rightarrow$2)Rha(1$\rightarrow$3)Rha(1$\rightarrow$4)Rha(1$\rightarrow$3)Gal(1- 3 ‖ 4 (on first Rha) Pyr	Bebault *et al.* [1978]

TABLE 1 (cont'd)

36	Rhamnose, glucose, galactose, glucuronic acid, pyruvate	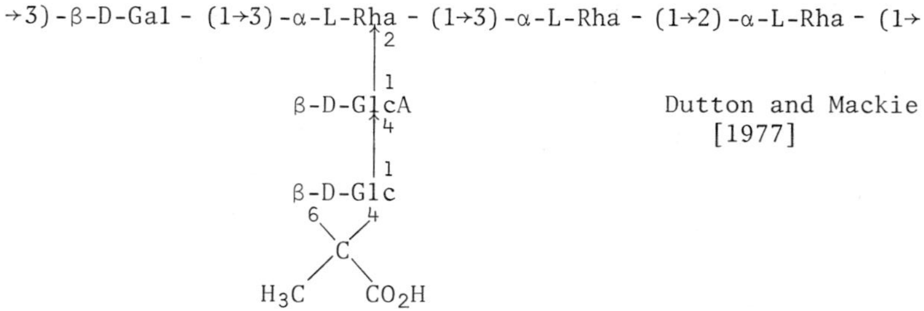 →3)-β-D-Gal-(1→3)-α-L-Rha-(1→3)-α-L-Rha-(1→2)-α-L-Rha-(1→ ↑2 1 β-D-GlcA ↑4 1 β-D-Glc 6 4 C H_3C CO_2H	Dutton and Mackie [1977]
37	Glucose, galactose, glucuronic acid	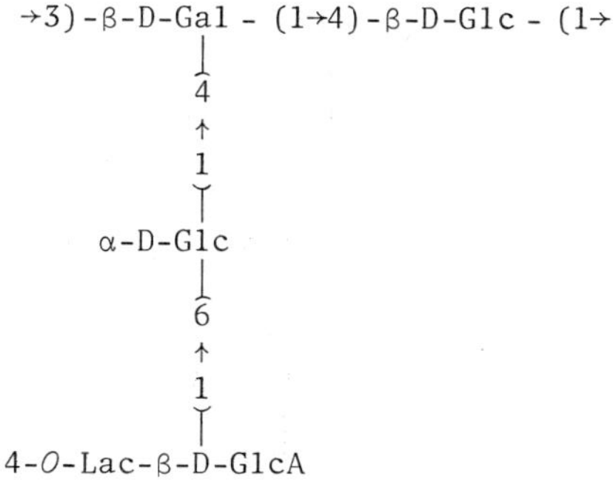 →3)-β-D-Gal-(1→4)-β-D-Glc-(1→ 4 ↑ 1 α-D-Glc 6 ↑ 1 4-*O*-Lac-β-D-GlcA	Lindberg *et al.* [1977]
47	Rhamnose, galactose, glucuronic acid	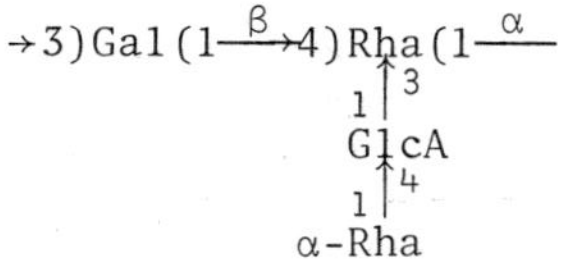 →3)Gal(1 —β→ 4)Rha(1 —α— ↑3 1 GlcA ↑4 1 α-Rha	Bjorndahl *et al.* [1973]

TABLE 1 (cont'd)

Type	Components	Structure	Reference
52	Rhamnose, galactose, glucuronic acid	→4)Rha(1→2)Gal(1— ↑3 1 GlcA ↑4 1 Rha ↑2 1 Gal ↑3 1 Gal	Bjorndahl *et al.* [1973]
54	Fucose, glucose, glucuronic acid (various acyl, no pyruvate)	→6)Glc(1→4)GlcA(1→3)Fuc(1— ↑4 1 β-Glc	Conrad *et al.* [1966] Sutherland [1967, 1970]
56	Rhamnose, glucose, galactose, pyruvate	→3)Glc(1 —β→ 3)Gal(1 —β→ 3)Gal(1 ——→ 3)Gal(1 —α— Glc: ‖ Pyr; Gal: ↑2 α-Rha	Choy and Dutton [1973]
59	Mannose, glucose, galactose, glucuronic acid, acetate	→3)Glc(1 —β→ 3)Gal(1 —β→ 2)Man(1 —α→ 3)Man(1 —β— Gal: ↑4 1 β-GlcA; Man: :6 O.Ac; Man: :6 O.Ac	Lindberg *et al.* [1975]
62	Mannose, glucose, galactose, glucuronic acid	→4)-α-D-Glc - (1→2)-β-D-GlcA - (1→2)-α-D-Man - (1→3)-β-D-Gal - (1→ Man: 3 ↑ 1 α-D-Man	Dutton and Yang [1977]
81	Rhamnose, galactose, glucuronic acid	→2)Rha(1 —α→ 3)Rha(1 —α→ 4)GlcA(1 —β→ 2)Rha(1 —α→ 3)Rha(1 ——→ 3)Gal(1 —β—	Curvall *et al.* [1975b]

Although relatively little direct work on the control and regulation of *Xanthomonas campestris* exopolysaccharide production has been reported in the scientific literature as opposed to confidential company reports, many of the findings made for other systems should be applicable to this and related species. The present paper therefore considers various aspects of control of exopolysaccharide synthesis studied in microbial and particularly bacterial systems and therefore likely to be applicable to those of industrial importance.

Control of Substrate Uptake

The first site at which control mechanisms can take effect in most exopolysaccharide producing microorganisms is that of nutrient uptake. The only exceptions to this are dextran-forming species of bacteria, which effectively achieve an extracellular cell-free synthesis of polysaccharide. As this differs from the production of other exopolysaccharides it will be considered separately.

Many exopolysaccharide-producing microbial species can synthesize some extracellular polymer in the absence of utilizable carbohydrate; if, however, optimal yields of exopolysaccharide are to be obtained, a carbohydrate substrate such as glucose, fructose or sucrose is added to the culture medium. If this substrate is to be effectively utilized by the microbial cells and converted to exopolysaccharide with high efficiency, transport mechanisms must be present in the cell membrane to permit substrate uptake. The carbohydrate may enter the cell through active transport (i.e. permease) systems or through group translocation processes. Both mechanisms require energy, either in the form of ATP and proton translocation in the former system or as phosphoenolypyruvate in the latter. (More detailed accounts of such uptake mechanisms can be found in reviews by Kornberg [1976] and by Saier [1977] and Postma and Roseman [1976], respectively.)

Glucose is commonly used as the carbohydrate substrate for exopolysaccharide synthesis in synthetic or semi-synthetic growth media. If non-carbohydrate components of the media provide precursors for cell material, conversion of the carbohydrate to exopolysaccharide can be high. In *Escherichia coli*, glucose is taken up by the phosphoenolpyruvate (PEP)-dependent phosphotransferase system [Kundig *et al.*, 1964]. Strains differ even in the glucose phosphotransferase, the enzyme being constitutive in one strain studied and inducible in another [Kornberg and Reeves, 1972]. It has been suggested that glucose transport is a rate-limiting step in the growth of *Escherichia coli* on glucose, using a wide range of growth rates [Herbert and Kornberg, 1976]. The strain of *Escherichia coli* which formed an inducible glucose phosphotransferase did not form sufficient enzyme to account for the observed growth rate

[Kornberg and Reeves, 1972] and may even be subject to catabolite repression during growth in media containing excess glucose [Herbert and Kornberg, 1976]. Studies on such mechanisms in exopolysaccharide-producing species are clearly necessary. Glucose is also transported in *Escherichia coli* by one of the seven known systems for galactose transport - the methyl galactoside transport system [Henderson *et al.*, 1977]. Even the control of substrate uptake in a strain of *Escherichia coli* is thus a complex phenomenon and it is probably not valid to extrapolate such results to strains or species in which the modes of carbohydrate transport are still unknown. The phosphotransferase type of system is certainly known to occur widely among bacteria, primarily in obligate and facultative anaerobes, but has not been demonstrated in fungi; even glucose does not always enter the cell by a phosphotransferase system under conditions where other sugars are transported by this type of mechanism [Postma and Roseman, 1976].

Xanthomonas campestris more closely resembles *Pseudomonas* species than *Escherichia coli*, but even in *Pseudomonas aeruginosa* glucose uptake is under several control mechanisms. Glucose may be phosphorylated directly at the expense of ATP, or it may be oxidized to gluconate and 2-oxogluconate then phosphorylated (Figure 3). Glucose uptake can be controlled through repression, effected by the concentration of gluconate, which is in turn formed within the periplasm of *Pseudomonas aeruginosa* cells by a glucose dehydrogenase [Whiting *et al.*, 1976]. In the

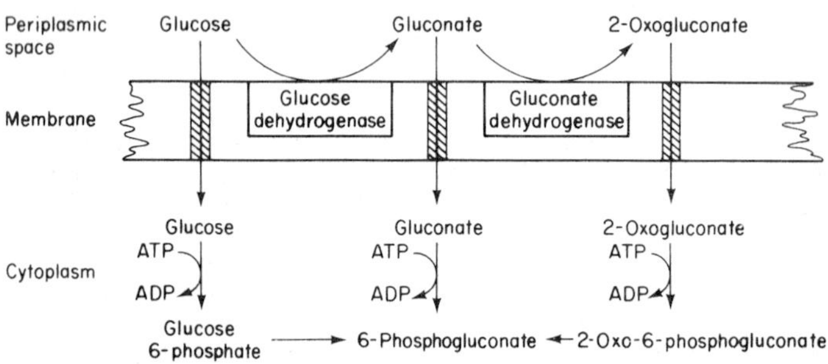

Fig. 3 Extracellular and intracellular pathways of glucose catabolism and related transport systems in *Pseudomonas aeruginosa*. Reproduced from Whiting *et al.* [1976] with the permission of the Journal and the authors.

presence of 30-mM citrate, glucose transport was induced by a concentration of 6-8 mM-glucose but was repressed by 20 mM-glucose. Added gluconate repressed the glucose transport system; this was also true of glucose dehydrogenase deficient mutants. Such mutants lacking endogenous gluconate are free of self-induced repression of glucose uptake at high substrate concentrations and are clearly an example of an 'improvement' which could be introduced into such a strain by mutagenesis. *Xanthomonas campestris* can use a number of carbohydrate substrates for polysaccharide production, but it is still not clear whether the uptake mechanisms are of the same complexity as those found in *Pseudomonas aeruginosa*. A further indication of the similarity between the two species has been the recent finding of a glucose dehydrogenase in a strain of *Xanthomonas campestris* [C. Whitfield, unpublished results].

Control of Polysaccharide Yield and Synthesis

Heteropolysaccharides

The yield of exopolysaccharide from microbial cultures is dependent on a number of different parameters. Some of these were determined several years ago using batch cultures of enteric bacteria. Aerobic culture of these facultatively anaerobic bacteria was essential for polysaccharide production; reduced incubation temperature also favoured synthesis. Highest yields were obtained when carbohydrate was present in excess and growth was limited by the available nitrogen, phosphorus or sulphur source [Duguid and Wilkinson, 1953; Wilkinson *et al.*, 1954]. The ratio of polysaccharide produced/cell nitrogen was highest (48) for nitrogen limitation and lowest (17) for sulphur limitation. The effect of potassium differed from that of other nutrient limitations; little stimulus of polysaccharide production occurred. This apparently anomalous result was elucidated by Dicks and Tempest [1967]. They demonstrated antagonism between K^+ and NH_4^+ uptake under conditions of K^+ limitation. Substrate uptake is also very dependent on the external pH and adequate control of the pH value is necessary both in batch and continuous culture. Results with other microorganisms vary, some species yielding results similar to *Escherichia coli* and *Enterobacter aerogenes* (synonyms: *Aerobacter* or *Klebsiella aerogenes*) while others produced little polysaccharide under these conditions [Sutherland, 1977b].

Recent studies on a pseudomonad producing large amounts of exopolysaccharide indicated that it too produced more polysaccharide in batch culture under nitrogen limitation when excess carbohydrate was present [Williams and Wimpenny, 1977]. The phosphate concentration had little effect on polysaccharide production in adequately buffered media. Yields of polysaccharide and bacteria increased with glucose concentration up to 2% (w/v) but did not increase when the carbohydrate concentration was higher than this. Similar results were obtained when sucrose was used as the growth substrate, but a number of other carbohydrates and non-carbohydrate substrates tested yielded less polysaccharide and smaller amounts of cells. Unlike the enteric bacteria and several other bacteria which have been studied, the pseudomonad produced polysaccharide mainly late in the growth phase and after growth had ceased.

There are obviously various constraints to which polysaccharide production in batch culture is prone and some of these may be removed through the judicious use of continuous culture [see Evans *et al.*, Chapter 3], but this technique has its limitations. For example, the conditions chosen may lead to the development of populations forming less polysaccharide or polysaccharide of different composition from that originally produced. Such problems have been encountered during studies of *Xanthomonas campestris* growth in continuous culture [Silman and Rogovin, 1972]. Thorough checking of the inoculum was also essential as mutants producing less or altered polysaccharide have been observed in *Xanthomonas campestris* stock cultures [Cadmus *et al.*, 1976].

Large scale fermentors have been used to study the characteristics of exopolysaccharide production by *Xanthomonas campestris* [Rogovin *et al.*, 1961]. Under the conditions employed, 2.5-3% (w/v) glucose in the medium yielded most polymer, but the efficiency of conversion of substrate to polymer was highest at 1% (w/v). Little cell material was formed from the added glucose when 'distillers solubles' were used as the growth substrate. Polysaccharide synthesis resembled that observed in enteric bacteria as it occurred throughout the fermentation, conversion of substrate to polymer being relatively constant during growth at optimum conditions [Moraine and Rogovin, 1966]. Unlike enteric species, sub-optimal growth temperatures did not appear to favour polysaccharide production.

The mechanisms of control under conditions yielding optimal amounts of polysaccharide are by no means clear. Substrate is committed to either anabolic processes or to catabolism. The extent to which substrate is converted into polymer suggests that under optimal growth conditions, little is catabolised or converted to cell material other than exopolysaccharide. In growing bacteria, glycogen is seldom formed; this is fortunate for the laboratory using bacteria for exopolysaccharide production which are also capable of glycogen synthesis. Under some conditions, glycogen may form up to 20% of the bacterial dry weight and could represent a considerable economic loss of substrate [Holme, 1957]. Glycogen synthesis is effectively separated from exopolysaccharide production through (i) the lack of involvement of isoprenoid lipid intermediates and (ii) the utilization of ADP-glucose as the glucosyl donor. Control of glycogen synthesis occurs through allosteric regulation of ADP-glucose synthesis [Preiss, 1969] and depends on the concentration of the precursor glucose-1-phosphate. In bacteria which are effectively in the logarithmic phase of growth, glycogen synthesis poses no threat to exopolysaccharide production, but knowledge of the mechanisms of control of synthesis and selection of mutants incapable of forming glycogen could significantly affect polysaccharide production in multistage systems.

As sugar nucleotides are the immediate precursors of the oligosaccharides which form the repeating units of exopolysaccharides, they represent an obvious site of control. UDP-glucose is frequently both a precursor of wall polysaccharides and of extracellular polymers. As well as being a glucosyl donor, it is capable of being epimerized to UDP-galactose or oxidised to UDP-glucuronic acid, both compounds often needed in the formation of exopolysaccharides [e.g. Norval and Sutherland, 1970], (Figure 4).

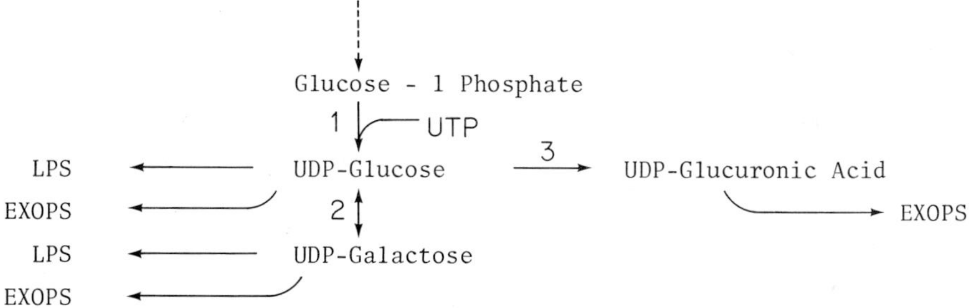

Fig. 4 The role of UDP-glucose in control. 1 UDP glucose pyrophosphorylase; 2 UDP galactose - 4 - epimerase; 3 UDP glucose dehydrogenase; LPS = lipopolysaccharide; EXOPS = exopolysaccharide.

Relatively high levels of UDP-glucose have been found in a number of exopolysaccharide-producing bacterial strains from the Enterobacteriaceae and the amounts in nucleotide pools of mutants incapable of forming exopolysaccharide were similar to those in wild type bacteria [Grant *et al.*, 1970]. Another sugar nucleotide functioning similarly to UDP-glucose is GDP-mannose. Not only is this the direct donor of D-mannose but it can be oxidized to GDP-mannuronic acid or undergo a complex oxidoreduction to GDP-L-fucose. Control over mannose and fucose synthesis has been examined in bacterial strains with polysaccharides containing (i) D-mannose, (ii) L-fucose, and (iii) D-mannose and L-fucose [Kornfeld and Ginsburg, 1966]. In the first group, GDP-mannose synthesis was controlled through the enzyme GDP-mannose pyrophosphorylase, whereas the level of GDP-fucose controlled both GDP-mannose pyrophosphorylase and GDP-mannose hydrolyase in those bacteria which contained only fucose in their polysaccharides. When the microorganisms produced polysaccharides containing both mannose and fucose, the levels of each sugar nucleotide controlled its own synthesis (Figure 5). *Xanthomonas campestris* probably falls into the first category, as fucose is absent from its

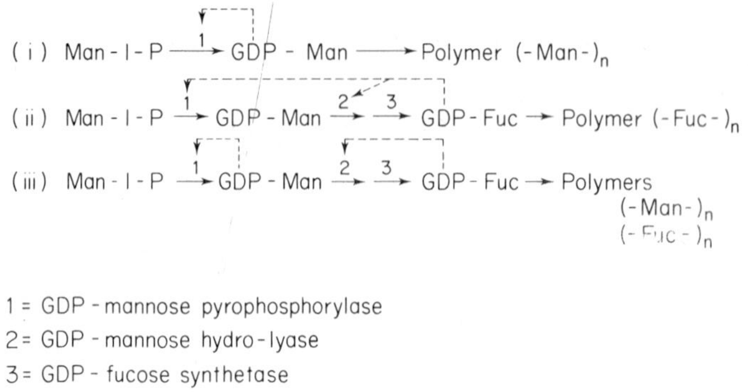

(i) Man-1-P $\xrightarrow{1}$ GDP-Man ⟶ Polymer $(\text{-Man-})_n$

(ii) Man-1-P $\xrightarrow{1}$ GDP-Man $\xrightarrow{2}$ $\xrightarrow{3}$ GDP-Fuc ⟶ Polymer $(\text{-Fuc-})_n$

(iii) Man-1-P $\xrightarrow{1}$ GDP-Man $\xrightarrow{2}$ $\xrightarrow{3}$ GDP-Fuc ⟶ Polymers $(\text{-Man-})_n$ $(\text{-Fuc-})_n$

1 = GDP-mannose pyrophosphorylase
2 = GDP-mannose hydro-lyase
3 = GDP-fucose synthetase

Fig. 5 Control of mannose and fucose synthesis [after Kornfeld and Glaser, 1966].

polysaccharides, but mannose may be needed both for lipopolysaccharide and for exopolysaccharide, while the same is true for glucose from UDP-glucose.

Many bacteria also contain sugar nucleotide hydrolases, some being apparently located in the periplasm [Beacham *et al.*, 1973]. The role of these enzymes is not clear, but it may represent a further means of regulating the nucleotide pool [Ward and Glaser, 1968; 1969]. Whether the enzymes have access to their substrates or are spatially separated from them in normal cells has not yet been determined. They also present problems in studies involving cell-free synthesis of polysaccharides from sugar nucleotide precursors, where they certainly can come into direct contact with the sugar nucleotides. Such enzymes are not confined to the Enterobacteriaceae, as studies on *Pseudomonas aeruginosa* and *Xanthomonas campestris* [N.L. Piggott and I.W. Sutherland, unpublished results] indicated the presence of one or more enzymes hydrolysing GDP-mannose; enzyme activity was partially inhibited by the presence of fluoride.

As well as regulation at the level of the enzyme activity, control can be exerted over enzyme synthesis. This is seen in the studies by Markovitz [1977] of colanic acid synthesis in various enterobacteria. Whether such complex genetic control systems may be found in other exopolysaccharide-forming bacteria is not clear. Some aspects of the regulation of alginate synthesis in *Pseudomonas aeruginosa* resemble those reported for the formation of colanic acid [J.R.W. Govan, unpublished results].

A key intermediate in exopolysaccharide synthesis, affording the possibility of control of synthesis, is the lipid-linked oligosaccharide. Both prokaryotes and eukaryotes contain isoprenoid lipids which are involved in glycosylation mechanisms. They provide a means of converting the hydrophilic sugars from their attachment to nucleoside diphosphate compounds to lipophilic substances capable of carrying the sugar into and through the cell membrane. In eukaryotic cells, the isoprenoid lipids are C_{80} - C_{100} compounds (dolichols) which are only present in very low concentrations. Hemming [1977] suggested that the concentration of dolichol phosphate is probably rate limiting for the glycosylation processes occurring in mammalian tissues; this may also be true for the shorter chain length lipids (C_{50} - C_{60}, but predominantly C_{55}) found in prokaryotes. The role of isoprenoid lipid intermediates was clearly demonstrated by Troy *et al.* [1971] in their elegant studies on the biosynthesis of an exopolysaccharide of *Enterobacter aerogenes*. They confirmed results using another strain of this species which had shown the involvement of lipid intermediates using a series of transferaseless mutants [Sutherland and Norval, 1970], (Figure 6). It has to be remembered that in Gram negative bacteria, the isoprenoid lipids are used to form intermediates in the formation of peptidoglycan, lipopolysaccharide and exopolysaccharide, all of which are polymers composed at least partially of repeating glycan units and which are located outside the cytoplasmic membrane. Good assay methods for the isoprenoid lipids and related compounds are still not available for Gram negative or indeed most types of cell, but a considerable amount of indirect evidence supports the suggestion that availability of isoprenoid lipid is a critical factor in exopolysaccharide synthesis.

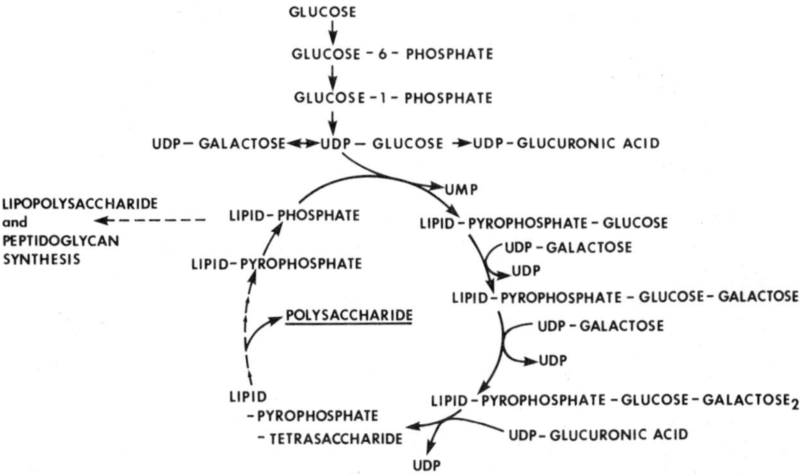

Fig. 6 Pathway for the biosynthesis of an exopolysaccharide.

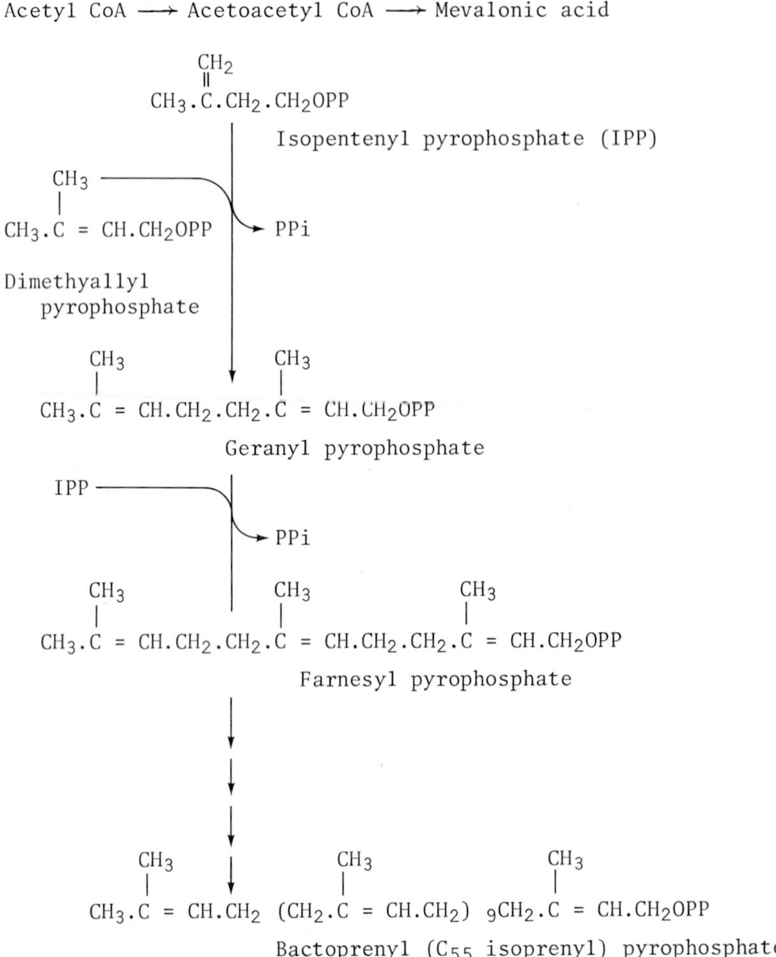

Fig. 7 Pathway for isoprenoid biosynthesis.

Isoprenoid lipids are synthesized from acetyl coenzyme A *via* isopentenyl pyrophosphate yielding a lipid pyrophosphate as shown in Figure 7. The enzymes responsible for this biosynthetic pathway in *Salmonella* species were partially soluble and partially membrane-bound [Christenson *et al.*, 1969]. Further enzymes capable of dephosphorylating and phosphorylating isoprenyl alcohols have been discovered in Gram positive bacteria [Higashi *et al.*, 1970; Goldman and Strominger, 1972; Willoughby *et al.*, 1972]. It seemed unlikely that such enzymes were confined to one group of bacteria and more recently, a kinase with similar properties although some physical differences was detected in

Gram negative bacteria [Poxton *et al.*, 1974]. The bacterial cell may thus control the amount of the active form, isoprenyl phosphate, through dephosphorylation of the pyrophosphate and through the phosphorylation of the metabolically inactive free alcohol.

It has to be assumed that in most bacteria there is sufficient isoprenoid lipid to permit the simultaneous synthesis of peptidoglycan, lipopolysaccharide and exopolysaccharide. Otherwise such bacteria would not be expected to form all three polymers while actively growing. In fact, the remarkable consistency of the rate of polysaccharide production under different nutrient limitations has been pointed out by Deavin *et al.* [1977] (Table 2). In some species, exopolysaccharide is only observed late in the log phase or early in the stationary phase of growth. Such a late 'switch on' of synthesis might be due to the limitation of the available isoprenoid lipid, preventing the concurrent synthesis of all three polymers. In many of the species tested, synthesis of exopolysaccharides can occur in non-growing cells under conditions where there should be no competition for isoprenoid lipid. It is, however, possible that some of the lipid is unavailable for exopolysaccharide production due to the attachment of some peptidoglycan or lipopolysaccharide intermediates. When mutants of *Salmonella* species unable to transfer the D-antigenic side-chains of lipopolysaccharides to the core structures were tested, they were unable to form exopolysaccharide [Sutherland, 1975]. Other types of mutant

TABLE 2

Exopolysaccharide production by Azotobacter vinelandii

Limiting nutrient	Cell mass (mg/ml)	Specific rate of polysaccharide production (mg/mg cell h^{-1})
MoO_4^{2-}	1.1	0.34
PO_4^{3-}	1.9	0.28
Fe^{3+}	1.4	0.25
C (as sucrose)	1.3	0.25
Ca^{2+}	1.2	0.20
K^+	1.9	0.16

$D = 0.15 \pm 0.01\ h^{-1}$

[Results taken from Deavin *et al.*, 1977.]

failing to form lipopolysaccharide were unaffected in exopolysaccharide synthesis. Exopolysaccharide production is frequently favoured by incubation at sub-optimal temperatures [see for example Sutherland, 1977a]. Under such conditions, it could be argued that less isoprenoid lipid is needed for peptidoglycan and lipopolysaccharide synthesis, which both occur at a reduced rate. More would then be available for exopolysaccharide formation. Bacitracin-resistant mutants may also possess higher levels of isoprenoid lipid which, in the absence of the physical constraints which limit peptidoglycan and lipopolysaccharide production and incorporation into walls, would become available for exopolysaccharide synthesis. Increases both in polysaccharide produced and in transfer of sugar to lipid-soluble material have been noted in bacitracin-resistant strains [Sutherland, 1977b].

An unusual type of mutant was isolated from several strains of *Enterobacter aerogenes* in our laboratory [Norval and Sutherland, 1969]. These mutants appear to have the attributes which might be expected from isoprenoid lipid-limited cells, when they are grown at lowered incubation temperature. They are autoagglutinable, contain normal peptidoglycan, less lipopolysaccharide than normal and form no exopolysaccharide until growth has ceased. The enzymes involved in exopolysaccharide synthesis are present at normal levels and indeed on transfer to washed cell suspension, near normal levels of exopolysaccharide are formed. At 37°C, the cells are indistinguishable from wild type. If these bacteria are indeed defective in isoprenoid lipid synthesis at lower temperatures, there appears to be a distinct system of priorities for the available lipid: (i) peptidoglycan, (ii) lipopolysaccharide and (iii) exopolysaccharide, once (i) and (ii) are satisfied. The role of isoprenoid lipid in control of synthesis of various polysaccharides has recently been reviewed [Sutherland, 1977a and b] but is summarised in Figure 8.

Added complications in the control mechanisms regulating polysaccharide synthesis are to be expected in those microbial strains which produce more than one extracellular polysaccharide. Relatively few such systems have been reported although they are known to occur in the Enterobacteriaceae in which a number of *Escherichia coli* strains can produce both the colanic acid type of polymer and other strain-specific polysaccharides [Ørskov *et al.*, 1963]. The ability to produce two exopolysaccharides was also found in *Alcaligenes faecalis* var *myxogenes* [Harada, 1977]. The polymers are a succinoglucan and a β 1,3-linked polyglucose respectively. The bacterial cultures undergo spontaneous mutation on subculture and the mutant colonies isolated were found to produce much more β glucan than succinoglucan, unlike the wild-type isolate in which the reverse was true [Amemura *et al.*, 1977]. The mechanisms by which this bacterial strain can form two exopolysaccharides simultaneously are not known, but it seems likely that formation of UDP-glucose may be particularly important as both polymers contain glucose and in addition galactose, formed from UDP-glucose by epimerization, is present in smaller quantities in the succinoglucan. The ability to

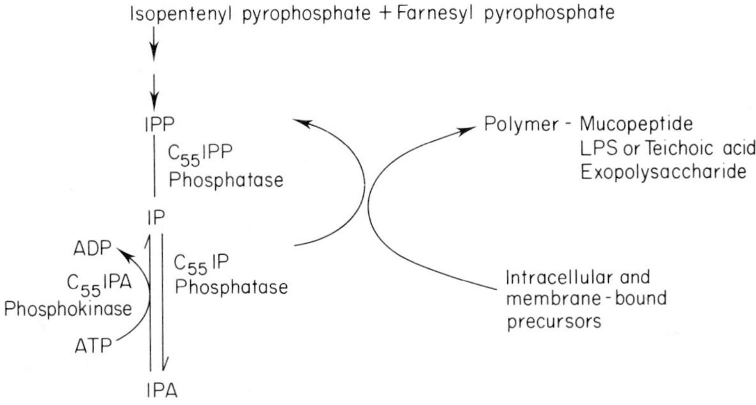

Fig. 8 Regulation of carrier lipids.

produce several exopolysaccharides is also known among bacterial species producing levans and dextrans.

Dextran

The production of dextrans by bacteria [see Brooker, Chapter 5] differs from the formation of most other exopolysaccharides in that the process is essentially extracellular and does not require the monosaccharide to be activated and attached to a donor of the nucleoside diphosphate type. Thus, although there is a stringent substrate requirement which is only satisfied by sucrose and a small number of related oligosaccharides, there are fewer control mechanisms which can be utilized by dextran-synthesising bacteria and relatively few reports on these. The extracellular enzyme systems involved in dextran synthesis have been partially purified then allowed to react with sucrose and the resultant polymer studied [Baird *et al.*, 1973; Newbrun and Sharma, 1976]. The product from cell-free synthesis may differ in some respects from that produced by whole cultures. In one study [Arnett and Mayer, 1975], *Streptococcus sanguis* dextrans from whole cells and from cell-free preparations differed in their content of α 1,3 linkages. The observed differences are probably due to the loss or inactivation of one of the enzymes involved, during the preparation of the cell-free material. There is thus the possibility of using a mixture of cell-free enzymes of known activities to produce a polymer containing the desired linkages in the correct proportions. Such cell-free preparations, partially purified by adsorption on to hydroxylapatite, have been used for periods up to 80 hours [Arnett and Mayer, 1975]. The products have been examined in respect of their chemical linkages but it is not entirely clear whether the use of cell-free preparations might yield products of differing chain length from those prepared using whole cultures.

The chain length and linkages obtained using the natural system probably depend on the strain used and on the physiological conditions and hence on the relative amounts and activities of the specific enzymes. Further work is needed, perhaps using insolubilised enzymes on solid matrices, before a clearer understanding of the dextran-forming system can be obtained. Using enzyme(s) immobilised on Biogel (agarose gel) Robyt and Taniguchi [1976] suggested that branching of the polymeric product did not necessarily require a distinct enzyme.

Two possible ways in which control could be exerted were through the availability of sucrose and the supply of a suitable dextran acceptor. Under glucose limitation, the maximum production of the glycosyl transferases occurred at low growth rates (14 hours mean generation time) and required a pH value close to 6.5 [Ellwood and Hunter, 1976].

Control of Polymer Release and of Physical Properties

Despite our knowledge relating to the early stages of exopolysaccharide synthesis, we know little about the final stages of polysaccharide production - the transfer of the oligosaccharide chains from their attachment to isoprenoid lipid to a possible surface receptor and extrusion from the cell surface. The electron microscopic studies of Brooker [1976; Chapter 5], Brown *et al.* [1976], Colvin and Leppard [1977] and Bayer and Thurow [1977] have revealed fibrous structures at the cell surface during polysaccharide synthesis and have indicated possible sites for extrusion of the polymers from the cells. However, some of the fibres visualised probably result from interaction between adjacent molecules with the correct orientation after extrusion, as is known to occur for several different polysaccharides in aqueous solution [see for example, Morris *et al.*, 1977]. There may even be differences in the release of polysaccharides from cells in different physiological states. Thus, peptidoglycan from the cell walls of logarithmic phase bacteria is less cross-linked than that obtained from stationary phase bacteria; some similar type of process might occur for extracellular polysaccharides. The outer membrane of Gram negative bacteria is also more readily lost from bacteria under some conditions than others. This is likely to affect the attachment and release of exopolysaccharides as well as affecting their final purity. It should also be remembered that in some microorganisms, exopolysaccharide is secreted continuously during growth, whereas in others exopolysaccharide production is a feature of the late log phase and stationary phase only. Examples of the latter are to be found in polysaccharide production by a pseudomonad [Williams and Wimpenny, 1977] and by *Aureobasidium (Pullularia) pullulans* [Catley, 1971].

Control of the molecular weight of exopolysaccharides is probably one of the ideals to be aimed for if there is a requirement for products with the same chemical structure but varying properties, as there is for alginates. The polysaccharides from different microorganisms vary widely in their molecular weight and are probably also dependent on the growth conditions used. In a few species, the size of the polymer molecule may be affected by the presence of polysaccharases in the culture fluid. This problem is of limited occurrence as few polysaccharides are degraded by enzymes from the cell which produces them. Exceptions are found in some alginate and dextran-producing bacteria and in hyaluronic acid-forming bacteria [Haug and Larsen, 1971; Dewar and Walker, 1975]. Molecular weight values as low as 8×10^4 have been quoted for an uncharacterised polymer from a strain of hydrocarbon-utilizing bacteria, but most exopolysaccharides have larger molecular weights. Typically, these are in the range 2×10^5 - 3×10^6 as can be seen from some of the preparations which have been studied and which are listed in Table 3. Variations can, however, occur between different strains of the same species.

Molecular weight is one of the parameters which affects the viscosity and other rheological properties of polysaccharides in solution. The ability to control molecular weight would thus be advantageous. Alteration of the physical conditions of growth or of the growth rate may be important. As Collins [1964] reported that increased growth rate resulted in shorter O-antigenic side-chains in the lipopolysaccharides of *Salmonella enteritidis* cell walls, it is possible that shorter chain extracellular polysaccharides may also be produced with increased growth rate. The result observed with the lipopolysaccharide side-chains is probably due to increased turnover of the isoprenoid lipid pyrophosphate-oligosaccharide intermediates, preventing accumulation of longer chains. Similar results might be expected for exopolysaccharide synthesis.

Exopolysaccharides are found in one of two forms. They may be attached to the cell as a discrete capsule or they may be secreted entirely as soluble slime unattached to the microbial cells from which they are produced. Capsulate cultures usually contain some slime formed from the disintegration of the capsules. The state of the polysaccharide depends on the microbial species and on the physiological state of the culture. In some cultures, the capsule

TABLE 3

Molecular weights of some exopolysaccharides

Strain		Molecular weight	Reference
Acetobacter xylinum		5.67×10^5	Brown [1962]
Xanthomonas campestris		2×10^6	Dintzis *et al.* [1970]
Streptococcus pneumoniae	Type 3	2.67×10^5	Koenig and Perrings [1955]
Escherichia coli K87		2.8×10^5	Tarcsay *et al.* [1971]
Klebsiella pneumoniae	Type 1	2.94×10^6	Wolf *et al.* [1978]
Enterobacter aerogenes	Type 5	1.29×10^6	
	Type 6	2.5×10^6	
	Type 8	1.13×10^6	
	Type 11	2×10^6	
	Type 56	1.7×10^5	
	Type 57	2.27×10^6	
Aureobasidium (Pullularia) pullulans		1.7×10^5	Taguchi *et al.* [1973]

is very firmly attached to the cell surface, perhaps by means of the knob-like structures recently demonstrated in electron micrographs of *Escherichia coli* exopolysaccharide preparations [Bayer and Thurow, 1977]. In other strains or species the capsule is so loosely attached that the turbulence of growth in a fermentor is sufficient to detach it; others again are consistently slime formers. It is an obvious disadvantage if a polysaccharide with industrial potential has first to be detached from the microbial cells. Fortunately, mutants no longer able to attach the polysaccharide in the form of capsules have been found regularly during studies of various capsulate microorganisms and such mutants are apparently stable [see for example Norval, 1969; Andreeson and Schlegel, 1974]. It is not clear whether the mutants have lost a transferase type of enzyme involved in the final stages of capsule formation or whether an attachment site on the cell surface has been lost. Both types of mutation might be expected to yield slime-forming (Sl) cultures. No chemical differences of significance have been reported between the capsule and slime polysaccharides prepared from wild type strains and their slime-forming mutants, but it was suggested that for an *Enterobacter aerogenes* strain, the slime polysaccharides had lower molecular weights than capsular material, 1.5×10^6 and 1×10^8 respectively [Wilkinson, 1958]. One of the problems in preparing slime-forming mutants is the lack of suitable selection procedures as is also true for other types of mutation affecting the physical properties of exopolysaccharides.

Detailed knowledge of the physical characteristics of polysaccharide solutions is slowly becoming available but it is unfortunate that many of these studies are being performed in industrial concerns and the results have not become widely available. Use of X-ray diffraction analysis to elucidate molecular structures of microbial exopolysaccharides has been of considerable value [Atkins *et al.*, 1977; Atkins, Chapter 7] but we still do not know all the parameters involved in the physical properties of polysaccharide solutions [see Powell, Chapter 6]. Thus we still do not know why polysaccharides produced by variants of the same strain possess different properties in solution. Changes in physical properties can certainly be expected when one strain forms more than one polysaccharide as is the case with *Alcaligenes faecalis* var. *myxogenes* [Harada, 1977]. Wild type cultures produce much more succinoglucan than curdlan (β 1,3 glucan). On the other hand, mutant strains selected on aniline blue agar yield mainly curdlan, a polysaccharide with characteristic gelling properties.

Viscosity changes in polysaccharide preparations are not always entirely due to changes in molecular weight. Alginate depends for its solution properties on the ratio of D-mannuronic acid to L-guluronic acid and on the proportion of the three possible types of structures derived from these two monomers: polymannuronic acid (-M-M-M-M-), polyguluronic acid (-G-G-G-G-), or alternating portions (-M-G-M-G-). By appropriate variations of the culture conditions, a range of alginates could be produced using continuous cultures of *Azotobacter vinelandii* [Deavin *et al.*, 1977]. These polymers had low, medium or high viscosities, thus corresponding to commercial alginates currently obtained from seaweed [see Jarman, Chapter 2]. Variations in the viscosity of continuous cultures of *Xanthomonas campestris* have also been reported, but a problem encountered in this work was the appearance of colonial variants which eventually failed to synthesize any polysaccharide [Silman and Rogovin, 1972]. Changes in the viscosity of solutions of this polymer may also be due to altered pyruvate content as discussed below.

An alternative approach, which probably depends on altered control of polysaccharide synthesis, may yield polymers of higher molecular weight and greatly increased solution viscosity. In this laboratory, we have started with polymers of known chemical composition and structure as opposed to the traditional approach of widespread screening of isolates from different environments. It is thus possible, to select a polysaccharide with the appropriate solution characteristics in respect of helical, ribbon or folded structure, and develop the culture until the polysaccharide is produced in the yield and physical form which is likely to be most useful. Starting with cultures of exopolysaccharide-producing bacteria such as *Escherichia coli* K12 and strains of *Enterobacter aerogenes*, both of which grow very well on synthetic media at 37°C, mutants were usually first isolated which produced polysaccharide slime only. From these, stable mutants were in turn obtained, the polymers from which differed considerably in their physical properties from those of the parent strains. The structure and the chemical composition appeared to be unaltered, while growth of mutant and parent strains proceeded at the same rate under comparable conditions. The polymers from the mutant and parent strains differ considerably in their viscosities as can be seen for two such polysaccharide families (Figures 9 and 10). The properties were the same for cultures grown on solid and liquid media and in continuous culture under several limitations. Polysaccharides from mutants showed much enhanced viscosity and different plots of viscosity against shear rate (Figures 11 and 12). Such viscous (V) mutants are isolated relatively easily although it has not yet proved possible to devise a selection procedure which will offer one of the several possible 'improvements' on wild-type capsulate bacteria (Figure 13). The lack of apparent chemical changes suggests that we still know little about the way in which microorganisms control properties such

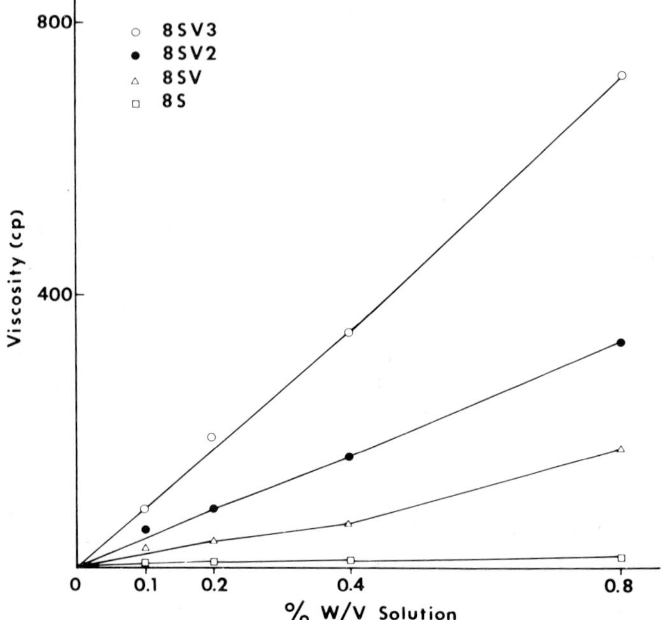

Fig. 9 Viscosity at low shear (10 sec^{-1}, 20°) for polysaccharides from *Enterobacter aerogenes* type 8 mutants.

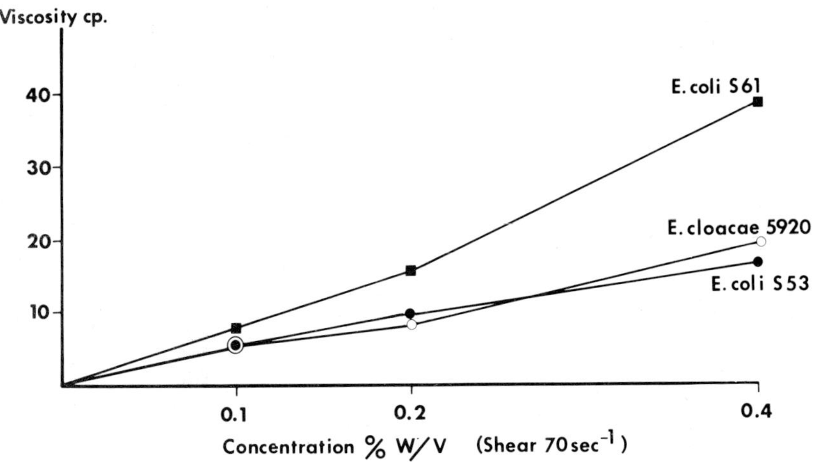

Fig. 10 Viscosity of colanic acid preparations (shear 70 sec^{-1}, Ferranti Viscometer, 20°C).

as the molecular weight of the exopolysaccharides which they produce and the alterations in biosynthetic mechanisms which bring about these changes.

Control of Acylation and Ketalation

Whereas, except in bacterial alginates, the carbohydrate composition of an exopolysaccharide is apparently constant, variations in the acyl and ketal groups are relatively common. This is revealed both from studies on a common polysaccharide from various bacterial strains and on the production of a polysaccharide under varying physiological conditions. Even bacterial alginates synthesized by *Pseudomonas aeruginosa* vary considerably in their acetyl content. A few preparations entirely lacked O-acetyl groups; several samples contained 4.1-4.7 %, but most of these polysaccharides had O-acetyl contents in the range 10.4-11.8 % [Evans and Linker, 1973].

Colanic acid (Figure 14) is produced by *Enterobacter cloacae, Escherichia coli* and numerous *Salmonella* strains [Grant *et al.*, 1969]. Although all preparations contained the same hexasaccharide repeating unit, several different types of ketal were detected [Garegg *et al.*, 1971b]. Another feature of particular interest was the discovery of a preparation from a *Salmonella typhimurium* mutant, which lacked O-acetyl groups [Garegg *et al.*, 1971a]. Thus, although polysaccharide synthesis requires all the sugar precursors and transferases, formation of the polymer can apparently proceed in the absence of the acetyl donor (presumably acetyl CoA) or more probably the acetyl trans-

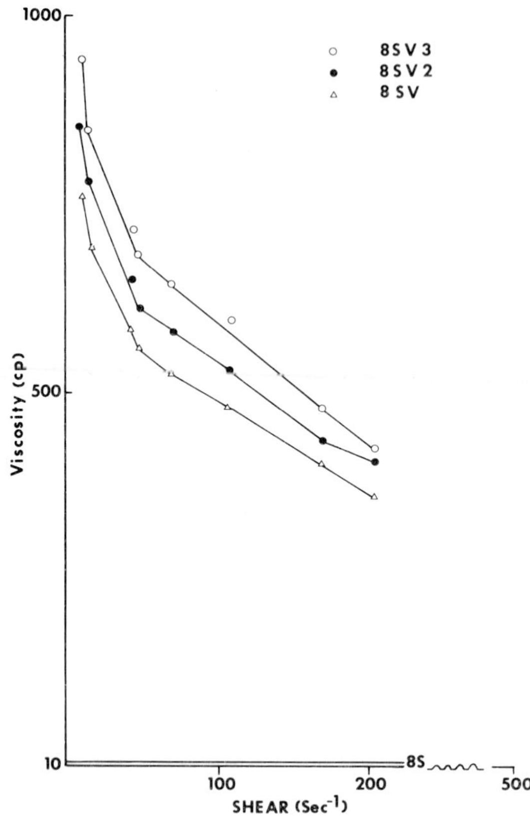

Fig. 11 The viscosity versus shear profile for mutants derived from an *Enterobacter aerogenes* type 8 strain. All polysaccharides were used at 0.8% (w/v) and 20°C.

ferase. Support for this view has recently come from a number of laboratories studying the production of exopolysaccharide by *Xanthomonas campestris* and related strains. Colony variants of *Xanthomonas campestris* yielded different amounts of polysaccharide and analysis showed that the contents of both O-acetyl groups and pyruvate were lower in polysaccharide from the small colony type (Table 4) [Cadmus *et al.*, 1976]. Studies on the proposed structure of the *Xanthomonas campestris* polysaccharide indicated an average of one pyruvate group per 2 or 3 repeat units and approximately one acetate group per repeat unit [Jansson *et al.*, 1975; Melton *et al.*, 1976]. The similarity of the values for the 'low pyruvate' polymer to a ratio of one pyruvate per 4 repeating units was noted by Sandford *et al.* [1977]. These authors also showed that the 'low pyruvate' polymer had altered physical properties, particularly when salts

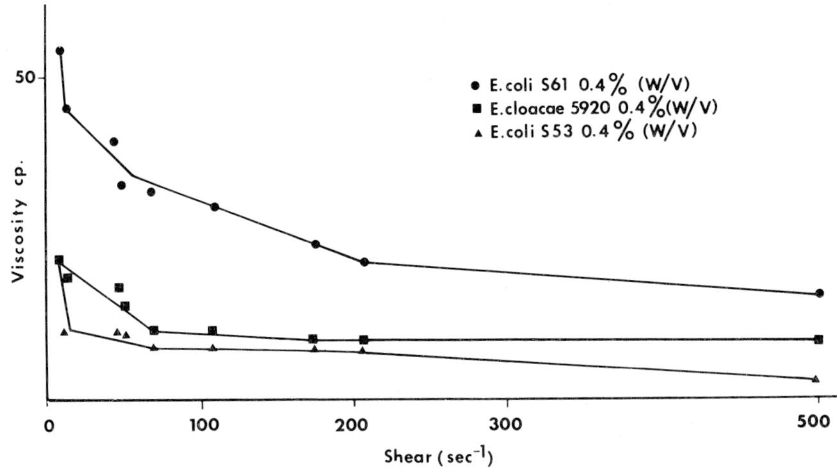

Fig. 12 Viscosity versus shear plots for colanic acid (all at 20°C, pH 7.0).

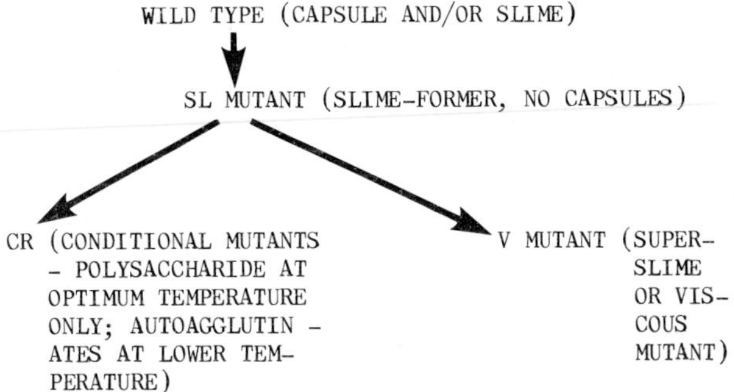

Fig. 13 Mutations affecting the physical state of a bacterial exopolysaccharide.

were present. The concept of one pyruvate residue attached to every fourth repeating unit is not the only possible interpretation of these results. One alternative structure is that in which some polysaccharide is pyruvylated on every second repeat unit and other on every fourth such unit. This would imply a somewhat loose control of pyruvylation in the biosynthetic process. The material used by Cadmus *et al.* [1976] was obtained by batch culture. As conditions vary during growth in batch culture, it is conceivable that pyruvylation of *Xanthomonas campestris* exopolysaccharide is a function of the physiological state of the culture, polymer at certain stages of growth being pyruvylated and not at other times; the final product

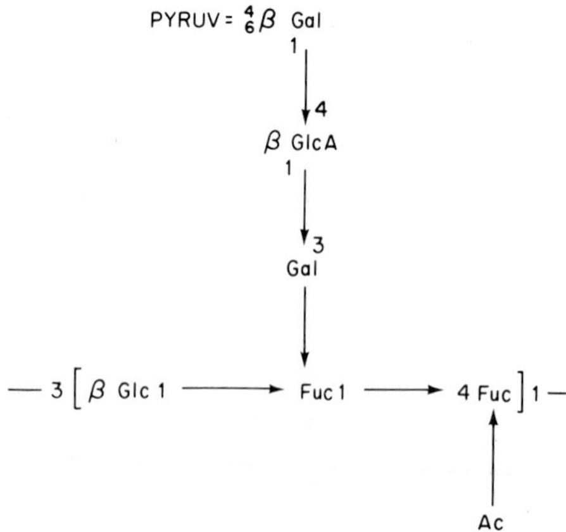

Fig. 14 The structure of colanic acid from *Escherichia coli*.

being a mixture of pyruvylated (perhaps on every repeat unit) and non-pyruvylated polysaccharide strands. Ideally one would like to have enzymes able to break the polysaccharide into its sub-units, the $(\text{oligosaccharide})_4$ or $(\text{oligosaccharide})_2$ each containing one mole of pyruvate. Use of such a technique demonstrated clearly that in the *Enterobacter aerogenes* type 54 strain examined, acetate was present on every alternate repeat unit [Sutherland and Wilkinson, 1968]. Unfortunately, enzymes of this general type capable of degrading *Xanthomonas campestris* polysaccharide have not yet been isolated. However, use of affinity techniques has indicated that batch preparations of *Xanthomonas campestris* polymer represent a mixture of strand types, some molecules being pyruvylated in a ratio close to 1 per repeat unit, while others lack pyruvate entirely [I.W. Sutherland, unpublished results]. At the same time, examination of polysaccharides from a wide range of strains of *Xanthomonas campestris* indicated that even in batch culture, several polymers were formed which lacked pyruvate entirely, while others resembled in all respects tested the structure shown in Figure 2. These results can be taken to imply that acetylation and pyruvylation of *Xanthomonas campestris* exopolysaccharides depend on (i) the strain and (ii) the physiological conditions of growth. In some strains (Table 4) it would appear that the repeat unit is fully acetylated and pyruvylated while in others, the enzymes for acetylation or for ketalation are apparently absent. Nevertheless, the carbohydrate structure is formed. In other strains, obviously including those used for the definitive studies on structure by Jansson *et al.* [1975] and Melton *et al.* [1976] polysaccharide composition depends on the physiological conditions, most probably on

TABLE 4

Composition of polysaccharides from Xanthomonas species

Material	Pyruvate	Acetyl	Glucose	Mannose	Glucuronic acid
Xanthomonas campestris HPX*, normal colony	4.4	4.5	37.0	43.4	19.5
Xanthomonas campestris LPX*, small colony	2.5	3.7	37.7	42.9	19.3
Xanthomonas fragaria 1469	7.0	3.9	42.7	37.5	19.8
Xanthomonas gummisudans 2182	4.7	10.0	42.4	37.4	20.1
Xanthomonas manihotis 1159	7.6	6.8	42.3	36.5	21.2
Xanthomonas vasculorum 796	5.4	8.8	44.3	35.0	20.4

* Results of Sandford *et al.* [1977]; other data from unpublished work (I.W. Sutherland).

Acetyl and pyruvate are calculated as % dry weight; each sugar as % total carbohydrate.

the availability under changing conditions of the presumed donors, acetyl CoA and phosphoenolypyruvate (PEP). It would be of particular interest to know more about the acetate and pyruvate content of polysaccharides obtained from the carefully defined conditions which can be realized using continuous culture. It might then be possible also to correlate acyl content with the conditions under which acyl donors are likely to accumulate or to be limited.

A possible approach to the study of pyruvylation and acetylation mechanisms is the use of inhibitors. These have been used in eukaryotic systems [Stubbe and Kenyon, 1972], and in studies on muramic acid precursor synthesis in bacteria [Cassidy and Kahan, 1973], but do not yet appear to have been tested in systems involving pyruvylated polysaccharides. Another possible analogue is the antibiotic fosfomycin [Woodruff *et al.*, 1977]. Susceptibility of strains producing pyruvylated polymers to this compound is not appreciably different from that of strains producing non-pyruvylated polysaccharides [I.W. Sutherland, unpublished results]. Under conditions where fosfomycin slightly reduced total growth but permitted some polysaccharide synthesis, the product appeared to have reduced pyruvate content. Further studies are, however, necessary using other possible antagonists and physiological conditions.

Whether the concentration of PEP and acetyl CoA plays an important part in regulating the pyruvylation and acetylation of exopolysaccharides is not clear. Pyruvylated polysaccharides are not confined to one group of bacteria (although they have not been commonly found in eukaryotic products). Thus, unlike group translocation mechanisms for substrate uptake, which are found in prokaryotes possessing the Embden-Meyerhof pathway (i.e. yielding two moles of PEP/mole of carbohydrate metabolised) and not in species using the Entner-Doudoroff scheme [Saier, 1977], both groups of bacteria may produce pyruvylated polymers.

Conclusions

In conclusion, the cell free synthesis of exopolysaccharides such as dextran can perhaps be regarded as the ideal situation. Currently production of extracellular heteropolysaccharides using whole microbial cells is still some way from this ideal. However, through a knowledge of the control mechanisms involved in substrate uptake, intermediary metabolism and polymer synthesis, it should be possible to improve the conversion rate of substrate to polymer, the rate at which polysaccharide is synthesised and perhaps even the physical properties of the final product.

References

Amemura, A., Hisamatsu, M. and Harada, T. (1977). Spontaneous mutation of polysaccharide production in *Alcaligenes faecalis* var. *myxogenes* 10C3. *Applied and Environmental Microbiology* **34**, 617-620.

Andreesen, M. and Schlegel, H.G. (1974). A new coryneform hydrogen bacterium: *Corynebacterium autotrophicum* strain 7C. *Archives of Microbiology* **100**, 351-361.

Arnett, A.T. and Mayer, R.M. (1975). Structural characteristics of native and enzymically formed dextran of *S. sanguis* ATCC 10558. *Carbohydrate Research* **42**, 339-345.

Atkins, E.D.T., Gardner, K.H. and Isaac, D.H. (1977). X-ray diffraction by bacterial capsular polysaccharides: trial conformations for *Klebsiella* polyuronides K5, K57 and K8. In *Cellulose Chemistry and Technology*, pp.56-72. Edited by A.C. Arthur. Washington: American Chemical Society.

Baird, J.K., Longyear, V.M.C. and Ellwood, D.C. (1973). Water insoluble and soluble glucans produced by extracellular glycosyl transferases from *Streptococcus mutans*. *Microbios* **8**, 143-150.

Bayer, M. and Thurow, H. (1977). Polysaccharide capsule of *Escherichia coli*: microscope study of its size, structure and site of synthesis. *Journal of Bacteriology* **130**, 911-936.

Beacham, I.R., Kahana, R., Levy, L. and Yagil, E. (1973). Mutants of *Escherichia coli* K12 "cryptic" or deficient in 5'-nucleotidase and 3'-nucleotidase activity. *Journal of Bacteriology* **116**, 957-964.

Björndal, H., Lindberg, B., Lönngren, J., Meszaros, M. Thompson, J.L. and Nimmich, W. (1973). Structural studies of the capsular polysaccharide of *Klebsiella* type 52. *Carbohydrate Research* **31**, 93-100.

Brooker, B.E. (1976). Surface coat transformation and capsule formation by *Leuconostoc mesenteriodes* NCDO 523 in the presence of sucrose. *Archives of Microbiology* **111**, 99-104.

Brown, A.M. (1962). The mechanism of cellulose biosynthesis by *Acetobacter acetigenum*. *Journal of Polymer Science* **59**, 155-163.

Brown, R.M., Willison, J.H.M. and Richardson, C.L. (1976). Cellulose biosynthesis in *Acetobacter xylinum*: visualization of the site of synthesis and direct measurement of the *in vivo* process. *Proceedings of the National Academy of Sciences, U.S.A.* **73**, 4565-4569.

Cadmus, M.C., Rogovin, S.P., Burton, K.A., Pittsley, J.E., Knutson, C.A. and Jeanes, A. (1976). Colonial variation in *Xanthomonas campestris* NRRL B-1459 and characterization of the polysaccharide from a variant strain. *Canadian Journal of Microbiology* **22**, 942-948.

Cassidy, P.J. and Kahan, F.M. (1973). A stable enzyme - phosphoenol pyruvate intermediate in the synthesis of uridine-5'-diphospho-N-acetyl-2-amino-2-deoxy glucose-3-O-enolpyruvyl ether. *Biochemistry* **12**, 1364-1373.

Catley, B.J. (1971). Utilization of carbon sources by *Pullularia pullulans* for the elaboration of extracellular polysaccharides. *Applied Microbiology* **22**, 641-649.

Chakraborty, A.K., Friebolin, H., Niemann, H. and Stirm, S. (1977). Primary structure of the *Klebsiella* serotype 16 capsular polysaccharide. *Carbohydrate Research* **59**, 525-530.

Choy, Y.M. and Dutton, G.G.S. (1973). The structure of the capsular polysaccharide from *Klebsiella* K type 21. *Canadian Journal of Chemistry* **51**, 198-207.

Choy, Y.M., Dutton, G.G.S. and Zanlunga, M. (1973). The structure of the capsular polysaccharide of *Klebsiella* K type 24. *Canadian Journal of Chemistry* **51**, 1819-1825.

Christenson, J.G., Gross, S.K. and Robbins, P.W. (1969). Enzymatic synthesis of the antigen carrier lipid. *Journal of Biological Chemistry* **244**, 5436-5439.

Collins, F.M. (1964). The effect of the growth rate on the composition of *S. enteritidis* cell walls. *Australian Journal of Experimental Biology and Medical Science* **42**, 255-262.

Colvin, J.R. and Leppard, G.G. (1977). The biosynthesis of cellulose by *Acetobacter xylinum* and *Acetobacter acetigenus*. *Canadian Journal of Microbiology* **23**, 701-709.

Conrad, H.E., Bamburg, J.R., Epley, J.D. and Kindt, T.J. (1966). The structure of the *Aerobacter aerogenes* A3 (SL) polysaccharide. II. Sequence analysis and hydrolysis studies. *Biochemistry* **5**, 2808-2817.

Curvall, M., Lindberg, B., Lönngren, J. and Nimmich, W. (1975a). Structural studies of the capsular polysaccharide of *Klebsiella* type 81. *Carbohydrate Research* **42**, 73-82.

Curvall, M., Lindberg, B., Lönngren, J. and Nimmich, W. (1975b). Structural studies of the capsular polysaccharide of *Klebsiella* type 28. *Carbohydrate Research* **42**, 95-105.

Deavin, L., Jarman, T.R., Lawson, C.J., Righelato, R.C. and Slocombe, S. (1977). The production of alginic acid by *Azotobacter vinelandii* in batch and continuous culture. In *Extracellular Microbial Polysaccharides*, pp.14-26. Edited by P.A. Sandford and A. Laskin. Washington: American Chemical Society.

Dewar, M.D. and Walker, G.J. (1975). Metabolism of the polysaccharides of human dental plaque. *Caries Research* **9**, 21-35.

Dicks, J.W. and Tempest, D.W. (1967). Potassium-ammonium antagonism in polysaccharide synthesis by *Aerobacter aerogenes* NCTC 418. *Biochimica et Biophysica Acta* **136**, 176-179.

Dintzis, F.R., Babcock, G.E. and Tobin, R. (1970). Studies on dilute solutions and dispersions of the polysaccharide from *Xanthomonas campestris* NRRL B-1459. *Carbohydrate Research* **13**, 257-267.

Duguid, J.P. and Wilkinson, J.F. (1953). The influence of cultural conditions on polysaccharide production by *Aerobacter aerogenes*. *Journal of General Microbiology* **9**, 174-189.

Dutton, G.G.S. and Mackie, K.L. (1977). Structural investigations of *Klebsiella* serotype K36 polysaccharide. *Carbohydrate Research* **55**, 49-63.

Dutton, G.G.S., Stephan, A.M. and Churms, S.C. (1974). Structural investigations of *Klebsiella* serotype 7 polysaccharide. *Carbohydrate Research* **38**, 225-237.

Dutton, G.G.S. and Yang, M.T. (1973). The structure of the capsular polysaccharide of *Klebsiella* K type 5. *Canadian Journal of Chemistry* **51**, 1826-1832.

Dutton, G.G.S. and Yang, M.T. (1977). Structural investigation of *Klebsiella* serotype 62 polysaccharide. *Carbohydrate Research* **59**, 179-192.

Ellwood, D.C. and Hunter, J.R. (1976). The mouth as a chemostat. In

Continuous Culture 6, pp.270-282. Edited by A.C.R. Dean, D.C. Ellwood, C.G.T. Evans and J. Melling. London: Society of Chemical Industry.

Erbing, C., Kenne, L., Lindberg, B., Lönngren, J. and Sutherland, I.W. (1976). Structural studies of the capsular polysaccharide from *Klebsiella* type 1. *Carbohydrate Research* **50**, 115-120.

Evans, L.R. and Linker, A. (1973). Production and characterization of the slime polysaccharide of *Pseudomonas aeruginosa*. *Journal of Bacteriology* **116**, 915-924.

Gahan, L.C., Sandford, P.A. and Conrad, H.E. (1967). The structure of the serotype 2 capsular polysaccharide of *Aerobacter aerogenes*. *Biochemistry* **6**, 2755-2766.

Garegg, P.J., Lindberg, B., Onn, T. and Holme, T. (1971a). Structural studies on the M-antigen from two mucoid mutants of *Salmonella typhimurium*. *Acta Chemica Scandinavica* **25**, 1185-1194.

Garegg, P.J., Lindberg, B., Onn, T. and Sutherland, I.W. (1971b). Comparative structural studies on the M-antigen from *Salmonella typhimurium*, *Escherichia coli* and *Aerobacter cloacae*. *Acta Chemica Scandinavica* **25**, 2103-2108.

Goldman, R. and Strominger, J.L. (1972). Purification and properties of C_{55}-isoprenyl-pyrophosphate phosphatase from *Micrococcus lysodeikticus*. *Journal of Biological Chemistry* **247**, 5116-5122.

Grant, W.D., Sutherland, I.W. and Wilkinson, J.F. (1969). Exopolysaccharide colanic acid and its occurrence in the Enterobacteriaceae. *Journal of Bacteriology* **100**, 1187-1193.

Grant, W.D., Sutherland, I.W. and Wilkinson, J.F. (1970). Control of colanic acid synthesis. *Journal of Bacteriology* **103**, 89-96.

Harada, T. (1977). Production, properties and application of curdlan. In *Extracellular Microbial Polysaccharides*, pp.265-283. Edited by P.A. Sandford and A. Laskin. Washington: American Chemical Society.

Haug, A. and Larsen, B. (1971). Biosynthesis of alginate. Part II. Polymannuronic acid C-5-epimerase from *Azotobacter vinelandii*. *Carbohydrate Research* **17**, 297-308.

Hemming, F.W. (1977). Dolichol phosphate, a coenzyme in the glycosylation of animal membrane-bound glycoproteins. *Biochemical Society Transactions* **5**, 1223-1331.

Henderson, P.J.F., Giddens, R.A. and Jones-Mortimer, M.C. (1977). Transport of galactose, glucose and their molecular analogues by *Escherichia coli* K12. *Biochemical Journal* **162**, 309-320.

Herbert, D. and Kornberg, H.L. (1976). Glucose transport as rate-limiting step in the growth of *Escherichia coli* on glucose. *Biochemical Journal* **156**, 477-480.

Higashi, Y., Siewert, G. and Strominger, J.L. (1970). Biosynthesis of the peptidoglycan of bacterial cell walls. XIX. Isoprenoid alcohol phosphokinase. *Journal of Biological Chemistry* **245**, 3683-3690.

Holme, T. (1957). Continuous culture studies on glycogen synthesis in *Escherichia coli* B. *Acta Chemica Scandinavika* **11**, 763-775.

Jansson, P.-E., Kenne, L. and Lindberg, B. (1975). Structure of the extracellular polysaccharide from *Xanthomonas campestris*. *Carbohydrate Research* **45**, 275-282.

Koenig, V.L. and Perrings, J.D. (1955). Sedimentation and viscosity studies on the capsular and somatic polysaccharides of *Pneumococcus* type III. *Journal of Biophysical and Biochemical Cytology* **1**, 93-98.

Kornberg, H.L. (1976). Genetics in the study of carbohydrate transport by bacteria. *Journal of General Microbiology* **96**, 1-16.

Kornberg, H.L. and Reeves, R.E. (1972). Inducible phosphoenolpyruvate-dependent hexose phosphotransferase activities in *Escherichia coli*. *Biochemical Journal* **128**, 1339-1344.

Kornfeld, R.H. and Ginsburg, V. (1966). Control of synthesis of guanosinediphosphate-D-mannose and guanosinediphosphate-L-fucose in bacteria. *Biochimica et Biophysica Acta* **117**, 79-87.

Kundig, W., Ghosh, S. and Roseman, S. (1964). Phosphate bound to histidine in a protein as an intermediate in a novel phosphotransferase system. *Proceedings of the National Academy of Sciences, U.S.A.* **52**, 1067-1074.

Lindberg, B., Lönngren, J., Ruden, U. and Nimmich, W. (1975). Structural studies of the capsular polysaccharide of *Klebsiella* type 59. *Carbohydrate Research* **42**, 83-93.

Lindberg, B., Lindquist, B. and Lönngren, J. (1977). Structural studies of the capsular polysaccharide of *Klebsiella* type 37. *Carbohydrate Research* **58**, 443-451.

Markovitz, A. (1977). Genetics and regulation of bacterial capsular polysaccharide biosynthesis and radiation sensitivity. In *Surface Carbohydrates of the Prokaryotic Cell*, pp.415-462. Edited by I.W. Sutherland. London, New York & San Francisco: Academic Press.

Melton, L.D., Mindt, L., Rees, D.A. and Sanderson, G.R. (1976). Covalent structure of the extracellular polysaccharide from *Xanthomonas campestris*: evidence from partial hydrolysis studies. *Carbohydrate Research* **46**, 245-257.

Moraine, R.A. and Rogovin, P. (1966). Kinetics of polysaccharide B1459 fermentation. *Biotechnology and Bioengineering* **8**, 511-524.

Morris, E.R., Rees, D.A., Walkinshaw, M.D. and Darke, A. (1977). Order-disorder transition for a bacterial polysaccharide in solution. *Journal of Molecular Biology* **110**, 1-16.

Newbrun, E. and Sharma, M. (1976). Further studies on extracellular glucans synthesized by glucosyltransferases of oral streptococci. *Caries Research* **10**, 255-272.

Norval, M. (1969). Exopolysaccharide Synthesis. Ph.D. Thesis, University of Edinburgh.

Norval, M. and Sutherland, I.W. (1969). A group of *Klebsiella* mutants showing temperature-dependent polysaccharide synthesis. *Journal of General Microbiology* **57**, 369-377.

Ørskov, I., Ørskov, F., Jann, B. and Jann, K. (1963). Acidic polysaccharide antigens of a new type from *E. coli* capsules. *Nature* **200**, 144-146.

Postma, P.W. and Roseman, S. (1976). The bacterial phosphoenolpyruvate: sugar phosphotransferase system. *Biochimica et Biophysica Acta* **457**, 213-257.

Poxton, I.R., Lomax, J.A. and Sutherland, I.W. (1974). Isoprenoid alcohol kinase - a third butanol-soluble enzyme in *Klebsiella aerogenes* membranes. *Journal of General Microbiology* **84**, 231-233.

Preiss, J. (1969). The regulation of the biosynthesis of α 1,4 glucans in bacteria and plants. In *Current Topics in Cellular Regulation*, Volume 1, pp.125-160. Edited by B.L. Horecker and E.R. Stadtman. New York: Academic Press.

Robyt, J.F. and Taniguchi, H. (1976). The mechanism of dextran sucrase action. *Archives of Biochemistry and Biophysics* **174**, 129-135.

Rogovin, S.P., Anderson, R.F. and Cadmus, M.C. (1961). Production of polysaccharide with *Xanthomonas campestris*. *Journal of Biochemical and Microbiological Technology and Engineering* **3**, 51-63.

Saier, M.H. (1977). Bacterial phosphoenolpyruvate:sugar phosphotransferase systems: structural, functional and evolutionary interrelationships. *Bacteriological Reviews* **41**, 856-871.

Sandford, P.A., Pittsley, J.E., Knutson, C.A., Watson, P.R., Cadmus, M.C. and Jeanes, A. (1977). Variation in *Xanthomonas campestris* NRRL B1459: characterization of xanthan products of varying pyruvic acid content. In *Extracellular Microbial Polysaccharides*, pp.192-210. Edited by P.A. Sandford and A. Laskin. Washington: American Chemical Society.

Silman, R.W. and Rogovin, P. (1972). Continuous fermentation to produce xanthan biopolymer: effect of dilution rate. *Biotechnology and Bioengineering* **14**, 23-31.

Stubbe, J.A. and Kenyon, G.L. (1972). Analogs of phosphoenol pyruvate substrate specificities of enolase and pyruvate kinase from rabbit muscle. *Biochemistry* **11**, 338-345.

Sutherland, I.W. (1967). Phage-induced fucosidases hydrolysing the exopolysaccharide of *Klebsiella aerogenes* type 54 [A3(SL)]. *Biochemical Journal* **104**, 278-285.

Sutherland, I.W. (1970). Structure of *Klebsiella aerogenes* type 8 polysaccharide. *Biochemistry* **9**, 2180-2185.

Sutherland, I.W. (1972). The exopolysaccharides of *Klebsiella* serotype 2 strains as substrates for phage-induced polysaccharide depolymerases. *Journal of General Microbiology* **70**, 331-338.

Sutherland, I.W. (1975). Lipids in the synthesis of lipopolysaccharides and exopolysaccharides. *Biochemical Society Transactions* **3**, 840-843.

Sutherland, I.W. (1977a). Microbial exopolysaccharide synthesis. In *Extracellular Microbial Polysaccharides*, pp.40-57. Edited by P.A. Sandford and A. Laskin. Washington: American Chemical Society.

Sutherland, I.W. (1977b). *Surface Carbohydrates of the Prokaryotic Cell*. London, New York & San Francisco: Academic Press.

Sutherland, I.W. and Norval, M. (1970). Synthesis of exopolysaccharide by *Klebsiella aerogenes* membrane preparations and the involvement of lipid intermediates. *Biochemical Journal* **120**, 567-576.

Sutherland, I.W. and Wilkinson, J.F. (1968). The exopolysaccharide of *Klebsiella aerogenes* A3 (SL) (type 54). *Biochemical Journal* **110**, 749-754.

Taguchi, R., Kikuchi, Y., Sakano, Y., and Kobayashi, T. (1973). Structural uniformity of pullulan produced by several strains of *Pullularia pullulans*. *Agricultural and Biological Chemistry* **37**, 1583-1588.

Tarcsay, L., Jann, B. and Jann, K. (1971). Immunochemistry of the K antigens of *Escherichia coli*. *European Journal of Biochemistry* **23**, 505-514.

Thurow, H., Choy, Y.M., Frank, N., Niemann, H. and Stirm, S. (1975). The structure of *Klebsiella* serotype 11 capsular polysaccharide. *Carbohydrate Research* **41**, 241-255.

Troy, F.A., Frerman, F.E. and Heath, E.C. (1971). The biosynthesis of capsular polysaccharide in *Aerobacter aerogenes*. *Journal of Biological Chemistry* **246**, 118-133.

Ward, J.B. and Glaser, L. (1968). An *E. coli* mutant with cryptic UDP-sugar hydrolase and altered metabolite regulation. *Biochemical and Biophysical Research Communications* **31**, 671-677.

Ward, J.B. and Glaser, L. (1969). Turnover of UDP-sugars in *E. coli* mutants with altered UDP-sugar hydrolase. *Archives of Biochemistry and Biophysics* **134**, 612-622.

Whiting, P.H., Midgley, M. and Dawes, E.A. (1976). The regulation of transport of glucose, gluconate, and 2.oxogluconate and of glucose catabolism in *Pseudomonas aeruginosa*. *Biochemical Journal* **154**, 659-668.

Wilkinson, J.F. (1958). The extracellular polysaccharides of bacteria. *Bacteriological Reviews* **22**, 46-73.

Wilkinson, J.F., Duguid, J.P. and Edmunds, P.N. (1954). The distribution of polysaccharide production in *Aerobacter* and *Escherichia* strains and its relation to antigenic character. *Journal of General Microbiology* **11**, 59-72.

Williams, A.G. and Wimpenny, J.W.T. (1977). Exopolysaccharide production by *Pseudomonas* NCIB 11264 grown in batch culture. *Journal of General Microbiology* **102**, 13-21.

Willoughby, E., Higashi, Y. and Strominger, J.L. (1972). Enzymatic dephosphorylation of C_{55}-isoprenoid alcohol pyrophosphate. *Journal of Biological Chemistry* **247**, 5113-5115.

Wolf, C., Elsässer-Beile, U., Stirm, S., Dutton, G.G.S. and Burchard, W. (1978). Conformational studies of bacterial polysaccharides. *Biopolymers* (in press).

Woodruff, H.B., Mata, J.M., Hernandez, S., Mochales, S., Rodriguez, A., Stapley, E.O., Wallick, H., Miller, A.K. and Hendlin, D. (1977). Fosfomycin: laboratory studies. *Chemotherapy* **23** (suppl. 1), 1-22.

Reference Added in Proof

Bebault, G.M., Dutton, G.G.S., Funnell, N.A. and Mackie, K.L. (1978). Structural investigation of *Klebsiella* serotype K32 polysaccharide. *Carbohydrate Research* **63**, 183-192.

Chapter 2

BACTERIAL ALGINATE SYNTHESIS

T.R. JARMAN

Tate & Lyle Ltd., Group Research and Development, Reading, UK

Introduction

Alginate has been obtained from certain species of marine algae, for example, *Laminaria digitata* and *Ascophyllum nodosum*, for many decades. It is well established as a commercially important polysaccharide being used predominantly as a gelling agent or a viscosifying agent in foods, textile printing and pharmaceuticals [Booth, 1975; McDowell, 1975]. The author's involvement with alginate is due to the finding that some bacteria produce an alginate-like exopolysaccharide, thus offering an alternative commercial source of this polymer. In this article studies on bacterial alginate synthesis conducted both in our own laboratories and elsewhere are reviewed.

Alginate-Producing Bacteria

Azotobacter vinelandii

Azotobacter vinelandii, in common with other species of the *Azotobacter* genus are predominantly mucoid when isolated from their natural soil habitat and retain this character on maintenance in the laboratory. Several strains of *Azotobacter vinelandii* have been shown to produce an alginate-like polysaccharide containing mannuronic and guluronic acid [Gorin and Spencer, 1966; Larsen and Haug, 1971; Page and Sadoff, 1975; Couperwhite and McCallum, 1975] but earlier investigations of *Azotobacter vinelandii* exopolysaccharide are at variance with these findings. Cohen and Johnstone [1964] identified galacturonic acid as the main component of the exopolysaccharide of three strains of *Azotobacter vinelandii* together with minor quantities of glucose, rhamnose, mannuronolactone, acetate and material reacting positively with thiobarbituric acid. Rigorous identification of galacturonic acid was not made and it is possible that glucose and rhamnose were derived from a second polysaccharide which was separated in later studies [Gorin and Spencer, 1966]. Rhamnose and 2-keto-3-deoxygalacturonic acid were proposed as the major constituents of the exopolysaccharide from *Azotobacter vinelandii* 1484 [Claus, 1965]. There is therefore some doubt about whether all strains of *Azotobacter vinelandii* produce alginate-like exopolysaccharide. Other species of *Azotobacter* have been shown to produce exopolysaccharides distinct from alginate [Parikh and Jones, 1963; Lawson and Stacy, 1954].

Pseudomonas aeruginosa

In contrast to the normally mucoid *Azotobacter vinelandii*, the majority of isolates of the ubiquitous *Pseudomonas aeruginosa* are not copious producers of exopolysaccharide [Doggett, 1969]. Mucoid strains of *Pseudomonas aeruginosa* have been found in association with pathological conditions, especially respiratory tract infections accompanying cystic fibrosis [Doggett and Harrison, 1969]. Mucoid strains can also be isolated from non-mucoid laboratory cultures on the basis of drug resistance [Govan and Fife, 1978], bacteriocin resistance [Govan, 1976] and phage resistance [Martin, 1973]. In addition a mucoid determinant was transferred from a mucoid strain by conjugation and conferred exopolysaccharide producing ability on a previously non-mucoid strain of *Pseudomonas aeruginosa* [Govan, 1976]. It has been suggested that a possible explanation for mucoid strains arising in the lungs of cystic fibrosis patients is *in vivo* selection based on drug resistance [Doggett and Harrison, 1969; Govan and Fife, 1978]. Both natural isolates and mucoid strains produced *in vitro* are unstable, rapidly changing to a non-mucoid form on repeated sub-culturing [Govan, 1975]. The rate of increase of non-mucoid organisms in an initially totally-mucoid, ammonia-limited continuous culture was consistent with mutation to and subsequent selective advantage of the non-mucoid form [Mian *et al.*, 1978]. Reversion to this form was curtailed by incorporation of surfactants into maintenance media [Govan, 1975] and by growth under sulphate, magnesium and iron limitation in continuous cultures [Jones *et al.*, 1977]. The exopolysaccharide produced both by mucoid *Pseudomonas aeruginosa* strains isolated from natural habitats and induced *in vitro* has been shown to be a partially acetylated alginate-like polymer [Linker and Jones, 1966; Carlson and Matthews, 1966; Evans and Linker, 1973; Govan, personal communication]. With the exception of the commercially important *Xanthomonas campestris* exopolysaccharide [Jansson *et al.*, 1975] the exopolysaccharides of other *Pseudomonas* species have received little attention. The exopolysaccharide produced by *Pseudomonas stutzeri* was shown to be composed mainly of glucose and mannose [Dellweg *et al.*, 1975] and that of *Pseudomonas fluorescens* entirely of mannose [Eagon, 1956]. Purified polysaccharide from an isolate identified as a *Pseudomonas* species [Williams and Wimpenny, 1977] has been shown to contain glucose, galactose, acetate and pyruvate in the ratio of 7:1:1:1 [Lawson and Symes, personal communication].

Structure and Biosynthesis

Algal alginate is a 1,4 linked unbranched co-polymer of β-D-mannuronic acid and its C-5 epimer α-L-guluronic acid (Figure 1) [Drummond *et al.*, 1962]. Partial acid hydrolysis has revealed that some regions of the polymer contain homopolymeric sequences or "blocks" which are almost exclusively polymannuronic acid or polyguluronic acid, with other sequences which contain both monomers (Figure 1) [Haug *et al.*, 1966]. It was thought by these authors that

MONOMERS

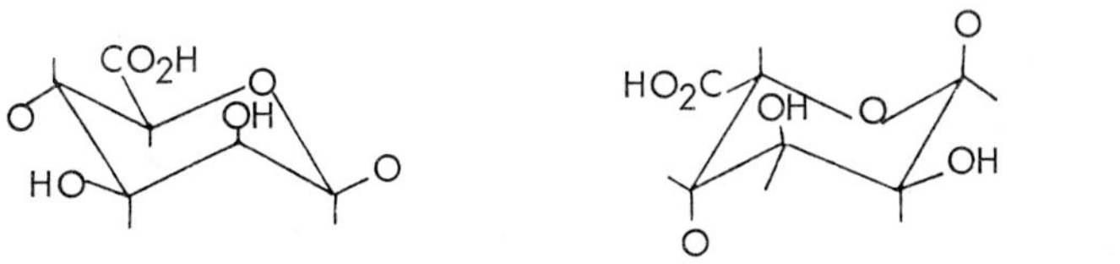

β-D-mannopyranosyluronic acid

α-L-gulopyranosyluronic acid

BLOCK STRUCTURE

....M-M-M-M-M-M....

....G-G-G-G-G-G....

....M-G-M-G-M-G....

Fig. 1 The structure of alginic acid.

in the mixed sequences the monomers were strictly alternating but more recent evidence obtained using alginate degrading enzymes suggests a more random sequence [Min *et al.*, 1977; Simonescu *et al.*, 1975]. Both the proportion of monomers in the polymer and their block arrangement can vary [Penman and Sanderson, 1972] thus altering the properties of the polymer [see Sutherland, Chapter 1]. For example, the greater the content of polyguluronic acid blocks, the stronger and more rigid are gels formed in the presence of calcium ions [Smidsrod, 1974]. The exopolysaccharides produced by *Azotobacter vinelandii* and *Pseudomonas aeruginosa* have the same basic structure as algal alginates but differ in that a proportion of the C2 and/or

C3 hydroxyl groups are usually acetylated. The degree of acetylation of the bacterial polymers is variable [see Sutherland, Chapter 1]. Gorin and Spencer [1966] found that 1 in 5.2 uronic acid residues of *Azotobacter vinelandii* exopolysaccharide were acetylated whereas Larsen and Haug [1971] found the acetate content to vary between 1 in 5.7 residues to very low levels (1 in 66 residues) depending on growth conditions. The O-acetyl content of *Pseudomonas aeruginosa* exopolysaccharide varied with bacterial strain between zero and 13.4% w/w [Evans and Linker, 1973], this higher limit representing one acetyl group per 1.5 uronic acid residues. Evidence obtained by degradation of bacterial alginates with lyase enzymes has shown that O-acetyl groups are exclusively associated with sequences of polymannuronic acid [Davidson *et al.*, 1977]. As in algal alginates the proportion of L-mannuronate to D-guluronate residues in bacterial polymers is variable and the monomers are in a block arrangement [Linker and Evans, 1976; Larsen and Haug, 1971].

Biosynthetic pathways

Exopolysaccharide is produced by *Azotobacter vinelandii* from a range of mono- and disaccharides including sucrose, glucose, fructose, lactose, maltose and mannitol [Deavin, personal communication]. A pathway for *Azotobacter* alginate synthesis from sucrose has been proposed on the basis of detectable enzymes (Figure 2) [Pindar and Bucke, 1976]. Available evidence indicates that sucrose is transported into the cell by an L-malate-linked system [Barnes, 1974] and cleaved to fructose and glucose by an intracellular invertase [Pindar and Bucke, 1976]. Under the majority of growth conditions little extracellular glucose or fructose is found in the culture broth during growth on sucrose. The alginate biosynthetic pathway proceeds from hexose-6-phosphates, formed and interconverted by the appropriate hexokinases and hexose-6-phosphate isomerases, via GDP-mannose and GDP-mannuronic acid. No evidence was obtained for the involvement of GDP-guluronic acid as an alginate precursor in *Azotobacter vinelandii*. This intermediate has been found in the marine algae [Lin and Hassid, 1966a] though its incorporation into alginic acid has not been demonstrated. It is likely therefore that in *Azotobacter vinelandii* polymannuronic acid is the initial polymeric product. The cellular location of the enzymes involved in polymannuronate synthesis has not been determined and it is not yet known whether membrane-bound polyisoprenyl phosphates act as cofactors in the polymerisation as has been found for some other bacterial exopolysaccharide and lipopolysaccharide synthesising systems [Sutherland, 1977]. A polymannuronic acid C-5-epimerase occurs extracellularly in cultures of *Azotobacter vinelandii* which catalyses the conversion of mannuronic acid residues to guluronic acid

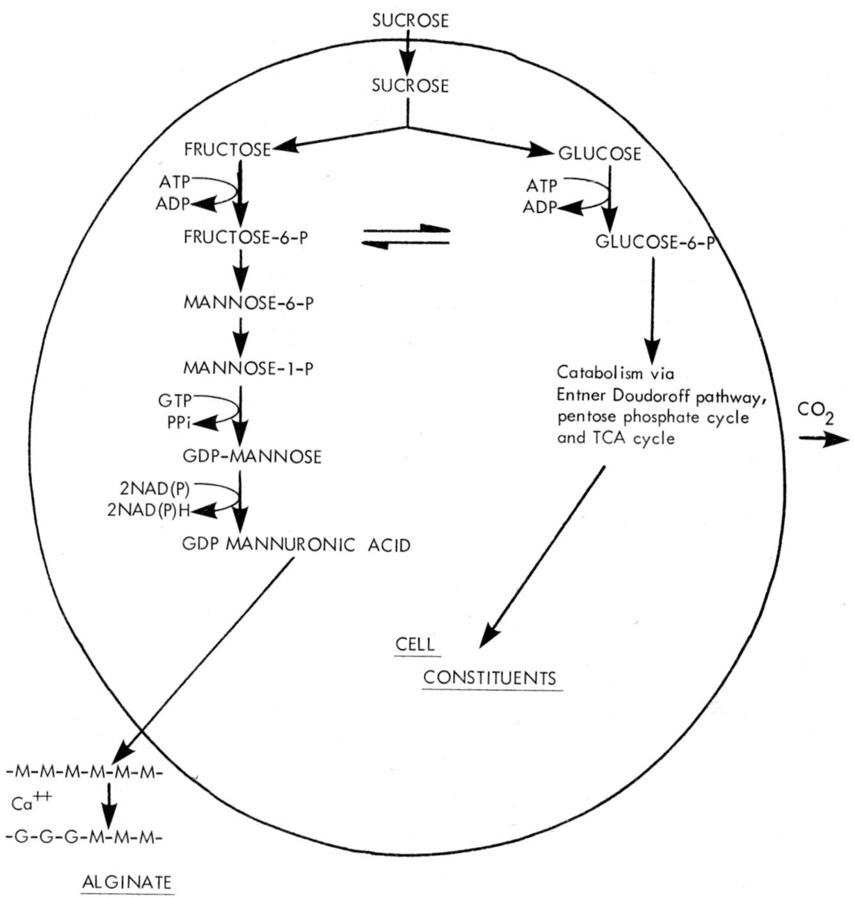

Fig. 2 Metabolism of *Azotobacter vinelandii* in relation to alginate synthesis.

residues in the polymer [Haug and Larsen, 1971]. This enzyme, which has also been found in the marine algae *Pelvetia canaliculata* [Madgwick *et al.*, 1973], is dependent on calcium ions for activity. By increasing the incubation time and the concentration of calcium in the growth medium of *Azotobacter vinelandii* it was possible to increase the guluronate content of the exopolysaccharide obtained from 18% to 67% of total uronic acids [Larsen and Haug, 1971] and thus to vary the properties of the product obtained in respect to its interaction with calcium and other divalent cations. This is additional evidence that polymannuronic acid is the initial polymeric product. The finding that O-acetyl groups are associated exclusively with mannuronic acid residues has led to the suggestion that O-acetyl groups protect mannuronic acid residues from epimerisation [Davidson *et al.*, 1977]. The variability in the extent of acetylation is consistent with the suggestion that as for other bacterial exopolysaccharides, insertion of acetyl groups is not a necessary part of alginate biosynthesis [Sutherland, 1977].

The pathway of alginate synthesis in *Azotobacter vinelandii* is therefore similar to that of marine algae [Lin and Hassid, 1966b] but whereas GDP-guluronic acid is possibly an intermediate in algae no evidence has been obtained for the involvement of this intermediate in *Azotobacter vinelandii*. Investigations currently in progress have determined that GDP-mannose pyrophosphorylase and GDP-mannose dehydrogenase are present in mucoid strains of *Pseudomonas aeruginosa* indicating that this organism possesses a similar biosynthetic pathway to *Azotobacter vinelandii* and marine algae [Piggot and Sutherland, personal communication].

Azotobacter alginate lyase

In addition to synthesising alginate *Azotobacter vinelandii* produces an alginate lyase which catalyses a cleavage of the polyuronide chain to yield one chain terminating at the reducing end with a uronic acid and one chain terminating at the non-reducing end with a 4,5-unsaturated uronic acid (Figure 3) [Haug and Larsen, 1971]. This enzyme is specific for mannuronic acid-containing regions of the polymer [Davidson, 1975] and is 'endo' in its mode of action, causing a very rapid decrease in the solution viscosity of the exopolysaccharide for a small increase in the presence of 4,5-unsaturated uronic acid residues [Jarman, unpublished data]. A bacterium identified as an *Azotobacter vinelandii* type has been shown to grow on its own exopolysaccharide [Proctor, 1959] although the nature of the polymer was not investigated. This observation, coupled with the presence of an alginate lyase enzyme, has led to the suggestion that alginate functions as a storage polymer in *Azotobacter vinelandii* [Couperwhite and McCallum, 1975]. However, we have failed to obtain growth of alginate lyase-producing strains of *Azotobacter vinelandii* on sodium alginate, a finding which contradicts that suggestion. The function of this enzyme therefore remains obscure but it might possibly be involved in release of the polysaccharide chain during biosynthesis or function in the germination of *Azotobacter vinelandii* cysts which contain alginate as a major component of the cyst exine [Sadoff, 1975].

The culture broth of mucoid *Pseudomonas aeruginosa* retains its viscosity on storage [Mian *et al.*, 1978] and thiobarbituric reacting material could not be detected following periodate oxidation of *Pseudomonas aeruginosa* culture broths which indicates that *Pseudomonas aeruginosa* does not produce alginate lyases.

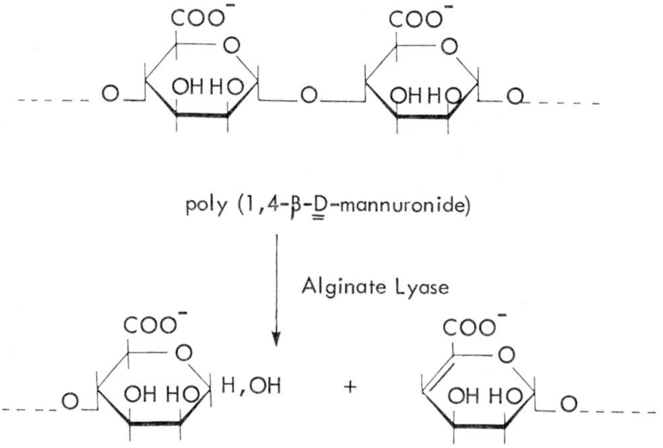

Fig. 3 Mode of action of alginate lyase acting on a region of β 1,4-polymannuronide to form one polymer containing 4,5-unsaturated uronic acid at the non-reducing end and one containing a mannuronate residue at the reducing end.

Physiological Studies on Bacterial Alginate Synthesis

As part of a development programme for a commercial process for alginate production from *Azotobacter vinelandii*, studies have been made on the physiology of the organism with respect to exopolysaccharide production and these have been considered in relation to studies on exopolysaccharide synthesis by other bacteria. The importance of using continuous culture techniques in such studies has been discussed elsewhere [Deavin *et al.*, 1976]. Inherent problems associated with the metabolic characteristics of *Azotobacter vinelandii* must be overcome before an efficient process can be developed. *Azotobacters* are nitrogen-fixing obligate aerobes and catabolise carbohydrate growth substrates via the Entner-Doudoroff pathway, pentose phosphate cycle and tricarboxylic acid to carbon dioxide [Still and Wang, 1964]. They respond to increases in dissolved oxygen tension by increasing their specific respiration rate and can thus convert a large proportion of the carbon and energy growth substrate to carbon dioxide in an energetically wasteful process [Downs and Jones, 1975]. It has been proposed that this feature of metabolism helps to maintain a low oxygen concentration environment thus protecting an oxygen-sensitive nitrogenase from inactivation

[Dalton and Postgate, 1969]. Conversely, under conditions where oxygen supply is restricted, especially at low growth rates, the intracellular storage compound poly-3-hydroxybutyrate is accumulated [Dawes, 1975] and can contribute up to 70% of cell mass. This also represents a diversion of carbohydrate substrate from alginate synthesis. For maximum yield of alginate from carbohydrate strict control of oxygen supply is therefore crucial.

Effect of oxygen availability

The effect of oxygen availability on exopolysaccharide production in phosphate-limited continuous cultures of *Azotobacter vinelandii* with a dilution rate of 0.15 h^{-1} was investigated by altering the fermenter agitator speed and thus the rate of oxygen transfer into the culture broth [Jarman *et al.*, 1978]. In the steady states obtained, cell mass was essentially constant and no significant accumulation of poly-3-hydroxybutyrate occurred. With increasing agitator speed the specific respiration rate increased (Figure 4) until at the highest agitator speed almost 70% of the sucrose used by the organism was converted to carbon dioxide. The conversion efficiency for sucrose into exopolysaccharide fell from an optimum value of 40% at the lowest respiration rate to 8% at very high specific respiration rates. The specific rate of exopolysaccharide synthesis, however, varied only two-fold for the tenfold increase in specific respiration rate which occurred.

Effect of growth rate

Variation of the dilution rate of phosphate-limited continuous cultures of *Azotobacter vinelandii* caused elevation of cell mass at dilution rates of 0.1 h^{-1} and below (Figure 5) [Jarman *et al.*, 1978]. This increase was found to be entirely due to intracellular accumulation of poly-3-hydroxybutyrate. "Active" cell mass, that is the measured cell mass minus poly-3-hydroxybutyrate, was essentially constant over the dilution rate range studied. The specific rate of exopolysaccharide synthesis, expressed on the basis of "active" cell mass was found to be independent of specific growth rate (which in continuous culture equals the dilution rate) within the range 0.05-0.25 h^{-1} [Jarman *et al.*, 1978]. Thus exopolysaccharide concentration increased with decreasing dilution rate. Re-examination of continuous culture studies on synthesis of the exopolysaccharide xanthan by *Xanthomonas campestris* and of a galactoglucan by a *Pseudomonas* species have revealed that the specific rate of exopolysaccharide synthesis was also independent of specific growth rate in these systems [Deavin *et al.*, 1976]. This finding is in conflict with the view, resulting from a batch culture experiment, that galactoglucan synthesis by *Pseudomonas* species resembles secondary

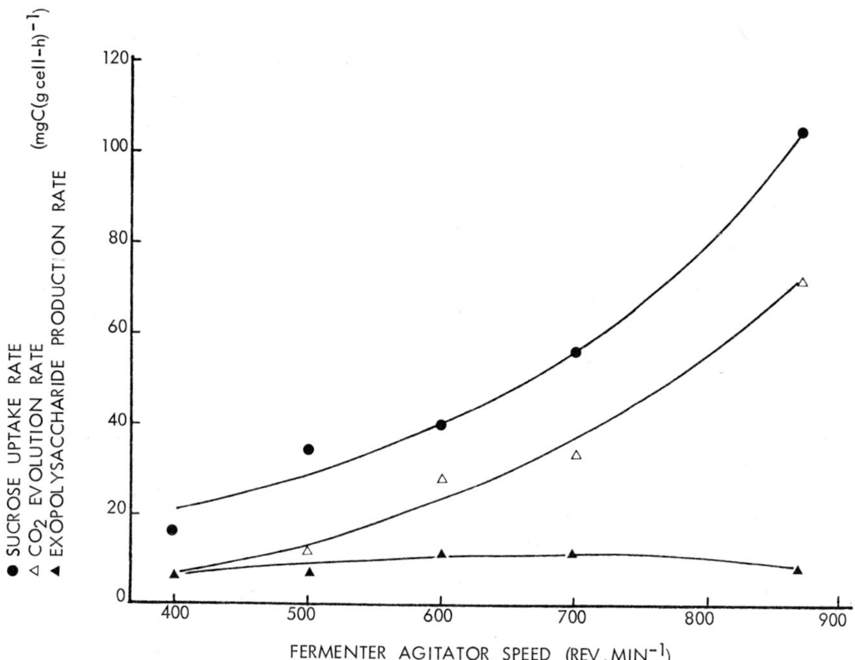

Fig. 4 Sucrose use and exopolysaccharide production in relation to respiration rate in phosphate-limited continuous cultures of *Azotobacter vinelandii*.

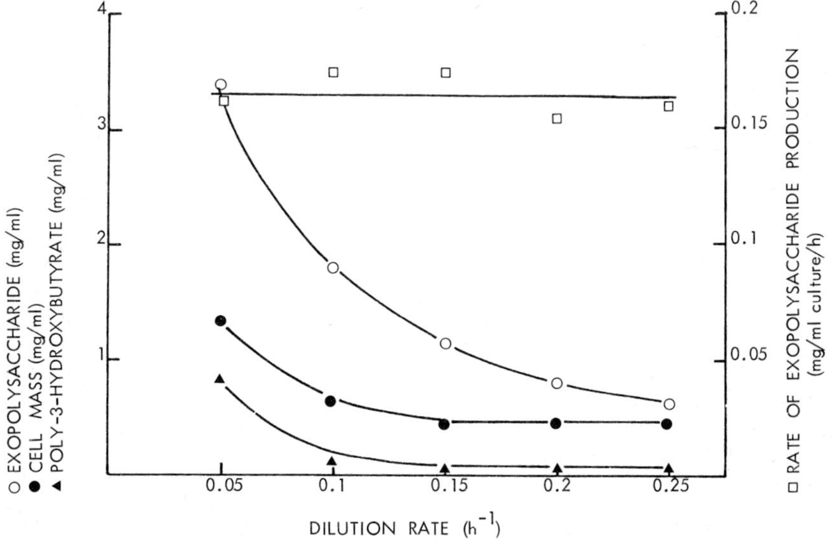

Fig. 5 Exopolysaccharide production in phosphate-limited continuous cultures of *Azotobacter vinelandii* as a function of dilution rate.

metabolite synthesis and that the polymer is only produced under conditions of very low growth rate or in stationary phase [Williams and Wimpenny, 1977]. In contrast to the systems mentioned above, in nitrogen-limited continuous cultures of a mucoid *Pseudomonas aeruginosa* strain grown at dilution rates of 0.05 h^{-1} and 0.1 h^{-1} both alginate concentration and cell mass were independent of dilution rate. The specific rate of alginate synthesis therefore increased with increased specific growth rate [Mian *et al*., 1978].

TABLE 1

Effect of growth-limiting nutrient on exopolysaccharide production by Azotobacter vinelandii grown in continuous culture at $D = 0.15\ h^{-1}$

Limiting nutrient	Cell mass ($g.l^{-1}$)	Polysaccharide produced mg C (g cell.h)$^{-1}$
$MoO_4{}^{2-}$	1.1	10.5
$PO_4{}^{3-}$	1.9	8.9
Fe^{2+}	1.4	8.1
C (sucrose)	1.3	7.7
$N(N_2)$	1.5	6.8
Ca^{2+}	1.2	6.2
$SO_4{}^{2-}$	1.7	7.8
K^+	1.9	5.0
O_2	1.2	1.9

Effect of growth-limiting nutrient

Exopolysaccharide, determined as isopropanol precipitated material, was produced from sucrose by *Azotobacter vinelandii* when growth was limited by a wide range of nutrients in continuous cultures with similar specific respiration rates (Table 1) [Jarman *et al*., 1978]. Except under conditions of oxygen limitation, in which cell mass would be expected to be boosted by the intracellular accumulation of poly-3-hydroxybutyrate, the specific rate of exopolysaccharide synthesis varied over a two-fold range. Molybdate limitation, followed by phosphate limitation, the conditions routinely used, gave the highest rates of exopolysaccharide synthesis. Exopolysaccharide was produced at rates comparable to those obtained under other limitations when sucrose, the carbon and energy substrate, was growth limiting. Although this was an unexpected finding it was also observed that alginate was produced by glucose-limited continuous cultures of *Pseudomonas aeruginosa* [Mian *et al*., 1978]. Some bacterial exopolysaccharides are not produced under carbon-limited conditions. A galacto-glucan was produced when a *Pseudomonas* species was continuously cultured under ammonia- and phosphate-limitation but not when grown under glucose-limitation [Williams and Wimpenny, 1978]. Ammonia-limitation was found to be most favourable for exopolysaccharide production from glucose by *Klebsiella aerogenes*. Production also occurred under sulphate- and phosphate-limitation but not under carbon-limited conditions [Neijssel and Tempest, 1975]. It was proposed that in *Klebsiella aerogenes* exopolysaccharide was produced as an 'overflow metabolite', its biosynthesis possibly functioning as a 'slip reaction' effecting the hydrolysis and turnover of ATP [Neijssel and Tempest, 1975; 1976]. The production of alginate by both *Azotobacter vinelandii* and *Pseudomonas aeruginosa* under carbon-limited conditions eliminates overflow metabolism as a possible function for exopolysaccharide production in these species.

Metabolic control of alginate biosynthesis in Azotobacter vinelandii

The specific rate of alginate synthesis from sucrose by *Azotobacter vinelandii* has proved to be relatively independent of nutrient limitation, specific respiration rate and specific growth rate. Nothing is known about how this constant rate of synthesis is achieved at the metabolic level. It is apparent, however, from the finding that alginate is produced in sucrose-limited conditions in continuous culture and at similar specific rates for a wide range of specific respiration rates in phosphate-limited conditions that the rate of alginate synthesis is largely independent of both the flux of sucrose into the cell and the flux of hexose intermediates through the catabolic pathways. In bacteria, exopolysaccharide synthesis shares common precursors (hexose nucleosides) and possible common cofactors (polyisoprenyl phosphates) with cell wall biosynthesis. Evidence has been presented which indicates that in some bacteria exopolysaccharide synthesis is dependent upon the availability of polyisoprenyl phosphate and is therefore in competition with cell wall synthesis for this cofactor [Sutherland, 1977]. The independence of rate of exopolysaccharide synthesis from specific growth rate in *Azotobacter vinelandii* and the increasing rate of alginate synthesis with specific growth rate in *Pseudomonas aeruginosa* suggests that in these bacteria competition between exopolysaccharide synthesis and cell wall synthesis for precursors or cofactors does not greatly influence the rate of exopolysaccharide synthesis.

Possible Functions of Bacterial Alginate

It is intriguing to speculate on why two bacteria as diverse as the soil-inhabiting, nitrogen-fixing *Azotobacter vinelandii* and strains of the opportunist pathogen *Pseudomonas aeruginosa* should, in common with marine algae, produce alginate or alginate-like polymers. In common with many other bacterial exopolysaccharides [Dudman, 1977], the role(s) of bacterial alginate remain obscure. Although loss of ability to produce exopolysaccharide does not affect the viability of laboratory strains of *Azotobacter vinelandii*, exopolysaccharide production under the wide range of environmental conditions described indicates a major role *in vivo*. Although expression of alginate synthesis by *Pseudomonas aeruginosa* is only selected for in specific conditions, the fact that most strains of *Pseudomonas aeruginosa* appear to contain the genetic information necessary for alginate synthesis would indicate that this is an important survival mechanism for this organism also.

Alginate has been identified as a cell wall component in the brown seaweeds [Vreeland, 1970; Evans and Holligan, 1972] and also occurs in the intracellular liquid of *Laminaria* and *Ascophyllum* species [Haug *et al*., 1969]. It has been suggested that it contributes to the tensile strength and flexibility demanded of these algae by their turbulent environment [Hellebust and Haug, 1972]. A structural function for alginate has also been demonstrated in metabolically dormant cysts of *Azotobacter vinelandii* [Sadoff, 1975]. There is evidence that it is an important component of both the cyst exine and intine, which are microscopically distinct regions outside the central body of the cyst. Mutants of *Azotobacter vinelandii* which do not produce exopolysaccharide or are depleted in mannuronate-C-5-epimerase were unable to encyst and addition of a phage-derived alginate lyase to an encysting culture of *Azotobacter vinelandii* inhibited encystment [Ekland *et al*., 1966]. This demonstrated structural function of alginate in cysts does not explain its copious extracellular production in metabolically active vegetative cells. Suggested roles for bacterial exopolysaccharides [see Dudman, 1977] include storage of carbon-energy source, involvement in uptake or detoxification of metal ions, acting as barriers to diffusion, participation in bacterial-plant and bacterial-animal interactions, protection against phagocytosis and phage attack and their synthesis functioning as 'overflow metabolism' in the presence of excess carbon source. As discussed previously, evidence suggests that bacterial alginate does not serve as a storage polymer or overflow metabolite. Some evidence has been presented to suggest a protective function for *Azotobacter vinelandii* alginate against heavy metal toxicity [Den Dooren de Jong, 1971] and as a diffusion barrier to oxygen [Postgate, 1974]. It is still not apparent, however, whether *Azotobacter vinelandii* exopolysaccharide serves a specific function or has a more general role in providing the bacterium with a hydrophilic, negatively charged coating which provides protection against attack and adverse environmental conditions. The findings that alginate production by *Pseudomonas aeruginosa* is related to resistance to antibiotics, phage and bacteriocins and protects against phagocytosis and antibody attachment [Costerton *et al.*, 1978] supports the suggestion that alginate might function as a general protective barrier.

References

Barnes, E.M. (1974). Glucose transport in membrane vesicles from *Azotobacter vinelandii*. *Archives of Biochemistry and Biophysics* **163**, 416-422.

Booth, E. (1975). Seaweeds in industry. In *Chemical Oceanography*, pp.219-268. Edited by J.P. Riley and G. Skirrow. London: Academic Press.

Carlson, D.M. and Matthews, L.W. (1966). Polyuronic acids produced by *Pseudomonas aeruginosa*. *Biochemistry* **5**, 2817-2822.

Claus, D. (1965). 2-keto-3-deoxygalacturonic acid as a constituent of an extracellular polysaccharide in *Azotobacter vinelandii*. *Biochemical and Biophysical Research Communications* **20**, 745-751.

Cohen, G.H. and Johnstone, D.B. (1964). Extracellular polysaccharides of *Azotobacter vinelandii*. *Journal of Bacteriology* **88**, 329-338.

Costerton, J.W., Geesey, G.G. and Cheng, K.-J. (1978). How bacteria stick. *Scientific American* **238**, 86-95.

Couperwhite, I. and McCallum, M.F. (1975). Polysaccharide production and the possible occurrence of GDP-D-mannose dehydrogenase in *Azotobacter vinelandii*. *Antonie van Leeuwenhoek Journal of Microbiology and Serology* **41**, 25-32.

Dalton, H. and Postgate, J.R. (1969). Effect of oxygen on growth of *Azotobacter chroococcum* in batch and continuous cultures. *Journal of General Microbiology* **54**, 463-473.

Davidson, I.W. (1975). Alginate lyases and their substrates. Ph.D. Thesis. University of Edinburgh.

Davidson, I.W., Sutherland, I.W. and Lawson, C.J. (1977). Localisation of O-acetyl groups of bacterial alginate. *Journal of General Microbiology* **98**, 603-606.

Dawes, E.A. (1975). The role and regulation of poly-β-hydroxybutyrate as a reserve in microorganisms. In *Proceedings of the International Symposium on Macromolecules*, pp.433-449. Edited by E.B. Mano. Amsterdam: Elsevier Scientific Publishing Company.

Deavin, L., Jarman, T.R., Lawson, C.J., Righelato, R.C. and Slocombe, S. (1977). The production of alginic acid by *Azotobacter vinelandii* in batch and continuous culture. In *Extracellular Microbial Polysaccharides*, pp.14-26. Edited by P.A. Sandford and A. Laskin. Washington: American Chemical Society.

Dellweg, H., John, M. and Foster, B. (1975). Characterisation of an extracellular polysaccharide from *Pseudomonas stutzeri*. *European Journal of Applied Microbiology* **1**, 307-312.

Den Dooren de Jong, L.W. (1971). Tolerance of *Azotobacter* for metallic and non-metallic ions. *Antonie van Leeuwenhoek Journal of Microbiology and Serology* **37**, 119-124.

Doggett, R.G. (1969). Incidence of mucoid *Pseudomonas aeruginosa* from clinical sources. *Applied Microbiology* **18**, 936-937.

Doggett, R.G. and Harrison, G.M. (1969). Significance of the pulmonary flora associated with chronic pulmonary disease in cystic fibrosis. *Proceedings of the 5th International Cystic Fibrosis Conference*, pp.175-188. Edited by D. Lawson. London: Cambridge Press.

Downs, A.J. and Jones, C.W. (1975). Respiration linked proton translocation in *Azotobacter vinelandii*. *FEBS Letters* **60**, 42-46.

Drummond, D.W., Hirst, E.L. and Percival, E. (1962). The constitution of alginic acid. *Journal of the Chemical Society (London)*, pp.1208-1216.

Dudman, W.F. (1977). The role of surface polysaccharides in natural environments. In *Surface Carbohydrates of the Prokaryotic Cell*, pp.357-414. Edited by I.W. Sutherland. London: Academic Press.

Eagon, R.G. (1956). Studies on polysaccharide formation by *Pseudomonas fluorescens*. *Canadian Journal of Microbiology* **2**, 673-676.

Ekland, C., Pope, L.M. and Wyss, O. (1966). Relationship of encapsulation and encystment in *Azotobacter*. *Journal of Bacteriology* **92**, 1828-1830.

Evans, L.R. and Linker, A. (1973). Production and characterisation of the slime polysaccharide of *Pseudomonas aeruginosa*. *Journal of Bacteriology* **116**, 915-923.

Evans, L.V. and Holligan, M.S. (1972). Correlated light and electron microscope studies on brown algae. 1. Localisation of alginic acid and sulphated polysaccharides in *Dictyota*. *New Phytologist* **71**, 1161-1172.

Gorin, P.A.J. and Spencer, J.F.T. (1966). Exocellular alginic acid from *Azotobacter vinelandii*. *Canadian Journal of Chemistry* **44**, 993-998.

Govan, J.R.W. (1975). Mucoid strains of *Pseudomonas aeruginosa*: the influence of culture medium on the stability of mucus production. *Journal of Medical Microbiology* **8**, 513-522.

Govan, J.R.W. (1976). Genetic studies on mucoid *Pseudomonas aeruginosa*. *Proceedings of the Society for General Microbiology* **3**, 187.

Govan, J.R.W. and Fife, J.A.M. (1978). Mucoid *Pseudomonas aeruginosa* and cystic fibrosis: Resistance of the mucoid form to carbenicillin, flucloxacillin and tobramycin and the isolation of mucoid variants *in vitro*. *Journal of Antimicrobial Chemotherapy* **4**, 233-240.

Haug, A. and Larsen, B. (1971). Biosynthesis of alginate. Polymannuronic acid C-5-epimerase from *Azotobacter vinelandii* (Lipman). *Carbohydrate Research* **17**, 297-308.

Haug, A., Larsen, B. and Baardseth, E. (1969). Comparison of the constitution of alginates from different sources. In *Proceeding of the Sixth International Seaweed Symposium*, pp.443-451. Edited by R.M. Margalef. Oxford: Pergamon Press.

Haug, A., Larsen, B. and Smidsrod, O. (1966). A study of the constitution of alginic acid by partial acid hydrolysis. *Acta Chemica Scandinavica* **20**, 183-190.

Hellebust, J.A. and Haug, A. (1972). Photosynthesis, translocation and alginic acid synthesis in *Laminaria digitata* and *Laminaria hyperborea*. *Canadian Journal of Chemistry* **50**, 169-176.

Jansson, P.E., Kenne, L. and Lindberg, B. (1975). Structure of the exopolysaccharide from *Xanthomonas campestris*. *Carbohydrate*

Research **45**, 275-282.

Jarman, T.R., Deavin, L., Slocombe, S. and Righelato, R.C. (1978). An investigation of the effect of environmental conditions on the rate of exopolysaccharide synthesis in *Azobacter vinelandii*. *Journal of General Microbiology* **107**, 59-64.

Jones, S.E., Brown, M.R.W. and Govan, J.R.W. (1977). The production of stable defined cultures of mucoid *Pseudomonas aeruginosa* in continuous culture. *Journal of Pharmacy and Pharmacology* **29**, 69.

Larsen, B. and Haug, A. (1971). Biosynthesis of alginate. Composition and structure of alginate produced by *Azotobacter vinelandii* (Lipman). *Carbohydrate Research* **17**, 287-296.

Lawson, G.J. and Stacey, M. (1954). Immunopolysaccharides. 1. Preliminary studies of a polysaccharide from *Azotobacter indicum* containing a uronic acid. *Journal of the Chemical Society (London)*, pp.1925-1931.

Lin, T.Y. and Hassid, W.Z. (1966a). Isolation of guanosine diphosphate uronic acids from a marine brown alga, *Fucus gardneri* Silva. *Journal of Biological Chemistry* **241**, 3282-3293.

Lin, T.Y. and Hassid, W.Z. (1966b). Pathway of alginic acid synthesis in the marine brown alga, *Fucus gardneri* Silva. *Journal of Biological Chemistry* **241**, 5284-5297.

Linker, A. and Evans, L.R. (1976). Unusual properties of glycuronans [poly(glycosyluronic)compounds]. *Carbohydrate Research* **47**, 179-187.

Linker, A. and Jones, R.S. (1966). A new polysaccharide resembling alginic acid isolated from Pseudomonads. *The Journal of Biological Chemistry* **241**, 3845-3851.

McDowell, R.H. (1975). New developments in the chemistry of alginates and their use in food. *Chemistry and Industry*, pp.391-395.

Madgwick, J., Haug, A. and Larsen, B. (1973). Polymannuronic acid 5-epimerase from the marine alga *Pelvetia canaliculata* (L.) Dene et Thur. *Acta Chemica Scandinavica* **27**, 3592-3594.

Martin, D.R. (1973). Mucoid variation in *Pseudomonas aeruginosa* induced by the action of phage. *Journal of Medical Microbiology* **6**, 111-118.

Mian, F., Jarman, T.R. and Righelato, R.C. (1978). Studies on the biosynthesis of exopolysaccharide by *Pseudomonas aeruginosa*. *Journal of Bacteriology* **134**, 418-422.

Min, H.E., Sasaki, S.F., Kashiwabara, Y., Umekawa, M. and Nisizawa, K. (1977). Fine structure of SMG alginate fragment in the light of its degradation by alginate lyases of *Pseudomonas* sp. *Journal of Biochemistry, Tokyo* **81**, 555-562.

Neijssel, O.M. and Tempest, D.W. (1975). The regulation of carbohydrate metabolism in *Klebsiella aerogenes* NCTC 418 organisms growing in chemostat culture. *Archives of Microbiology* **106**, 251-259.

Neijssel, O.M. and Tempest, D.W. (1976). Bioenergetic aspects of aerobic growth of *Klebsiella aerogenes* NCTC 418 in carbon-limited and carbon-sufficient chemostat culture. *Archives of Microbiology* **107**, 215-221.

Page, W.J. and Sadoff, H.L. (1975). Relationship between calcium and uronic acids in the encystment of *Azotobacter vinelandii*. *Journal of Bacteriology* **122**, 145-151.

Parikh, V.M. and Jones, J.K.N. (1963). The structure of the extracellular polysaccharide of *Azotobacter indicum*. *Canadian Journal of Chemistry* **41**, 2826-2835.

Penman, A. and Sanderson, G.R. (1972). A method for the determination of uronic acid sequences in alginates. *Carbohydrate Research* **25**, 273-282.

Pindar, D.F. and Bucke, C. (1975). The biosynthesis of alginic acid in *Azotobacter vinelandii*. *Biochemical Journal* **152**, 617-622.

Postgate, J.R. (1974). Evolution within nitrogen-fixing systems. *Symposia of the Society for General Microbiology* **24**, 263-292.

Proctor, M.H. (1959). A function for the extracellular polysaccharide of *Azotobacter vinelandii*. *Nature* **184**, 1934-1935.

Sadoff, H.L. (1975). Encystment and germination in *Azotobacter vinelandii*. *Bacteriological Reviews* **39**, 516-539.

Simonescu, R.I., Popa, V.I., Liga, A. and Rusan, V. (1975). Researches in the field of seaweed chemistry. IV. Molecular structure of alginic acid. *Cellulose Chemistry and Technology* **9**, 547-554.

Smidsrod, O. (1974). Molecular basis for some physical properties of alginates in the gel state. *Faraday Discussions of the Chemical Society* **57**, 263-274.

Still, G.C. and Wang, C.H. (1964). Glucose catabolism in *Azotobacter vinelandii*. *Archives of Biochemistry and Biophysics* **105**, 126-132.

Sutherland, I.W. (1977). Bacterial exopolysaccharides - their nature and production. In *Surface Carbohydrates of the Prokaryotic Cell*, pp.22-96. Edited by I.W. Sutherland. London: Academic Press.

Vreeland, V. (1970). Localisation of a cell wall polysaccharide in a brown alga with labelled antibody. *Journal of Histochemistry and Cytochemistry* **18**, 371-373.

Williams, A.G. and Wimpenny, J.W.T. (1977). Exopolysaccharide production by *Pseudomonas* NCIB 11264 grown in batch culture. *Journal of General Microbiology* **102**, 13-21.

Williams, A.G. and Wimpenny, J.W.T. (1978). Exopolysaccharide production by *Pseudomonas* NCIB 11264 grown in continuous culture. *Journal of General Microbiology* **104**, 47-57.

Chapter 3

CONTINUOUS CULTURE STUDIES ON THE PRODUCTION OF EXTRACELLULAR POLYSACCHARIDES BY *XANTHOMONAS JUGLANDIS*

C.G.T. EVANS, R.G. YEO and D.C. ELLWOOD

Microbiological Research Establishment, Porton Down, Nr. Salisbury, U.K.

Introduction

Microbial exopolysaccharides - "gums" - are produced during the growth of various genera of bacteria and yeasts, and following the pioneer work of the group at the Northern Regional Research Laboratory, Peoria, many of these gums, particularly those produced by *Xanthomonas* species [see Sutherland, Chapter 1], have been shown to be useful as thickeners and emulsifiers in a wide variety of applications which includes the food and oil industries [Wells, 1977; Andrew, 1977; Gabriel, Chapter 8]. Production of xanthan gum is now an important commercial undertaking with demand increasing rapidly. In the oil industry alone, the annual demand for xanthan is anticipated to reach 50,000 tonnes per year by 1990 [Wells, 1977].

In 1973 we started to investigate the possibility of using continuous culture to produce gums of the xanthan type since, by analogy with Single Cell Protein production [Herbert, 1976], economic supply of material sufficient to meet the growing demand must require the development of continuous production processes. However, reports in the literature have indicated that the ability of *Xanthomonas campestris* to produce useful amounts of gum is rapidly lost during continuous culture in a single stage chemostat [see e.g. Silman and Rogovin, 1972], and attempts have been made by some workers to prolong useful production by the use of two or more stages [Lindblom and Patton, 1967].

At this point it is as well to emphasize a fundamental difference between batch and continuous culture methods: the environment of cells growing in batch culture is continually changing throughout the "growth cycle". For example, provided that other adverse conditions such as toxic products or extreme pH have not arisen, at the stationary phase the supply of a nutrient that was originally in excess will be exhausted. In a continuous culture the growth medium is continuously supplied to a culture vessel of constant volume and is so formulated that one specific chosen component, the "limiting nutrient", is present in a concentration low enough to ensure that the other components are present in excess. The cells' environment is therefore constant, and growth is controlled by the rate at which the medium containing the limiting nutrient is supplied; the medium feed pump consequently regulates the rate at which the cells can grow. Only in such a system can nutrient-limited cells be continuously produced for long periods.

Most of the early work on continuous culture was done, with the carbon source as the limiting nutrient, to investigate the effect of growth rate upon the physiology of cells. Subsequently, chemostat studies - particularly at Porton - have shown that the choice of limiting nutrient can have profound effect upon the composition and properties of, *inter alia*, some of the constituent cellular polymers [Ellwood and Tempest, 1969; Tempest and Neijssel, 1976]. This is well illustrated by the replacement of the phosphorus-containing teichoic acid cell wall polymer by a non-phosphorus-containing teichuronic acid polymer when growth of some bacilli is limited by phosphate [Ellwood and Tempest, 1969]. That such changes are not confined to structural or intracellular polymers was shown by San Blas and Cunningham [1974] in an experiment with *Hansenula holstii* (NRRL 2448), which had been shown [Jeans *et al.*, 1961] to produce an extracellular phosphomannan; when growth in batch culture was restricted by phosphate, the polysaccharide produced was a non-phosphorylated mannan. A similar effect has been demonstrated in phosphate-limited chemostat cultures of this organism [D.C. Ellwood and B. Johnson, unpublished]. The choice of limiting nutrient has been found also to affect the culture stability of some organisms [C.G.T. Evans and B. Capel, unpublished observations].

Despite general statements to the contrary [Dye and Lelliott, 1974], all the *Xanthomonas* strains tested by us were able to grow well in simple, chemically defined media with glucose as the carbon source. This enabled us to study the effects of growth-limitation by selected nutrients upon culture stability and the production and properties [see Powell, Chapter 6] of the extracellular polysaccharides produced. We report here experiments which show that, with suitable growth conditions, good yields of polysaccharides can be maintained for more than 2000 h, and that the natures and properties of the resultant polysaccharides can be affected by the growth conditions [Ellwood *et al.*, 1975].

Growth Experiments

The strains were scored as shown in Table 1. Although three strains of *Xanthomonas juglandis* and the *Xanthomonas campestris* strain were equally the most mucoid, we decided, in view of the extensive patent coverage of *Xanthomonas campestris*, to concentrate first on *Xanthomonas juglandis* and arbitrarily selected XJ 107 for our preliminary experiments.

TABLE 1

Estimates of the mucoid appearance on solid medium of 10 strains of Xanthomonas juglandis and one strain of Xanthomonas campestris (ATCC 13951)

Culture collection number	Mucoid appearance
NCPPB 411 and 1659	M-
NCPPB 362	M
NCPPB 412, 415	M+
NCPPB 413, ATCC 11329	M++
NCPPB 414, 1447, ICPB XJ 107, ATCC 13951	M+++

M- = non mucoid; M to M+++ indicates increasingly mucoid appearance. Inocula from all the primary seed stocks were streaked onto Tryptone Soya Agar (Oxoid/England) plates and examined for apparent polysaccharide production. This involved making subjective estimate of the variation in glistening appearance of obliquely lit colonies.

Polysaccharide production in complex nitrogen-limited medium

To establish a baseline level of polysaccharide yield, XJ 107 was grown batchwise in a fermenter in the complex organic medium. After 93 h the yield of acetone precipitable material (APM) was 20.5 g/l. The culture was then put on flow at a dilution rate (D) of 0.03 h^{-1}. A maximum level of acetone precipitated material of 22 ± 2 g/l was maintained for 5 days, but thereafter diminished steadily until at 400 h it had decreased to 1 g/l. During that time the glucose consumption also decreased from 37 to 11 g/l. While there was no evidence of contamination, colonial variation was early apparent. This result was similar to those reported for *Xanthomonas campestris*; Silman and Rogovin [1972] quoted maximum "Q" values of 6.5-8.7, where Q value = 1 was defined as a volume of feed medium equal to the fermenter culture volume. In the above experiments the Q value was only about 5. The physical properties of the polysaccharide were similar, but not identical, to those of commercial xanthan gum (Keltrol, Kelco, San Diego) from *Xanthomonas campestris*; a 1% (w/v) solution of Keltrol has an apparent viscosity at zero shear (K value) of about 5,500 cps whereas our material had a K value of 11,000 cps. The results of this and other experiments on the effects of other growth-limiting nutrients are shown in Table 2.

TABLE 2

*The effect of growth-limiting nutrient upon polysaccharide production by Xanthomonas juglandis XJ 107 in the chemostat**

Limiting element and medium*	Time on flow (h)	Dilution rate (h^{-1})	Amount of glucose used (g l^{-1})	APM yield (g l^{-1})	Max. culture K value (cps)	Culture variation (wks)†
N, complex	400	0.03	37-11	22	11,500	1
N, defined	900	0.03	37	23	15,000	-
C, defined	300	0.045	20	9	500	1
K, defined	1,000	0.03	25	9	360	1
Mg, defined	500	0.03	25	9	270	3
P, defined	1,500	0.03	44	18	5,000	3
S, defined	2,000	0.035	44	27	18,000	-

†Time on flow before appearance of variants. APM: Acetone precipitated material.

*Chemostat culture media: (a) Complex organic medium (g/100 ml): Malt extract (Oxoid), 0.3; Yeast Extract (Oxoid), 0.3; Mycological Peptone (Oxoid), 0.5; KH_2PO_4, 0.2; Glucose, 4.5. (b) Defined simple salts media: These were as described by Evans *et al.* [1970] and chosen to ensure that growth was limited by the carbon, nitrogen, potassium, magnesium or sulphur content. For a carbon-limited culture medium the glucose concentration was 2% (w/v), but for other limitations, the glucose concentration was increased to 4.5% (w/v), unless otherwise stated, to ensure its presence in excess. With complex organic medium, antifoam addition (P.2000, Shell Chemicals, England) was necessary during the first 48 h - that is, until just after the start of polysaccharide formation. The use of antifoam was often avoided by starting with a low impeller speed (250 rev. min^{-1}) gradually increased to 1000 rev. min^{-1} over 48 h as the culture viscosity increased. Culture pH was controlled using NaOH and HCl. Continuous culture vessels were of 2.5 l. working volume, similar to that described by Evans *et al.* [1970] but with two six bladed impellers mounted on a single shaft. Sterile air (1 vol/vol culture min^{-1} was supplied through a single hole sparger beneath the lower impeller; there were four equi-spaced pierced baffles. The temperature was controlled at 30°C ± 0.1°C. Medium and culture glucose determinations were by the glucose oxidase method. Bacterial dry weights were determined from diluted culture samples by the method of Ford [1967]. Apparent viscosities were measured over a range of shear rates with a

Wells-Brookfield "cone and plate" viscometer, model HBT, at a temperature of 25°C using the 0.8° cone and 0.5 ml samples from a hypodermic syringe. The apparent viscosity at no shear (the K value) was found by extrapolating to zero a log-log plot of apparent viscosities over the range of shear rates used. Due to the high viscosities of the samples, and the entrainment of air bubbles, culture samples were measured by weight for cell dry weight and polysaccharide determinations. Polysaccharide yields were obtained by adding three volumes of acetone to one volume of culture, decanting the supernatant and drying the precipitated material to constant weight. Subtracting the dry weight of bacteria from a replicate sample gave the yield of cell-free polysaccharide produced. Samples for sugar analysis of the polysaccharide were first diluted with 9 parts of normal saline to reduce their viscosity and then centrifuged at 9000 g for two hours to remove bacterial cells. The cell-free polysaccharide was precipitated from the decanted supernatant with acetone but in the presence of normal saline. The precipitate was then washed thrice by re-dissolving it in normal saline, followed by re-precipitation with acetone.

Polysaccharide production in defined salts nitrogen-limited medium

Whereas it is difficult to be certain what the limiting nutrient really is in complex media, in defined salts media there need be no doubt. The formation of intracellular polysaccharide (glycogen) and extracellular polysaccharide [Williams and Wimpenny, 1978] is promoted by high carbon/nitrogen ratios. We therefore expected that growth-limitation by nitrogen would encourage the production of high yields of polysaccharide.

After 48 h batch growth, flow of the same medium was started and maintained at a flow rate designed to give a dilution rate of 0.03 h^{-1}. Polysaccharide was produced from about the middle of the batch culture period and, after three days of continuous flow, reached a steady state level of 23 g/l which was maintained for at least 900 h (Q = >30). At the end of this period there was no evidence of culture variation.

Polysaccharide production in defined salts carbon-limited medium

Although there are instances of carbon-limited cultures forming storage compounds, such as poly-β-hydroxybutyric acid [Wilkinson and Munroe, 1967; Harder *et al.*, 1977], we believed that the production of polysaccharides of the xanthan type was an expression of "over-flow" metabolism. However, we felt that growth in a carbon-limited environment might exert a selection pressure for a constitutive producing strain, or, if a plasmid was involved in polysaccharide synthesis, might force its incorporation into the genome. In either case, this should result in a more stable culture. To test these ideas a carbon-limited culture was grown with the glucose content of the medium reduced to 2% (w/v). At no time did the culture broth viscosity exceed 500 cps, and after 240 h at a dilution rate of 0.045 h^{-1}, the residual glucose level was 0.1 mg/ml and the culture viscosity was that of water. Culture variation by colonial morphology was also much greater than that seen in nitrogen-limited cultures. While this seemed to be a vindication of our belief, at the end of the experiment we were surprised to find that, despite the water-like viscosity of the culture broth, it yet contained 9 g/l of acetone precipitable material. This suggested that while some polysaccharide production is constitutive, the material is different from that produced when glucose is present in excess. Subsequent cultures of mucoid isolates from this and similar experiments showed no increased culture stability or increased polysaccharide.

Polysaccharide production in defined salts potassium-limited medium

Potassium is an important cell constituent. We therefore thought that if the acidic polysaccharide produced by this organism was involved in cation uptake, then potassium limitation might encourage production of polysaccharide. However, experiments showed, as did those of Duguid and Wilkinson [1954], that potassium limitation suppressed polysaccharide formation almost completely and that which was produced was of very low viscosity. During more than 2,000 h growth at dilution rates of 0.03-0.07 h^{-1}, the acetone precipitable material remained at about 9 g/l with only 50% utilization of the glucose supplied; the culture broth K value varied between 190 and 360 cps. Culture variation appeared early and became considerable.

Polysaccharide production in defined salts magnesium-limited medium

The potassium-limited experiments suggested that the polysaccharide does not bind monovalent cations; but if it were involved in the uptake of divalent cations such as magnesium, then an enhanced production might be expected if the limiting nutrient were a divalent cation. However, it is well known that magnesium, through *inter alia* its involvement in ribosome and membrane structure, plays an important part in controlling the activity of enzymes concerned in extracellular synthesis; if then its concentration were low, polysaccharide synthesis could be at least reduced, or even absent. Experiments at dilution rates of 0.02-0.03 h^{-1} showed that, as with potassium-limited cultures, the yield of polysaccharide was only 9 g/l with only 50% utilization of the glucose supplied. However, culture variations were reduced and the K value of a 1% (w/v) solution of cell-free polysaccharide was 6,500 cps. The broth

K value, however, was only about the same as the potassium-limited culture - 270 cps.

Polysaccharide production in defined salts phosphorus-limited medium

Phosphorus is known to be involved in RNA and cell wall structure and function. As an anionic limiting nutrient it therefore seemed unlikely to encourage the formation of an anionic polymer. Furthermore, Neijssel and Tempest [1975] with *Klebsiella aerogenes* obtained less than half the polysaccharide yield in phosphate-limited conditions that they had obtained in nitrogen-limited conditions, and Williams and Wimpenny [1978] reported that "phosphate limitation did not increase polysaccharide production".

At dilution rates between 0.02-0.03 h^{-1} the yield of acetone precipitable material was, surprisingly, more than 17.5 g/l with 85% utilization of the glucose supplied, and broth K values were maintained at more than 5,000 cps for 1,500 h. Culture variation was slight, but there was a fairly rapid takeover of the culture by a poorly pigmented variant which has continued to breed true on subsequent subculture.

Polysaccharide production in defined salts sulphur-limited medium

Restriction of an organism's sulphur supply has been shown to modify the protein composition of the envelope [Pardee, 1968]; a reduction of membrane -SH groups could increase permeability of the membrane to proteins such as extracellular polysaccharide synthesizing enzymes. Alternatively, or in addition, sulphur restriction might, as suggested by Neijssel and Tempest [1975], reduce the synthesis of those sulphur-containing and essential components of the cell, such as coenzyme A, thiamine pyrophosphate and lipoic acid. Such restriction could reduce the normal efficiency of glucose oxidation, making available the unoxidized products and excess glucose for production of polysaccharide which could serve as receptors for unwanted pyruvate and acetate. The data of Neijssel and Tempest [1975] show that, of the conditions examined by them, sulphur-limited growth on glucose produced the greatest amount of pyruvate, although with much less polysaccharide than was formed under nitrogen-limited conditions. It has also been shown by Sandford *et al.* [1977] that viscosity increases with pyruvate content; we therefore hoped that sulphur-limited growth on glucose might encourage, not only an increased production of polysaccharide, but also of a polysaccharide of increased viscosity. Our first sulphur-limited growth experiment showed that although our theory might be incorrect, our hopes were largely fulfilled: at a dilution rate of 0.035 h^{-1}, with an initial glucose concentration of 5% (w/v), we obtained a yield of acetone precipitated material of 27 g/l after about three days on flow, and this was maintained, ± 2 g, for 12 weeks which represents a "Q" value of about 98. Throughout this experiment the culture broth K value remained at 18,500 ± 1,000 cps. The most remarkable feature of this experiment was that as well as these conditions producing the highest yield of polysaccharide, they also virtually abolished culture variation. We therefore wondered if we had perhaps chosen conditions that encouraged the selection of a stable, producing variant, or, if a plasmid was involved, that the conditions had forced its incorporation into the genome. However, when cells from this culture were grown in complex nitrogen-limited medium there was no apparent change in the previously observed rate of variant production. It therefore seemed that the observed stabilization of polysaccharide production was the result of metabolic direction resulting from sulphur limitation. The polysaccharide produced did not however show the anticipated increased viscosity.

The effect of oxygen availability

We were concerned that the very viscous culture broths with which we were working might impose oxygen deficient conditions upon the cells, and, since polysaccharides are believed to be oxidation products, that this could affect the rate and efficiency of their production. To investigate this possibility we grew a culture of XJ 107 in sulphur-limited medium with approximately 5% (w/v) glucose; the dilution rate was adjusted to 0.06 h^{-1} to increase the oxygen demand of the culture. In the steady state we obtained a yield of acetone precipitable material of 2.4% (w/v). The medium was then diluted with an equal volume of deionised water and the experiments continued. The yield of acetone precipitable material from the half-strength medium was 1.3% (w/v). Since, even under these rather extreme conditions, the yields were almost in proportion to the strengths of the media, we concluded that under our conditions oxygen availability was not limiting the production of polysaccharide and that the oxygen level of the culture is less critical than had been supposed; perhaps being required for growth more than polysaccharide production.

The effect of sulphur-limitation upon polysaccharide production and the stability of other strains of Xanthomonas juglandis

Having found that sulphur-limited growth conditions maintained stable polysaccharide production of strain XJ 107, we wished to discover if other strains were similarly influenced. To this end, we grew the remaining nine

TABLE 3

The effect of sulphur-limited growth upon polysaccharide production and stability of various strains of Xanthomonas juglandis in the chemostat

Type culture collection number	Time on flow (h)	Dilution rate (h^{-1})	Amount of carbon used (g/l)	APM yield (g/l)	Culture K value (cps)	Culture variation
NCPPB 411	215	0.038	42.5	10.5	ND	None seen
NCPPB 412	200	0.026	44.5	14	5,300	None seen
NCPPB 413	205	0.026	44.5	19	5,600	None seen
NCPPB 414	480	0.026	35.5	17	9,200	None seen
NCPPB 415	820	0.032	37	23	7,700	None seen
NCPPB 1447	500	0.032	45	27.5	15,500	None seen
NCPPB 1659	300	0.038	27	19.5	11,000	None seen
NCPPB 362	380	0.026	45	33	4,400	None seen
ATCC 11329	400	0.020	44	18	15,000	None seen

strains in sulphur-limited chemostat culture at a variety of dilution rates (Table 3). An obvious but surprising feature of these results is the lack of agreement between the apparent polysaccharide production of some strains on plates (see Table 1) and their performance in continuous culture; for example NCPPB 1659 was scored as a poor producer on plates but produced a reasonable amount of polysaccharide and had a good culture viscosity; NCPPB 362 was scored low on plates but produced the greatest yield of acetone precipitable material, but had the lowest culture viscosity; and NCPPB 414 which had seemed the equal of XJ 107 on plates was inferior in the chemostat both for yield and viscosity. Another unexpected feature of the results was the variation between strains of acetone precipitable material and culture broth viscosity: for example, NCPPB 362 produced the highest yield (33g) but had the lowest culture viscosity (4,400 cps). This could be explained by postulating that the material extracted contained a significant amount of protein, since Neijssel and Tempest [1975] reported the excretion of protein by *Klebsiella aerogenes* growing on glucose in a defined sulphur-limited medium. Although this type of result was recognized with only one of our strains, it is a further indication of the unreliability of a naive screening programme like that with which we started. While all our results indicate that sulphur-limitation provides conditions for the maintenance of culture stability, care must be taken to select those strains that do not excrete significant amounts of low viscosity material.

The effect of dilution rate upon the production of polysaccharide

Changes in dilution rate are known to affect not only the efficiency of substrate utilization [Herbert *et al.*, 1956]; they also influence the production of polymeric storage compounds, such as poly-β-hydroxybutyrate and glycogen [Wilkinson and Munroe, 1967] and the structure and composition of some polymers of the bacterial cell wall. Moreover, Cadmus *et al.* (quoted by Sandford *et al.* [1977]) showed that the viscosity of *Xanthomonas campestris* gum produced in batch culture increases with its pyruvate content which increases during the period of culture. Sandford *et al.* [1977] have confirmed the correlation of viscosity with pyruvate content.

To investigate the effect of dilution rate upon the efficiency and rate of polysaccharide production, we grew a series of sulphur-limited cultures of XJ 107 at a range of dilution rates from 0.03-0.06 h^{-1} in a medium containing 24 mg S/1 and 50 g/1 glucose (Figure 1).

A surprising feature of the results was the quite small variation of the amount of polysaccharide in the culture broth over the range of dilution rates from 0.03 to 0.05 h^{-1}. At higher rates, however, there was a dramatic decrease. The efficiency with which glucose was converted to polysaccharide was low and increased only slowly up to a dilution rate of 0.035 h^{-1}. This we attribute to the effect of the organisms' standing maintenance requirement [see Powell, 1972], the effect of which diminishes with increasing growth rate. Indeed, at dilution rates from 0.04 to 0.06 h^{-1} the efficiency of glucose utilisation rose steadily to nearly 50%. This gradual increase compensated for the decreasing glucose utilisation and maintained the level of polysaccharide in the culture virtually constant to a dilution rate of 0.05 h^{-1}. Beyond that point the increased efficiency was small compared with the large decrease in glucose used.

The difficulties encountered when working with solutions of high viscosity are well known and must be acknowledged. It is, for example, hard to be sure that dilution rates are real and not merely apparent, since if mixing is not always homogeneous, the stirred volume of the culture vessel may be less than the volume of liquid that it contains. To avoid this dilemma, Williams and Wimpenny [1978] worked with culture broths of low viscosity containing less than 4 g/1 polysaccharide. We felt that for practical purposes we should use conditions as nearly practical as are

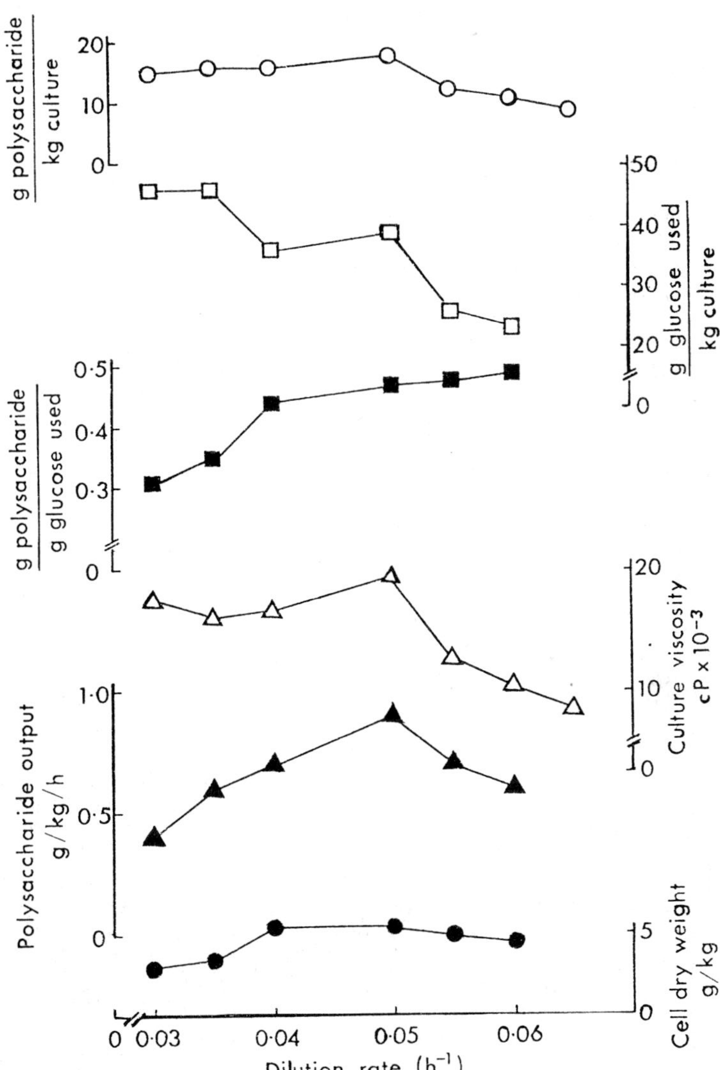

Fig. 1 The effect of dilution rate upon polysaccharide production (○), cell dry weight (●), glucose consumption (□), polysaccharide/glucose used (■), culture viscosity (K value) (△) and polysaccharide output (▲) by sulphur-limited *Xanthomonas juglandis* cultures.

possible on a small laboratory scale, and the parallelism of our results with diluted and normal medium gave us some assurance that our results were credible.

The opposite effects produced by increasing dilution rate upon glucose conversion efficiency and the amount of glucose used, posed something of a dilemma when we tried to determine the optimum economic dilution rate. However, at a dilution rate of 0.05 h^{-1}, the utilization of glucose is at a maximum (78%) as is the level of polysaccharide in the culture (18 g/l); the conversion efficiency of glucose to polysaccharide is also approaching its maximum. When the output rate of polysaccharide is plotted against the dilution rate it also reaches a sharp maximum at this flow rate. Since the dilution rate of 0.05 h^{-1} represents a mean residence time of 20 h, this compares favourably with the batch culture time, in nitrogen "limited" complex medium, of 93 h for this organism [Ellwood *et al.*, 1975] and 96 h for *Xanthomonas campestris* [Rogovin *et al.*, 1965]. Moreover, the weight of acetone precipitable material from a continuous culture (29 g/l) was greater than that from a batch culture (20.5 g/l). A similar, but reduced, effect was also noticed when batch and continuous cultures of *Xanthomonas campestris* grown in defined media were compared [Evans and Yeo, unpublished observations]; culture viscosities were much higher, as were the efficiencies of conversion of glucose to polysaccharide.

The effect of dilution rate upon the physical properties of polysaccharide samples

We expected that the physical properties of the polymer produced at different dilution rates might not be the same, and that if there were differences they could result from changes in structure such as reduced substitution with pyruvate, reduced side chains, shorter chain lengths and less cross-linking. Such changes should be shown by comparing the absolute viscosities of different concentrations of the samples in normal saline (a similar method has been used by Jeanes *et al.* [1961]). Kenis [1968] showed that linear, high molecular weight polymers from bacteria can greatly reduce the frictional drag of a liquid, and using a turbulent flow rheometer [Hoyt, 1965] he showed that a solution containing 200 mg/l of xanthan gum - Kelzan - could reduce drag by 62%. Parker and Hedley [1972] have indicated that drag reduction results from bridging, by the molecules, of the turbulent vortices and that the properties of the molecules most important in this are length and flexibility. Differences in drag reduction properties should, therefore, indicate differences in structure. Drag reduction was measured using a 0.46 m length of 1 mm bore tube [Parker and Hedley, 1972] at a Reynolds number of 14,000.

The results shown in Table 4 suggest that the nature of

the polysaccharide is unaffected by dilution rates between 0.04 and 0.06 h^{-1} and that the relation between viscosity and concentration of the samples is generally similar to that of both the commercial samples from *Xanthomonas campestris*. There is, however, a significant difference in drag reducing abilities which suggests that the *juglandis* material might be composed of longer, less branched molecules. The material produced at the lowest dilution rate, on the other hand, does seem to be different from the material produced at the higher dilution rates and to be composed of longer unbranched molecules.

TABLE 4

Comparison of apparent viscosities at zero shear of 3 concentrations, in normal saline, of cell-free polysaccharides produced at various dilution rates in sulphur-limited defined salts medium, and of the drag reducing properties of the samples at a concentration of 200 mg l^{-1} (w/v). Results with two samples of commercial xanthan gum, "Kelzan" and "Keltrol", are included for comparison

Dilution rate (h^{-1})	Concentration of polysaccharide (% w/v) 0.5	1.0	2.0	Drag reduction (%)
0.03	2,800	6,700	9,600	65
0.04	1,400	5,400	15,800	58
0.05	1,800	7,500	10,300	61
0.055	1,800	6,500	10,500	58
0.06	1,600	7,000	13,000	64
Keltrol	3,000	6,600	16,000	54
Kelzan	2,500	9,800	19,500	48

The effect of growth limitation upon the chemical composition of polysaccharide samples

The acetone-precipitated material produced by growth in complex nitrogen-limited medium was found by cetavalon fractionation to contain two polysaccharides, one having a composition similar to the well known xanthan gum from *Xanthomonas campestris*, and the other composed of glucose and rhamnose [Lawson and Symes, 1977]. The ratio of the two polysaccharides varied with culture conditions [see Table 5], being highest when the cells were grown carbon-limited and lowest when sulphur-limited; indeed, material from sulphur-limited cultures seemed very similar to xanthan gum. When growth was potassium-limited, the material precipitated with acetone was only poorly soluble in water and was largely composed of glucose. It is tempting to speculate that this is a cellulosic backbone material.

TABLE 5

Carbohydrate composition of the polysaccharide produced by Xanthomonas juglandis XJ 107, grown under different limitations

		% Composition of crude polysaccharide					
Limiting medium	Dilution rate (h^{-1})	Glucose	Mannose	Glucuronic acid	Rhamnose	Acetate	Pyruvate
N complex	0.03	36.3	23.2	18.0	5.0	3.0	1.6
N defined	0.03	34.0	26.3	18.0	5.7	3.0	1.7
C defined	0.045	38.0	24.1	15.2	15.0	0.5	1.8
K defined	0.03	75.0	5.6	2.0	7.5	ND	ND
Mg defined	0.03	63.3	10.7	7.1	7.5	ND	ND
P defined	0.03	29.9	29.0	17.0	2.0	3.0	3.7
S defined	0.035	34.3	30.8	18.1	1.4	1.6-3.2	1.5-5.4

Samples of polysaccharides were redissolved in water, dialysed for three days against constant changes of distilled water, and dried to constant weight over phosphorus pentoxide. When required, further purification was carried out using the cetyl trimethyl ammonium bromide precipitation procedure described by Scott [1965]. Routinely, samples of polysaccharide (5 mg) were dissolved in 0.5M H_2SO_4 (0.5 ml) and heated to 105°C for 16 h. The hydrolysate was neutralized to pH 7.0 with barium hydroxide. The precipitate of barium sulphate was removed by centrifugation (5,000 x g, 10 min), washed once with water and recentrifuged. The supernatant and washings were combined and lyophilized. The sugars present were converted to alditol acetates by the method of Blake and Richards [1970] and quantitatively analysed by gas chromatography using the method of Sawardeker *et al.* [1965]. Total carbohydrate was estimated by the phenol-sulphuric acid method of Dubois *et al.* [1956]. Acetate and pyruvate contents were determined by the methods of Hestrin [1949] and Sloneker and Orentas [1962] respectively.

Conclusion

There has been much discussion of the question "why do organisms produce polysaccharide?", but the answer remains unclear. The question is not of merely academic interest since, if answered, it could provide a clue as to how production might be encouraged. Explanations of the purpose of polysaccharide production have included: improved infectivity resulting from protection of wind-borne organisms from desiccation [Hedges, 1926] or the effects of ultra violet light [Leach *et al.*, 1957]; or their subsequent adhesion to the surface of the new host with the additional possibility of exploiting the ion concentrating properties of surfaces [Brown *et al.*, 1977]; increased phytotoxicity by encouraging the collapse (wilt) of the infected plant's protective structure [Husain and Kelman, 1958]; protection of the invader from the action of inhibitors [Misaki *et al.*, 1962]; facilitating the spread of infection through the tissues [Schluchterer and Stacey, 1945]. The polysaccharide could also act as a receptor for unwanted metabolites such as acetate and pyruvate, or even as a defence against ingestion by amoebae in the soil [Wilkinson, 1958; Jarman, Chapter 2].

In view of the instability of strains of, for example, *Xanthomonas campestris*, it is not easy to see how most of these possible advantages to the organism of polysaccharide production could be exploited on a practical scale. However, the fact that some polysaccharides can incorporate pyruvate and acetate [Sandford *et al.*, 1977] indicates one way in which a selection pressure might be exerted in favour of continued polysaccharide production since Neijssel and Tempest [1975] have shown that, in the presence of excess glucose, nitrogen- and sulphur-limited cultures of *Klebsiella aerogenes* produce increased amounts of pyruvate and acetate, and of acetate in phosphorus-limited conditions.

There remains the problem of the reduced stability of nitrogen-limited cultures growing in complex media. The virtual absence of variation in cultures growing on or in defined salts media can be explained by the absence of other potential carbon sources such as amino acids and peptides. In complex media, deamination of these makes available carbon sources which, within the cell, could be metabolised more readily than glucose. It is significant that the glucose utilization of chemostat cultures growing in complex medium always diminishes with time, and that this is associated with an increasing variant population and reduced polysaccharide production. It is also possible that growth on the carbon sources derived from amino compounds might inhibit production of, or repress polysaccharide synthesizing enzymes.

Our results show that it is possible to produce polysaccharide continuously for periods of at least 2000 h with yields per g of glucose at least equal to those obtained in batch culture, but with an output per volume of fermenter space that is considerably greater. The use of a defined mineral salts medium should also make easier the recovery of clean product.

Acknowledgements

It is with pleasure that we acknowledge the skilled and devoted assistance of W. Robinson with the maintenance of cultures and the isolation and physical testing of polysaccharide samples, and of T. Collinge in operating the fermenters. We also wish to thank V.M.C. Longyear for the sugar analyses.

References

Andrew, J.R. (1977). Application of xanthan gum in foods and related products. *American Chemical Society Symposium Series* **45**, 231-241.

Blake, J.D. and Richards, G.N. (1970). A critical re-examination of problems inherent in compositional analysis of hemicelluloses by gas-liquid chromatography. *Carbohydrate Research* **14**, 375-387.

Brown, C.M., Ellwood, D.C. and Hunter, J.R. (1977). Growth of bacteria at surfaces: influence of nutrient limitation. *FEMS Microbiology Letters* **1**, 163-166.

Cadmus, M.C., Rogovin, S.P., Burton, K.A., Pittsley, J.E., Knutson, C.A. and Jeanes, A. (1976). Colonial variation in *Xanthomonas campestris* NRRL B-1459 and characterization of the polysaccharide from a variant strain. *Canadian Journal of Microbiology* **22**, 942-948.

Dubois, M., Gilles, K.A., Hamilton, J.K., Rebers, P.A. and Smith, F. (1956). Colorimetric method for determination of sugars and related substances. *Analytical Chemistry* **28**, 350-356.

Duguid, J.P. and Wilkinson, J.F. (1954). Note on the influence of potassium deficiency upon production of polysaccharide by *Aerobacter aerogenes*. *Journal of General Microbiology* **11**, 71-73.

Dye, D.W. and Lelliott, R.A. (1974). Xanthomonas. In Bergey's *Manual of Determinative Bacteriology*, 8th edn., pp.243-249. Edited by R.E. Buchanan and N.E. Gibbons. Baltimore: Williams and Wilkins.

Ellwood, D.C., Evans, C.G.T. and Yeo, R.G. (1975). British Patent Application 46784/75.

Ellwood, D.C. and Tempest, D.W. (1969). Control of teichoic acid and teichuronic acid biosynthesis in chemostat cultures of *Bacillus subtilis* var. *niger*. *Biochemical Journal* **111**, 1-5.

Evans, C.G.T., Herbert, D. and Tempest, D.W. (1970). Continuous cultivation of microorganisms 2. Construction of a chemostat. In *Methods in Microbiology* **2**, pp.277-327. Edited by J.R. Norris and D.W. Ribbons. London and New York: Academic Press.

Ford, J.W.S. (1967). Measurement of bacterial dry weight using an infrared oven. *Chemistry and Industry* 1556-1557.

Harder, W., Kuenen, J.G. and Matin, A. (1977). A review: microbial selection in continuous culture. *Journal of Applied Bacteriology* **43**, 1-24.

Hedges, F. (1926). Bacterial wilt of beans (*Bacterium flaccumfaciens* Hedges), including comparisons with *Bacterium phaseoli*. *Phytopathology* **15**, 1-22.

Herbert, D. (1976). Stoicheiometric aspects of microbial growth. In *Continuous Culture 6: Applications and New Fields*, pp.1-30. Edited by A.C.R. Dean, D.C. Ellwood, C.G.T. Evans and J. Melling. Chichester, England: Ellis Horwood Ltd.

Herbert, D., Elsworth, R. and Telling, R.C. (1956). The continuous culture of bacteria; a theoretical and experimental study. *Journal of General Microbiology* **14**, 601-622.

Hestrin, S. (1949). The reaction of acetyl choline and other carboxylic acid derivatives with hydroxylamine and its analytical application. *Journal of Biological Chemistry* **180**, 249-261.

Hoyt, J.W. (1965). A turbulent-flow rheometer. In *Symposium on Rheology, American Society of Mechanical Engineering*, pp.71-82. Edited by A.W. Marris and J.T.S. Wang. New York: American Society of Mechanical Engineering.

Husain, A. and Kelman, A. (1958). Relation of slime production to mechanism of wilting and pathogenicity of *Pseudomonas solanacearum*. *Phytopathology* **48**, 155-165.

Jeanes, A., Pittsley, J.E. and Senti, F.R. (1961). Polysaccharide B-1459: a new hydrocolloid polyelectrolyte produced from glucose by bacterial fermentation. *Journal of Applied Polymer Science* **5**, 519-526.

Jeanes, A., Pittsley, J.E., Watson, P.R. and Dimler, R.J. (1961). Characterization and properties of the phosphomannan from *Hansenula holstii* NRRL Y-2448. *Archives of Biochemistry and Biophysics* **92**, 343-350.

Kenis, P. (1968). Drag reduction by bacterial metabolites. *Nature, London* **217**, 940-942.

Lawson, C.J. and Symes, K.C. (1977). Xanthan gum - acetolysis as a tool for the elucidation of structure. *American Chemical Society Symposium Series* **45**, 183-191.

Leach, J.G., Lilly, V.G., Wilson, H.A. and Purvis, M.R. (1957). Bacterial polysaccharides: the nature and function of the exudate produced by *Xanthomonas phaseoli*. *Phytopathology* **47**, 113-120.

Lindblom, G.P. and Patton, J.T. (1967). *United States Patent* 3 328 262.

Misaki, A., Kirkwood, S., Scaletti, J.V. and Smith, F. (1962). Structure of the extracellular polysaccharide produced by *Xanthomonas oryzae*. *Canadian Journal of Chemistry* **40**, 2204-2213.

Neijssel, O.M. and Tempest, D.W. (1975). The regulation of carbohydrate metabolism in *Klebsiella aerogenes* NCTC 418 organisms growing in continuous culture. *Archives of Microbiology* **106**, 251-258.

Pardee, A.B. (1968). Membrane transport proteins. *Science*, New York **162**, 632-637.

Parker, C.A. and Hedley, A.H. (1972). Drag reduction and molecular structure. *Nature, London* **236**, 61-62.

Powell, E.O. (1972). Hypertrophic growth. *Journal of Applied Chemistry and Biotechnology* **22**, 71-78.

Rogovin, P., Albrech, W. and Sohns, V. (1965). Production of industrial-grade polysaccharide B-1459. *Biotechnology and Bioengineering* **7**, 161-169.

San Blas, G. and Cunningham, W.L. (1974). Structure of cell wall and exocellar mannans from the yeast *Hansenula holstii*; mannans produced in phosphate-limited medium. *Biochimica et Biophysica Acta* **354**, 247-253.

Sandford, P.A., Pittsley, J.E., Knutson, C.A., Watson, P.R., Cadmus, M.C. and Jeanes, A. (1977). Variations in *Xanthomonas campestris* NRRL-1459: Characterization of Xanthomonas products of differing pyruvic acid content. *American Chemical Society Symposium Series* **45**, 192-210.

Sawardeker, J.S., Sloneker, J.H. and Jeanes, A. (1965). Quantitative determination of monosaccharides as their alditol acetates by gas liquid chromatography. *Analytical Chemistry* **37**, 1602-1604.

Schluchterer, E. and Stacey, M. (1945). The capsular polysaccharide of *Rhizobium radicicolum*. *Journal of the Chemical Society*, 776-783.

Scott, J.E. (1965). Fractionation by precipitation with quaternary ammonium salts. *Methods in Carbohydrate Chemistry V*, 38-44.

Silman, R.W. and Rogovin, P. (1972). Continuous fermentation to produce xanthan biopolymer: effect of dilution rate. *Biotechnology and Bioengineering* **14**, 23-31.

Sloneker, J.H. and Orentas, D.G. (1962). Pyruvic acid a unique component of an exocellular bacterial polysaccharide. *Nature, London* **194**, 478-479.

Tempest, D.W. and Neijssel, O.M. (1976). Microbial adaptation to low nutrient environments. In *Continuous Culture 6: Applications and New Fields*, pp.283-296. Edited by A.C.R. Dean, D.C. Ellwood, C.G.T. Evans and J. Melling. Chichester, England: Ellis Horwood Ltd.

Wells, J. (1977). Extracellular microbial polysaccharides - a critical overview. *American Chemical Society Symposium Series* **45**, 299-313.

Wilkinson, J.F. (1958). The extracellular polysaccharides of bacteria. *Bacteriological Reviews* **22**, 46-73.

Wilkinson, J.F. and Munroe, A.L.S. (1967). The influence of growth limiting conditions on the synthesis of possible carbon and energy storage polymers in *Bacillus megaterium*. In *Microbial Physiology and Continuous Culture*, pp.173-184. Edited by E.O. Powell, C.G.T. Evans, R.E. Strange and D.W. Tempest. England: Her Majesty's Stationery Office.

Williams, A.G. and Wimpenny, J.W.T. (1978). Exopolysaccharide production by *Pseudomonas* NCIB 11264 grown in continuous culture. *Journal of General Microbiology* **104**, 47-57.

Chapter 4

PULLULAN SYNTHESIS BY *AUREOBASIDIUM PULLULANS*

B.J. CATLEY

Department of Brewing and Biological Sciences, Heriot-Watt University, Edinburgh, UK

Introduction

It is now twenty years since Bernier [1958] observed that extracellular polysaccharides were produced by *Pullularia* (now *Aureobasidium*) *pullulans*; an observation that was followed a year later by publication of the work of Bender *et al.* [1959] who characterized the neutral glucan component and termed it pullulan. Since that time a number of groups have taken up the task of examining the diverse aspects of the elaboration of this polymer and it would appear appropriate to take stock of what has, and has not, been achieved in the intervening years. A title such as the above may well suggest a particular area of investigation to one reader, but an entirely different one to another. In attempting to produce a review for the microbial physiologist on the one hand and a survey for the biochemist on the other it is hoped that in this chapter there will be found an overview of investigations that have been, and are being, conducted into the environmental conditions and intracellular controls that enable *Aureobasidium pullulans* to transform low molecular weight sugars into the extracellular polysaccharide pullulan.

Pullulan structure

Examining the nature of the extracellular polysaccharides produced by *Pullularia pullulans* grown in a modified Czapek-Dox medium containing glucose [Bernier, 1958], Bender *et al.* [1959] isolated a neutral glucan, which on the basis of optical rotation, partial acidic hydrolysis, methylation analysis and its infra-red spectrum, was shown to have a predominance of α 1,4- and α 1,6-glucosidic linkages. This polysaccharide was termed pullulan. Further study [Wallenfels *et al.*, 1961] indicated that, by treatment with hot formic acid, it could be converted into "Restpullulan" or limit-dextrin, in 90% yield. This treatment was subsequently shown [Wallenfels *et al.*, 1965] to inactivate a glucamylase adsorbed to the precipitated pullulan, and not to produce any chemical modification of the polysaccharide. Periodate oxidation and methylation analysis [Wallenfels *et al.*, 1961, 1965] revealed that pullulan contains α 1,4- and α 1,6-glucosidic linkages in the ratio 2:1. Pullulanase (EC 3.2.1.41), an enzyme specific for catalysing the hydrolysis of α 1,6-glucosidic linkages [Bender and Wallenfels, 1961], converted the polysaccharide almost quantitatively into maltotriose, and since little or no 2,3,4,6-tetra-O-methyl glucoside was afforded by methylation analysis, the maltotriose was presumed to be polymerized through α 1,6 linkages on the outer glucosyl units. From these observations a structure of pullulan was proposed (Figure 1). Careful investigation by paper chromatography showed that pullulan that had been exhaustively subjected to pullulanolysis contained traces of a tetrasaccharide. This minor component was thought to have a branched structure and to constitute the terminal sequences of the polysaccharide [Wallenfels *et al.*, 1965].

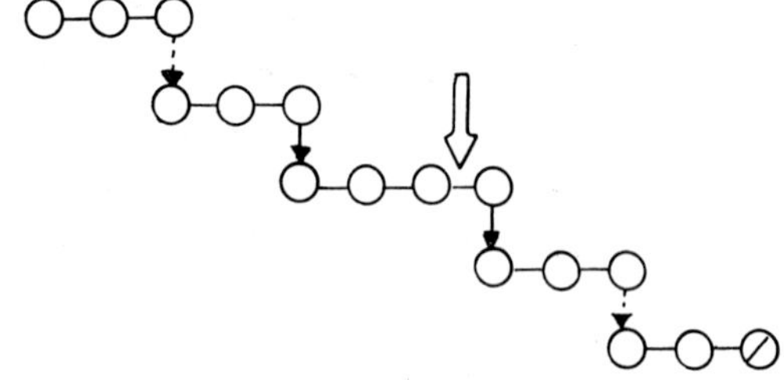

Fig. 1 The basic polymaltotriose structure of pullulan incorporating the minor maltotetraose component. α 1,6 linkages are represented by ↓, α 1,4 linkages by —. Non-reducing D-glucosyl units are shown by ○, and the reducing unit as ⊘. The site of the proposed endo-amylase attack is indicated by the arrow: ⇩.

Later work [Catley and Whelan, 1971] suggested that the terminally located branched tetrasaccharide originated as an intramolecular linear maltotetraose, and that it is the cleavage of these portions of the glucosyl chain by a postulated endo-amylase that give rise to the branched termini (Figure 1). However, polymaltotriose is not the only structure that has been proposed for the neutral glucan elaborated by *Aureobasidium pullulans* and other classes of polysaccharide, in addition to the neutral α glucans, have been characterized.

Bouveng *et al.* [1962], using cetyl trimethylammonium hydroxide (CTA-OH), fractionated the extracellular polysaccharides into neutral and acidic glycans. The neutral fraction, a glucan, was readily soluble in water but the other was less so and appeared strongly adherent to the cell walls. Using sucrose as carbon source [Bouveng *et al.*, 1962; 1963a] the major elaborated polysaccharide was found to be the neutral glucan which, on analysis, was shown to be pullulan-like. The acidic glycan appeared to be composed of both water-soluble and water-insoluble components, the former rich in D-galactose and D-mannose, the latter composed mainly of D-glucose. Polysaccharide formation occurred using sucrose, D-glucose, maltose, D-mannose, D-xylose and L-arabinose as carbon sources, but could not be detected when either D-galactose or mannitol was used. In traversing this series from sucrose to L-arabinose the efficiency of polysaccharide formation falls and the proportion of D-galactose, D-mannose, and D-glucuronic acid occurring in the polysaccharide increases [Bouveng *et al.*, 1962]. Further investigation [Bouveng *et al.*, 1963b] of extracellular polysaccharides afforded by 5% D-xylose as sole carbon source revealed the presence of a β glucan containing 1,3 and 1,6 linkages as well as the α glucan and heteroglycan.

Other investigations of the neutral glucan structure have led to reports of 6% of 1,3 linkages [Sowa *et al.*, 1963]. Ueda *et al.* [1963] examined the products of sixteen strains of *Pullularia* grown on sucrose and, of the five strains examined in detail, found that the ratio of 1,4 to 1,6 linkages varied from 1.1 to 2.7:1. In contrast a comparison of extracellular polysaccharides produced by three different strains showed that whereas there may be differences in the ratio of neutral to hetero-glycans produced, there are no differences in the composition of the neutral glucan, even the ratio of maltotriose to maltotetraose remaining constant [Taguchi *et al.*, 1973a]. This work also produced additional evidence regarding pullulan structure in that the hydrolysis products of pullulan 4-glucanohydrolase (EC 3.2.1.57) supported the proposed linear structure, devoid of multiple branching [Sakano *et al.*, 1971; Taguchi *et al.*, 1973a].

A proposed structure for the neutral glucan that differs markedly from the polymaltotriose format is that proposed by Elinov and Matveeva [1972] and Elinov *et al.* [1975]. The principal saccharide chain is composed of β 1,3-linked glucosyl residues attached to which, at every third residue through an α 1,6 linkage, are α 1,4-linked oligosaccharides that vary in length according to the conditions of culturing. X-ray diffraction patterns obtained from both oriented and unoriented preparations did not permit a calculation of the unit cell characteristics, but did suggest the formation of ordered crystalline regions.

It would appear that, historically, the term pullulan serves only as a generic term describing the neutral glucan produced by *Aureobasidium pullulans*. Nevertheless the present drift of nomenclature, used for instance by the chemical supply companies in describing a polysaccharide substrate for the estimation of pullulanase, appears to favour the term as a particular name for a polymer of polymaltotriose structure.

Molecular weight measurements of pullulan have been achieved in a number of ways. Light scattering methods afford a value of 2.5×10^5 [Ueda *et al.*, 1963]; a figure which is close to that derived from the hydrodynamic techniques of Wallenfels *et al.* [1965], namely 2.35×10^5. Lower values of 4.5×10^4 have been determined through osmotic pressure measurements of the polysaccharide methyl ether [Bouveng *et al.*, 1963a], and that of the neutral glucan examined by Elinov *et al.* [1975] appears to be considerably higher with an estimated value of 10^7. The polydisperse structure of polysaccharides, arising from their mode of synthesis, may be modified by being subjected to post-polymerization hydrolysis. Gel-filtration profiles of pullulan, isolated at different points of the growth cycle, show a decline in molecular weight with an increased duration of culture [Catley, 1970]. Initially of a maximum molecular weight of 2×10^6 the polysaccharide is converted to a species with one-tenth of the original size. Labelled high-molecular weight pullulan, isolated from a five-day culture, was incorporated into a three-day culture which was allowed to grow for a further six days. Whereas there was no incorporation of label into the cells there was a decrease in the molecular weight of the labelled polymer. Suspecting that the occasional maltotetraose incorporated into the polymaltotriosyl structure might be the point of an extracellular endo-amylase attack, evidence was sought that it had been hydrolysed. Analysis of products from a pullulanase catalysed hydrolysis of the low-molecular weight labelled polysaccharide revealed the presence of a branched tetrasaccharide appearing in place of the linear, thus suggesting that hydrolysis had indeed taken place at these points (Figure 1) [Catley, 1970; Catley and Whelan, 1971]. Clearly this explanation is incomplete since, assuming a regular distribution of the tetrasaccharide throughout the molecule, cleavage of all the maltotetraose residues would reduce the pullulan to a molecular weight of approximately 7×10^3. Claims have been made that, through control of culturing conditions, the molecular weight of pullulan can be altered [Yuen, 1974]. Whether this reflects a control of the extent of synthesis of individual molecules, subsequent degradation, or both is not presently clear.

Descriptions of an extracellular microbial polysaccharide that is set free in the culture medium, must, at some point, be accompanied by consideration of a possible relationship with the carbohydrate structures of the cell wall. Opinions appear to differ as to whether generalizations are appropriate to this situation. On the one hand, an appraisal of the probably different functions that capsule and cell wall serve would suggest that different structures should have evolved to fulfil these roles [Wilkinson, 1958]. On the other hand, experimental observation arising from comparative studies of the *Aspergillus* and *Candida* cell wall and their associated exocellular polysaccharides suggest that a common enzymatic process might be used for the assembly of carbohydrates destined for different locations [Bull, 1972]. However, data are often insufficient to justify these conclusions.

Analyses of the cell wall of *Aureobasidium pullulans* suggest that pullulan is not a component of that assembly. Brown and Nickerson [1965] detected D-glucose, D-galactose, D-mannose, L-rhamnose and D-glucosamine in the cell walls of both filamentous and yeast-like growths. X-ray diffraction studies showed that D-glucosamine was not present as chitin or chitosan. Three β glucans have been extracted from the yeast-like form [Brown and Lindberg, 1967 a]. Though differing in solubility in alkali they all possess a β 1,3-linked structure with occasional substitutions by D-glucosyl residues at position 6 of the glucosyl ring. An accompanying heteropolysaccharide consisted of an α 1,6-linked mannan containing side chains of D-glucose, D-galactose and D-glucuronic acid [Brown and Lindberg, 1967b]. Kikuchi *et al.* [1973], examining heteropolysaccharides present in the growth medium and in phenol extracts of cell walls, suggested that the extracellular glycan might be derived from cell wall or intracellular polysaccharides through autolysis of the cell. The same studies revealed that some polysaccharides, the product of aqueous extraction of cell walls at 100°, were susceptible to pullulanase catalysed hydrolysis thus suggesting the presence of α 1,6-glucosidic bonds. The extraction of 10g of fresh cell with hot water yielded 1.2g of polysaccharide, which, upon pullulanolysis, gives 17% conversion into maltotriose. Whether these are equivalent maltotriosyl residues or were identified as liberated maltotriose is not clear. A more thorough investigation of cells as chlamydospores revealed D-glucose present in linear chains of β 1,3 and β 1,6-linkages, with occasional branch points, as the major constituent of the cell-wall polysaccharides [Brown *et al.*, 1973]. Moreover, the same components were present in the yeast-like cells, but with increased amounts of bound lipid, D-galactose, and D-mannose, but less D-glucose. The architecture but not the chemical composition of the yeast cell-wall has been described [Ramos and García Acha, 1975a].

Nomenclature of producer organism

The microorganism producing pullulan suffers from a surfeit of descriptions. *Aureobasidium pullulans* (De Bary) Arnaud appears to be the currently acceptable term for this deuteromycete, though as Durrell [1968] remarked, both *aureo* and *basidium* appear inappropriate in view of the fact that it is both classified with the black yeasts and is imperfect. Other designations have been *Pullularia pullulans* (De Bary) Berkhout and *Dermatium pullulans*. Revisions of the genus *Pullularia* have been proposed [Wynne and Gott, 1956] and a comprehensive survey of the morphological, physiological and biochemical attributes, as well as the nomenclature, of the organism has been compiled [Cooke, 1959].

Pullulan Production, Cell Morphology and the Environment

Considerable attention has been given to the determination of conditions in which assimilated carbon sources are elaborated as pullulan; an impetus to which has been given by the possible commercial uses of the polysaccharide [Yuen, 1974; Jeanes, 1977]. It is apparent from the preceeding review of the structure and composition of this polymer that polymaltotriose may not be the only exocellular macromolecule to be elaborated and it is therefore necessary to employ an assay wherein specific degradation products of polymaltotriose may be measured and identified. Such an assay might incorporate pullulanase [Bender and Wallenfels, 1961] yielding maltotriose or isopullulanase (pullulan 4-glucanohydrolase) yielding isopanose [Taguchi *et al.*, 1973a]. The less specific assays such as the phenol-sulphuric acid procedure [Dubois *et al.*, 1956], may be a convenient method for the approximate estimation of total carbohydrates but cannot be relied upon to provide an accurate measure of pullulan.

A consideration of the efficiency with which sugars are transformed into pullulan may be divided into two headings; the ability of a particular carbon source to be used in the donation of glucosyl residues for assembly into polysaccharides, and the influence of constituents, other than the carbon source, that are present in the growth medium and which may directly and/or indirectly modulate the mechanisms of biosynthesis or elaboration. In a practical sense a distinction must also be made with regard to the duration of the investigation, particularly with respect to the lengths of cell and growth cycles.

Some of the most successful conversions of supplied carbon source to polysaccharide have been obtained with partial hydrolysates of starch [Yuen, 1974], a yield of 76% being reported from a 10% hydrolysate of a starch syrup, (dextrose equivalent 45) when cells were grown for seven days at 28°. Maltose, sucrose, and D-glucose under the same conditions yielded 51%, 35% and 31% conversions respectively. Other strains gave different yields. Thus a comparison of three strains [Taguchi *et al.*, 1973a] showed that whereas one converts a 5% sucrose carbon source into 43% CTA-OH-soluble material (presumably pullulan) and 5% CTA-OH-insoluble polysaccharide, other strains under the same conditions (30°; 96 h culture period) produce very little pullulan (0.8% yield) and proportionately larger amounts of heteropolysaccharide (1.7% yield). Similar observations have been made by Ueda *et al.* [1963]; out of sixteen strains examined, five produced little or no pullulan and the yield from the remainder varied from 25.4% to 5.9%. The cells were grown for five days at 24° on a 10% sucrose carbon source. Our own analyses of composition and yield of extracellular polysaccharide afforded by a culture grown for 100 h at 25-27° on 2.5% carbon sources indicated that 60% extracellular polymer is produced from sucrose [Catley, 1971a]. Of this 73% was polysaccharide containing 76% polymaltotriose representing an overall yield of 33% pullulan; thus illustrating the errors arising from the use of non-specific assays. Approximately the same analyses were obtained over the whole growth cycle indicating that the proportion of pullulan to non-pullulan material remained unchanged. Cultures provided with an equivalent amount of a combination of D-glucose and D-fructose in place of sucrose exhibit similar growth characteristics. However, the use of [^{14}C]-labelled monosaccharides in these experiments indicated a preferential use of D-glucose both for cell growth and for exocellular polymer production. Examination of the labelled polymer produced from [^{14}C]-D-fructose in this combination showed that 80% of the polymeric product is pullulan; an analysis to be contrasted with a value of 29% when D-fructose is the sole carbon source.

Carbon sources other than D-glucose, D-fructose, sucrose or maltose have been reported as capable of supporting growth of *Aureobasidium pullulans*. Thus, constitutive enzymes for the utilization of D-galactose, D-mannose, raffinose and trehalose, but not lactose, have been identified [Clark and Wallace, 1958a, b] and the production of both pullulan and heteropolysaccharides from D-mannose, D-xylose and L-arabinose have been documented [Bouveng *et al.*, 1962, 1963b]. Growth on 2.5% glycerol occurs but no extracellular polysaccharides are elaborated [Catley, 1971a]. Addition of D-glucose to a concentration of 2.5% restored production of polysaccharide and, by the use of [^{14}C]-glycerol in this two carbon substrate culture it was established that 4.3% of the polysaccharide, a little over half of which is pullulan, is derived from the triol. Growth on acetate, even when supplemented with D-glucose was not conducive to the synthesis of extracellular polymers, though the rise in pH during the course of growth may have inhibited production.

Pullulan is not produced by cells supplied with a reduced level of 0.5% D-glucose, and since the maximum cell density under these conditions is about half of that when growth is on 2.5% D-glucose it might be assumed that the lower concentration of D-glucose presents conditions of carbon limitation [Catley, 1971a]. Growth in continuous culture under D-glucose limiting conditions has confirmed this [G. Carolan, B.J. Catley and T.R. Jarman, unpublished data].

On entering the logarithmic phase of cell growth there

is a temporal lag between the increase of cell mass and the onset of pullulan production [Catley, 1971a, b]. Moreover the accumulation of polysaccharide appears linear with time during a phase of the life-cycle where growth is rapid. If the ability of the cell to elaborate pullulan, once initiated, remains constant then extracellular polysaccharide should accumulate at an increasing rate. *Aureobasidium pullulans* does not normally metabolize pullulan [Catley, 1970] and the constant rate of accumulation cannot therefore be explained by proposing a steady state maintained by cellular synthesis and utilization. Examining the assimilation patterns of [^{14}C]-D-glucose, the rate of extracellular polysaccharide synthesis was seen to rise and fall; the sequence distinctly following, and not being concurrent with, the rate of cellular incorporation of ^{14}C [Catley, 1973]. The maximum proportion of assimilated D-glucose that is diverted to extracellular polysaccharide is about 65%, though the overall yield is considerably less. That pullulan is not synthesized throughout the growth cycle, but is produced in conditions where there is an excess of a suitable carbon source together with some other necessary nutrient that has become rate-limiting to growth, is seen as an indication that the polysaccharide may be classed as a secondary metabolite associated with the storage and maintenance phase of the growth cycle.

The factors that control the rate of synthesis remain obscure, though a number of parameters have been identified that influence production, either by directly acting on the synthesis or excretion mechanism or by modifying cellular metabolism. Though it might be fairly assumed that the pH of the intracellular milieu is constrained to lie between narrow limits, the same cannot be claimed for the extracellular environment. Typical buffering ions in the medium are those of orthophosphate and these are not effective between pH 6 and 3. Most cultures commencing growth around neutrality and using ammonium ion as a nitrogen source rapidly increase in acidity once they cease to be buffered by the phosphate ion. Cells actively elaborating pullulan at pH 5 produce little or no polysaccharide when the medium is adjusted to neutrality [Catley, 1971b]. These observations have been corroborated and extended by Ono *et al.* [1977b] who have shown that whereas cell growth appears adequate at pH 2.5-2.0 there is little production of pullulan. Their data also show that a better yield of polysaccharide (66%) is obtained from cultures initially at pH 6 and allowed to become more acidic than that (48%) from cultures otherwise identical but in which the pH is maintained at 6. Moreover, the proportion of pullulan was better (95.4%) in the acidifying culture than that in the controlled (69.1%). Differences were also noted between shaken cultures and those actively aerated, the latter being more productive of polysaccharide.

Other factors that might govern the accumulation of

polysaccharide in the medium have not been extensively studied, but of those which have the concentration of ammonium ions and protons have a pronounced effect [Catley, 1971b]. The rationale of the observed inhibition of pullulan synthesis effected by ammonium ion is not at all clear and there may be several individual contributions to the phenomenon. On the one hand, there is the possibility of the general consequence of nitrogen limitation signalling onset of secondary metabolism, one result of which is the biosynthesis of pullulan. On the other hand, a more specific step in the carbon flow of metabolic pathways may be the target of control. Evidence for both of these effects has been produced though their relationship to one another, if any, has not been investigated. Experiments in shake culture demonstrated that the addition of eight-fold the amount of ammonium sulphate normally present at the commencement of growth maintains biomass production in the log-phase for a longer period and inhibits polysaccharide production [Catley, 1971b]. Later work using continuous culture with ammonium limited media [Catley and Kelly, 1975; Kelly, 1976] indicated that pullulan elaboration only commences when the nitrogen source is at almost zero concentration. A more specific effect of ammonium ion on glucose catabolism has been proposed [Rothman and Cabib, 1969]. Addition of 50 mM ammonium ion to log-phase suspensions of *Saccharomyces cerevisiae* suppresses glycogen accumulation; an effect that is explained by the stimulation of glycolysis through the activation of phosphofructokinase. Similar effects have been observed with *Aureobasidium pullulans* isolated in the mid-log phase of growth; the addition of 50 mM ammonium ion reduces the elaboration of pullulan by 30% [W. Anderson and B.J. Catley, unpublished observations]. Whether these observations gathered from investigation over the period of an hour are related to those studies of the growth-cycle lasting a few days is not known.

Suspecting that pullulan synthesis is not constitutive, the role of sugars as inducers was investigated [Catley, 1972]. Starvation of washed cell suspensions for an hour before the addition of cycloheximide followed by D-glucose inhibits the production of pullulan whereas the omission of the protein synthesis inhibitor leads to the recommencement of elaboration. Addition of the inhibitor after exposure to D-glucose does not appreciably inhibit production over the subsequent hour. The naive interpretation of these results is that messenger RNA involved in pullulan production must have a half-life of considerably less than 1 hour and that the proteins responsible for the appearance of the polysaccharide must, at least *in vivo*, be inducible and quite unstable.

Whatever the rationale for activating the production of pullulan there appears to be a relationship between that process and the morphological changes exhibited by the organism. Precisely what this link might be is still a matter for conjecture but in general terms it seems that extracellular polysaccharide production is associated with the yeast-like form of the microorganism. The fungus, formerly described as being dimorphic [Cooke, 1959], exhibits a varied, often complex, morphology [Ramos and García-Acha, 1975b; Ramos, García-Acha and Peberdy, 1975; Pechak and Crang, 1977]. Studies of a number of dimorphic fungi, for example with *Mucor* and *Mycotypha*, have demonstrated relationships between morphology and the alternative metabolic states of respiration and fermentation [Hall and Kolankaya, 1974]. Environmental conditions which lead to an impairment of respiration function, for example high D-glucose concentration, the presence of cyanide ion or oligomycin, give rise to an increase in the number of yeast-like organisms. The isolation of spontaneous and ultra-violet light induced mutants show that inhibition of electron flow down the respiratory chain affords an organism growing almost entirely in the yeast-like phase [Storck and Morrill, 1971; Schultz *et al*., 1974].

The first suggestion that pullulan synthesis might be associated with the yeast-like form of the organism was the observation of the coincidence of onset of polysaccharide elaboration and the budding of hyphae [Catley, 1973]. Separation of the two morphological forms on a mannitol gradient confirmed this observation. Even taken together these two lines of evidence do not constitute proof that mycelia are incapable of polysaccharide production and it is likewise apparent from this review that the single-celled form of *Aureobasidium pullulans* does not produce pullulan continuously. Nevertheless, the isolation of "yeasty" mutants that exhibit increased synthesis of polysaccharide provides additional evidence of a possible correlation between morphology and the appearance of pullulan. In the belief that this correlation exists mutagenesis was attempted with ethidium bromide (2,7-diamino-10-ethyl-9-phenylphenanthridinium bromide). The mutagen is thought to have a selective action on circular DNA; just such a form being present in mitochondria. Any modification that occurred to this DNA might inhibit mitochondrial function and promote a non-respiratory metabolism. Acridine dyes, and especially ethidium bromide, act preferentially on the mitochondrial genome in yeast cells affording "petite" mutants that possess an altered cytochrome chain and grow fermentatively. Ethidium bromide can effect an almost complete conversion of *Saccharomyces cerevisiae* to these phenotypes [Slonimski *et al*., 1968]. However, though ethidium bromide is known to act by intercalation with the DNA helix it has been shown to be capable of other interactions: binding covalently to mitochondrial DNA and causing alterations in the structure of transfer RNA [Jones and Kearns, 1975].

In making a reduction of the time-scale for experimentation from a typical 100 h that might be the extent of the growth-cycle to a 10 h span that might encompass the cell-cycle, the question of a periodic synthesis of pullulan must be considered. Investigations into the production of yeast cell-wall components of *Saccharomyces cerevisiae* [Sierra *et al.*, 1973] demonstrate a continuous synthesis of mannan and the alkali- and acid-insoluble glucans. This contrasts with the periodic synthesis of chitin [Cabib and Farkas, 1971], which is effective during a limited portion of the cycle, the enzyme being activated for the assembly of the septum separating daughter and parent cells. Whether or not extracellular polysaccharides, and in particular pullulan, are produced in a cyclic fashion is not known with any certainty though there is some evidence that the major elaboration rate is just before cell-division [B.J. Catley and O. Whiteoak, unpublished observations].

The free energy that is required for the synthesis of a glycosidic bond is usually derived *in vivo* either from the clevage of an existing glycosidic bond or the clevage of a pyrophosphate bond. Examples of the former pathway are dextran synthesis from sucrose by *Leuconostoc mesenteroides* [see Brooker, Chapter 5] and the formation of the α 1,6-glucosidic bond at the expense of the α 1,4-glucosidic bond in the synthesis of glycogen. Examples of the latter route are the polymerization of adenine diphosphoglucose (ADP-D-glucose) to form the linear α 1,4 glucan precursor of bacterial glycogen and the polyprenyl pyrophosphate saccharides participating in assembly of the bacterial cell-wall. Both these mechanisms have been reported as playing a role in the biosynthesis of oligo- and polysaccharides of *Aureobasidium pullulans*. Thus, using a preparation of acetone-dried cells, a process of trans-glucosylation affording a range of oligosaccharides from a maltose donor has been observed [Ueda and Kono, 1965]. The initial concentration of disaccharide was high (0.29M), the maximum degree of polymerization (DP) was 5 and the oligosaccharides characterized were maltotriose, panose (6^2-α-glucosylmaltose), maltotetraose and maltopentaose. The reaction could be displaced in favour of synthesis by the addition of glucose oxidase whereby a maximum DP of 9 was observed. In the absence of glucose oxidase the yield of oligosaccharides was of the order of 20%. Cell-free supernatant preparations from cells in the mid logarithmic phase of growth have likewise been reported as exhibiting transferase activity in their action on disaccharides [Catley, 1971a]. A sucrose substrate (0.03M) afforded a major pattern of hydrolysis to D-glucose and D-fructose with a minor portion (5%) involved in transfer reactions. Similar supernatant preparations from maltose-grown cultures demonstrated maltase activity but appreciably more (40%) transfer reaction products were generated from the 0.03M maltose substrate. That glucosidic bonds other than the α 1,4 are produced was seen from the chromatographic mobilities of the products. In no case involving transglucosylation reactions has there been evidence to suggest the formation of polysaccharides. More recently [Taguchi *et al.*, 1973b] there has been a report of pullulan synthesis from maltose and sucrose by acetone-dried cells. The results, however, were somewhat variable and five times more dried cells were required to produce similar yields to those of the washed cells. With the above possible exception it must be concluded that pullulan at least on present evidence, is not assembled due to transferase activity.

Turning to the possibility of involvement with nucleotide and polyprenyl pyrophosphate sugars as donors there is limited but more promising evidence for resolution of the *in vivo* assembly process. Using high concentrations (160mM) of uridine diphosphoglucose a 54% conversion of nucleotide sugar into pullulan has been reported [Taguchi *et al.*, 1973b]. An absolute requirement for the synthesis is adenosine triphosphate, possibly as the magnesium adenylate, present at the equally high concentration of 160mM. ADP-D-glucose, even in the presence of ATP, is not a donor. Seeking the participation of a lipid intermediate, acetone-dried cells were administered with [^{14}C]-sucrose. Analysis of the subsequently labelled products indicated that there might well be a glucosylated polyprenyl pyrophosphate but the evidence, though suggestive of the presence of lipid intermediates, is not unequivocal. With the rationale that whereas pullulan is not produced at pH 2 the precursors, possibly lipid in nature, might be, an analysis of glycolipids extracted from mycelia at this acidic pH has been reported [Ono *et al.*, 1977a]. Lipids extracted from water-washed mycelia contain glucolipid wherein the carbohydrate was a single glucose moiety. On the other hand the carbohydrate moiety present in glucolipids extracted from acetone-treated mycelia was of an oligosaccharyl nature though seemingly not present as simple α 1,4 or α 1,6 glucans.

Conclusions

Whereas it is apparent that, at the physiological level, information about the conditions for growth of *Aureobasidium pullulans* leading to a major diversion of assimilated carbon sources to the production of pullulan is now available, very little is known about the mechanism of assembly at the molecular level. The past successes of structural analysis of the polysaccharide and growth studies of the organism producing it must now be matched with an elucidation of the biosynthetic mechanism.

References

Bender, H., Lehmann, J. and Wallenfels, K. (1959). Pullulan, ein extracelluläres Glucan von *Pullularia pullulans*. *Biochimica et Biophysica Acta* **36**, 309-316.

Bender, H. and Wallenfels, K. (1961). Untersuchungen an Pullulan. II. Spezifischer Abbau durch ein bakterielles Enzym. *Biochemische Zeitschrift* **334**, 79-95.

Bernier, B. (1958). The production of polysaccharides by fungi active in the decomposition of wood and forest litter. *Canadian Journal of Microbiology* **4**, 195-204.

Bouveng, H.O., Kiessling, H., Lindberg, B. and McKay, J. (1962). Polysaccharides elaborated by *Pullularia pullulans*. Part I. The neutral glucan synthesised from sucrose solutions. *Acta Chemica Scandinavica* **16**, 615-622.

Bouveng, H.O., Kiessling, H., Lindberg, B. and McKay, J. (1963a). Polysaccharides elaborated by *Pullularia pullulans*. Part II. The partial acid hydrolysis of the neutral glucan synthesised from sucrose solutions. *Acta Chemica Scandinavica* **17**, 797-800.

Bouveng, H.O., Kiessling, H., Lindberg, B. and McKay, J. (1963b). Polysaccharides elaborated by *Pullularia pullulans*. Part III. Polysaccharides synthesised from xylose solutions. *Acta Chemica Scandinavica* **17**, 1351-1356.

Brown, R.G. and Nickerson, W.J. (1965). Composition and structure of cell walls of filamentous and yeastlike forms of *Aureobasidium (Pullularia) pullulans*. *Proceedings of the American Society for Microbiology*, p.26.

Brown, R.G. and Lindberg, B. (1967a). Polysaccharides from cell walls of *Aureobasidium (Pullularia) pullulans*. Part I. Glucans. *Acta Chemica Scandinavica* **21**, 2379-2382.

Brown, R.G. and Lindberg, B. (1967b). Polysaccharides from cell walls of *Aureobasidium (Pullularia) pullulans*. Part II. Heteropolysaccharide. *Acta Chemica Scandinavica* **21**, 2383-2389.

Brown, R.G., Hanic, L.A. and Hsiao, M. (1973). Structure and chemical composition of yeast chlamydospores of *Aureobasidium pullulans*. *Canadian Journal of Microbiology* **19**, 163-168.

Bull, A.T. (1972). Environmental factors influencing the synthesis and excretion of exocellular macromolecules. *Journal of Applied Chemistry and Biotechnology* **22**, 261-292.

Cabib, E. and Farkas, V. (1971). The control of morphogenesis: an enzymatic mechanism for the initiation of septum formation in yeast. *Proceedings of the National Academy of Sciences, U.S.A.*, **68**, 2052-2056.

Catley, B.J. (1970). Pullulan, a relationship between molecular weight and fine structure. *F.E.B.S. Letters* **10**, 190-193.

Catley, B.J. (1971a). Utilization of carbon sources by *Pullularia pullulans* for the elaboration of extracellular polysaccharides. *Applied Microbiology* **22**, 641-649.

Catley, B.J. (1971b). Role of pH and nitrogen limitation in the elaboration of the extracellular polysaccharide pullulan by *Pullularia pullulans*. *Applied Microbiology* **22**, 650-654.

Catley, B.J. (1972). Pullulan elaboration, an inducible system of *Pullularia pullulans*. *F.E.B.S. Letters* **20**, 174-176.

Catley, B.J. (1973). The rate of elaboration of the extracellular polysaccharide, pullulan, during growth of *Pullularia pullulans*. *Journal of General Microbiology* **78**, 33-38.
Catley, B.J. and Kelly, P.J. (1975). Metabolism of trehalose and pullulan during the growth cycle of *Aureobasidium pullulans*. *Biochemical Society Transactions* **3**, 1079-1081.
Catley, B.J. and Whelan, W.J. (1971). Observations on the structure of pullulan. *Archives of Biochemistry and Biophysics* **143**, 138-142.
Clark, D.S. and Wallace, R.H. (1958a). Carbohydrate metabolism of *Pullularia pullulans*. *Canadian Journal of Microbiology* **4**, 43-54.
Clark, D.S. and Wallace, R.H. (1958b). Oxidation of compounds in the Kreb's Cycle by *Pullularia pullulans*. *Canadian Journal of Microbiology* **4**, 125-139.
Cooke, W.B. (1959). An ecological life history of *Aureobasidium pullulans* (de Bary) Arnaud. *Mycopathologia et Mycologia Applicata* **12**, 1-45.
Dubois, M., Gilles, K.A., Hamilton, J.K., Rebers, P.A. and Smith, F. (1956). Colorimetric method for determination of sugars and related substances. *Analytical Chemistry* **28**, 350-356.
Durrell, L.W. (1968). Studies of *Aureobasidium pullulans* (de Bary) Arnaud. *Mycopathologia et Mycologia Applicata* **35**, 113-120.
Elinov, N.P. and Matveeva, A.K. (1972). Extracellular glucan produced by *Aureobasidium pullulans*. *Biokhimiya* **37**, 255-257.
Elinov, N.P., Marikhin, V.A., Dranishnikov, A.N., Myasnikova, L.P. and Maryukhta, Y.B. (1975). Peculiarities of the glucan produced by a culture of *Aureobasidium (Pullularia) pullulans*. *Doklady Akademii Nauk S.S.S.R.* **221**, 213-216.
Hall, M.J. and Kolankaya, H. (1974). The physiology of mould-yeast dimorphism in the genus *Mycotypha* (Mucorales). *Journal of General Microbiology* **82**, 25-34.
Jeanes, A. (1977). Dextrans and Pullulans: Industrially significant α D-glucans. In *American Chemical Society Symposia*, Vol. 45, pp. 284-297. Edited by P.A. Sandford and A. Laskin.
Jones, C.R. and Kearns, D.R. (1975). Identification of a unique ethidium bromide binding site on yeast t-RNA^{Phe} by high resolution (300 MHz) nuclear magnetic resonance. *Biochemistry* **14**, 2660-2665.
Kelly, P.J. (1976). Carbohydrates elaborated by *Aureobasidium pullulans*. Ph.D. thesis, Heriot-Watt University, Edinburgh.
Kikuchi, Y., Taguchi, R., Sakano, Y. and Kobayashi, T. (1973). Comparison of extracellular polysaccharide produced by *Pullularia pullulans* with polysaccharides in the cells and cell wall. *Agricultural and Biological Chemistry* **37**, 1751-1753.
Ono, K., Kawahara, Y. and Ueda, S. (1977a). Effect of pH on the content of glycolipids in *Aureobasidium pullulans* S-1. *Agricultural and Biological Chemistry* **41**, 2313-2317.
Ono, K., Yasuda, N. and Ueda, S. (1977b). Effect of pH on pullulan elaboration by *Aureobasidium pullulans* S-1. *Agricultural and Biological Chemistry* **41**, 2113-2118.
Pechak, D.G. and Crang, R.E. (1977). An analysis of *Aureobasidium pullulans* developmental stages by means of scanning electron microscopy. *Mycologia* **69**, 783-792.
Ramos, S. and García-Acha, I. (1975a). Cell wall enzymatic lysis of the yeast form of *Pullularia pullulans* and wall regeneration by

protoplasts. *Archives of Microbiology* **104**, 271-277.

Ramos, S. and García-Acha, I. (1975b). A vegetative cycle of *Pullularia pullulans*. *Transactions of the British Mycological Society* **64**, 129-135.

Ramos, S., García-Acha, I. and Peberdy, J.F. (1975). Wall structure and the budding process in *Pullularia pullulans*. *Transactions of the British Mycological Society* **64**, 283-288.

Rothman, L.B. and Cabib, E. (1969). Regulation of glycogen synthesis in the intact yeast cell. *Biochemistry* **8**, 3332-3341.

Sakano, Y., Masuda, N. and Kobayashi, T. (1971). Hydrolysis of pullulan by a novel enzyme from *Aspergillus niger*. *Agricultural and Biological Chemistry* **35**, 971-973.

Schultz, B.E., Kraepelin, G. and Hinkelmann, W. (1974). Factors affecting dimorphism in *Mycotypha* (Mucorales): a correlation with the fermentation/respiration equilibrium. *Journal of General Microbiology* **82**, 1-13.

Sierra, J.M., Sentandreu, R. and Villaneuva, J.R. (1973). Regulation of wall synthesis during *Saccharomyces cerevisiae* cell cycle. *F.E.B.S. Letters* **34**, 285-290.

Slonimski, P.P., Perrodin, G. and Croft, J.H. (1968). Ethidium bromide induced mutation of yeast mitochondria: complete transformation of cells with respiratory deficient non-chromosomal "petites". *Biochemical and Biophysical Research Communications* **30**, 232-239.

Sowa, W., Blackwood, A.C. and Adams, G.A. (1963). Neutral extracellular glucan of *Pullularia pullulans* (de Bary) Berkhout. *Canadian Journal of Chemistry* **41**, 2314-2319.

Storck, R. and Morrill, R.C. (1971). Respiratory-deficient yeast-like mutant of *Mucor*. *Biochemical Genetics* **5**, 467-479.

Taguchi, R., Kikuchi, Y., Sakano, Y. and Kobayashi, T. (1973a). Structural uniformity of pullulan produced by several strains of *Pullularia pullulans*. *Agricultural and Biological Chemistry* **37**, 1583-1588.

Taguchi, R., Sakano, Y., Kikuchi, Y., Sakuma, M. and Kobayashi, T. (1973b). Synthesis of pullulan by acetone-dried cells and cell-free enzyme from *Pullularia pullulans*, and the participation of lipid intermediate. *Agricultural and Biological Chemistry* **37**, 1635-1641.

Ueda, S., Fujita, K., Komatsu, K. and Nakashima, Z. (1963). Polysaccharide produced by the genus *Pullularia*. I. Production of polysaccharide by growing cells. *Applied Microbiology* **11**, 211-215.

Ueda, S. and Kono, H. (1965). Polysaccharide produced by the genus *Pullularia*. II. Trans-α-glucosidation by acetone cells of *Pullularia*. *Applied Microbiology* **13**, 882-885.

Wallenfels, K., Bender, H., Keilich, G. and Bechtler, G. (1961). Über Pullulan, das Glucan der Schleimhülle von *Pullularia pullulans*. *Angewandte Chemie* **73**, 245-246.

Wallenfels, K., Keilich, G., Bechtler, G. and Freudenberger, D. (1965). Untersuchungen an Pullulan. IV. Die Klärung des Strukturproblems mit physikalischen, chemischen und enzymatischen Methoden. *Biochemische Zeitschrift* **341**, 433-450.

Wilkinson, J.F. (1958). The extracellular polysaccharides of bacteria. *Bacteriological Reviews* **22**, 46-73.

Wynne, E.S. and Gott, C.L. (1956). A proposed revision of the genus

Pullularia. *Journal of General Microbiology* **14**, 512-519.
Yuen, S. (1974). Pullulan and its applications. *Process Biochemistry* (November) **22**, 7-9.

Chapter 5

ELECTRON MICROSCOPY OF THE DEXTRANS PRODUCED BY LACTIC ACID BACTERIA

B.E. BROOKER

National Institute for Research in Dairying, Shinfield, Reading, UK

Introduction

Many lactic acid bacteria possess the ability to produce extracellular dextran when grown in the presence of sucrose. The enzyme involved in dextran formation, dextransucrase, is a member of a general class of glucosyltransferases which catalyses the synthesis of chains of glucopyranosyl residues from these groups in sucrose [see Sutherland, Chapter 1]. In some species, for example *Streptococcus mutans*, *Streptococcus bovis*, this enzyme is constitutive since it is present in cultures supplemented with sugars other than sucrose [Bailey, 1959; Guggenheim and Newbrun, 1969] but in others, for example *Leuconostoc mesenteroides* it is induced only by the presence of sucrose [Neely and Nott, 1962]. The dextran produced by a given species or strain of bacterium may be a water-soluble polymer which disperses and increases ambient viscosity or it may be water-insoluble capsular dextran. In some cases, as will be seen below, soluble and insoluble dextrans are produced simultaneously.

To date, the dextrans whose structure has been most thoroughly investigated are those produced by *Leuconostoc mesenteroides* and *Streptococcus mutans*. There is a good reason for this. Those produced by *Leuconostoc mesenteroides* are commercially important and there is a voluminous literature devoted not only to their industrial production but also to their chemical structure [see Powell, Chapter 6] and biosynthesis [Neely, 1960; Jeanes, 1966; Sidebotham, 1974]. However, there have been relatively few light or electron microscopic studies of this important organism or of the dextrans it produces. In the case of *Streptococcus mutans*, the dextrans are of interest because they appear to promote dental caries by facilitating the colonization of tooth surfaces by cariogenic bacteria and by contributing to the formation of dental plaque.

Thus, there is growing interest in the chemical structure of the dextrans produced by oral streptococci and already most of the available information on dextran ultrastructure is derived from studies of the insoluble extracellular polysaccharides produced by *Streptococcus mutans*.
This chapter briefly reviews this information and presents the results of some recent ultrastructural studies of the insoluble dextrans produced by *Leuconostoc mesenteroides*.

Methods of Preparation

Since the nature of the information obtained from any ultrastructural study depends to a large extent on the method of specimen preparation, it is relevant to consider briefly some of the techniques that are available or have been used to prepare dextrans for electron microscopy.

One of the simplest methods of examining dextran involves negative staining or metal shadowing molecules which have been dried onto a support film. This approach, used by Holzwarth and Prestridge [1977] to examine xanthan, and by Simionescu and Palade [1971] to determine the size distribution of particles in dextran solutions, may be useful for the study of soluble dextrans but it is less satisfactory in the case of insoluble dextrans because of the difficulty in achieving and maintaining an adequate dispersion [Newbrun *et al.*, 1971]. Freeze fracturing offers one of the few satisfactory methods for the examination of soluble dextrans *in situ* since these polymers are not normally immobilized during fixation and are lost during the processing steps before embedding. Thus, in a study of dextran production by *Streptococcus bovis*, Cheng *et al.* [1976] demonstrated an extracellular, filamentous network of water-soluble glucan which had separated from the aqueous phase during freezing. The method can also be used to advantage in those cases where dextrans prove difficult or impossible to embed. Embedding and thin sectioning, however, remain the most widely used methods of preparing dextrans for electron microscopy although, as mentioned above, it suffers from the disadvantage that soluble dextrans are lost during processing. Moreover, dextrans are not intrinsically electron dense and it is often difficult to confer sufficient electron density using either conventional fixatives or the lead and uranium salts normally used to stain sections.

To overcome this problem, i.e. to render the dextran visible, and at the same time to localize periodate-reactive carbohydrate, many investigators have used cytochemical methods derived from the periodic acid-Schiff (PAS) technique in which a metal is deposited at the sites of aldehydes produced by the periodate oxidation of adjacent hydroxyl groups in the dextran. There are several closely related methods by which this can be done [reviewed by Thiéry and Rambourg, 1974] each of which can be applied to the dextran in two ways. In the method of Critchley *et al.* [1967], which has been used to examine the extracellular polysaccharides produced by cariogenic streptococci [Guggenheim and Schroeder, 1967; Newman *et al.*, 1976], periodic acid, thiosemicarbazide and osmium tetroxide (PA-TSC-OsO_4) are applied in turn directly to the material before embedding. The other approach which uses the same or similar reagents on thin sections of conventionally fixed and embedded material, is easier to perform and is less likely to produce the artefacts caused by inadequate washing between reagents. Although giving good contrast, the resolution obtained by these methods is limited by the relatively large grain size of the deposited metal. As will be seen below, more uniform, finer grained staining of insoluble native dextran can be achieved by the use of ruthenium red during fixation [Brooker, 1976, 1977]. Introduced to electron microscopy largely through the work of Luft [1964, 1971], the current extensive use of this cationic dye in ultrastructural studies of microorganisms stems from its ability to stain extracellular material associated with the surface of cell walls that is otherwise difficult or impossible to demonstrate. Thus, it was used by Pate and Ordal [1965] to demonstrate surface filaments on myxobacteria, by Springer and Roth [1973] to show the capsular structure of *Diplococcus pneumoniae* and *Klebsiella pneumoniae* and by Costerton *et al.* [1974] to reveal the surface layers of some rumen bacteria. Fletcher and Floodgate [1973] and Jones *et al.* [1969] have shown that the surface of many aquatic bacteria is covered with ruthenium red-positive material which is probably involved in their adhesion to solid surfaces; similar material is also found on the surface of some bacteria that adhere to the epithelia of the alimentary canal [Brooker and Fuller, 1975].

Luft [1971] and Blanquet [1976] showed that the essential step in ruthenium red staining was its exposure to specimens in the presence of osmium tetroxide. Although ruthenium red in aqueous or buffered solutions will bind to substrates, simultaneous or prior exposure to osmium tetroxide is necessary to impart electron density. However, some authors claim that exposure of unfixed cells to ruthenium red prior to the method advocated by Luft [1971] gives improved staining of extracellular polysaccharide [Cagle *et al.*, 1972; Cagle, 1975]. In another approach, bacteria are treated with ruthenium red before fixation and dehydrated in the presence of the cationic dye [Patterson *et al.*, 1975; Cheng *et al.*, 1976] but neither technique is used widely.

Two factors which appear to be important in determining the affinity of a molecule for ruthenium red are its anionic charge density and its degree of polymerization. Hence polyanions with high charge density react strongly with ruthenium red but highly polymerized substances with very low charge density were found by Luft [1971] not to react. The ability to stain a neutral polysaccharide such as dextran with ruthenium red is therefore somewhat surprising and requires some explanation. Although the polysaccharides produced by lactic acid bacteria are assumed not to contain charged groups, some aspects of their behaviour can be explained in terms of ionic interactions [Kelstrup and Funder-Nielsen, 1972; Melvaer *et al.*, 1974]. Kelstrup and Funder-Nielsen [1972] found that the soluble and insoluble polysaccharides produced by *Streptococcus mutans* contained negatively charged sites which probably resulted from the uptake of organic or inorganic anions from the surrounding medium. Melvaer *et al.* [1974] came to a similar conclusion and were able to demonstrate the incorporation of labelled phosphate by the polysaccharides produced by *Streptococcus mutans* and *Streptococcus salivarius*. Rorem [1955] showed the uptake of labelled rubidium and phosphate ions by dextrans of *Leuconostoc mesenteroides* and so cations as well as anions can be bound by these polymers. There are two possible explanations for the staining of dextrans by ruthenium red. Either the ruthenium red is bound to sites of high charge density conferred by absorbed anions or there is direct binding of the ruthenium to the dextran similar to that reported by Rorem [1955] for labelled rubidium. Either of these possibilities may also explain the affinity of dextrans for lead ions reported by Simionescu and Palade [1971] in their study of the use of clinical dextrans as particulate tracers. All that can be said at present is that the insoluble dextrans produced by *Leuconostoc mesenteroides* stain with ruthenium red yet they cannot be demonstrated to possess negatively charged sites using cationic ferritin or colloidal iron hydroxide at low pH nor do they adhere to poly-L-lysine coated glass surfaces as perhaps would be expected if a significant number of anionic groups were present [Tsutsui *et al.*, 1976]. The formation of ruthenium complexes with free hydroxyl groups seems a more likely mechanism.

A method for the ultrastructural visualization of dextrans that produces results comparable to those obtained with ruthenium red has recently been described by Ainsworth [1977] and involves secondary fixation with osmium tetroxide which has been partially reduced by the addition of potassium ferrocyanide. The intense electron density of dextrans produced by this method is probably due to the formation of an osmium black complex with the dextran and may, if necessary, be enhanced by counterstaining sections with bismuth subnitrate chelated by tartrate.

Using a primary aldehyde fixative containing tris-(1-aziridinyl) phosphine oxide (TAPO), Cassone and Garaci [1977] have shown that the capsule of *Klebsiella pneumoniae* appears as a uniform, finely fibrillar structure, quite different from that shown by Springer and Roth [1973] using ruthenium red. Although its mechanism of action is unknown, the successful use of aldehyde-TAPO fixation in demonstrating polysaccharide and glycoprotein-rich cell wall components [Djaczenko and Cassone, 1972; Cassone, 1973] suggests that the method may be used to advantage in future ultrastructural studies of dextrans.

Dextrans Produced by *Streptococcus mutans*

Although there have been many ultrastructural studies of the extracellular polysaccharides produced by mixed populations of bacteria in dental plaque [Newman, 1976], comments here will be restricted to the glucans produced in cultures of known organisms. In one of the earliest and most detailed studies of the native, insoluble dextrans produced by *Streptococcus mutans*, Guggenheim and Schroeder [1967] distinguished three different structural elements: (1) homogeneous, periodate-reactive material closely associated with single cells or groups of streptococci, (2) fibrils (2 nm thick protofibrils), often arranged in parallel, forming a loose network and representing the major portion of the dextran and (3) structureless, heavily electron-dense particles scattered throughout the meshwork of fibrillar material. Identification of the protofibrillar elements as dextran was established using PA-TSC-OsO_4 staining and dextranase digestion. Newbrun *et al.* [1971] examined purified polysaccharides obtained from three species of oral streptococci. The glucan and dextran obtained from *Streptococcus mutans* and *Streptococcus sanguis* respectively, were found to be in the form of ellipsoidal particles which sometimes aggregated to produce beaded fibres reminiscent of the protofibrils reported by Guggenheim and Schroeder [1967]. Using different strains of *Streptococcus mutans*, several workers have confirmed that the principal dextran produced by this species in the presence of sucrose is fibrillar in nature [Johnson *et al.*, 1974; Nalbandian *et al.*, 1974; Newman *et al.*, 1976] and that this material is dextranase sensitive [Nalbandian *et al.*, 1974]. Johnson *et al.* [1974] and Nalbandian *et al.* [1974] showed that the production of fibrillar dextran was important in determining the ability of *Streptococcus mutans* to adhere to solid surfaces. Thus, in mutants of *Streptococcus mutans* whose ability to adhere was lost or greatly diminished, globular or homogeneous extracellular polysaccharides were produced with little or none of the fibrillar material synthesized by wild-type strains. The dextran produced by one of these mutants was found by Johnson *et al.* [1977] to contain significantly fewer 1,3-glucosidic linkages than that produced by a wild-type strain thus supporting the suggestion originally made by Guggenheim and Schroeder [1967] that the protofibrillar elements contained 1,3-linked glucopyranosyl residues.

Insoluble Dextrans Produced by *Leuconostoc mesenteroides*

Before considering ultrastructural aspects of dextran formation by *Leuconostoc mesenteroides*, some light microscope observations relevant to this phenomenon must be described.

The eleven strains that will be referred to here can be divided into two groups. In one, cells grown in the presence of sucrose produce only soluble dextran and by light microscopy do not appear grossly different from cells of the same strain grown in glucose. The other group includes those strains which, when grown in the presence of sucrose, contain a mixture of acapsulate cells producing soluble dextran and cells that produce a capsule of insoluble dextran visible by light microscopy. The proportion of capsulate cells is variable and depends to some extent on the culture medium used, but they always constitute the minority (5-20%) of the cells present. Although the ability of some strains to produce water-soluble (S) and water-insoluble (L) dextrans simultaneously is well established [Jeanes *et al.*, 1954], it is not generally recognized that in many cases they are produced by different populations of cells.

It is interesting to note that a microscope is not needed to tell if a particular strain is producing capsulate cells. This can be done by examining colonies after 48 h growth on MRS [deMan *et al.*, 1960] agar plates supplemented with 3.6% sucrose. If the colonies contain distinct white specks, capsulate cells are present; if the colonies are relatively uniform in density, capsulate cells are probably absent. Each of the white specks is an aggregate of numerous capsulate cells.

Strains that contain two variants (capsulate and acapsulate cells) when grown in the presence of sucrose do not contain genetically distinct populations with respect to capsule formation. If this was the case, colonies grown on agar media would be expected to consist exclusively of capsulate or acapsulate cells and, assuming that soluble and insoluble dextrans differ in chemical structure [Wilham *et al.*, 1955], one would expect to find differences between the dextrans they produce. In reality, we have observed that colonies always contain both types of cell and it is already known that colonies of any given strain are uniform with respect to the chemical properties of their dextrans [Jeanes *et al.*, 1954; Wilham *et al.*, 1955]. It would appear therefore that both variants can arise from a common parent cell. This can be clearly demonstrated using a dilute suspension of MRS-grown cells mounted on a microscope slide thinly coated with MRS agar containing sucrose. The daughter cells derived from a single, isolated bacterium can then be observed for several hours by phase contrast light microscopy. A colony of cells produced in this way always contains a minority of capsulate cells (Figure 1). It is frequently observed that one of the daughter cells produced by division of an acapsulate cell will start producing insoluble dextran whereas the other does not and may subsequently divide to give two acapsulate cells. However, the daughters of capsulate cells always appear to produce capsules. The mechanism of this interesting phenomenon is not known and deserves further study.

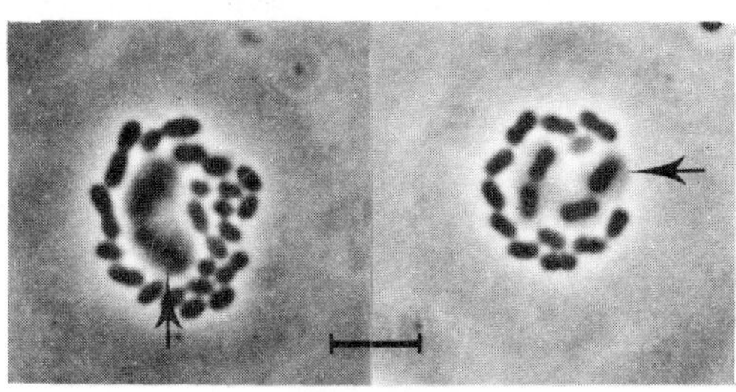

Fig. 1 *Leuconostoc mesenteroides* NCDO 1875. Small colonies of cells derived from a single bacterium. Grown for 7 h on a microscope slide coated with MRS agar supplemented with 3.6% sucrose. Capsulate (arrows) and acapsulate cells are present. Phase contrast microscopy. Bar = 10 μm.

Surface Coat Formation

Exposure of *Leuconostoc mesenteroides* to sucrose leads to the induction of dextransucrase and the initiation of dextran synthesis. In all strains that have been examined by electron microscopy, these events are accompanied by dramatic morphological changes in the surface of the cell wall. It should be noted that these surface changes are easily demonstrated using ruthenium red [Luft, 1971] or reduced osmium tetroxide [Ainsworth, 1977] during fixation but they cannot be followed satisfactorily using conventional methods of fixation. Surface changes are detectable within minutes of exposure to sucrose and always precede the appearance of insoluble dextran in those strains that produce capsules. The morphology of the surface changes is identical for all cells of a given strain even though some may be destined to form capsules and others are not.

All of the strains fall into one of two groups which differ in the details of their surface transformation. In some strains (Group 1 in Table 1, p.99) the cell wall of MRS-grown cells is covered with a layer of material which often has the appearance of a mass of tangled filaments (Figures 2 & 4). Sections stained with periodic acid-thiosemicarbazide-silver proteinate (PA-TSC-Ag) [Thiéry, 1967] show that this surface coat contains periodate-reactive carbohydrate (Figure 7). When transferred to medium containing sucrose, the surface coat increases in thickness and under-

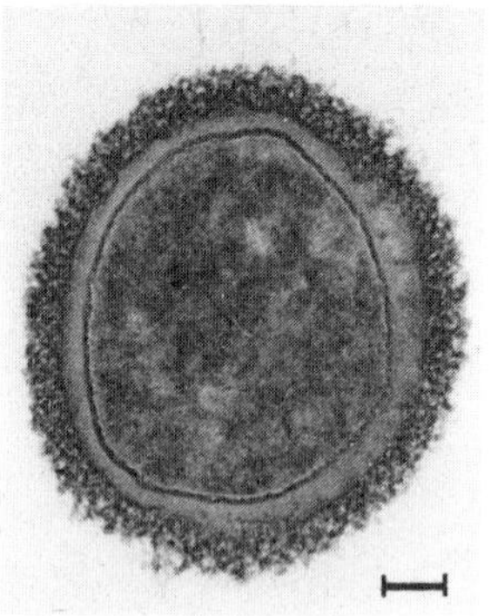

Fig. 2 *Leuconostoc mesenteroides* NCDO 553. Appearance of the cell coat when grown in MRS broth.*

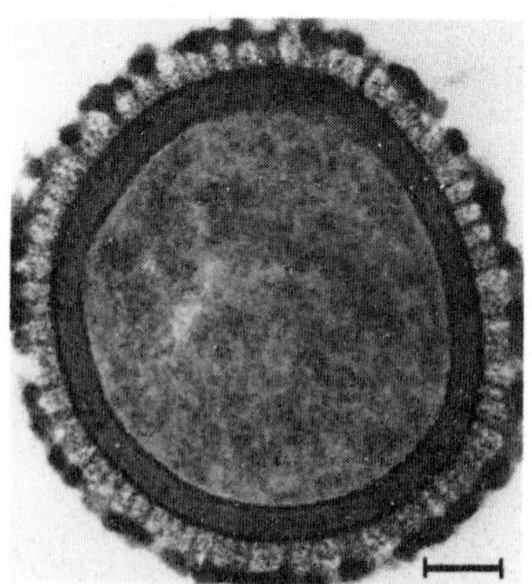

Fig. 3 *Leuconostoc mesenteroides* NCDO 553. Appearance of the cell coat when grown in MRS + sucrose. Filamentous and radial components of the coat are visible.

goes internal reorganization. The result is a uniform layer of very fine filamentous material in which radially arranged rods are visible (Figures 3 & 6). In most strains of this group, the filamentous component of the surface coat of some cells may disappear or become less conspicuous and the radial rods become particularly easy to see (Figure 5). The staining reaction produced by the method of Thiéry suggests that both components of the coat contain periodate-reactive carbohydrate (Figure 8).

In other strains (Group 2), the cell wall of MRS-grown cells has the same appearance as that found in many other

*Figures 2 - 34. Unless otherwise stated, all micrographs are of specimens which have been postfixed in the presence of 0.15% ruthenium red. Bar = 0.1 μm except where stated.

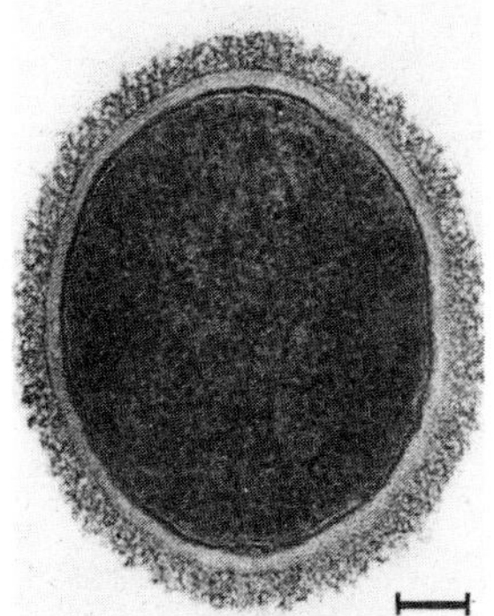

Fig. 4 *Leuconostoc mesenteroides* NCDO 523. Appearance of the cell coat when grown in MRS broth.

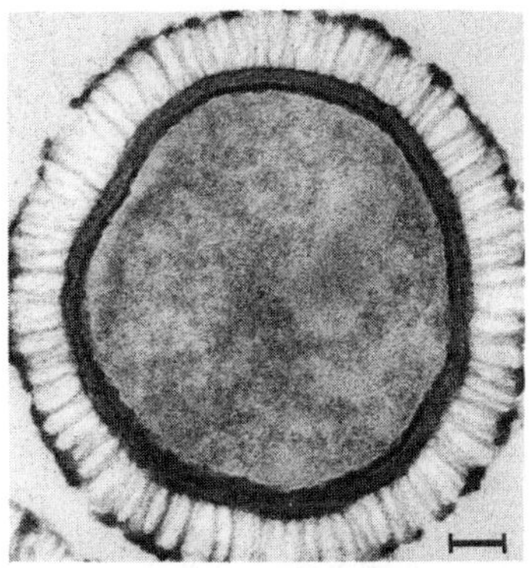

Fig. 5 *Leuconostoc mesenteroides* NCDO 523. Appearance of the cell coat when grown in MRS + sucrose.

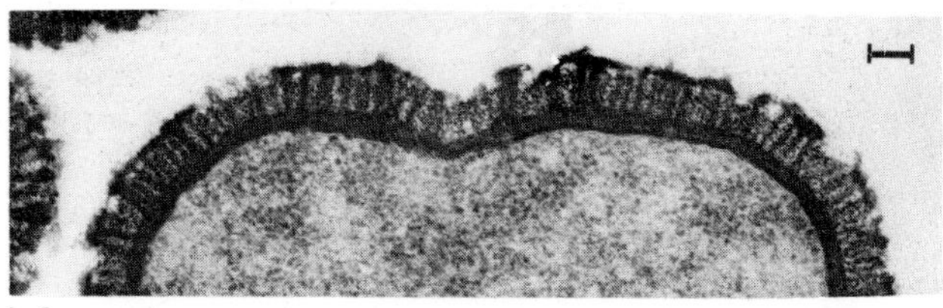

Fig. 6 As for Fig. 5 but showing filamentous and radial components of the cell coat.

Gram positive bacteria and there is no evidence of an additional surface layer (Figure 9). However, when exposed to sucrose, a uniform layer of very electron dense material appears on the surface of the cell wall and increases in prominence until it is 100-150 nm thick (Figures 10, 11). Depending on the strain, this surface coat is two to five times thicker than the cell wall, has a fine filamentous

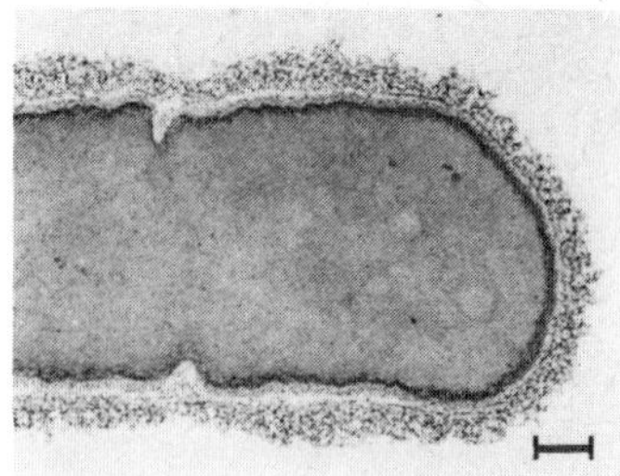

Fig. 7 *Leuconostoc mesenteroides* NCDO 523. MRS-grown cell stained with PA-TSC-Ag. The cell coat contains periodate-reactive carbohydrate.

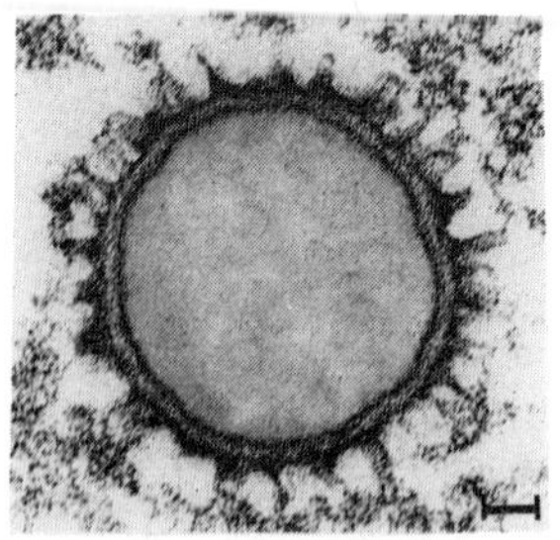

Fig. 8 *Leuconostoc mesenteroides* NCDO 523. Cell grown in MRS + sucrose stained with PA-TSC-Ag. The radial rods of the surface coat contain periodate-reactive carbohydrate.

Fig. 9 *Leuconostoc mesenteroides* NCDO 1593. Cell grown in MRS broth. There is no cell coat.

internal structure and can be shown by cytochemical methods to contain periodate-reactive carbohydrate (Figure 12). It is very interesting that the thickness of the coat can be controlled by manipulating the concentration of sucrose to which the cells are exposed. In strain NCDO 1875 (NRRL 1299), the surface coat produced by exposure to 0.01% sucrose for 6 h is barely detectable (Figure 13). When the concentration of sucrose is increased to 0.1%, the coat (50 nm thick) becomes clearly visible (Figure 14) but it is much thicker (140 nm thick) in cells exposed to 1% sucrose

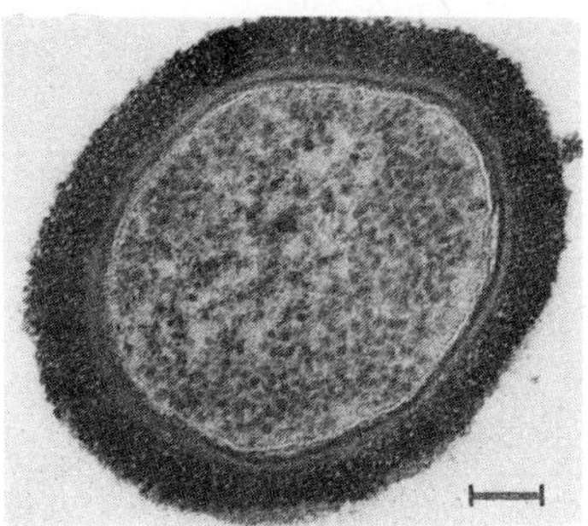

Fig. 10 *Leuconostoc mesenteroides* NCDO 1593. Cell grown in MRS + sucrose showing a thick, uniform cell coat.

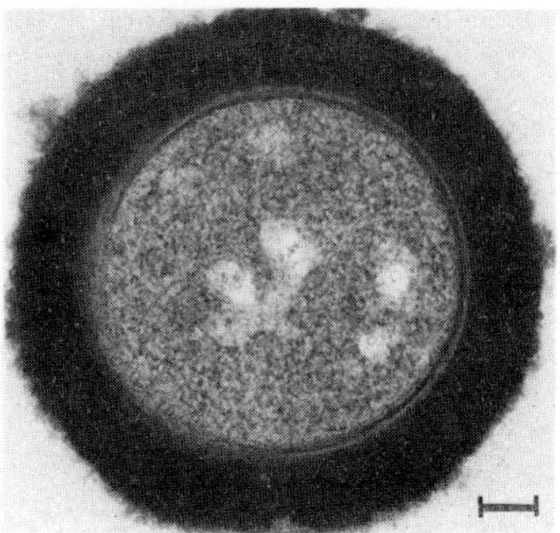

Fig. 11 *Leuconostoc mesenteroides* NCDO 1875. Cell grown in MRS + sucrose and postfixed in osmium tetroxide reduced with potassium ferrocyanide. A thick cell coat is clearly visible.

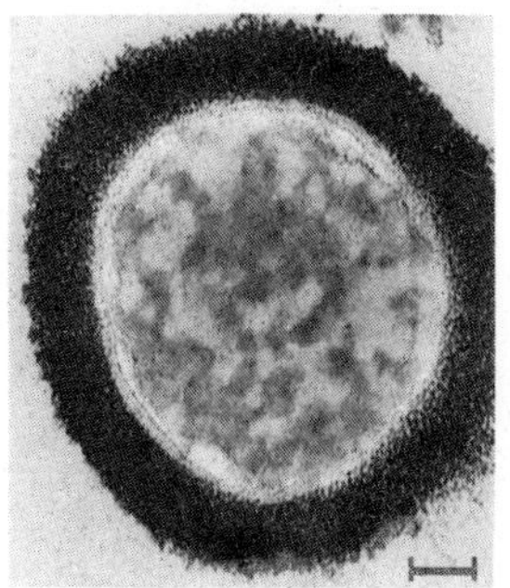

Fig. 12 *Leuconostoc mesenteroides* NCDO 1875. Cell grown in MRS + sucrose whose cell coat contains abundant periodate-reactive carbohydrate after staining with PA-TSC-Ag.

(Figure 15). Concentrations of sucrose greater than 1% cause no increase in the dimensions of the coat.

Since dextran synthesis and the formation of a thick carbohydrate-rich surface coat are both sucrose-dependent,

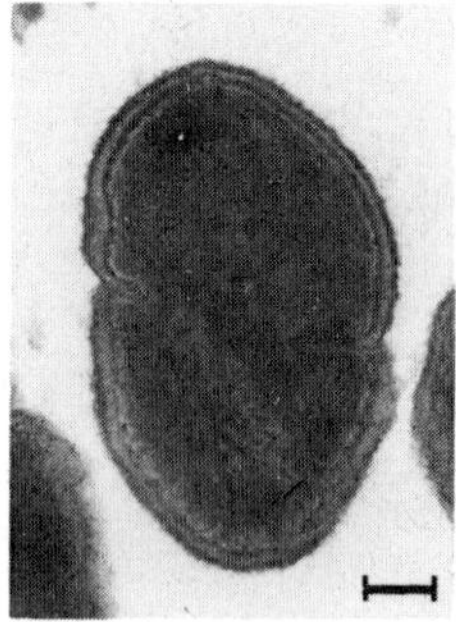

Fig. 13 *Leuconostoc mesenteroides* NCDO 1875. Cells exposed to MRS + 0.01% sucrose. The surface coat is barely visible.

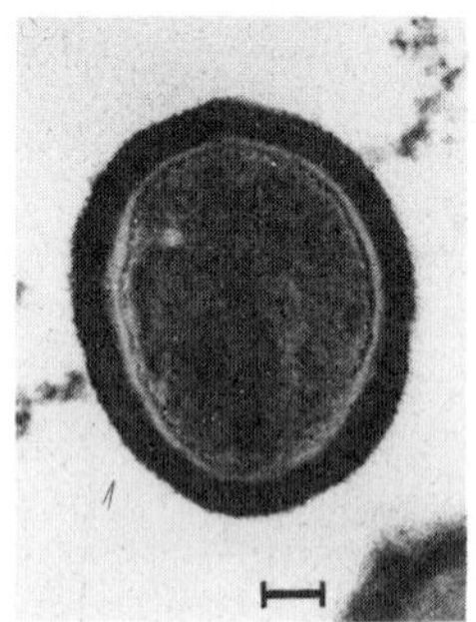

Fig. 14 As for Fig. 13 but exposed to 0.1% sucrose.

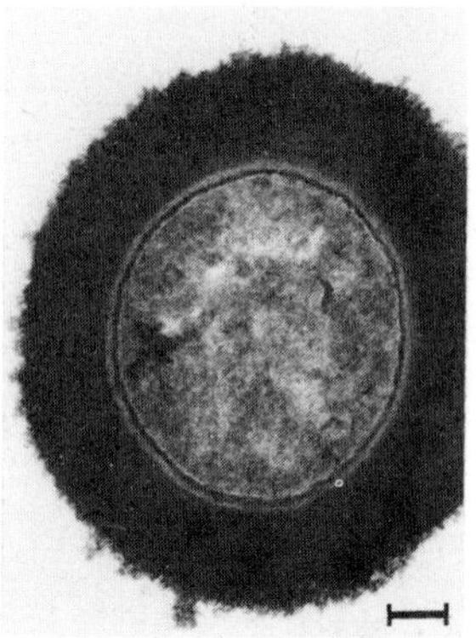

Fig. 15 As for Fig. 13 but exposed to 1.0% sucrose. The cell coat is much thicker than that shown in Figures 13 and 14.

it has been suggested elsewhere [Brooker, 1976, 1977], that the cell coat of *Leuconostoc mesenteroides* contains dextran which is held in position by its association with a cell-bound dextransucrase. Accordingly, the coat may be regarded as the structural equivalent of the cell-bound dextransucrase-dextran complex suggested by the work of Smith [1970] or the covalently linked enzyme-dextran complex visualized by Ebert and Schenk [1968] in their 'insertion type' mechanism of dextran synthesis. Since previous studies suggest that the level of dextransucrase activity is directly related to the concentration of sucrose [Neely and Nott, 1962], the effect of low levels of sucrose on coat thickness referred to above can be attributed to reduced levels of enzyme activity and/or dextran synthesis in the cell coat.

Direct evidence that the carbohydrate component of the surface coat is dextran can be obtained using a dextranase (Worthington Biochemical Corp.) that specifically hydrolyses the α 1,6-D-glucopyranosyl linkages of dextrans. This can be done by transferring MRS-grown cells of NCDO 1875 to the same medium containing 3.6% sucrose to induce surface coat formation. If the medium contains dextranase (10 units ml^{-1}), the surface coat is much thinner or absent after 5 h (Figure 16) whereas in controls which contain the same amount of inactivated enzyme (by boiling) the coat is of normal dimensions. The use of larger or smaller amounts of the enzyme results in thinner or thicker cell coats respectively. It appears therefore that under these conditions coat thickness is determined by the relative rates of dextran synthesis and digestion. Exposure to dextranase completely inhibits capsule formation.

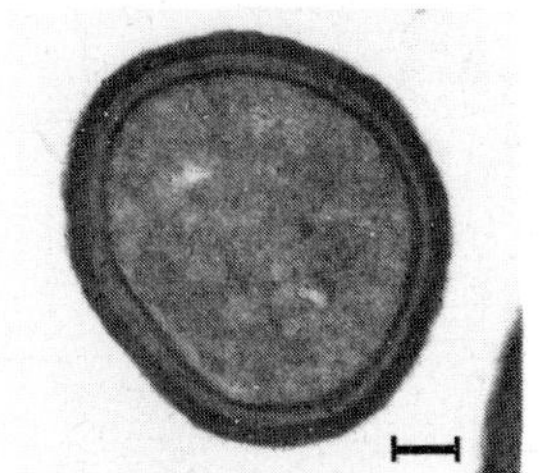

Fig. 16 *Leuconostoc mesenteroides* NCDO 1875. Cell grown in MRS + sucrose + 10u ml^{-1} dextranase. Control cells treated with inactivated dextranase are identical to that shown in Figure 15. Postfixed in reduced osmium tetroxide.

Some information on the types and percentages of different glucosidic linkages in the dextrans produced by strains of *Leuconostoc mesenteroides* that have been examined by electron microscopy is given in Table 1. Periodate oxidation analysis, from which this information is derived, collectively determines certain differently linked D-glucopyranosyl residues and other methods can provide more precise data. However, for some of the strains under study, more precise information is not available and the data given here are probably sufficient for the present purpose of attempting to relate ultrastructural features of dextrans to their chemical structure. Most of the information in Table 1 is derived from soluble dextrans since there is little or no comparable information on insoluble capsular dextrans.

Strains belonging to Group 1 generally produce dextrans in which the predominant non-1,6-D-glucosidic linkages are 1,3 in nature whereas in Group 2 they are chiefly of the 1,3 or 1,4 type (Table 1). Such an attempt to establish a relationship between ultrastructural results and those obtained by chemical analysis is made difficult by inadequate information on dextran structure but it may be that the existence of two morphologically distinct types of sucrose-induced surface coats is the consequence of structural differences between their dextran moieties. However, it should be noted that in strains producing soluble and insoluble dextrans which differ considerably in chemical structure (e.g. NCDO 798), the capsulate and acapsulate cells produce identical surface coats.

Although a surface coat of filamentous (or fuzzy) material has also been described from some strains of *Streptococcus mutans*, it is not a constant feature of the species (Figure 25). In the strains used by Guggenheim and Schroeder [1967] (OMZ 176 and 61), the formation of the coat was sucrose-dependent whereas in a wild-type strain (WT 6715-13) and several mutants, Nalbandian *et al.* [1974] showed that this was not the case since a conspicuous coat was also present on cells grown in equimolar glucose and fructose. However, both groups of workers suggested that the 'fuzzy' coat represented, or was the site of, the surface bound dextransucrase. Guggenheim and Schroeder [1967] were unable to demonstrate the surface coat with PA-TSC-OsO_4 and Nalbandian *et al.* [1974] did not find the coat to be glucanase labile. It appears from these observations therefore that the surface coat of *Streptococcus mutans*, unlike that of *Leuconostoc mesenteroides*, does not contain a dextran component.

Cell Wall Growth

Immunofluorescent, radioactive and morphological markers of the cell wall of chain-forming cocci have been used by several workers to study the sites at which new

TABLE 1

Percentages of differently linked D-glucopyranosyl residues in the dextrans produced by 11 strains of Leuconostoc mesenteroides and the incidence of capsule formation

NCDO	NRRL	Insoluble capsular dextran present (+)/absent (-) when grown in sucrose	Percentages of differently linked D-glucopyranosyl residues as determined by periodate oxidation analysis*		
			(1→6)	(1→4)	(1→3)
		Group 1.			
523	1118	+	76	3	21
797	523	+	66	10	24
519	1121	+	65	2	33
518	1120	+	85	0	15
808	1149	+	52	8	40
553	512	-	95	5	0
		Group 2.			
1875	1299	+	58	36	6 (L)
1585	1399	+	65	35	0
1593	1193	-	95	2	3
798	742	-	81	19	0 (L)
2068	1397	+	75	25	0

*Data taken from Jeanes *et al.* [1954]. (L) = data for least soluble dextran component.

wall material is added to the surface during cell growth. These studies [reviewed by Higgins and Shockman, 1971; Rogers *et al.*, 1978] show that new cell wall appears symmetrically in the equatorial or septal region of the cell. Extensive ultrastructural studies of *Streptococcus faecalis* by Higgins *et al.* [1971] and Higgins and Shockman [1970, 1971, 1976], have shown that in this equatorial addition to the surface, discrete areas of new cell wall are separated from the older cell wall by bands of raised material. Although these bands are also found in strains of *Leuconostoc mesenteroides*, they are generally less clearly preserved than those of *Streptococcus faecalis* and are probably unreliable as an indicator of cell wall growth. However, the surface coat of sucrose-grown *Leuconostoc mesenteroides* provides a cell wall marker which allows the sites of new growth to be studied.

The surface coat of *Leuconostoc mesenteroides* grown in the presence of sucrose always covers the entire surface of the cell wall irrespective of the rate of cell division. Thus the formation of coat material under these conditions keeps pace with the appearance of new cell wall in cells which are increasing their surface area prior to division. However, if cells which have been grown for 8-12 h in the presence of sucrose are washed and transferred to a medium lacking sucrose, cell coat formation appears to cease abruptly, for within 15 min, clearly defined areas of cell wall devoid of cell coat mark the sites at which new cell wall growth has occurred. The appearance of cell wall without coat does not result from the loss of coat from the surface on transfer to a different medium since cells transferred to phosphate buffer containing 0.1% tryptone (pH 7.0) remain viable, do not divide yet retain a complete surface coat for several hours. The surface coat is a natural marker of the cell wall formed during growth in the presence of sucrose and areas without coat thus appear to represent cell wall produced after transfer to sucrose-free medium. Naturally, this process is easier to follow in strains belonging to Group 2 which have no surface coat when grown in the absence of sucrose.

Profiles such as that in Figure 17 show that in dividing cells of *Leuconostoc mesenteroides*, as in streptococci [Higgins and Shockman, 1971], new cell wall is formed in the equatorial zone adjacent to the origin of the growing cross wall. The area of new cell wall increases as the septum completes its growth and divides the cell into two daughters. In cells which started division at the time they were transferred to sucrose-free medium, the area of cell wall possessing surface coat is approximately equal to that without coat (Figure 18). However, in cells which had started division before transfer, the proportion of new, coat-free cell wall is correspondingly smaller. Thus, one of the cells in Figure 19 in which a septum is just beginning to form has a little more than half its surface

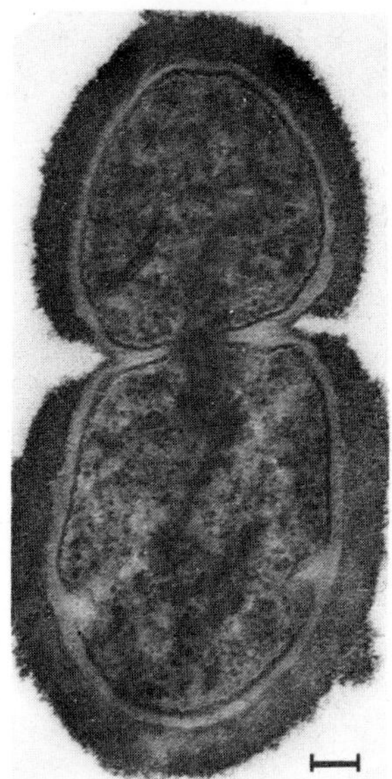

Fig. 17 *Leuconostoc mesenteroides* NCDO 1875. Cell grown in MRS + sucrose and transferred to MRS only. In this dividing cell, new cell wall added to the surface in the region of the cross wall is free of surface coat.

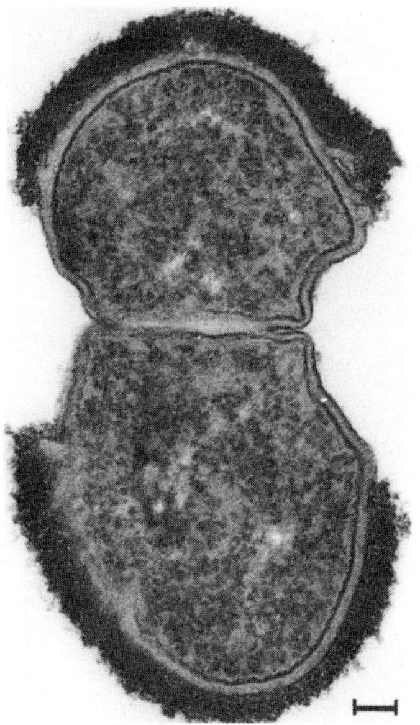

Fig. 18 As for Figure 17, but with the cross wall completed, approximately half of the cell wall surface of each daughter cell is free of cell coat.

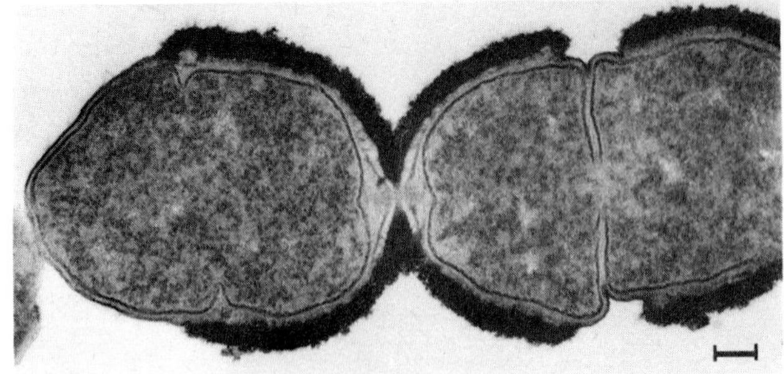

Fig. 19 As for Figure 17. Part of a chain of cells. When the cell on the left divides, the cell wall of one daughter (on the left) will have no surface coat except for a narrow band encircling the cell.

covered with coat material. It can be seen from the presumptive plane of division that when this cell divides, the insertion of new cell wall will separate the coat into two unequal zones; one daughter cell will have about one half of its surface covered with coat but in the other daughter, the coat will exist only as a narrow band encircling the cell (Figure 20). Quantitative aspects of cell wall formation in *Leuconostoc mesenteroides* have yet to be studied.

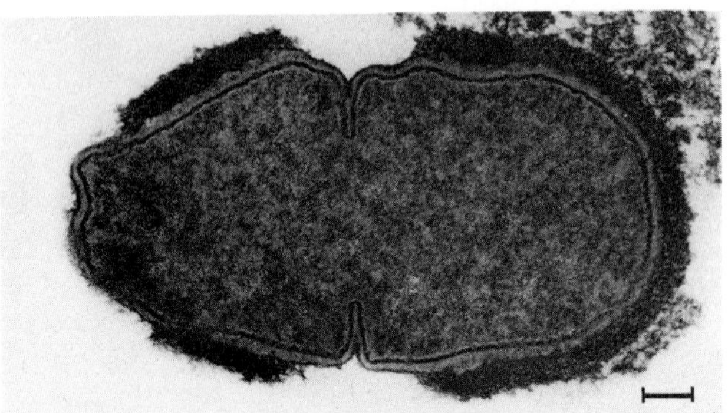

Fig. 20 As for Figure 17, showing a dividing cell. One daughter will have a band of surface coat encircling the cell.

Insoluble Dextran

In cells exposed to sucrose, the completion of surface transformation is marked by the appearance of insoluble dextran on the surface coat of those cells destined to form capsules. In all stages of capsule formation, dextran is in direct contact with the surface of the cell coat. Although there is considerable strain variation in the gross appearance and internal structure of the capsules, those produced by the strains in Group 1 (see Table 1), except

NCDO 808, are indistinguishable and will be considered first.

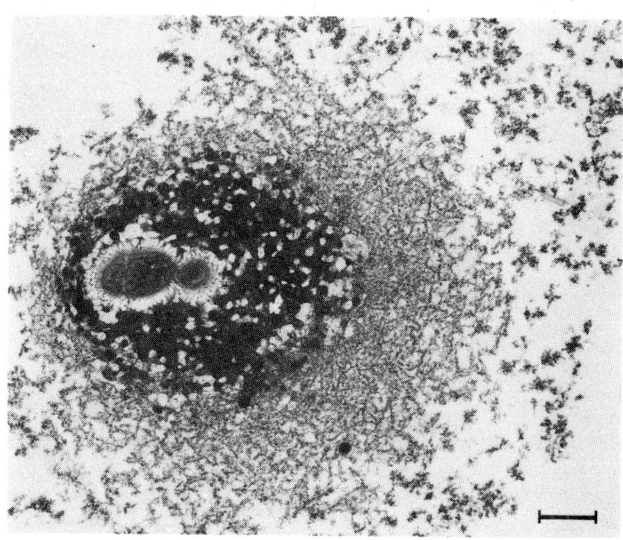

Fig. 21 *Leuconostoc mesenteroides* NCDO 523 grown in MRS + sucrose showing the two layers of the capsule. The outer zone of fibrils surrounds the central region of densely packed globular, filamentous and fibrillar dextran. Bar = 0.5 μm.

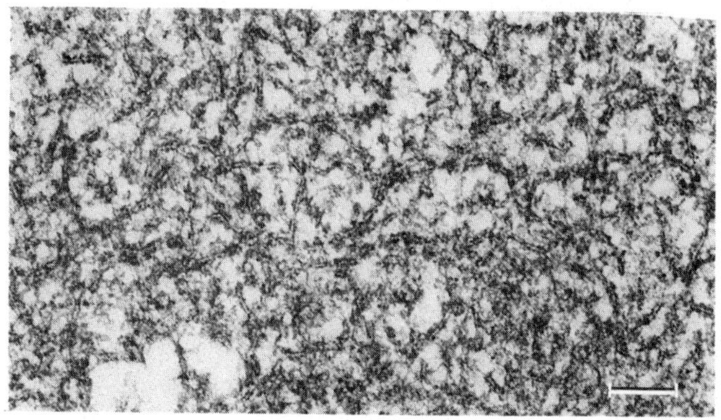

Fig. 22 Part of the fibrillar zone of a capsule of *Leuconostoc mesenteroides* NCDO 523. Spaces between the fibrils contain filamentous dextran.

a) Group 1 At least three morphologically distinct types of dextran can be recognized in each capsule. The major component is a fibril (9-10 nm diameter) whose centre is completely electron lucent but whose surface is covered with fine, electron dense material (Figures 21, 22). This description implies that the fibrils are solid but it may well be that they should be regarded as hollow tubules since, as will be seen below, only the surface can be shown to contain periodate-reactive carbohydrate using PA-TSC-Ag staining. The fibrils sometimes branch, are randomly orientated and are always associated with filamentous material (Figure 22). Individual fibrils are most clearly seen at the periphery of the capsule where they are less densely packed together (Figure 21). Although some capsules consist entirely of this fibrillar material, in most cases there is a clearly defined zone adjacent to the bacterium in which two other dextran components are visible (Figure 21).

One of these is referred to here as globular dextran since it takes the form of spherical aggregates (100-110 nm diameter) of tightly packed material (Figure 24). A salient feature of this dextran component is that the axis of each globule is traversed by a fibril of the type described above. The tubular nature of these fibrils is clearly demonstrated by their circular profiles in transverse sections (Figure 27). The other dextran that can be distinguished appears as a tightly packed reticulum of filaments and may be regarded as a matrix in which the globular and fibrillar dextrans are embedded (Figure 24). It seems very likely, for reasons which have been given elsewhere [Brooker, 1976], that the densely packed filamentous and globular dextrans disperse as they move towards the periphery of the capsule and that the resulting filaments are equivalent to those associated with the fibrils near the surface of the capsule.

This complex central region of the capsule is not a solid mass of dextran, for it is pervaded by a labyrinth of spaces which extends from the surface coat of the bacterium to the periphery of the zone (Figures 21 & 24). The presence of these spaces indicates that dextran is not constantly produced from the entire surface of the cell coat and may facilitate the exchange of metabolites and waste products with the surrounding medium across what might otherwise be an impermeable barrier of densely packed dextran.

A peculiarity of strains NCDO 518 and 519 is that many of the acapsulate cells produce, at the surface of the cell coat, a thick layer of very dense material (Figure 28). This probably represents an accumulation of the material that is found in much smaller quantities on the surface of acapsulate cells in most of the strains in Group 1.

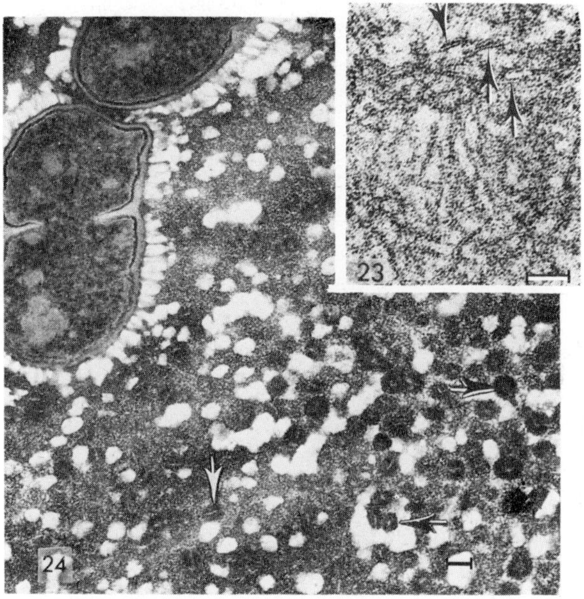

Fig. 23 (inset, top right) Fibrillar dextran similar to that in Figure 22 but stained with PA-TSC-Ag. Only the surface (arrows) of the fibrils stains.

Fig. 24 Dense central zone of the capsule produced by *Leuconostoc mesenteroides* NCDO 523. Densely packed filamentous dextran forms a matrix which is pervaded by a labyrinth of spaces. Globular dextran (arrowed) contains axial dextran fibrils.

The gross appearance of capsules produced by NCDO 808 (Figure 26) is quite different from that seen in all other strains of Group 1, but the same three dextran components can be recognized. This difference in appearance is due not only to the much larger proportion of filamentous dextran but also to the larger size of the globular component (160-180 nm diameter) and the smaller number of fibrils (Figure 27). The capsules produced by NCDO 808 appear in-

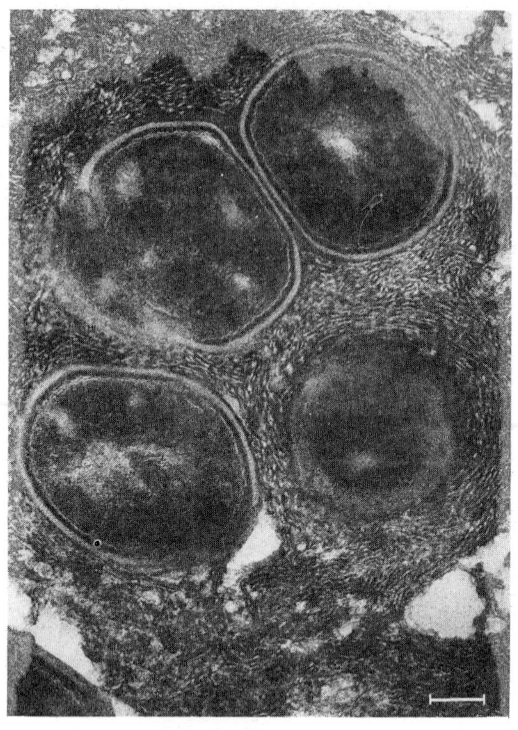

Fig. 25 *Streptococcus mutans* NCDO 2062 grown in the presence of sucrose and producing large amounts of extracellular fibrillar dextran.

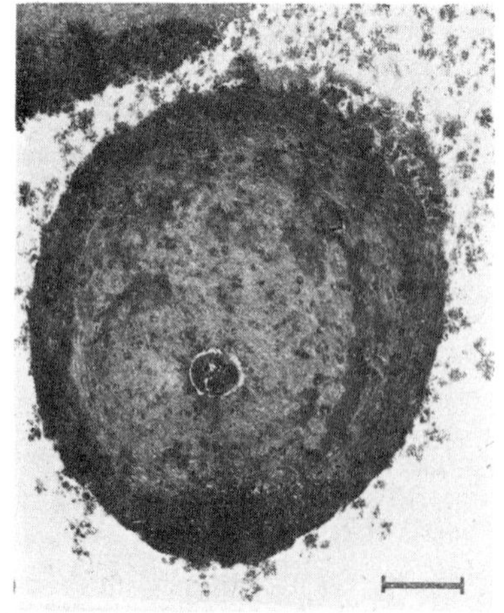

Fig. 26 Capsule of *Leuconostoc mesenteroides* NCDO 808. Bar = 1 μm.

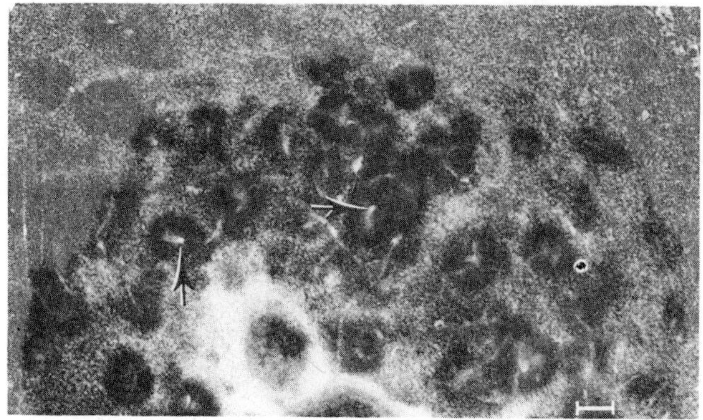

Fig. 27 Capsule matrix of NCDO 808. Circular sections of the fibrils (arrows) which penetrate the large dextran globules are clearly visible. Most of the matrix is of tightly packed filaments.

Fig. 28 Acapsulate cells of NCDO 518 showing aggregates of material on the surface coat.

termediate in structure between those produced by strains in Group 1 and those in Group 2 (see below).

All three dextran components of the capsules produced by strains in Group 1 contain reaction product after staining with PA-TSC-Ag. However, in the case of the fibrils, staining is restricted to the surface and their centre remains electron lucent (Figure 23).

In *Streptococcus mutans*, much of the insoluble extracellular glucan produced in the presence of sucrose appears as fibrils which are identical in appearance and diameter to those produced by *Leuconostoc mesenteroides* (Figure 25). Since the insoluble dextran produced by *Streptococcus mutans* contains a large proportion of 1,3 glucosidic linkages [Ceska *et al.*, 1972; Baird *et al.*, 1973] and the strains of *Leuconostoc mesenteroides* in Group 1 which produce abundant fibrillar dextran are also known to produce dextrans containing a high proportion of 1,3 or 1,3-like linkages (Table 1), it appears that there is a relationship between the presence of large amounts of fibrillar dextran and a high proportion of 1,3 glucosidic linkages. A similar conclusion was reached by Johnson *et al.* [1977] who found that the dextran produced by a mutant of *Streptococcus mutans* contained only 3% 1,3-like linkages and did not contain a fibrillar component. Although it is probably too simple to regard the fibrils as solid structures containing large numbers of 1,3 linkages and with reactive sites only at their surface, such a model would explain why their centre does not stain with PA-TSC-Ag, since this reaction depends on the generation of aldehyde groups by the periodate oxidation of non-1,3-linked residues. An alternative interpretation is that the fibrils are hollow tubules (or, more precisely, tightly coiled helices) of dextran which are rich in 1,3 linkages but which, nevertheless, contain sufficient 1,6-linked residues to produce reaction product when treated with PA-TSC-Ag.

The protofibrils of dextran which have often been described from *Streptococcus mutans* almost certainly correspond to one wall of the structures that have been referred to here as fibrils. Hence, previous references to double protofibrils [Newman *et al.*, 1976] or protofibrils arranged parallel to each other [Guggenheim and Schroeder, 1967; Johnson *et al.*, 1974] almost certainly correspond to the tubular fibrils shown in Figure 25.

b) Group 2 The capsules of three of the strains in Group 2 are of quite different internal structure from that seen in Group 1. Although the three dextran components described above also constitute the bulk of the capsular material, there is an unmistakable shift in the relative proportions of each. Although fibrillar and globular dextrans are constantly present in capsules of NCDO 1875, they constitute only a small part of the total insoluble dextran (Figures 29, 30). In strains NCDO 2068 and 1585, these same components are also found but they are not a regular feature of the capsules. In all of these strains, the major part of the insoluble dextran consists of a compact reticulum of fine filaments whose mesh size varies from one strain to another (Figures 31-34). From Table 1 it can be seen that the dextrans produced by these strains contain much higher proportions of 1,4-like linkages than 1,3-like linkages. However, more recent work has shown that the majority of the linkages formally described as 1,4-like are 1,2 linkages [Allen and Kabat, 1959; Bourne *et al.*, 1972; Miyaji *et al.*, 1973]. It can be argued therefore that the filamentous material, as the major component of the cap-

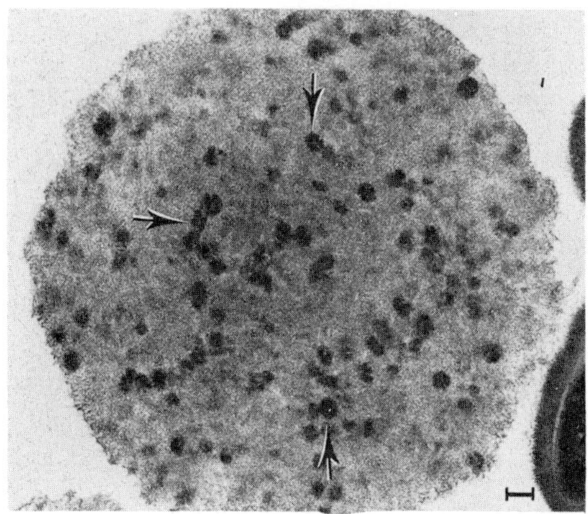

Fig. 29 Capsule of *Leuconostoc mesenteroides* NCDO 1875. Most of the capsule is composed of filamentous dextran although globular dextran (arrowed) is also visible.

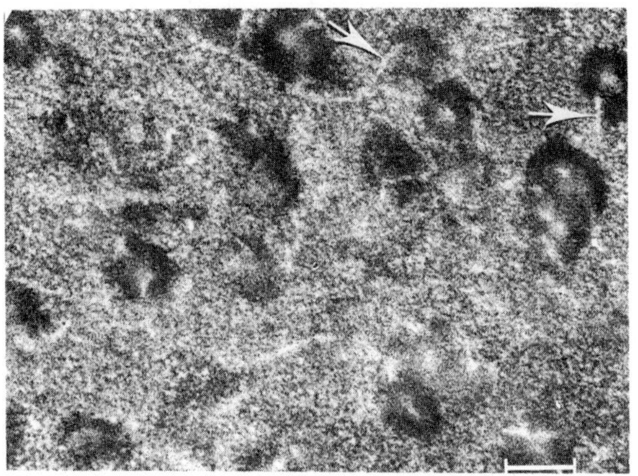

Fig. 30 Capsule matrix of NCDO 1875 consisting of globular, fibrillar (arrowed) and filamentous dextran.

sules produced by these three strains, represents a dextran containing most of the 1,2 linkages. Moreover, the relationship between the incidence of fibrillar dextran and the percentage of 1,3 linkages derived from a study of the dextrans in Group 1 is supported by the observation that Group 2 insoluble dextrans, with a relatively small proportion of 1,3 linkages [Bourne *et al.*, 1972; Miyaji *et al.*, 1973], contain fibrillar dextran but as a minor component.

It has been mentioned already that most of the information available on the types and percentages of different glucosidic linkages in *Leuconostoc mesenteroides* dextrans has been derived from studies of soluble dextrans. In most

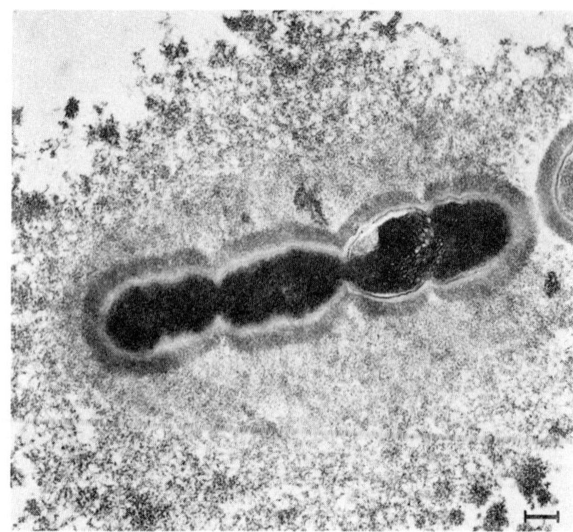

Fig. 31 Cells and capsule of *Leuconostoc mesenteroides* NCDO 1585. The capsule is composed almost entirely of filamentous dextran. Bar = 0.2 μm.

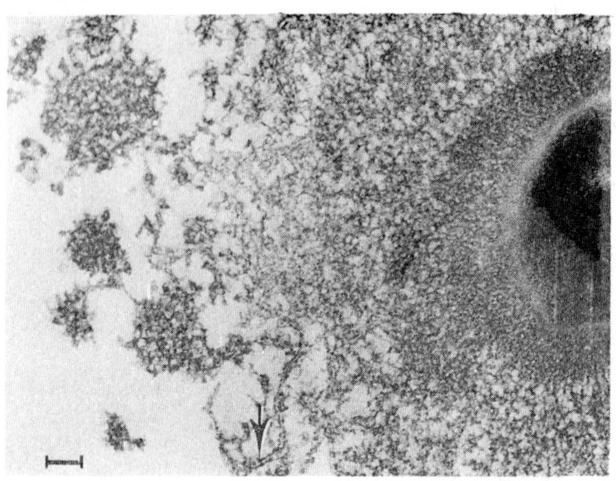

Fig. 32 Fibrillar dextran matrix of the capsule produced by NCDO 1585. Some fibrils are visible (arrow).

cases, the preparation of material for analysis has involved centrifugation of the culture medium to remove bacteria and insoluble material and precipitation of soluble dextran from the supernatant. However, it has been observed that small particles of insoluble dextran are constantly lost from the surface of capsules into the surrounding medium (Figures 21, 26, 31, 33). The particle size and density of this dextran are such that it is unlikely to be removed completely from the supernatant by the sedimentation regimes that have been generally used. It would

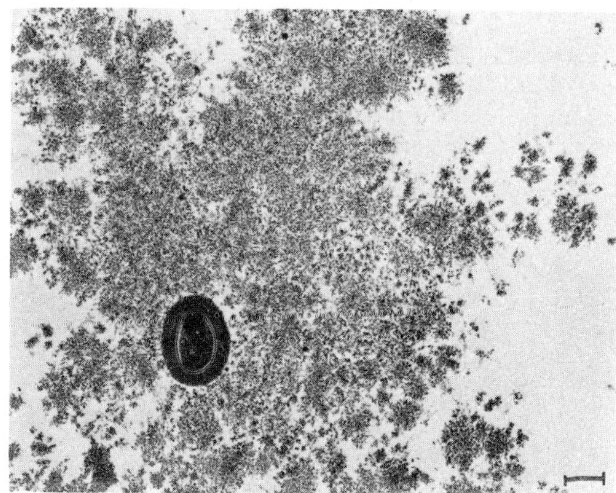

Fig. 33 Loosely packed dextran in the capsule produced by *Leuconostoc mesenteroides* NCDO 2068. Globular and fibrillar dextrans are absent. Bar = 0.5 μm.

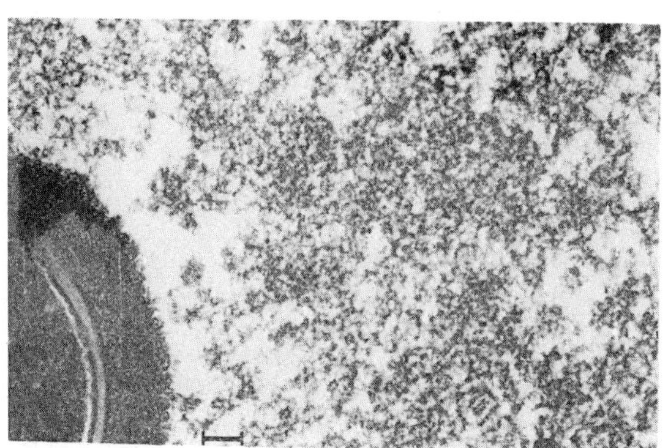

Fig. 34 Reticulum of dextran filaments in the capsule of NCDO 2068.

appear therefore that this material has, in past studies, not only contributed to the reported structural heterogeneity of dextrans [Wilham *et al.*, 1955; Kobayashi *et al.*, 1973], but also may have affected the results of more detailed structural studies. Moreover, in view of the ultrastructural heterogeneity of insoluble capsular dextran demonstrated by electron microscopy, it seems unlikely that heterogeneity with respect to chemical structure is limited to the soluble dextrans.

Acknowledgements

The author is indebted to Dr. E.I. Garvie for providing the strains of bacteria used in this study and for much useful discussion.

References

Ainsworth, S.K. (1977). An ultrastructural method for the use of polyvinylpyrrolidone and dextrans as electron opaque tracers. *Journal of Histochemistry and Cytochemistry* **25**, 1254-1259.

Allen, P.Z. and Kabat, E.A. (1959). Immunochemical studies on dextrans. II. Antidextran specificities involving the α 1→3 and the α 1→2 linked glucosyl residues. *Journal of the American Chemical Society* **81**, 4382-4386.

Bailey, R.W. (1959). Transglucosidase activity of rumen strains of *Streptococcus bovis*. 2. Isolation and properties of dextransucrase. *Biochemical Journal* **72**, 42-49.

Baird, J.K., Longyear, V.M.C. and Ellwood, D.C. (1973). Water insoluble and soluble glucans produced by extracellular glycosyltransferases from *Streptococcus mutans*. *Microbios* **8**, 143-150.

Blanquet, P.R. (1976). Ultrahistochemical study on the ruthenium red surface staining: I. Processes which give rise to electron-dense marker. *Histochemistry* **47**, 63-78.

Bourne, E.J., Sidebotham, R.L. and Weigel, H. (1972). Studies on dextrans and dextranases. Part X. Types and percentages of secondary linkages in the dextrans elaborated by *Leuconostoc mesenteroides* NRRL-B1299. *Carbohydrate Research* **22**, 13-22.

Brooker, B.E. (1976). Surface coat transformation and capsule formation by *Leuconostoc mesenteroides* NCDO 523 in the presence of sucrose. *Archives of Microbiology* **111**, 99-104.

Brooker, B.E. (1977). Ultrastructural surface changes associated with dextran synthesis by *Leuconostoc mesenteroides*. *Journal of Bacteriology* **131**, 288-292.

Brooker, B.E. and Fuller, R. (1975). Adhesion of lactobacilli to the chicken crop epithelium. *Journal of Ultrastructural Research* **52**, 21-31.

Cagle, G.D. (1975). Fine structure and distribution of extracellular polymer surrounding selected aerobic bacteria. *Canadian Journal of Microbiology* **21**, 395-408.

Cagle, G.D., Pfister, R.M. and Vela, G.R. (1972). Improved staining of extracellular polymer for electron microscopy: examination of *Azotobacter, Zoogloea, Leuconostoc* and *Bacillus*. *Applied Microbiology* **24**, 477-487.

Cassone, A. (1973). Improved visualization of cell wall structure in *Saccharomyces cerevisiae*. *Experientia* **29**, 1303-1306.

Cassone, A. and Garaci, E. (1977). The capsular network of *Klebsiella pneumoniae*. *Canadian Journal of Microbiology* **23**, 684-689.

Ceska, M., Granath, K.A., Norrman, B. and Guggenheim, B. (1972). Structural and enzymatic studies on glucans synthesized with glucosyltransferases of some strains of oral streptococci. *Acta Chemica Scandinavica* **26**, 2223-2230.

Cheng, K.J., Hironaka, R., Jones, G.A., Nicas, T. and Costerton, J.W. (1976). Frothy feedlot bloat in cattle: production of extracellular polysaccharides and development of viscosity in cultures of *Streptococcus bovis*. *Canadian Journal of Microbiology* **22**, 450-459.

Costerton, J.W., Damgaard, H.N. and Cheng, K.J. (1974). Cell envelope morphology of rumen bacteria. *Journal of Bacteriology* **118**, 1132-1143.

Critchley, P., Wood, J.M., Saxton, C.A. and Leach, S.A. (1967). The polymerisation of dietary sugars by dental plaque. *Caries Research* **1**, 112-129.

deMan, J.C., Rogosa, M. and Sharpe, M.E. (1960). A medium for the cultivation of lactobacilli. *Journal of Applied Bacteriology* **23**, 130-135.

Djaczenko, W. and Cassone, A. (1972). Visualization of new ultrastructural components in the cell wall of *Candida albicans* with fixatives containing TAPO. *Journal of Cell Biology* **52**, 186-190.

Ebert, K.H. and Schenk, G. (1968). Mechanisms of biopolymer growth: the formation of dextran and levan. *Advances in Enzymology* **30**, 179-221.

Fletcher, M. and Floodgate, G.D. (1973). An electron-microscopic demonstration of an acidic polysaccharide involved in the adhesion of a marine bacterium to solid surfaces. *Journal of General Microbiology* **74**, 325-334.

Guggenheim, B. and Newbrun, L. (1969). Extracellular glucosyltransferase activity of an HS strain of *Streptococcus mutans*. *Helvetica Odontologica Acta* **13**, 84-97.

Guggenheim, B. and Schroeder, H.E. (1967). Biochemical and morphological aspects of extracellular polysaccharides produced by cariogenic streptococci. *Helvetica Odontologica Acta* **11**, 131-152.

Higgins, M.L., Pooley, H.M. and Shockman, G.D. (1971). Reinitiation of cell wall growth after threonine starvation of *Streptococcus faecalis*. *Journal of Bacteriology* **105**, 1175-1183.

Higgins, M.L. and Shockman, G.D. (1970). Model for cell wall growth of *Streptococcus faecalis*. *Journal of Bacteriology* **101**, 643-648.

Higgins, M.L. and Shockman, G.D. (1971). Procaryotic cell division with respect to wall and membranes. *Critical Reviews of Microbiology* **1**, 29-72.

Higgins, M.L. and Shockman, G.D. (1976). Study of a cycle of cell wall assembly in *Streptococcus faecalis* by three-dimensional reconstructions of thin sections of cells. *Journal of Bacteriology* **127**, 1346-1358.

Holzwarth, G. and Prestridge, E.B. (1977). Multistranded helix in xanthan polysaccharide. *Science*, New York **197**, 757-759.

Jeanes, A. (1966). Dextran. In *Encyclopaedia of Polymer Science and Technology* 4, pp.805-824. Edited by H.F. Mark. New York: J. Wiley.

Jeanes, A., Haynes, W.C., Wilham, C.A., Rankin, J.C., Melvin, E.H., Austin, M.J., Cluskey, J.E., Fisher, B.E., Tsuchiya, H.M. and Rist, C.E. (1954). Characterization and classification of dextrans from ninety-six strains of bacteria. *Journal of the American Chemical Society* **76**, 5041-5052.

Johnson, M.C., Bozzola, J.J. and Shechmeister, I.L. (1974). Morphological study of *Streptococcus mutans* and two extracellular polysaccharide mutants. *Journal of Bacteriology* **118**, 304-311.

Johnson, M.C., Bozzola, J.J., Shechmeister, I.L. and Shklair, I.L. (1977). Biochemical study of the relationship of extracellular glucan to adherence and cariogenicity in *Streptococcus mutans* and an extracellular polysaccharide mutant. *Journal of Bacteriology* **129**, 351-357.

Jones, H.C., Roth, I.L. and Sanders, W.M. (1969). Electron microscopic study of a slime layer. *Journal of Bacteriology* **99**, 316-325.

Kelstrup, J. and Funder-Nielsen, T.D. (1972). Molecular interactions between the extracellular polysaccharides of *Streptococcus mutans*. *Archives of Oral Biology* **17**, 1659-1670.

Kobayashi, M., Shishido, K., Kikuchi, T. and Matsuda, K. (1973). Fractionation of the *Leuconostoc mesenteroides* NRRL B-1299 dextran and preliminary characterization of the fractions. *Agricultural and Biological Chemistry* **37**, 357-365.

Luft, J.H. (1964). Electron microscopy of cell extraneous coats as revealed by ruthenium red staining. *Journal of Cell Biology* **23**, 54A-55A.

Luft, J.H. (1971). Ruthenium red and violet. I. Chemistry, purification, methods of use for electron microscopy and mechanism of action. *Anatomical Record* **171**, 347-368.

Melvaer, K.L., Helgeland, K. and Rölla, G. (1974). A charged component in purified polysaccharide preparations from *Streptococcus mutans* and *Streptococcus sanguis*. *Archives of Oral Biology* **19**, 589-595.

Miyaji, H., Misaki, A. and Torii, M. (1973). The structure of a dextran produced by *Leuconostoc mesenteroides* NRRL B-1397: the linkages and length of the branches. *Carbohydrate Research* **31**, 277-287.

Nalbandian, J., Freedman, M.L., Tanzer, J.M. and Lovelace, S.M. (1974). Ultrastructure of mutants of *Streptococcus mutans* with reference to agglutination, adhesion and extracellular polysaccharide. *Infection and Immunity* **10**, 1170-1179.

Neely, W.B. (1960). Dextran: structure and synthesis. *Advances in Carbohydrate Chemistry* **15**, 341-369.

Neely, W.B. and Nott, J. (1962). Dextransucrase, an induced enzyme from *Leuconostoc mesenteroides*. *Biochemistry* **1**, 1136-1140.

Newbrun, E., Lacy, R. and Christie, T.M. (1971). The morphology and sizes of the extracellular polysaccharides from oral streptococci. *Archives of Oral Biology* **16**, 863-872.

Newman, H.N. (1976). Dental plaque. In *Microbial Ultrastructure*, pp.223-263. Edited by R. Fuller and D.W. Lovelock. London: Academic Press.

Newman, H.N., Donoghue, H.D. and Britton, A.B. (1976). Effect of glucose and sucrose on the survival in batch culture of *Streptococcus mutans* C67-1 and a non-cariogenic mutant C67-25. Morphological studies. *Microbios* **15**, 113-125.

Pate, J.L. and Ordal, E.J. (1967). The fine structure of *Chondrococcus columnaris*. III. The surface layers of *Chondrococcus columnaris*. *Journal of Cell Biology* **35**, 37-51.

Patterson, H., Irvin, R., Costerton, J.W. and Cheng, K.J. (1975). Ultrastructure and adhesion properties of *Ruminococcus albus*. *Journal of Bacteriology* **122**, 278-287.

Rogers, H.J., Ward, J.B. and Burdett, I.D.J. (1978). Structure and growth of the walls of Gram-positive bacteria. In *Relations between Structure and Function in the Prokaryotic Cell*. 28th Symposium of the Society for General Microbiology. pp.139-175. Edited by R.Y. Stanier, H.J. Rogers and B.J. Ward. Cambridge: Cambridge University Press.

Rorem, E.S. (1955). Uptake of rubidium and phosphate ions by polysaccharide-producing bacteria. *Journal of Bacteriology* **70**, 691-701.

Sidebotham, R.L. (1974). Dextrans. *Advances in Carbohydrate Chemistry and Biochemistry* **30**, 371-444.
Simionescu, N. and Palade, G.E. (1971). Dextrans and glycogens as particulate tracers for studying capillary permeability. *Journal of Cell Biology* **50**, 616-624.
Smith, E.E. (1970). Biosynthetic relation between the soluble and insoluble dextrans produced by *Leuconostoc mesenteroides* NRRL B-1299. *FEBS Letters* **12**, 33-37.
Springer, E.L. and Roth, I.L. (1973). The ultrastructure of the capsules of *Diplococcus pneumoniae* and *Klebsiella pneumoniae* studied with ruthenium red. *Journal of General Microbiology* **74**, 21-31.
Thiéry, J.P. (1967). Mise en évidence des polysaccharides sur coupes fines en microscopie électronique. *Journal de Microscopie* **6**, 987-1018.
Thiéry, J.P. and Rambourg, A. (1974). Cytochimie des polysaccharides. *Journal de Microscopie* **21**, 225-232.
Tsutsui, K., Kumon, H., Ichikawa, H. and Tawara, J. (1976). Preparative method for suspended biological materials for SEM by using of polycationic substance layer. *Journal of Electron Microscopy* **25**, 163-168.
Wilham, C.A., Alexander, B.H. and Jeanes, A. (1955). Heterogeneity of dextran preparations. *Archives of Biochemistry and Biophysics* **59**, 61-75.

Chapter 6

STRUCTURE, SOLUTION PROPERTIES AND BIOLOGICAL INTERACTIONS OF SOME MICROBIAL EXTRACELLULAR POLYSACCHARIDES

D.A. POWELL

Unilever Research, Colworth House, Bedford, UK

Introduction

Many bacteria produce extracellular polysaccharides under certain culture conditions. These may take the form of discrete capsules or may be free of any attachment to the cell and are then described as slimes. The exact state of the polysaccharide seems to depend on the particular bacterial species and the age and conditions of culture [Sutherland, 1977].

Early studies of extracellular polysaccharides were stimulated by the recognition that the antigenic specificities of bacterial cells often reside in their outermost layers, but more recently interest has been renewed by the discovery that some bacteria produce extracellular materials which display solution properties of great industrial importance [see, for example, Lawson, 1976]. Early work was involved mainly with covalent structural characterization and suggested that these polysaccharides embraced a large range of very complex structures; in more recent years a clearer picture has emerged with the use of more sophisticated techniques of structural analysis and most extracellular polysaccharides are recognized to consist of simple regular repeating units [Sutherland, 1977].

Little is known about the biological roles or functions of extracellular polysaccharides [see Brooker, Chapter 5; Evans *et al.*, Chapter 3] but with the realization that many have regular structures they may now be recognized as close relations of polysaccharides from plant and animal sources. The primary roles of many of these polymers are as producers of structure in biological tissue, and their ability to act in this way is dependent on their adoption of ordered shapes in hydrated environments [Rees and Welsh, 1977]. Various levels of molecular organization have now been identified for a limited number of bacterial polysaccharides; some specific interactions between them and other polysaccharides have also been characterized and found to involve ordered chain conformations. Sufficient selectivity has been found in interacting systems to suggest that microbial extracellular polymers may be involved in recognition processes such as host-pathogen relationships.

Studies on materials isolated from plant and animal systems show that most polysaccharides adopt ordered molecular conformations in the solid state and that such ordered shapes may persist under more hydrated conditions; they are favoured by non-covalent interactions of various types between conformationally regular chains. Chain-chain interactions have considerable effects on solution properties and may result in gel formation when they are long-lived, or produce interesting viscous effects when they are transient [Rees, 1973; Rees and Welsh, 1977]. In this chapter the composition of bacterial extracellular polysaccharides will be reviewed briefly to illustrate the types of structure that have been identified. Following some discussion of the way in which aspects of covalent structure (sugar type and linkage pattern) can be used to predict the type of conformation adopted by polysaccharide chains, some examples of bacterial extracellular polysaccharides, for which solution properties have been studied, will be considered. These systems are classified according to the types of chain conformations and interchain interactions important in determining their solution properties. Finally, some clues as to the biological roles of some extracellular polysaccharides have become apparent from the study of interactions between conformationally ordered polysaccharides in solution and these will be discussed for individual systems under the heading "biological interactions".

Levels of Polysaccharide Structure

Until recently, descriptions of polysaccharides have been confined to their primary structures (the nature and sequence of their constituent sugars and the way in which they are joined together). Recent work has indicated the importance of considering higher levels of polysaccharide structure [Rees and Welsh, 1977]. Covalent linkages between adjacent sugars are not completely flexible but restrict the residues to a narrow range of relative orientations and thus isolated polysaccharide chains are capable of adopting only certain shapes (or secondary structures) which are dependent on their primary sequence [Rees and Welsh, 1977; Morris *et al.*, 1977a]. At the next level of organisation, energetically favourable interactions between chains may result in ordered specific structures (tertiary structure). Finally, such compact structures may interact amongst themselves, or with other polymers, to give still higher levels of organization which are described as quaternary structures.

Primary Structure of Bacterial Extracellular Polysaccharides

Structurally, extracellular polysaccharides are distinct from other classes of bacterial cell envelope polymers in that they do not conform to a simple general model as do, for example, the lipopolysaccharides from Gram-negative bacteria. Other than their common disposition with respect to the cell there seem to be insufficient common features in the structures of extracellular materials for them all to be classified under the same heading. For the same reasons, advances in the study of the solution properties of extracellular polysaccharides are likely to be slow since it is not easy to extrapolate from the behaviour of one polymer to another even when detailed primary structures are known. Studies of solution properties have so far been dictated by the discovery of extracellular polysaccharides which show interesting or potentially commercially useful properties and not by any systematic effort. Hence only a limited number are understood in any detail.

From a primary structural viewpoint an exhaustive discussion of extracellular polysaccharides will not be given here, but a sufficient number of materials will be considered to demonstrate the extreme diversity of known types.

Composition

Many different sugars have been identified as components of extracellular polysaccharides, the commonest being the simple hexoses D-glucose, D-galactose and D-mannose. Other neutral sugars found include the 6-deoxy-hexoses L-rhamnose (6-deoxy-L-mannose) and L-fucose (6-deoxy-L-galactose). Pentoses are comparatively rare. Some extracellular polysaccharides resemble the teichoic acids found normally in cell walls and membranes of Gram-positive bacteria [Baddiley, 1972; Duckworth, 1977] and hence contain the glycitols glycerol or ribitol together with phosphate. The common 2-deoxy-2-amino sugars, D-glucosamine and D-galactosamine, are found as their *N*-acetyl derivatives and many extracellular polymers contain uronic acids. D-glucuronic acid is the commonest charged sugar, but D-galacturonic acid, D-mannuronic acid and L-guluronic acid are also found. It seems to be a fairly common feature of extracellular polysaccharides to possess a net negative charge, and this is conferred not only by uronic acids but also by non-sugar substituents. These may be the already mentioned phosphate, or acyl groups such as succinate [Harada, 1965] or pyruvic acid ketal substituents. The latter is of fairly widespread occurrence and has been identified in the capsular polysaccharide of *Xanthomonas campestris* [Sloneker and Orentas, 1962]. Identification of ketal substituents is facilitated by their stability to alkali and lability to dilute acid. Other acyl substituents are found; formyl residues have been identified in extracellular materials from various *Klebsiella* strains [Sutherland, 1971] and *O*-acetyl residues are probably amongst the most common non-sugar components. Acyl residues are extremely labile in dilute alkali and thus can be easily distinguished from ketal group substituents. Ether-linked lactic acid has been found as a substituent of D-glucuronic acid in the *Klebsiella* type 37 capsular polysaccharide [Lindberg *et al.*, 1976] and joined to D-glucose in the extracellular polysaccharide from *Aerococcus viridans* var. *homari* [Kenne *et al.*, 1976].

Although many extracellular polysaccharides have been studied, rigorous structures have not been determined for all of them. Often only sugars and their approximate molar ratios have been determined. There is widespread structural variation between polysaccharides even from different strains within the same bacterial genera. Any attempt at structural classification is thus impossible on a species level and polysaccharide structures will therefore be discussed on the basis of their monomer composition. Both homopolysaccharides and heteropolysaccharides have been found to be of widespread occurrence.

Early structural work on extracellular polysaccharides was often hampered by the techniques then available for fragmenting polymers and identifying and quantitating the resultant monomer units. As a result very complex structures have been proposed for many materials and it is only in recent years, with the advent of modern techniques, that many of these suggestions have been revised to more simple structures. Present methods of methylation analysis coupled with specific degradation techniques, ^{1}H-and ^{13}C-n.m.r., optical rotation and identification of reaction products by gas liquid chromatography-mass spectrometry mean that much structural information is now available from only a few milligrammes of polysaccharide material [Lönngren and Svensson, 1974; Lindberg *et al.*, 1975; Aspinall, 1977; Lindberg and Lönngren, 1978]. On biosynthetic grounds also, many of the previously suggested complex structures seem unlikely and it is now being regularly demonstrated that extracellular polysaccharides consist of fairly simple regular repeating units. The largest of these so far conclusively demonstrated is the octasaccharide repeating unit from the capsular material of *Rhizobium meliloti* [Jansson *et al.*, 1977].

Homopolysaccharides

Homopolysaccharides, although at first sight fairly simple since they contain only one monomeric unit, may present difficult problems in structural determinations. Acid hydrolysis of methylated homopolysaccharides will provide information on the types and amounts of various sugar-sugar linkages present but identification of linkage sequence and consequent determination of repeating structure may prove difficult through the lack of starting points for specific degradations. For heteropolysaccharides sequence information may come from the occurrence of particularly acid stable or labile glycosidic linkages due to the presence of uronic acids and amino sugars or 6-deoxyhexoses respectively. For homopolymers such techniques are not applicable and suggestions for repeating units may come only from the occurrence in simple ratios of derivatives indicative of a particular linkage in the methylation analysis. Most homopolysaccharides identified so far consist of D-glucose. Cellulose, a β 1,4 D-glucan, is produced as fibrils by *Acetobacter* species and appears similar to the plant polysaccharide [Hestrin *et al.*, 1954]. Chain lengths for the polymer of about 600 residues have been claimed.

Of more widespread occurrence, and the subject of much study due to their industrial importance, are the dextrans [Brooker, Chapter 5]. These branched chain glucans are synthesised usually from sucrose by *Leuconostoc mesenteroides*, *Leuconostoc dextranicum* and *Streptococcus viridans* or from starch derived dextrins by *Acetobacter capsulatum* and *Acetobacter viscosum* [Hehre and Hamilton, 1951]. Dextran is the collective name [Jeanes, 1966] for a large class of polymers composed mainly of α 1,6-linked D-glucopyranosyl units but also having branch points at positions 2, 3 or 4. Structures are determined largely by the producing strain and polymer chain lengths may vary from 40 to greater than 500 monosaccharide units. Some unbranched dextrans are known [Bailey, 1959] but commonly structures can best be described in terms of 60-90% α (1,6) linkages and 5-30% α 1,4 branch points [Larm *et al.*, 1971]. Other extracellular homopolysaccharides based on glucose are produced by *Alcaligenes faecalis* [Harada *et al.*, 1966; Misaki *et al.*, 1969]; a mutant strain produces a linear β 1,3-linked polymer [Harada *et al.*, 1968]. The latter polysaccharide, known as curdlan, has gelling properties and is finding application in the food industry. Its solution properties will be discussed later. The β 1,2-linked D-glucans are produced by various strains of *Agrobacterium* [Gorin *et al.*, 1961] and similar polymers, also containing β 1,3 and β 1,6 linkages, are products of *Rhizobium japonicum* [Dedonder and Hassid, 1964].

Levans (poly-D-fructans) are produced by many plant pathogenic bacteria of the genera *Pseudomonas* and *Xanthomonas* and also the Gram-positive *Streptococcus salivarius* and the genus *Bacillus*. Levans often have very high molecular weights (one million or greater) and β 2,6 linkages predominate [Cooper and Preston, 1935].

Heteropolysaccharides

Very many heteropolysaccharides have been identified among bacterial extracellular materials and it is only possible to discuss a sufficient number of them here to demonstrate the large diversity of their structures. It is particularly amongst the heteropolysaccharides that complex structures previously reported have been revised to more simple repeating units; these are usually confined to less than six sugars.

Structures for capsular polysaccharides from many types of the organisms *Streptococcus pneumoniae* [Larm and Lindberg, 1976] and *Klebsiella* [see Atkins *et al.*, Chapter 7] species have been determined since these materials are the principal type specific antigenic determinants of the organisms. Immunologically the many types of *Streptococcus pneumoniae* are well understood, and many antigenic cross-reactions are now reflected in similarities of capsular structures. Amongst the *Streptococcus pneumoniae* extracellular polysaccharides there are simple structures such as the disaccharide repeating unit of Type III material [Adams *et al.*, 1941] and more complex units containing sugars, alditols and phosphate which resemble the cell wall and membrane teichoic acids, for example Types XA and XIA [Rao *et al.*, 1966a; Rao *et al.*, 1966b; Kennedy *et al.*, 1969]. An example of a complex structure being revised to a more simple one is found for the Type XIV capsular material; original work [How *et al.*, 1964; references cited in Larm and Lindberg, 1976] had suggested dodecasaccharide or hexasaccharide repeating units whereas more recent results are consistent with a tetrasaccharide repeating structure which incorporates many of the original structural features [Lindberg *et al.*, 1977]. Table 1 shows some of the structures identified and illustrates the diversity of structural types.

In the genus *Klebsiella* wide variations in structural complexity are again apparent. Among the many types of repeating units found are linear structures (with from 2 to 6 sugars), and single and double branched units (with from mono- to tetrasaccharide side chains). In addition, substitution by *O*-acetyl and ketallically linked pyruvic acid groups provides additional complexity. Commonly the *Klebsiella* capsular materials are negatively charged, mainly through containing glucuronic acids or pyruvic acid substituents. Some 81 *Klebsiella* serotypes have been identified all having different capsular structures. Around 30 of these have been elucidated and some examples are given in Table 2.

A capsular polysaccharide of great industrial importance is that from *Xanthomonas campestris* [see Evans *et al.*, Chapter 3; Gabriel, Chapter 8]. Early studies indicated a repeating unit containing 16 sugar units [Sloneker *et al.*, 1964] but this has recently been revised to a penta-

TABLE 1

Structure of some pneumococcal capsular polysaccharides. All sugars have the pyranose ring form unless otherwise shown; all sugars are D except glycerol and rhamnose

Type	Structure	Reference
II	→3)Rha(1→3)Rha(1→3)Rha(1→4)Glc(1→ α α 2 β α α ↑ 1 Glc 6 α ↑ 1 GlcA	Kenne *et al.* (1975)
III	→3)GlcA(1→4)Glc(1→ β β	Adams *et al.* (1941)
XA	O⁻ | →O-P-O-6Gal f(1→3)Gal(1→4)GalNAc(1→3)Gal(1→2)Ribitol(5→ ‖ 6 O ↑ 1 Gal f	Rao (1966a,b)
XIA	OAc |2/3 →3)Gal(1→4)Glc(1→6)Glc(1→4)Gal(1→ β α ↑4 α α O | O=P-O⁻ | O ↑1 Glycerol	Kennedy *et al.* (1969)
XIV	→6)GlcNAc(1→3)Gal(1→4)Glc(1→ 4 β β β β | 1 Gal	Lindberg *et al.* (1977)

TABLE 2

Some Klebsiella extracellular polysaccharides conforming to different structural patterns. All sugars have the pyranose ring form. All sugars are D except rhamnose

Type	Structural pattern	Structure	Reference
32	Uronic acid absent	→3)Gal(1→2)Rha(1→3)Rha(1→4)Rha(1→ α ↑ α β | α 3.4 pyr 3.4 pyr	Bebault *et al.* (1978)
81	Uronic acid in chain (i) linear	→2)Rha(1→3)Rha(1→4)GlcA(1→2)Rha(1→3)Rha(1→3)Gal(1→ α α β α α β	Curvall *et al.* (1975a)
57	(ii) branch on uronic acid	→3)Gal(1→3)GalA(1→2)Man(1→ β 4 α α α | 1 Man	Kamerling *et al.* (1975)
52	(iii) branch on neutral sugar	→3)Gal(1→2)Rha(1→4)GlcA(1→3)Gal(1→4)Rha(1→ 2 ↑ 1 Gal	Björndal *et al.* (1973)
8	Uronic acid in side-chain (i) single unit side-chain	→3)Gal(1→3)Glc(1→3)Glc(1→ β ↑4 α β α |1 GlcA	Sutherland (1970)

28	(ii) two unit side-chain	see structure below	Curvall *et al.* (]975b)

```
→2)Gal(1→3)Man(1→2)Man(1→3)Glc(1→
    α    2    α        α       β
       β ↑
         1
        GlcA
         3
       β ↑
         1
        Glc
```

saccharide repeat [Jansson *et al.*, 1975; Melton *et al.*, 1976] (Figure 1). The structure is based on a cellulose backbone having alternate glucosyl residues substituted by trisaccharide side-chains. The detailed structure is discussed along with the solution properties of this material later. An interesting feature of this polysaccharide is that it is variably substituted by pyruvic acid ketal groups, depending on the growth conditions under which it is produced [Cadmus *et al.*, 1976]. The largest repeating unit conclusively demonstrated to occur in extracellular polysaccharides contains eight sugars. The capsular material from *Rhizobium meliloti* has a backbone containing one D-galactosyl residue and three D-glucosyl units, one of which is substituted by a tetrasaccharide side-chain containing only D-glucose. The polysaccharide is negatively charged since each side-chain is terminated by a pyruvic acid ketal substituent. This group was the starting point for the very elegant structural characterization of this material which involved step-wise degradation of the side-chain. The structure determination clearly demonstrates the power of modern techniques for characterizing complex polysaccharides of this type [Jansson *et al.*, 1977].

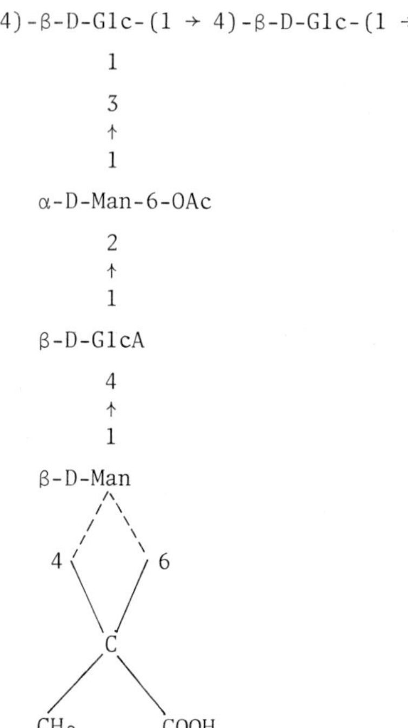

Fig. 1 The repeating unit structure of the extracellular polysaccharide from *Xanthomonas campestris*.

Bacterial cellulose has been mentioned as an example of a polysaccharide produced by both prokaryotic and eukaryotic organisms. Such dual occurrence is not unique to homopolymers. Hyaluronic acid, a polysaccharide having a disaccharide repeating structure containing D-glucuronic acid and *N*-acetyl-D-glucosamine, is found in the connective tissues of animals and is also found associated with various strains of *Streptococcus* [Kendall *et al.*, 1937]. Similarly an algal polysaccharide, alginic acid, is also formed extracellularly by the Gram-negative bacteria *Azotobacter vinelandii* and *Pseudomonas aeruginosa* [Gorin and Spencer, 1966; Linker and Jones, 1966; Carlson and Matthews, 1966; Jarman, Chapter 2]. Alginic acid contains D-mannuronic acid and L-guluronic acid. The proportions of which may vary with the growth conditions of the microorganism. The bacterial polymer is also acetylated. The polysaccharide does not have a simple repeating unit but has a block structure (see below).

Determinants of Polysaccharide Shape

Building blocks

Fundamental constraints on polysaccharide shapes are imposed by the adoption of certain favoured conformations by their monosaccharide constituents [Rees, 1977]. In this respect ring size is important, and for all of the bacterial polysaccharides to be considered here it is sufficient to consider six-membered rings since the building blocks are all pyranose forms of hexose or derivatised hexose units. These units can be regarded as geometrically rigid for chain conformational purposes since fixed carbon-carbon and carbon-oxygen bond lengths and angles dictate that only a small number of unstrained pyranose ring shapes are likely to exist. Boat conformations (Figure 2) are destabilized due to bond eclipsing and steric crowding of substituents, and the two possible chair conformations for a particular sugar are usually energetically distinguishable since one can minimize steric repulsions between crowded axial substituents. For the D-series of sugars the $^4C_1$

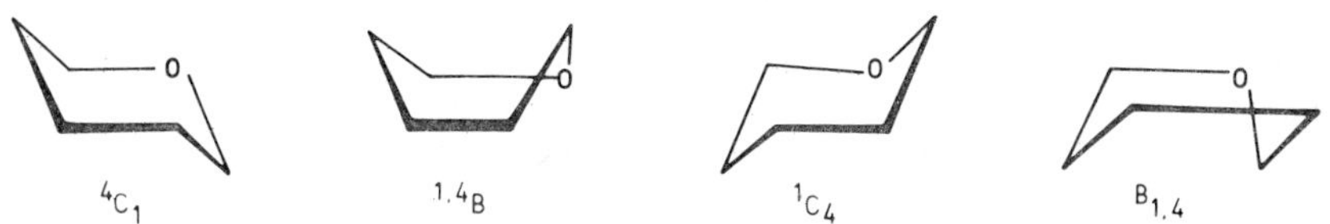

Fig. 2 Pyranose ring forms. Chair conformations allow all substituents to be staggered, whereas for both boat forms bonds at C(2) and C(3) are eclipsed. Repulsions between groups on C(1) and C(4) also cause destabilization of the boat forms.

conformation is usually preferred since the bulky $-CH_2OH$ group at C-5 is orientated in the least crowded equatorial position. For the same reason L sugars usually adopt the $^1C_4$ conformation.

Chain types

Since the shape of a particular sugar unit is fixed, its influence on chain geometry is determined only by its linkage pattern. The complexity of a polysaccharide's shape or higher ordered structure is hence dependent on the nature of its monosaccharide constituents and their linkage configurations, i.e. the primary structure of the chain. For a brief consideration of chain conformations it is convenient to classify the numerous possible polysaccharide types on the basis of their covalent structures [Rees, 1973; Rees, 1977]; this will be done in a slightly different way to the division into homopolysaccharides and heteropolysaccharides used below. Three types can be recognized:
(i) Periodic sequences;
(ii) Interrupted sequences;
(iii) Aperiodic sequences.

Periodic chains, the simplest structural types, have sugars arranged in a repeating pattern and hence may be found among both homo- and hetero-polysaccharides. Many extracellular bacterial polysaccharides fall into this category.

Interrupted sequences also have repeating units but these are upset by departures from regularity. Examples of this structural type are less common amongst bacterial polysaccharides but alginic acid is an example and will be discussed here.

Aperiodic sequences are characterized by irregular sequences of sugar units and can be highly complex; examples of this type have not been definitively characterized for capsular materials.

Linkage conformation

Since the overall shapes of all chain types are a result of the summation of the relative orientations of adjacent sugars the problem of defining higher level structures reduces ultimately to a study of disaccharide shapes. The relative orientation of adjacent sugars may be described by the angles of rotation Φ, Ψ about the two bonds to the glycosidic bridging oxygen atom between them (Figure 3). For a particular disaccharide the conformational energy associated with a particular pair of Φ and Ψ angles can be estimated taking into account van der Waals, polar, torsional and H-bonding interactions across the glycosidic bond [Rees, 1973].

When this is done, for example for cellobiose (the re-

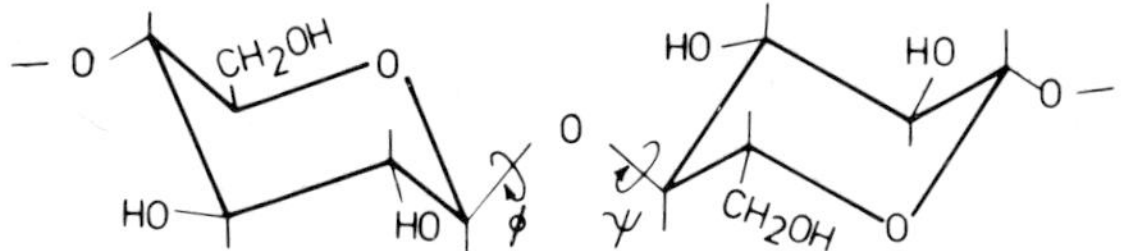

Fig. 3 Rotational angles between adjacent residues. The relative orientations of adjacent residues joined by a glycosidic linkage are defined by two rotational angles Φ and Ψ.

peat disaccharide of cellulose), most orientations are found to be extremely unfavourable and allowed conformations are restricted to a narrow range of Φ and Ψ. These angles agree well with those determined by X-ray diffraction for various cellobiose derivatives; confident extrapolation to solution shapes is indicated from optical rotation considerations which show that, when hydrated, cellobiose retains a shape corresponding to its solid state conformation no doubt with some fluctuation by bond oscillation and rotation. Hence the approach of using energy calculations for conformational predictions is vindicated for this disaccharide.

Order versus disorder

Although extension of favoured disaccharide orientations to longer chains will predict regular conformations for polysaccharides in the solid state, there are further considerations for aqueous solutions. Ordered conformations for highly hydrated chains cannot be assumed since the conformational entropy drive arising from continuous small fluctuations of polysaccharides about a large number of internal glycosidic linkages will be towards a disordered random coil state [Rees, 1973; 1977]. The fact that polysaccharides can, under particular circumstances, adopt ordered shapes in solution indicates that the entropic drive to disorder can be overcome by energy terms [Rees, 1973; 1975; 1977].

For this to occur, favourable non-covalent interactions (for example, hydrogen bonding, ionic interactions, solvent terms) must act co-operatively to fix the polysaccharide in an ordered conformation. For many of the polysaccharides from plant and animal sources which have so far been studied, transitions from a disordered to an ordered state are encouraged by interactions between chains. Stabilization of the ordered state is by alignment and co-operative association of long structurally regular sequences of carbohydrate chains into regions of ordered tertiary structure [Rees and Welsh, 1977]. Intermolecular associations are usually stable only above certain critical chain-lengths (typically around 20 residues) and are often terminated by covalent interruptions to the regular primary structure which are not compatible with ordered chain conformations. A result of these interruptions is that each chain can form junctions with several others so setting up a network which may result in gel formation. Similar arguments will apply to bacterial polymers.

The next stage in predicting higher levels of structure is therefore a consideration of the conformations likely to be adopted by individual polysaccharide chains of given primary structural types.

Conformational families

For this discussion of bacterial extracellular polysaccharides it will be sufficient to concentrate on the way in which possible ordered states can be predicted for chains having periodic sequences. This treatment will also be relevant to polysaccharides with interrupted sequences since they can be considered to have regions of periodic structure interrupted by other sequences which may themselves be irregular or have a different kind of regularity.

The approach can only predict the types of order particular chains are able to adopt; whether these actually exist can be investigated in the solid state by X-ray diffraction. Characterization of chain conformations under hydrated conditions usually relies on the X-ray evidence for the solid state to suggest models around which suitable experimental tests can be designed. It is usually found that the shapes adopted by polysaccharides in highly hydrated environments do correspond to a solid state conformation which has been characterized. Determination of solid state conformations by X-ray diffraction will be discussed in a later chapter of this book.

Simple periodic chain sequences When describing overall shapes for chains containing one type of sugar with a single linkage type the preceding discussion of linkages suggests that when these chains do adopt ordered shapes the shapes will be regular and periodic. Further it is found that a regularly repeating linkage type can generate only a limited number of conformational shapes all of which may be described mathematically as helices.

Helices can be described completely by two parameters n and h which refer respectively to the number of monomer residues per turn of helix and the projected length of each monomer unit on the helical axis [Rees, 1977]. These parameters are geometrically related to the glycosidic bond angles Φ and Ψ. Distinct ranges of n and h are predicted for different homopolymers depending on their covalent structures and these ranges correspond to different types of chain shape. Four chain families are distinguishable (Figure 4): Type A - extended ribbons, have $2 \leq n \leq 4$, and h close to the length of a sugar residue, meaning that each unit must lie nearly parallel to the helix axis. A

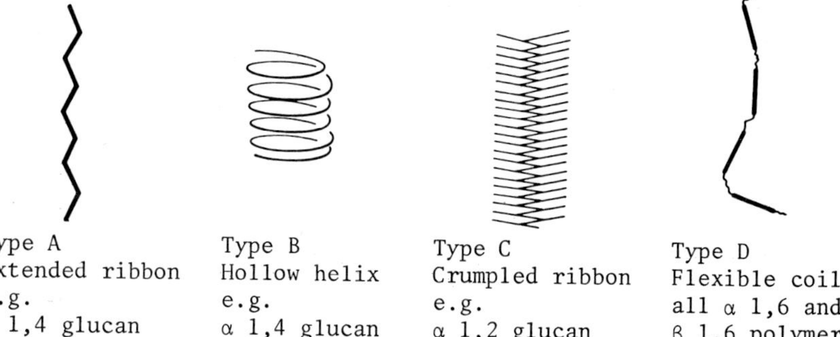

Fig. 4 Schematic representation of the regular conformations predicted for homopolysaccharides by conformational analysis. Examples for glucose polymers are given.

typical example is cellulose, a β 1,4 glucan, which adopts a 'bent-chain' conformation having a 2-fold symmetry axis. Adjacent residues are H-bonded between O(3) and O(5). The effect of this chain shape is manifested in the ability of polysaccharides of this type to pack together efficiently into dense insoluble arrays having efficient H-bonding within and between organized layers. Type B - hollow-helix family, resemble coiled springs in various states of extension. Values of $2 \leq n \leq 10$ are found with h very small, meaning that residues are nearly perpendicular to the helix axis. Amylose, an α 1,4 glucan, is of this type and is known to exist in many forms [Wu and Sarko, 1978a, 1978b; Murphy and French, 1975]. Polymers of this type normally exist in solution as random coils since, in the ordered state, they would have an empty void along their helical axis [Rees, 1977]. They may however form stable ordered conformations in solution if the void is filled by formation of an inclusion complex with a small molecule or, if in a more extended state, they combine with another chain to form a double helix by mutual twisting. For amylose ordered conformations are stabilized by both these methods. Other helical polysaccharides can fill their voids by 'nesting' with a partner; this is close packing without mutual chain twisting. Type C - crumpled ribbons, are not normally found in nature as repeated sequences of the required linkage type (for example an α 1,2 glucan) would cause extensive clashes between non-consecutive units. Ordered conformations are thus difficult for this chain type. Type D - flexible coils; polysaccharides having 1,6 linkages have an extra bond between residues. This leads to an extra freedom of rotation both because of the extra bond and because the sugar rings, being further apart, are less likely to clash. Such polysaccharides have enhanced conformational entropies and are unlikely to adopt an ordered conformation; they tend to exist in solution as flexible, random coils.

The adoption of a particular helical shape by a polysaccharide seems not to depend on interactions between sugar residues but rather on the geometric relationship between glycosidic and aglycone bonds on either side of each sugar unit [Rees, 1973] (Figure 5). Thus for the β 1,4 glucosyl linkage the bonds to a sugar from both its bridging oxygen atoms define a zig-zag ('Z') form and the polymer adopts an extended ribbon (Type A) conformation. For a β 1,3 glucosyl linkage the bonds show a 'U' turn and chains will have a coiled spring (Type B) shape. The primary influence then that determines the type of overall chain shape adopted by a particular homopolysaccharide is the geometric relationship within each sugar unit.

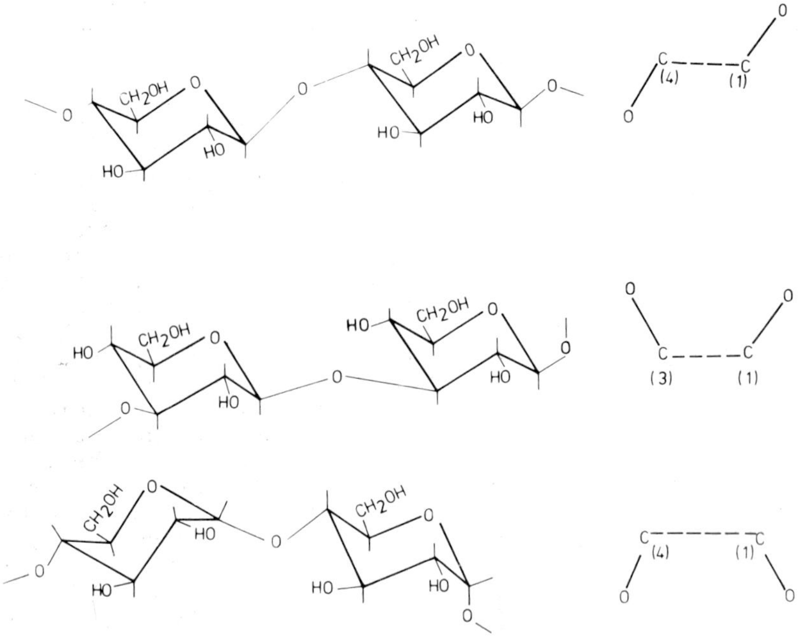

Fig. 5 Bonding arrangements for β 1,4, β 1,3 and α 1,4-linked glucans. Trans ('Z') bonding arrangements across sugar rings give rise to Type A (ribbon-like) shapes whilst cis ('U') bonding results in Type B (helical) chain conformations.

Complex periodic chain sequences For more complex periodic structures the conformational type is not immediately apparent from an inspection of covalent structures. The approach of relating primary structure to likely conformation is not totally without value however. This will be illustrated here by reference to some alternating polysaccharides of non-bacterial origin.

A large family of polymers from plant and animal tissues are known in which sugar units are alternately linked equatorially 1,3 and 1,4. Sugars linked in this way would, if themselves making up homopolymers, generate different chain conformations (types A and B respectively). Computer model building calculations [reviewed in Rees, 1973] predict that possible conformations range from undulating ribbons to extended hollow helices; actual X-ray diffraction analysis of shapes found in the condensed phase shows that they are all variants of an extended hollow helix, each differing in the degree of helix extension. This influences the way in which chains pack as already shown for the simple Type A and B shapes. Agarose (a polymer of alternating 3-linked β-D-galactose and 4-linked 3,6-anhydro-α-L-galactose) and iota carrageenan (3-linked β-D-galactose 4-sulphate and 4-linked 3,6-anhydro-α-D-galactose 2-sulphate) are polysaccharides of this type and their helical chains are not very extended and pack by double helix formation [reviewed in Rees and Welsh, 1977].

The important conclusion is that it is not possible to make clear divisions into structural types as can be done for simple homopolysaccharides. For mixed types A and B alternating polysaccharides the resultant conformations are the results of mixing to different extents for different polymers the shapes imposed by the constituent sugars. Thus even for these materials primary structures are helpful in identifying the type of ordering to be found in the solid state. Extrapolation to ordering under hydrated conditions requires the same considerations as already discussed for simple periodic polysaccharides.

Solution Properties of Bacterial Extracellular Polysaccharides

The solution properties of only a limited number of extracellular polysaccharides have been studied in detail; these will be discussed according to the types of chain shape and chain-chain interactions important for each particular system. Enough polymers have been studied to demonstrate that the ground rules laid for prediction of ordered polysaccharide shapes can be confidently used for polymers of bacterial origin.

Rigid rod type polysaccharides

The extracellular polysaccharide from Xanthomonas campestris The extracellular polysaccharide from *Xanthomonas campestris* was first isolated and examined in the early 1960's [Sloneker *et al.*, 1964]. Aqueous solutions of xanthan are highly viscous and suspend particles well and additionally, under appropriate conditions, the viscosity is fairly insensitive to temperature and ionic strength whilst being highly sensitive to rate of shear [Glicksman, 1970]. This unusual combination of properties has proved industrially attractive and for this reason much effort has been expended in explaining them at a molecular level [Morris, 1977; Morris *et al.*, 1977b]. The properties have been traced back to the adoption of rigid ordered molecular conformations by the polysaccharide chains in solution. Evidence for molecular order has come from optical rotation and n.m.r. relaxation measurements; these techniques have revealed a thermally induced order-disorder transition coinciding with changes in solution viscosity [Morris *et al.*, 1977b] (Figure 6). Xanthan is the first polysaccharide of bacterial origin for which convincing correlations between polymer conformation and solution behaviour have been made.

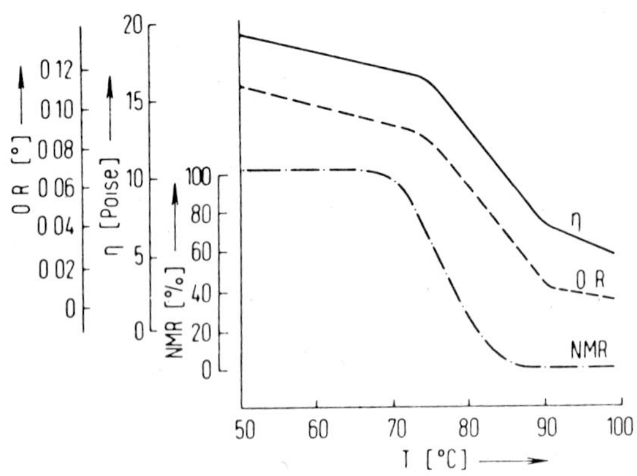

Fig. 6 Transitions of optical rotation, n.m.r. peak area and low-shear viscosity (η) with temperature for a 1% (w/v) aqueous solution of xanthan.

Xanthan is a heteropolysaccharide of the complex periodic type (see above) having a pentasaccharide repeating unit structure [Jansson *et al.*, 1975; Melton *et al.*, 1976]. For conformational purposes the polymer has a cellulose (β 1,4 D-glucose) backbone substituted on alternate residues by trisaccharide side chains (Figure 1). The *O*-acetyl and pyruvic acid ketal substituents have proved to be useful probes of polymer conformation.

At high temperatures aqueous solutions of xanthan give well resolved ^{1}H n.m.r. spectra; peaks of δ 1.5 ppm and δ 2.1 ppm have been assigned to the methyl protons of the pyruvate and acetate substituents. On cooling these peaks at first collapse and finally disappear altogether from the spectrum, evidence for the polysaccharide chain adopting an ordered conformation (Figure 6). In the n.m.r. experiment decay of magnetization can occur by loss of phase of individual precessing nuclei and is characterized by the spin-spin relaxation time, T_2. Since thermal motion interferes with this process, the relaxation rate is inversely related to the degree of molecular mobility. Relaxation data can be measured indirectly from n.m.r. spectra as T_2 is inversely proportional to the signal linewidth. Small molecules moving freely in solution have sharp narrow spectral lines whilst signals for solids are so broad that high resolution spectra cannot be recorded. Between these extremes typical random-coil polysaccharides in solution have T_2 values of around 50 msec and conformationally rigid species, for example, the carrageenan double helix, have values around 50 μsec [S. Ablett and A. Darke, unpublished work]. For the former high resolution peaks are discernable clearly but for the latter the linewidth is so great that all peaks are flattened into the base-line. Hence disappearance of a particular peak from a polymer spectrum can be used quantitatively to monitor disorder/order transitions. For xanthan (Figure 6) a sigmoidal curve results for a plot of n.m.r. peak area against temperature.

Optical rotation changes frequently accompany changes in polysaccharide conformation [Rees, 1970]. For xanthan, optical rotation also shows a sigmoidal dependence on temperature [Morris *et al.*, 1977b]; the transition mid-point coincides with that of the n.m.r. transition (Figure 6) but shifts to higher temperature with increase in ionic strength. Above ~ 0.15M salt no discontinuity is seen up to 100°C suggesting that the ordered chain conformation is completely stable at this temperature. Circular dichroism (C.D.) has been developed for use in polysaccharide systems to monitor the chiroptical behaviour of suitable chromophores and assess their effects on overall optical activity [Morris and Sanderson, 1972; Morris *et al.*, 1977b]. The technique is a sensitive index of the local environment of chromophores and as such is a useful probe of conformational changes. Most commonly $n \rightarrow \pi^*$ transitions are observed for the carboxyl groups of uronic acids, acyl (e.g. acetyl) and ketal (e.g. pyruvate) substituents. At room temperature the C.D. spectrum of xanthan (Figure 7) shows two main bands, a peak (at ~ 205 nm) and a trough (220 nm). On heating, the peak shows no cooperativity but diminishes linearly in height in amounts comparable to the changes in spectra of model compounds such as uronic acids. The trough however, increases in amplitude sigmoidally in a manner similar to that shown for optical rotation and n.m.r. peak area and can be attributed to *O*-acetate groups since it is diminished in size by de-*O*-acetylation of the polymer. These results indicate that the *O*-acetate groups located at the interior of the side-chain enter a more disymmetric environment when the polymer adopts its ordered form whereas the terminally located carboxylate groups do not. Since overall the C.D. change with heating is negative it must contribute negatively to the higher wave-

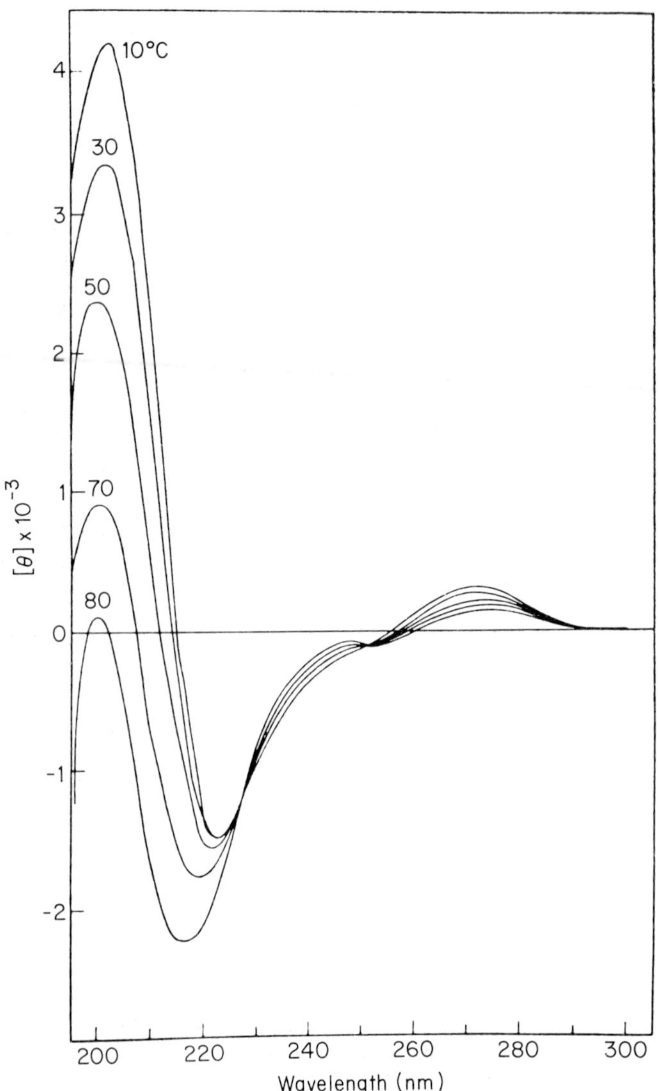

Fig. 7 Temperature dependence of the circular dichroism of xanthan; θ = molar ellipticity in degree cm^2 $decimole^{-1}$.

length optical rotation shift. However the optical rotation shift itself is positive and must therefore be dominated by large positive changes in the far ultra-violet region of the C.D. spectrum where electronic transitions of the polymer backbone occur. Such changes are consistent with, for example, a helix-coil transition since the alteration of conformational angles in such a transition is known to be capable of producing such an effect [Rees, 1970].

Ordered transitions of polysaccharides are frequently stabilized by intermolecular associations into rigid structures [Rees and Welsh, 1977], but for xanthan the transition midpoints of the indexes of order adoption (n.m.r. and optical rotation) are concentration independent suggesting that the order-disorder transition is either unimolecular or extremely cooperative. The latter, although observed for DNA, is unlikely for xanthan since the transition is rather broad.

X-ray diffraction of xanthan fibres shows that in the solid state the molecule exists as a 5-fold helix with a pitch of about 5 n.m. Thus, although having a Type A backbone, the polymer does not adopt the 2-fold ribbon conformation of cellulose but its structure is less extended, presumably being influenced by the trisaccharide side-chains which fold down in alignment with the main chain stabilizing the conformation by non-covalent interactions. The backbone conformation can be regarded as tending towards a very extended Type B hollow helix whose void is partially filled by the side-chains.

The spectroscopic evidence is in agreement with the existence of such an ordered structure in solution and the stabili ation of the ordered conformation with increasing ionic strength is consistent with reduction in electrostatic repulsions between the charged trisaccharide side-chains. A schematic representation of the order-disorder transition believed to occur on cooling xanthan solutions is given in Figure 8.

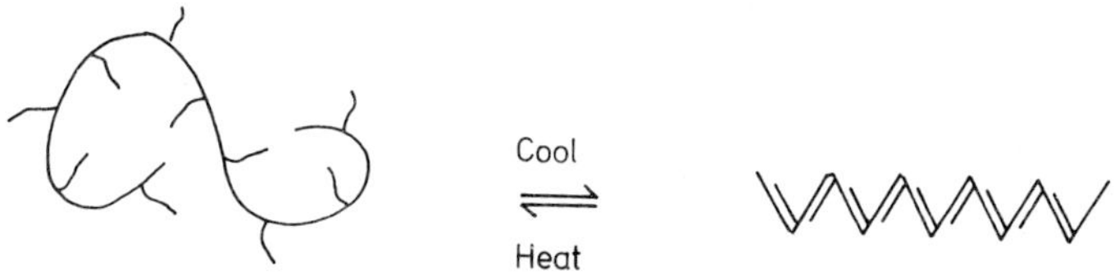

Fig. 8 Schematic representation of the disorder-order transition which occurs on cooling xanthan solutions. Trisaccharide side-chains fold down to stabilise an ordered helical main-chain conformation.

The unusual solution properties of xanthan can be correlated with its rigid-rod like conformation in solution. Intermolecular associations of rigid structures are to be expected even if interchain associations are only slight since entropy losses are small compared with those of a system involving associations of flexible polymer chains [Flory, 1956]. Weak associations of rigid rods into a three-dimensional network explain xanthan's birefringence and particle-suspending ability whilst progressive breakdown of the network with shear provides an explanation for its pseudoplastic behaviour [Jeanes, 1973]. Membrane partition chromatography of xanthan solutions and electron microscopy of samples prepared by drying down solutions indeed provide evidence for aggregation of ordered xanthan molecules [Holzwarth and Prestridge, 1977].

Typical polyelectrolytes adopt expanded conformations in low ionic strength solutions but collapse on addition of salt due to charge screening. Polymer solution rheology is dependent on molecular shape and hence variations in coil dimensions are reflected in solution viscosity changes [Smidsrød and Haug, 1971]. For xanthan, increase in ionic strength only further stabilizes the rigid rod form and viscosity is largely unaffected; the increased stability of the rigid rod is reflected by the maintenance of solution viscosity on heating [Morris *et al.*, 1977b].

At low ionic strengths, solution viscosity behaviour is more complex but can be correlated with chain conformation, although exact behaviour depends critically on shear rate (Figure 9). At low shear rates breakdown of the helix structure is paralleled by an accelerated loss of viscosity with temperature as expected for an order-disorder transition; at higher shear rates breakdown in viscosity is interrupted by an anomalous increase in viscosity which may be due to non-specific molecular entanglements of the disordered form [Morris *et al.*, 1977b]. Understanding of the conformational state of xanthan has thus provided considerable insight into its unusual rheological properties. In a later section of this chapter some biological interactions of the polysaccharide will be discussed in a similar light.

Extracellular polysaccharides from the genus Arthrobacter

Extracellular polysaccharides from three members of the genus of soil-borne bacteria *Arthrobacter* have solution properties in many ways analogous to those described for xanthan [Jeanes, 1974a, b]. The origin of the properties has, in a similar way, been traced to conformational ordering of polysaccharide chains [Darke *et al.*, 1978]. Unlike xanthan, where a knowledge of primary structure helped considerably in solution and condensed phase conformational studies, complete information on the covalent structure of the *Arthrobacter* polysaccharides is not yet available. It is therefore more difficult to trace secondary and tertiary structures to their primary structural origins. The general composition of the *Arthrobacter* polysaccharides is given in Table 3. For *Arthrobacter viscosus* B1973 a linear trisaccharide repeating unit has been established [Sloneker *et al.*, 1968]:

$$\rightarrow 4)\text{-D-manA-}\beta\text{-}(1 \rightarrow 4)\text{-D-glc-}\beta(1 \rightarrow 4)\text{-D-gal }\beta(1 \rightarrow$$

The acetate content would correspond to acylation of four of the eight available hydroxyls in the repeating unit but their positions have not been identified. Preliminary X-

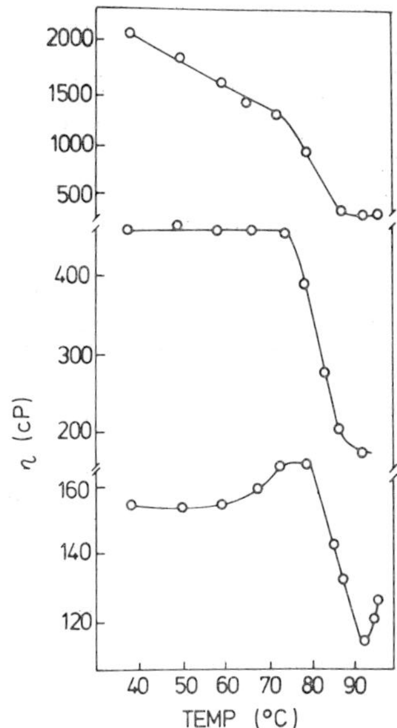

Fig. 9 The thermally induced order-disorder transition of xanthan as monitored by solution viscosity at various rates of shear: upper curve, low shear (14 s^{-1}); middle curve, intermediate shear (127 s^{-1}); bottom curve, high shear (571 s^{-1}).

ray diffraction data on the condensed phase indicates a 3-fold helical structure having a repeat of 4.25 n.m. [Darke *et al.*, 1978]. On this model *O*-acetyl substituents would be on O(2) of mannuronic acid and on O(3) and O(6) of the glucose and galactose units. Conformationally the polysaccharide can be classed as a mixed Type A (1,4-linked β-D-manA and β-D-glc), Type B (1,4-linked β-D-gal); as discussed above the predicted shape would be an extended hollow helix.

Aqueous solutions of all three polysaccharides show thermally reversible gelation behaviour at concentrations above around 1% [Darke *et al.*, 1978]. For the polysaccharide from *Arthrobacter stabilis* B3225, gelation is accompanied by a sigmoidal optical rotation transition, the mid-point of which corresponds to the gelation temperature. By analogy with earlier discussion this is characteristic of a gelation mechanism involving cooperative association of chains into conformationally regular ordered regions. The optical rotation transition is also observed at sub-gelling concentrations confirming that it is of genuine molecular origin and not a bulk artefact of gelation. The transition shifts to higher temperature in the presence of salt and to lower temperature (by about 10°C) on deacylation of the polysaccharide suggesting that the acyl groups,

TABLE 3

Composition of Arthrobacter extracellular polysaccharides
(Darke et al., 1978)

Bacterium	NRRL strain	Composition	Substituents (%)
Arthrobacter stabilis	B3225	Glucose : Galactose 2 : 1	*O*-Acetate 5.0 *O*-Succinate 3-5 Pyruvate 5.1
Arthrobacter viscosus	B1973	Glucose : Galactose : Mannuronic acid 1 : 1 : 1	*O*-Acetate 25
Arthrobacter viscosus species n	B1797	Glucose : Galactose : Glucuronic acid 3 : 1 : 1	*O*-Acetate 8.0 Pyruvate 5.5

whilst not appreciably altering the geometry of the ordered conformation discourage chain-chain interactions; the effect of salt is consistent with stabilization of the ordered state by charge screening and consequent prevention of internal electrostatic repulsions.

Further evidence for the existence of an ordered molecular conformation for the *Arthrobacter stabilis* polysaccharide comes from n.m.r. studies; again analogously to xanthan the high resolution spectrum observed at 95°C (disordered form) collapses on cooling (ordered form) [Darke *et al.*, 1978] and shows a sigmoidal dependence on temperature (Figure 10).

The two *Arthrobacter viscosus* extracellular materials show order-disorder transitions in solution using optical rotation and n.m.r. as probes (Figure 11). For the *Arthrobacter viscosus* species n polysaccharide the transition again occurs around 60°C although overall being broader than that for the *Arthrobacter stabilis* material. Melting out of structure is over an even wider range still for the *Arthrobacter viscosus* material, and is incomplete at 100°C. Conformational ordering is destroyed for both these polysaccharides by deacylation since they both then have high resolution n.m.r. spectra at ambient temperatures as expected for random-coil polymers. Gelation behaviour is also destroyed by this treatment.

The evidence for the adoption by these polysaccharides of ordered shapes in solution is attributable to ordered, although as yet unidentified, molecular conformations and provides rationalization for their rheological properties. Like xanthan aqueous solutions of the polysaccharides show enhanced viscosity with increased ionic strength, a property explicable in terms of the reduction of electrostatic repulsion between extended rigid molecular species leading overall to an enhancement of intermolecular association. The gain in stability of the ordered species in the presence of added salt, as shown by the raised optical rotation transition for the *Arthrobacter stabilis* polysaccharide, is consistent with this, since addition of salt to a normal random-coil polyelectrolyte causes viscosity reduction as a result of coil collapse [Pasika, 1977]. Further evidence for intermolecular association comes from considerations of solution rheology. Whereas normal polysaccharide solutions show continuous deformation under stress, solutions of these ordered bacterial polysaccharides have a yield point [Jeanes, 1974b] and flow at high shear rates; at low shear rates they show particle suspending properties and behave essentially as gels. The yield point thus corresponds to a breakdown of tenuous intermolecular associations between ordered species.

The closely analogous rheological properties of the *Arthrobacter* polysaccharides and xanthan can therefore be rationalized in terms of similarity in molecular shapes rather than from direct similarities in primary structures.

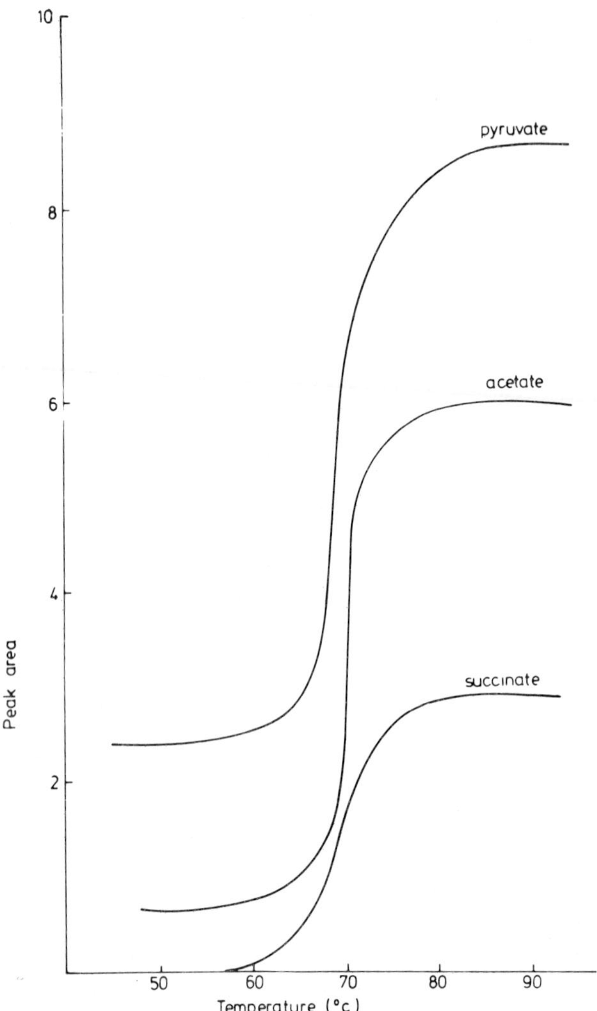

Fig. 10 Loss of peak area for resonances in the n.m.r. spectrum of the extracellular polysaccharide from *Arthrobacter stabilis* as an index of a thermally induced order-disorder transition.

Curdlan - a triple helical ordered polysaccharide

Curdlan, a β-D-glucan, is produced by *Alcaligenes faecalis* var *myxogenes* and some strains of *Agrobacterium radiobacter* [Harada *et al.*, 1966]. Wild-type strains of *Alcaligenes faecalis* produce a succinylated β-D-glucan containing mixed 1,3 and 1,4 links [Misaki *et al.*, 1969]. Some genetically stable mutant strains have been found to

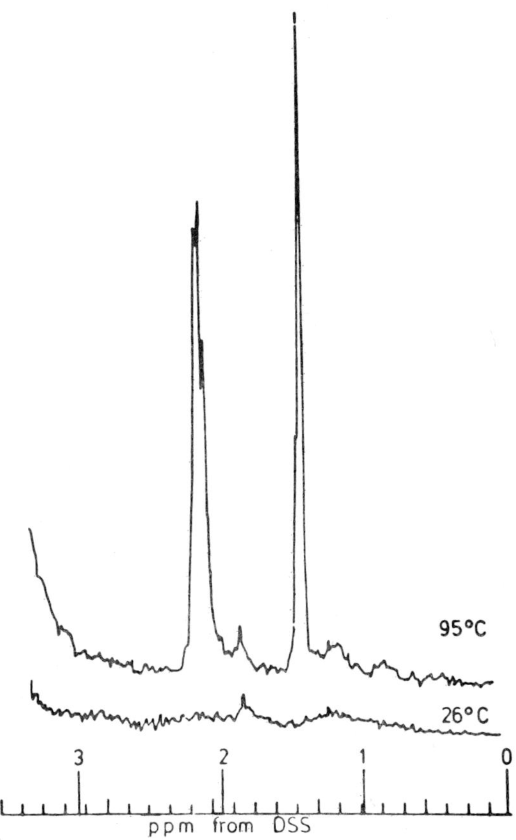

Fig. 11 High resolution ^{1}H-n.m.r. spectra of the ordered (26°C) and disordered (95°C) forms of the extracellular polysaccharide from *Arthrobacter viscosus* species n.

produce only curdlan in high yield [Harada *et al.*, 1968]. Curdlan is a linear D-glucan containing only β 1,3 linkages as shown chemically and by ^{1}H and ^{13}C high resolution n.m.r. [Bluhm and Sarko, 1977]. The polysaccharide is insoluble in cold water, but aqueous suspensions form firm resilient gels on heating and the polymer may find use as a texturiser in the food industry.

β 1,3 D-Glucosyl units present in homopolymeric sequences will give rise to chains of the hollow helix type since they have the 'U turn' relationship between the bridging oxygen atoms on either side of each sugar unit and the ability of curdlan to form gels under appropriate conditions is indicative of its ability to form an ordered conformation. X-ray diffraction of annealed curdlan fibres [Marchessault *et al.*, 1977] suggests the ordered structure to be triple helical.

Many experimental observations provide evidence of a disorder-order transition on gel formation. On heating a curdlan suspension to above 54°C a clear solution forms which subsequently gels at higher temperature [Harada, 1977]. This behaviour contrasts gel formation by marine polysaccharides, e.g. agar, where gels form on cooling hot aqueous solutions [Rees and Welsh, 1977]. Presumably curdlan's behaviour reflects some breakdown of ordering on solution formation followed by a build-up of ordered structure on gelation.

Curdlan dissolves in dilute alkali, but on dialysis against water a strong gel forms; Ogawa *et al.* [1972] studied the behaviour of curdlan in such a system by monitoring optical rotatory dispersion (O.R.D.), viscosity and flow birefringence as a function of alkali concentration. Large changes of these parameters occurred at around 0.2M sodium hydroxide and this was concluded to reflect a transition from an ordered form to a random coil with increase in alkali concentration. This was confirmed by ^{13}C n.m.r. studies; a high resolution spectrum of the random coil could be recorded but the spectral lines broadened and almost disappeared on gel formation [Harada, 1977]. Residual ^{13}C peaks were ascribed to a region of random coil whereas the peak-loss portion is due to multiple helical junction zone formation. Conformational ordering can also be induced for curdlan on adding a non-solvent such as water to a solution of the polysaccharide in dimethylsulphoxide [Harada, 1974].

Adoption of an ordered conformation by curdlan seems to depend on chain length [Ogawa *et al.*, 1973]. Using optical rotation as a probe, water soluble curdlan fragments of chain length (D.P.) below about 25 were shown to have a disordered form in both neutral and alkaline solution whereas glucans of higher DP had an ordered structure in dilute alkali (0.1M). The proportion of ordered structure increased to reach a maximum at around DP 200, which may be the lower limit of chain length for gel formation in neutral solution [Ogawa *et al.*, 1973]. It appears therefore that gels of curdlan result from cooperative intermolecular association of chains into triple helical junctions which are stable above a chain length of around 25 units; these zones may be linked by short single chain regions. The model is similar to that proposed for agar and carageenan-type gels [Rees and Welsh, 1977].

Electron microscopy has been used to examine the supramolecular structure of different molecular weight fractions of curdlan [Koreeda *et al.*, 1974]. Microfibrils (10-20 n.m. wide), themselves composed of many elementary fibrils were detected in original (DP 400) and partly degraded (DP 140-260) curdlan whilst smaller fragments (DP 13-36) showed no such structure. Only the higher chain length glucans were capable of gel formation, thus there appears to be a relationship between gel forming ability and the existence of microfibrils. These may represent aggregates of triple helical ordered regions [Koreeda *et al.*, 1974].

Alginate - an ion mediated ordered conformation

Alginates are a group of structurally related polysaccharides of great commercial importance. They demonstrate, under favourable conditions, gelling and viscosity properties which have been exploited in the food, textile, pharmaceutical and paper industries. Once again these solution properties can be traced to adoption of ordered molecular conformations by the polysaccharide chains.

Alginates were originally isolated from the brown seaweeds (*Laminaria* sp., *Macrocystis* sp.) where they occur as the major structural polysaccharides, but in the mid 1960's certain bacteria were found to produce extracellular polysaccharides conforming to the same structural type. Bacterial alginate is produced by the organisms *Pseudomonas aeruginosa* and *Azotobacter vinelandii* during their growth cycles probably as a protection against dehydration [Gorin and Spencer, 1966; Linker and Jones, 1966; Carlson and Matthews, 1966]. The term alginate refers to a group of polysaccharides having a linear 1,4-linked structure composed of β D-mannuronate and α L-guluronate residues. The sugars are arranged in homopolymeric sequences of each type and also in regions which approximate to a disaccharide repeating structure, although recent enzymic structural evidence suggests some deviation from an idealised regular alternating sequence [see Morris *et al.*, 1978]. Alginates from different sources, both plant and bacterial, vary within this structural framework, and in addition bacterial products may be partially acetylated.

In contrast to the extracellular polysaccharides already discussed alginates do not show thermally induced order-disorder transitions in solution, but rather adopt ordered tertiary structures by interactions with divalent cations, typically calcium. Thus in contrast to xanthan, which can exist in ordered chain conformations that can be further stabilized by high ionic strength conditions, the presence of a counterion is a prerequisite for the adoption of ordered chain conformations by alginate.

X-ray fibre diffraction studies [Atkins *et al.*, 1973a] on alginate samples having high proportions of mannuronic acid have shown that such chain sequences adopt flat ribbon-like 2-fold chain conformations in the solid state similar to those found for β 1,4 diequatorially linked polymers such as cellulose [Rees, 1977] (Figure 12). Polyguluronic acid, being 1,4 diaxially linked adopts a buckled 2-fold chain conformation in the acid form [Atkins *et al.*, 1973b] (Figure 12). From various lines of evidence it seems that poly-L-guluronate sequences are of prime importance in the calcium mediated gelation of alginate. When

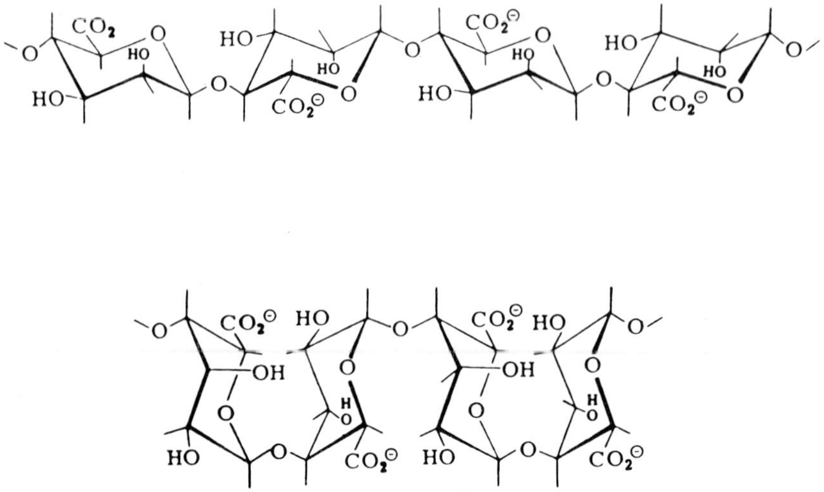

Fig. 12 Ribbon-like and buckled chain conformations for 1,4-linked β D-mannuronate (above) and α L-guluronate (below) sequences.

the various block sequences are isolated from intact alginate by partial acid hydrolysis [Haug *et al.*, 1966] polyguluronate shows an enhanced binding of calcium above a chain length of around 20 residues [Kohn, 1975]. This threshold for strong calcium binding suggests a cooperative mechanism in which binding sites exist in an ordered array and binding of one ion facilitates binding of the next. Similar effects were not observed for poly-mannuronate or alternating chain sequences [Kohn, 1975].

Model building [Grant *et al.*, 1973] suggests a plausible model for calcium binding in which poly-guluronate sequences adopt a chain conformation similar to that characterized in the solid state for the acid form. An array of calcium binding sites is formed by the alignment of two poly-guluronate chains (Figure 13); the buckled chains produce a linear array of carboxylate and hydroxyl lined cavities which have the correct size to bind calcium ions. Examination of poly-mannuronate sequences shows that the cavities formed between pairs of extended ribbon-like chains are too "shallow" to provide similar favourable binding sites. This model for alginate gelation is known as the "egg-box" [Grant *et al.*, 1973]. Further evidence for the model comes from C.D. spectroscopy [Rees and Welsh, 1977; Morris *et al.*, 1978]. When poly-guluronate sequences, or alginates rich in these, bind calcium large changes are seen in the $n \rightarrow \pi^*$ region of the C.D. spectrum, consistent with the n orbitals of the guluronate residues being involved with chelation. The egg-box model indeed places the cations in the correct position for coordina-

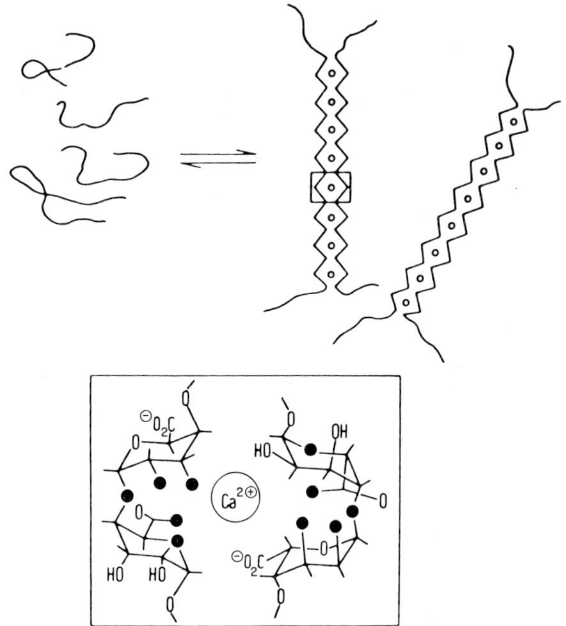

Fig. 13 Schematic representation of the calcium mediated association of poly-guluronate sequences in alginate into "egg-boxes". Below: oxygen atoms (•) coordinated to calcium ions.

tion by carboxylate n orbitals (Figure 13), and the association of poly-guluronate sequences by cation binding is believed to be the principal mechanism of cross-linking in alginate gelation [Morris *et al.*, 1973].

Since poly-guluronate sequences have 2-fold conformations in their ordered state, egg-box junctions could involve two chains or extend to infinite sheets of calcium guluronate. The primary gelation event appears, however, to involve only chain dimerisation as shown by equilibrium dialysis experiments where one cooperatively bound calcium ion is bound to every four guluronate residues; for the infinite sheet model the ratio would be 1:2 [Morris *et al.*, 1978].

Competitive inhibition experiments provide further evidence for the mode of alginate gelation (Figure 14). Poly-guluronate block sequences, when added to gelling calcium alginate systems, cause marked reduction in gel strength by occupying binding sites on intact alginate chains without contributing to inter-molecular cross-linking [Morris *et al.*, 1977a]; alternating blocks have less effect and poly-mannuronate sequences have none at all. These results again indicate the importance of poly-guluronate sequences in alginate gelation and support the evidence for a limited number of chains per junction zone since more extensive aggregation could clearly include such blocks with no loss of cross-linking.

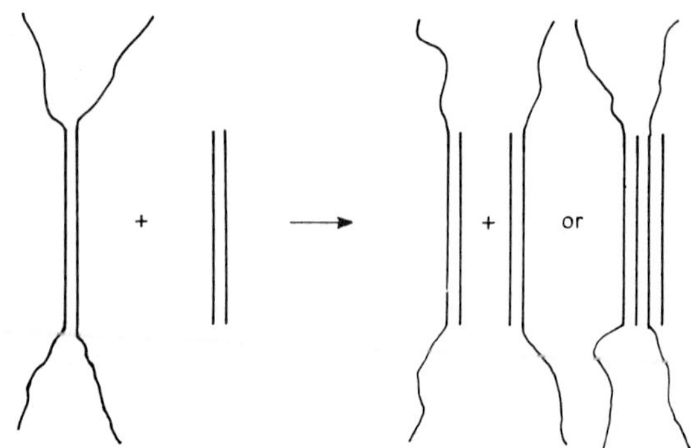

Fig. 14 Possible effects of the incorporation of poly-guluronate block sequences into alginate junction zones: left, dimerisation inhibited by loss of chain-chain cross-linking; right, block sequences incorporated into aggregated network without loss of cross-linking.

The role of poly-mannuronate and alternating sequences in gelation appears to be to provide primary covalent interruptions which are incompatible with the ordered conformation of poly-guluronate involved in junction zone formation. Hence a gel network may be set up rather than an insoluble precipitate by allowing each alginate chain to participate in ordered associations with several different partners. The extent to which ordered solution conformations can occur is thus limited by the number and lengths of poly-guluronate sequences. These variables appear to be under enzymic control at the polymer level; for *Azotobacter vinelandii* polymer chains are first synthesised as poly-mannuronate and poly-guluronate sequences are then introduced by epimerisation at carbon-5 by appropriate enzymes [Larsen and Haug, 1971; Haug and Larsen, 1971].

Dextran - a random coil polysaccharide

Dextran was the first microbial polysaccharide to be produced and used on an industrial scale [Jeanes, 1966]. Nowadays most dextran is produced from a single strain (NNRL B 512) of *Leuconostoc mesenteroides* [Jeanes, 1974a] and its use depends largely on the molecular weight distribution of the component molecules; important uses of dextran fractions are as blood plasma substitutes [Jeanes, 1966; Grönwall, 1957] and in the manufacture of Sephadex, a cross-linked dextran used widely for molecular sieve chromatography.

Dextran is usually highly soluble in water, those polymers having the highest proportion of α 1,6 linkages being the most readily soluble [Jeanes, 1966]. The industrially produced dextran has an α 1,6-linked backbone having side-chains either one or two D-glucose units long [Larm *et al.*, 1971]. Polymers that are 1,6-linked, as compared to 1,3- and 1,4-linked polysaccharides, have an extra degree of flexibility since residues are linked through three rather than two covalent bonds and are not so restricted in their rotation [Rees, 1977; Tvaroska *et al.*, 1978].

Weight average molecular weights in the range 40-50 x 10^6 have been reported for industrially produced dextrans using light scattering techniques [Jeanes, 1965] and the extremely low values of intrinsic viscosity [Jeanes, 1966] of these shows that a consequence of the 1,6 linkage flexibility is that they have small coil dimensions. Further evidence for chain flexibility is from ^{1}H n.m.r. relaxation data. Polysaccharides adopting ordered conformations in solution, such as xanthan or iota-carrageenan in its double helical form [S. Ablett and A. Darke, unpublished work; Dea *et al.*, 1977], have a very short (micro-second) spin-spin relaxation time (T_2) whereas random coil polymers have T_2 values around three orders of magnitude higher than this reflecting their greater conformational mobility. For dextran the extra flexibility of the 1,6 linkage is demonstrated clearly by T_2 values being even higher than those for 1,4- and 1,3-linked polysaccharides in their random coil states. Dextran chains do not tend to adopt ordered conformations in solution due to both their loosely jointed backbones and because chain-chain packing would be inhibited by their branched structure. The absence of intermolecular interactions is apparent from solution viscosity behaviour. Dextran solutions at 5% concentration show essentially Newtonian-type viscosity behaviour [Jeanes, 1974a] (viscosity is independent of shear rate) and comparison with the solution properties of polysaccharides having only two bonds in their glycosidic linkages, for example, plant galactomannans, is revealing. These are linear polymers of α 1,4-linked D-mannose with D-galactose side-chains; they are random coil polysaccharides but are somewhat "stiffer" than dextran [Rees, 1977] and hence more likely to pack and interact intermolecularly. Solutions of locust bean gum show plastic characteristics at concentrations around 1%, but similar properties were not observed for dextran until concentrations of 15% were used [Jeanes, 1974a; Patton, 1969]. Work in this laboratory has shown that if the viscosities of solutions of guar and dextran having comparable intrinsic viscosities (dextran therefore having a higher molecular weight) are compared, the onset of intermolecular coupling for guar (as monitored by a sharp viscosity increase) occurs at approximately 5-fold lower concentrations than for dextran.

Dextran is hence a good example of how primary structure effects solution behaviour. Its inability to form ordered molecular conformations is manifest on comparison of its solution properties with those of polysaccharides able to adopt ordered shapes and interact intermolecularly.

Biological Interactions

Little is understood of the function of bacterial extracellular polysaccharides and whilst many trivial roles have been suggested [Dudman, 1977] it would be surprising if such structurally complex materials and the necessary genetic machinery for their biosynthesis would be required to fulfil them [Sutherland, 1972; Sutherland and Norval, 1970]. The situation of the extracellular polymers at the surface of the bacterial cell suggests that they may be involved in interactions of the cell with its surroundings. Recent demonstrations that extracellular polysaccharides can interact specifically with other biologically important polysaccharide chains gives credence to this view [Morris *et al.*, 1977b; Dea *et al.*, 1977; Darke *et al.*, 1978]. Furthermore the best characterized of these interactions have been shown to occur under physiological conditions and to involve extracellular polysaccharides in ordered conformations.

The *Xanthomonas campestris* extracellular polysaccharide, xanthan, shows evidence of intermolecular interactions in solution [see above and Morris *et al.*, 1977b] but does not gel. On mixing with the plant galactomannan locust bean gum however, firm gels form providing evidence of strong specific interactions [Dea and Morrison, 1975; Morris *et al.*, 1977b; Dea *et al.*, 1977]. Other plant galactomannans also interact with xanthan [Dea *et al.*, 1977]. These polysaccharides all have β 1,4-linked main chains of D-mannose and have extended 2-fold ribbon-like conformations in the solid state [Dea and Morrison, 1975; Rees, 1977]; they are hence similar in secondary structure to cellulose although somewhat more flexible [Rees, 1977]. Substitution of the mannan backbone by α 1,6 D-galactose "stubs" occurs to varying extents and in different patterns for galactomannans from different sources [Dea and Morrison, 1975; Baker and Whistler, 1975; Hoffman *et al.*, 1976].

The strength of the interactions between xanthan and these polysaccharides is closely correlated with the pattern of substitution of the mannan chain [Dea and Morris, 1977; Morris *et al.*, 1977b; Dea *et al.*, 1977]. Thus guaran, the principal polysaccharide of guar gum (mannose (M): galactose (G) ~ 2:1) shows slight viscous interactions, tara gum (M:G ~ 3:1) gives soft fairly weak gels and locust bean gum (M:G ~ 4:1) shows the strongest interaction forming strong rigid gels. Even stronger gels result by using certain fractions of locust bean gum having higher M:G ratios. Thus unsubstituted regions of the mannan backbone are clearly implicated in these interactions. Xanthan-galactomannan gels may be formed by cooling a mixed solution of the polysaccharides and optical rotation and X-ray diffraction studies of this process show that the native ordered conformation of xanthan is present in the gel [Dea *et al.*, 1977; Dea and Morris, 1977] (Figure 15). The optical rotation transition normally accompanying the disorder-order transition of xanthan occurs in the presence of galactomannan and is complete at temperatures above the gel point of the mixture.

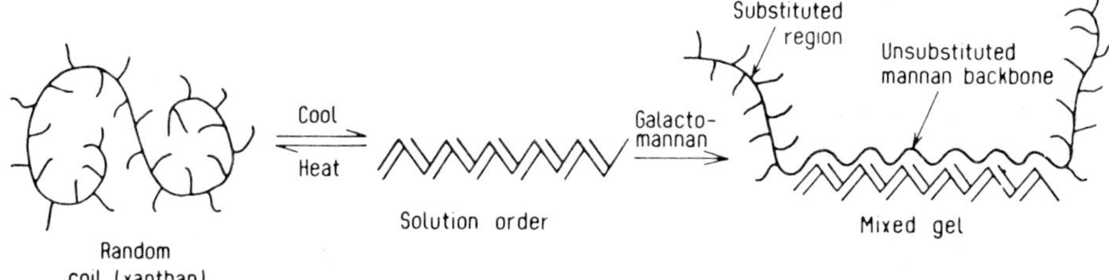

Fig. 15 Proposed model for the synergistic interaction of the ordered form of xanthan with the unsubstituted mannan backbone region of galactomannans.

The synergistic interactions occurring in these systems are examples of polysaccharide quaternary structure and involve cooperative association of regular ribbon-like unsubstituted mannan regions with xanthan rigid rods. The substituted parts of the galactomannan chains presumably act as "solubilising" regions preventing complete aggregation and precipitation of the system. A proposed mechanism [Dea *et al.*, 1977] for the interaction is illustrated schematically in Figure 15. The biological significance of this synergism is that *Xanthomonas* species are pathogenic organisms responsible for blight diseases of a number of plants, such as peas, beans, cabbage, whose cell walls consist of cellulose and hemicellulosic galactoglucomannans, polymers having similar backbones to the galactomannans. The specific binding or recognition of bacterial extracellular polysaccharides and cell-wall type components may suggest a role in the host-pathogen relationship [Morris *et al.*, 1977b]. Xanthan does in fact interact more strongly with polysaccharides bearing a closer structural resemblance to plant cell-wall polymers. Mixtures of xanthan and Konjac mannan (a partially acetylated β 1,4-linked linear copolymer of D-mannose and D-glucose derived from tubers of *Amorphophallus konjac*) form gels which have melting points 20°C higher than comparable xanthan-locust bean gum gels [Dea *et al.*, 1977]. There is some evidence that xanthan itself can interact with the cell walls of living plants [Leach *et al.*, 1975; Lesley and Hochster, 1959] and thus it might act as an anchor for bacterial cells or even play a more subtle role in identifying particular sites for invasion in particular plants by recognizing characteristic polysaccharide compositions in the plant cell wall. The synergistic interaction of xanthan and galactomannans therefore mimics this natural association.

Extracellular polysaccharides from certain *Arthrobacter* species have ordered solution conformations and also show synergistic interactions with galactomannans [Darke *et al.*, 1978]. Similar interpretation of these properties suggests a role for the polymers in the known interactions of these soil-borne bacteria with components of plant root systems [Sperber and Rovira, 1959]. There is some evidence that aggregation of the plant polysaccharides is a prerequisite for extensive synergistic interactions to occur [Darke *et al.*, 1978].

Further evidence that extracellular polysaccharides are involved in lectin mediated recognition processes comes from work on *Rhizobium* species. All *Rhizobium* species can form nitrogen-fixing nodules on the roots of leguminous plants but the process is highly specific. A particular species of bacterium can only interact with its particular host plant and the recognition process is thought to involve plant lectins in reactions similar to antigen-antibody binding [Hamblin and Kent, 1973; Bohlool and Schmidt, 1974]. There is some divergence of evidence on whether the bacterial antigen in the lectin binding process resides in the lipopolysaccharide or extracellular polysaccharide of the nodulating *Rhizobium* species [Dazzo and Hubbell, 1975; Wolpert and Albersheim, 1976; Planqué and Kijne, 1976; Albersheim *et al.*, 1977]. Recently a mutant of *Rhizobium leguminosarum* has been isolated which produces diminished amounts of extracellular polysaccharide and fails to nodulate pea, its normal host. This suggests that the presence, at least, of the extracellular material is essential for nodulation to occur [Sanders *et al.*, 1978]. The mutant, otherwise identical to its parent, can produce spontaneous revertants which renew production of extracellular material and regain their ability to nodulate the host [Sanders *et al.*, 1978]. The exact role of extracellular polysaccharides in the host-symbiont recognition process is unclear [Beringer, 1978], but it could well involve adoption of ordered conformations by these known covalently regular polymers [Jansson *et al.*, 1977; Jansson *et al.*, 1978].

The importance of ordered conformations of extracellular polysaccharides in determining some immunological properties of bacteria is suggested by recent work on the capsular polysaccharide from a serotype of *Escherichia coli* [Moorhouse *et al.*, 1977]. The K-specific capsular polysaccharide from *Escherichia coli* serotype 29, K29, has a regular covalent structure consisting of hexasaccharide repeating units [Choy *et al.*, 1975] and is the receptor for *Escherichia coli* K phage 29. The bacteriophage is highly specific for this capsule and for capsular material synthesized by two mutants, M13 and M41, of the same strain [Moorhouse *et al.*, 1977]. All three polysaccharides have identical ordered helical conformations as determined by X-ray diffraction of orientated fibres [Moorhouse *et al.*, 1977]. In the mutant M13 the capsular polysaccharide is attached to the exterior of the cell in the form of radial strands or coiled structures [Bayer and Thurow, 1976] and it is reasonable to assume that the ordered molecular structure determined by X-ray diffraction for highly hydrated fibres of the polysaccharide will persist for these strands *in vivo*. Absorption of phage and direction of it to a particular cell surface receptor site may therefore involve some particular conformational feature of the capsular polysaccharide [Moorhouse *et al.*, 1977].

Recent indications that some of the K antigenic capsular polysaccharides from *Klebsiella* can adopt ordered conformations under hydrated conditions [Wolf *et al.*, 1978] have led to the suggestion that ordered shapes for antigens may sometimes be important in the antigen-antibody reaction. The specificity of this reaction is known to depend on the antibody having a particular conformation [Pauling, 1940; Edelmann *et al.*, 1969] and so it is possible that the interacting capsular polysaccharide (the antigen in this case) may adopt a complementary conformation for complex formation. Alternatively the capsular polysaccharide may initially be disordered, and during interaction with the antibody any tendency to adopt an ordered shape may be strengthened by induced fit with the antibody; i.e. adoption of an ordered conformation may be a result of binding.

Acknowledgements

I would like to express my thanks to my colleagues, Dr. E.R. Morris and Dr. D.A. Rees, for their valuable help and constructive criticism during the preparation of this chapter.

References

Adams, M.H., Reeves, R.E. and Goebel, W.F. (1941). The Synthesis of 2,4 dimethyl-β-methyl glucoside. *Journal of Biological Chemistry* **140**, 653-661.

Albersheim, P., Ayers, A.R., Valent, B.S., Ebel, J., Hahn, M., Wolpert, J. and Carlson, R. (1977). Plants interact with microbial polysaccharides. *Journal of Supramolecular Structure* **6**, 599-616.

Aspinall, G.O. (1977). The selective degradation of carbohydrate polymers. *Pure and Applied Chemistry* **49**, 1105-1134.

Atkins, E.D.T., Niedusynski, I.A., Mackie, W., Parker, K.D. and Smolko, E.E. (1973a). Structural components of alginic acid. I. The crystalline structure of poly-β-D-mannuronic acid. Results of X-ray diffraction and polarised infrared studies. *Biopolymers* **12**, 1865-1878.

Atkins, E.D.T., Niedusynski, I.A., Mackie, W., Parker, K.D. and Smolko,

E.E. (1973b). Structural components of alginic acid. II. The crystalline structure of poly-α-L-guluronic acid. Results of X-ray diffraction and polarised infrared studies. *Biopolymers* **12**, 1879-1887.

Baddiley, J. (1972). Teichoic acids. *Essays in Biochemistry* **8**, 35-77.

Bailey, R.W. (1959). Trans-glucosidase activity of rumen strains of *Streptococcus bovis*. *Biochemical Journal* **72**, 42-49.

Baker, C.W. and Whistler, R.L. (1975). Distribution of D-galactosyl groups in guaran and locust bean gum. *Carbohydrate Research* **45**, 237-243.

Bayer, M.E. and Thurow, H. (1977). The polysaccharide capsule of *Escherichia coli*: microscope study of its size, structure and sites of synthesis. *Journal of Bacteriology* **130**, 911-936.

Bebault, G.M., Dutton, G.G.S., Funnell, N. and Mackie, K.L. (1978). Structural investigation of *Klebsiella* serotype K32 polysaccharide. *Carbohydrate Research* **63**, 183-192.

Beringer, J.E. (1978). *Rhizobium* recognition. *Nature, London* **271**, 206-207.

Björndal, H., Lindberg, B., Lönngren, J., Mészáros, M., Thompson, J.L. and Nimmich, W. (1973). Structural studies of the capsular polysaccharide of *Klebsiella* type 52. *Carbohydrate Research* **31**, 93-100.

Bluhm, T.L. and Sarko, A. (1977). The triple helical structure of lentinan, a linear β-(1 → 3)-D-glucan. *Canadian Journal of Chemistry* **55**, 293-299.

Bohlool, B.B. and Schmidt, E.L. (1974). Lectins, a possible basis for specificity in the *Rhizobium*-legume nodule symbiosis. *Science*, New York **185**, 269-271.

Cadmus, M.C., Rogovin, S.P., Burton, K.A., Pittsley, J.E., Knutson, C.A. and Jeanes, A. (1976). Colonial variation in *Xanthomonas campestris* NRRL B1459 and characterisation of the polysaccharide from a variant strain. *Canadian Journal of Microbiology* **22**, 942-948.

Carlson, D.M. and Matthews, L.W. (1966). Polyuronic acids produced by *Pseudomonas aeruginosa*. *Biochemistry* **5**, 2817-2828.

Choy, Y.M., Fehmel, F., Frank, N. and Stirm, S. (1975). The *Escherichia coli* serotype 29 capsular polysaccharide. *Journal of Virology* **16**, 581-590.

Cooper, E.A. and Preston, J.F. (1935). Enzyme formation and polysaccharide synthesis by bacteria. *Biochemical Journal* **29**, 2267-2275.

Curvall, M., Lindberg, B., Lönngren, J. and Nimmich, W. (1975a). Structural studies of the capsular polysaccharide of *Klebsiella* type 81. *Carbohydrate Research* **42**, 73-82.

Curvall, M., Lindberg, B., Lönngren, J. and Nimmich, W. (1975b). Structural studies of the capsular polysaccharide of *Klebsiella* type 28. *Carbohydrate Research* **42**, 95-105.

Darke, A., Morris, E.R., Rees, D.A. and Welsh, E.J. (1978). Spectroscopic characterisation of order-disorder transitions for extracellular polysaccharides of *Arthrobacter* species. *Carbohydrate Research* **66**, 133-144.

Dazzo, F.B. and Hubbell, D.H. (1975). Cross-reactive antigens and lectins as determinants of symbiotic specificity in the *Rhizobium*-clover association. *Applied Microbiology* **30**, 1017-1033.

Dea, I.C.M. and Morris, E.R. (1977). In *Extracellular Microbial Polysaccharides*. *American Chemical Society Symposium Series No. 45*.

pp.174-182. Edited by P.A. Sandford and A. Laskin. Washington D.C.: American Chemical Society.

Dea, I.C.M. and Morrison, A. (1975). Chemistry and interactions of seed galactomannans. *Advances in Carbohydrate Chemistry and Biochemistry* **31**, 241-312.

Dea, I.C.M., Morris, E.R., Rees, D.A., Welsh, E.J., Barnes, H.A. and Price, J. (1977). Association of like and unlike polysaccharides: mechanism and specificity in galactomannans, interacting bacterial polysaccharides and related systems. *Carbohydrate Research* **57**, 249-272.

Dedonder, R. and Hassid, W.Z. (1964). The enzymatic synthesis of a β (1 → 2) linked glucan by an extract of *Rhizobium japonicum*. *Biochimica et Biophysica Acta* **90**, 239-248.

Dudman, W.F. (1977). The role of surface polysaccharides in natural environments. In *Surface Carbohydrates of the Prokaryotic Cell*, pp.357-414. Edited by I.W. Sutherland. London: Academic Press.

Duckworth, M. (1977). Teichoic acids. In *Surface Carbohydrates of the Prokaryotic Cell*, pp.177-208. Edited by I.W. Sutherland. London: Academic Press.

Edelmann, G.M., Cunningham, B.A., Gall, W.E., Gottlieb, P.D., Rutishauser, U. and Waxdal, M. (1969). The covalent structure of an entire γG immunoglobulin molecule. *Proceedings of the National Academy of Sciences, U.S.A.* **63**, 78-85.

Flory, P.J. (1956). Statistical thermodynamics of semi-flexible chain molecules. *Proceedings of the Royal Society Series A* **234**, 50-73.

Glicksman, H. (1970). *Polysaccharide Gums in Food Technology*. New York: Academic Press.

Gorin, P.A.J. and Spencer, J.F.T. (1966). Exocellular alginic acid from *Azotobacter vinelandii*. *Canadian Journal of Chemistry* **44**, 993-998.

Gorin, P.A.J., Spencer, J.F.T. and Westlake, D.W. (1961). The structure and resistance to methylation of 1,2 β glucan from species of *Agrobacteria*. *Canadian Journal of Chemistry* **39**, 1067-1073.

Grant, G.T., Morris, E.R., Rees, D.A., Smith, P.J.C. and Thom, D. (1973). Biological interactions between polysaccharides and divalent cations: the egg-box model. *FEBS Letters* **32**, 195-198.

Grönwall, A. (1957). *Dextran and its Uses in Colloidal Infusion Solutions*. New York: Academic Press.

Hamblin, J. and Kent, S.P. (1973). Possible role of the phytohaemagglutinin from *Proteus vulgaris* L. *Nature New Biology* **245**, 28-30.

Harada, T. (1965). Succinoglucan 10C3: A new polysaccharide of *Alcaligenes faecalis* var. *myxogenes*. *Archives of Biochemistry and Biophysics* **112**, 65-69.

Harada, T. (1974). Succinoglucan and gel-forming beta-1,3-glucan. *Process Biochemistry* **9**, 21-27.

Harada, T. (1977). In *Extracellular Microbial Polysaccharides. American Chemical Society Symposium Series No. 45*, pp.265-283. Edited by P.A. Sandford and A. Laskin. Washington D.C.: American Chemical Society.

Harada, T., Masada, M. and Fujimori, K. (1966). Production of a firm resilient gel-forming polysaccharide by a mutant of *Alcaligenes* var. *myxogenes* 10C3. *Agricultural and Biological Chemistry* **30**,

196-202.

Harada, T., Misaki, A. and Saito, H. (1968). Curdlan: a bacterial gel-forming β 1 → 3 glucan. *Archives of Biochemistry and Biophysics* **124**, 292-298.

Haug, A. and Larsen, B. (1971). Biosynthesis of alginate - Part II. Polymannuronic acid C-5-epimerase from *Azotobacter vinelandii*. *Carbohydrate Research* **17**, 297-308.

Haug, A., Larsen, B. and Smidsrød, O. (1966). A study of the constitution of alginic acid by partial acid hydrolysis. *Acta Chemica Scandinavica* **20**, 183-190.

Hehre, E.J. and Hamilton, D.M. (1951). The biological synthesis of dextran from dextrins. *Journal of Biological Chemistry* **192**, 161-174.

Hestrin, S., Avineri-Shapiro, S. and Aschner, M. (1954). The enzymic production of levan. *Biochemical Journal* **37**, 450-456.

Hoffman, J., Lindberg, B. and Painter, T. (1976). The distribution of the D-galactose residues in guaran and locust bean gum. *Acta Chemica Scandinavica* **B30**, 365-366.

Holzwarth, G. and Prestridge, E.B. (1977). Multistranded helix in xanthan polysaccharide. *Science*, New York **197**, 757-759.

How, M.J., Brimacombe, J.S. and Stacey, M. (1964). The pneumococcal polysaccharides. *Advances in Carbohydrate Chemistry* **19**, 303-358.

Jansson, P.E., Kenne, L. and Lindberg, B. (1975). Structure of the extracellular polysaccharide from *Xanthomonas campestris*. *Carbohydrate Research* **45**, 275-282.

Jansson, P.E., Kenne, L., Lindberg, B. and Ljunggren, H. (1978). Structural studies of the *Rhizobium trifoli* capsular polysaccharide. Abstract G21, *IXth International Symposium on Carbohydrate Chemistry*. International Union of Pure and Applied Chemistry. London.

Jansson, P.E., Kenne, L., Lindberg, B., Ljunggren, H., Lönngren, J., Rudén, U. and Svensson, S. (1977). Demonstration of an octasaccharide repeating-unit in the extracellular polysaccharide of *Rhizobium meliloti* by sequential degradation. *Journal of the American Chemical Society* **99**, 3812-3815.

Jeanes, A. (1965). In *Methods in Carbohydrate Chemistry* Vol. V, pp. 118-127. Edited by R.L. Whistler and J.N. BeMiller. New York and London: Academic Press.

Jeanes, A. (1966). In *Encyclopaedia of Polymer Science and Technology* Vol. 4, pp.805-824. Edited by H.F. Mark. New York: J. Wiley.

Jeanes, A. (1973). In *Proceedings of the American Chemical Society Conference on Water Soluble Polymers*, pp.227-242. Edited by N.M. Bikales. New York: Plenum Press.

Jeanes, A. (1974a). Extracellular microbial polysaccharides. *Food Technology* **28**, 34-40.

Jeanes, A. (1974b). Applications of extracellular microbial polysaccharide polyelectrolytes: review of literature including patents. *Journal of Polymer Science, Symposium No. 45*, 209-227.

Kamerling, J.P., Lindberg, B., Lönngren, J. and Nimmich, W. (1975). Structural studies of the *Klebsiella* type 57 capsular polysaccharide. *Acta Chemica Scandinavica* (B) **29**, 593-598.

Kendall, F.E., Heidelberger, M. and Dawson, M.H. (1937). A serologically inactive polysaccharide elaborated by mucoid strains of

Group A haemolytic streptococci. *Journal of Biological Chemistry* **118**, 61-69.

Kenne, L., Lindberg, B. and Svensson, S. (1975). The structure of capsular polysaccharide of the *Pneumococcus* type II. *Carbohydrate Research* **40**, 69-75.

Kenne, L., Lindberg, B., Lindqvist, B., Lönngren, J., Arie, B., Brown, R.G. and Stewart, R.E. (1976). 4-O-[(S)-1-carboxyethyl]-D-glucose: a component of the extracellular polysaccharide material from *Aerococcus viridans* var. *homari*. *Carbohydrate Research* **51**, 287-290.

Kennedy, D.A., Buchanan, J.G. and Baddiley, J. (1969). The type-specific substance from pneumococcus Type 11A (43). *Biochemical Journal* **115**, 37-45.

Kohn, R. (1975). Ion binding on polyuronates-alginate and pectin. *Pure and Applied Chemistry* **42**, 371-397.

Koreeda, A., Harada, T., Ogawa, K., Sato, S. and Kasai, N. (1974). Study of the ultrastructure of gel-forming (1 → 3)-β-D-glucan (curdlan-type polysaccharide) by electron microscopy. *Carbohydrate Research* **33**, 396-399.

Larm, O. and Lindberg, B. (1976). The pneumococcal polysaccharides - a re-examination. *Advances in Carbohydrate Chemistry* **19**, 303-358.

Larm, O., Lindberg, B. and Svensson, S. (1971). Studies on the length of the side-chains of the dextran elaborated by *Leuconostoc mesenteroides* NRRL B-512. *Carbohydrate Research* **20**, 39-48.

Larsen, B. and Haug, A. (1971). Biosynthesis of alginate - Part I. Composition and structure of alginate produced by *Azotobacter vinelandii*. *Carbohydrate Research* **17**, 287-296.

Lawson, C.J. (1976). Microbial polysaccharides. *Chemistry and Industry*, 258-261.

Leach, J.G., Lilly, V.G., Wilson, H.A. and Purvis, M.R. Jr. (1975). Bacterial polysaccharides: the nature and function of the exudate produced by *Xanthomonas phaseoli*. *Phytopathology* **47**, 113-120.

Lesley, S.M. and Hochster, R.M. (1959). The extracellular polysaccharide of *Xanthomonas phaseoli*. *Canadian Journal of Biochemistry and Physiology* **37**, 513-529.

Lindberg, B., Lindqvist, B., Lönngren, J. and Nimmich, W. (1976). 4-O-[(S)-1-carboxyethyl]-D-glucuronic acid: a component of the *Klebsiella* Type 37 capsular polysaccharide. *Carbohydrate Research* **49**, 411-417.

Lindberg, B. and Lönngren, J. (1978). Methylation analysis of complex carbohydrates, general procedure and application in sequence analysis. *Methods in Enzymology*, in press.

Lindberg, B., Lönngren, J. and Powell, D.A. (1977). Structural studies on the specific Type 14 pneumococcal polysaccharide. *Carbohydrate Research* **58**, 177-186.

Lindberg, B., Lönngren, J. and Svensson, S. (1975). Specific degradation of polysaccharides. *Advances in Carbohydrate Chemistry and Biochemistry* **31**, 185-239.

Linker, A. and Jones, R.S. (1966). A new polysaccharide resembling alginic acid isolated from pseudomonads. *Journal of Biological Chemistry* **241**, 3845-3851.

Lönngren, J. and Svensson, S. (1974). Mass spectrometry in structural analysis of natural carbohydrates. *Advances in Carbohydrate Chemistry and Biochemistry* **29**, 41-106.

Marchessault, R.H., Deslandes, Y., Ogawa, K. and Sundararajan, P.R. (1977). X-ray diffraction data for β-(1 → 3)-D-glucan. *Canadian Journal of Chemistry* **55**, 300-303.

Melton, L.D., Mindt, L., Rees, D.A. and Sanderson, G.R. (1976). Covalent structure of the extracellular polysaccharide from *Xanthomonas campestris*: evidence from partial hydrolysis studies. *Carbohydrate Research* **46**, 245-257.

Misaki, A., Saito, H., Ito, T. and Harada, T. (1969). Structure of succinoglucan, an exocellular acidic polysaccharide of *Alcaligenes faecalis* var. *myxogenes*. *Biochemistry* **8**, 4645-4650.

Moorhouse, R., Winter, W.T., Arnott, S. and Bayer, M.E. (1977). Conformation and molecular organisation in fibres of the capsular polysaccharide from *Escherichia coli* M41 mutant. *Journal of Molecular Biology* **109**, 373-391.

Morris, E.R. (1977). In *Extracellular Microbial Polysaccharides, American Chemical Society Symposium Series No. 45*, pp.81-89. Edited by P.A. Sandford and A. Laskin. Washington D.C.: American Chemical Society.

Morris, E.R., Rees, D.A., Boyd, J. and Turvey, J.R. (1978). Chiroptical and stoichiometric evidence of a specific primary dimerisation process in alginate gelation. *Carbohydrate Research* **66**, 145-154.

Morris, E.R., Rees, D.A. and Thom, D. (1973). Characterisation of polysaccharide structure and interactions by circular dichroism: order-disorder transition in the calcium alginate system. *Journal of the Chemical Society Chemical Communications*, 245-246.

Morris, E.R., Rees, D.A., Thom, D. and Welsh, E.J. (1977a). Conformation and intermolecular interactions of carbohydrate chains. *Journal of Supramolecular Structure* **6**, 259-274.

Morris, E.R., Rees, D.A., Young, G., Walkinshaw, M.D. and Darke, A. (1977b). Order-disorder transition for a bacterial polysaccharide in solution. A role for polysaccharide conformation in recognition between *Xanthomonas* pathogen and its plant host. *Journal of Molecular Biology* **110**, 1-16.

Morris, E.R. and Sanderson, G.R. (1972). In *New Techniques in Biophysics and Cell Biology*, pp.113-147. Edited by R.H. Paine. London: John Wiley.

Murphy, V.G. and French, A.D. (1975). The structure of V amylose dehydrate; a combined X-ray and stereochemical approach. *Biopolymers* **14**, 1487-1501.

Ogawa, K., Tsurugi, J. and Watanabe, T. (1973). The dependence of the conformation of a (1 → 3)-β-D-glucan on chain length in alkaline solution. *Carbohydrate Research* **29**, 397-403.

Ogawa, K., Watanabe, T., Tsurugi, J. and Ono, S. (1972). Conformational behaviour of a gel-forming (1 → 3)-β-D-glucan in alkaline solution. *Carbohydrate Research* **23**, 399-405.

Pasika, W.M. (1977). In *Extracellular Microbial Polysaccharides, American Chemical Society Symposium Series No. 45*, pp.128-143. Edited by P.A. Sandford and A. Laskin. Washington D.C.: American Chemical Society.

Patton, T. (1969). Viscosity profile of typical polysaccharides in the ultra-low shear rate range. *Cereal Science Today* **14**, 178-183.

Pauling, L. (1940). A theory of the structure and process of formation of antibodies. *Journal of the American Chemical Society* **62**,

2643-2657.

Planqué, K. and Kijne, J.W. (1977). Binding of pea lectins to a glycan-type polysaccharide in the cell walls of *Rhizobium leguminosarum*. *FEBS Letters* **73**, 64-66.

Rao, E.V., Buchanan, J.G. and Baddiley, J. (1966a). The type-specific substance from *Pneumococcus* Type 10A (34) - (structure of the dephosphorylated repeating-unit). *Biochemical Journal* **100**, 801-810.

Rao, E.V., Buchanan, J.G. and Baddiley, J. (1966b). The type-specific substance from *Pneumococcus* Type 10A (34) - (the phosphodiester linkages). *Biochemical Journal* **100**, 811-814.

Rees, D.A. (1970). Conformational analysis of polysaccharides - Part IV. *Journal of The Chemical Society* (B), 877-884.

Rees, D.A. (1973). In *Carbohydrates, MTP International Review of Science. Organic Chemistry Series One*, **7**, pp.251-283. Edited by G.O. Aspinall. London: Butterworths.

Rees, D.A. (1975). In *Biochemistry of Carbohydrates, MTP International Review of Science. Biochemistry Series One*, **5**, pp.1-42. Edited by W.J. Whelan. London: Butterworths.

Rees, D.A. (1977). *Polysaccharide Shapes*, second edition. London: Chapman and Hall.

Rees, D.A. and Welsh, E.J. (1977). Secondary and tertiary structure of polysaccharides in solutions and gels. *Angewandte Chemie (international edition in English)* **16**, 214-224.

Sanders, R.E., Carlson, R.W. and Albersheim, P. (1978). A *Rhizobium* mutant incapable of nodulation and normal polysaccharide secretion. *Nature, London* **271**, 240-242.

Sloneker, J.H. and Orentas, D.G. (1962). Pyruvic acid: a unique component of a bacterial exocellular polysaccharide. *Nature, London* **194**, 478-479.

Sloneker, J.H., Orentas, D.G. and Jeanes, A. (1964). Exocellular bacterial polysaccharide from *Xanthomonas campestris* NRRL 1459; Part III. Structure. *Canadian Journal of Chemistry* **42**, 1261-1269.

Sloneker, J.H., Orentas, D.G., Knutson, C.A., Watson, P.R. and Jeanes, A. (1968). Structure of the extracellular bacterial polysaccharides from *Arthrobacter viscosus* NRRL B-1973. *Canadian Journal of Chemistry* **46**, 3353-3361.

Smidsrød, O. and Haug, A. (1971). Estimation of the relative stiffness of the molecular chain in polyelectrolytes from measurements of viscosity at different ionic strengths. *Biopolymers* **10**, 1213-1227.

Sperber, J.I. and Rovira, A.D. (1959). A study of the bacteria associated with the roots of subterranean clover and Wimmera Rye Grass. *Journal of Applied Bacteriology* **22**, 85-92.

Sutherland, I.W. (1970). Structure of *Klebsiella aerogenes* type 8 polysaccharide. *Biochemistry* **9**, 2180-2185.

Sutherland, I.W. (1971). The occurrence of acyl groups in *Klebsiella* exopolysaccharides. *Journal of General Microbiology* **65**, v-vi.

Sutherland, I.W. (1972). Extracellular polysaccharides. *Advances in Microbial Physiology* **8**, 143-213.

Sutherland, I.W. (1977). Enzymes acting on bacterial surface carbohydrates. In *Surface Carbohydrates of the Prokaryotic Cell*, pp. 27-96. Edited by I.W. Sutherland. London: Academic Press.

Sutherland, I.W. and Norval, M. (1970). The synthesis of exopolysaccharides by *Klebsiella aerogenes* membrane preparations and the involvement of lipid intermediates. *Biochemical Journal* **120**, 567-576.

Tvaroska, I., Pérez, S. and Marchessault, R.H. (1978). Conformational analysis of (1 → 6)-α-D-glucan. *Carbohydrate Research* **61**, 97-106.

Wolf, Ch., Elsässer-Beile, U., Stirm, S., Dutton, G.G.S. and Burchard, W. (1978). Conformational studies of bacterial polysaccharides. II. Optical activity of some bacterial capsular polysaccharides. *Biopolymers* **17**, 731-748.

Wolpert, J.S. and Albersheim, P. (1976). Host-symbiont interactions. I. The lectins of legumes interact with the O-antigen containing lipopolysaccharides of their symbiont *Rhizobia*. *Biochemical and Biophysical Research Communications* **70**, 729-737.

Wu, H.C. and Sarko, A. (1978a). The double-helical molecular structure of crystalline B-amylose. *Carbohydrate Research* **61**, 7-25.

Wu, H.C. and Sarko, A. (1978b). The double-helical molecular structure of crystalline A-amylose. *Carbohydrate Research* **61**, 27-40.

Chapter 7

CONFORMATIONS OF MICROBIAL EXTRACELLULAR POLYSACCHARIDES BY X-RAY DIFFRACTION: PROGRESS ON THE *KLEBSIELLA* SEROTYPES

E.D.T. ATKINS, D.H. ISAAC and H.F. ELLOWAY

H.H. Wills Physics Laboratory, University of Bristol, Bristol, UK

Introduction

The past few years have seen a notable advance in our knowledge of polysaccharide conformation and structure, in particular of those containing uronic acid residues (polyuronides). This rapid progress is mainly a result of the development of improved crystallization technique during the early seventies [Atkins and Mackie, 1972; Atkins and Sheehan, 1972]. Initially these techniques were applied successfully to the simpler plant polymonosaccharides and to the polydisaccharides of animal connective tissue. More recently an even larger field of study has been opened up with the crystallization of the more complex polysaccharides, which occur as microbial capsules, and contain up to eight residues in their chemical repeating sequence. A further important advance in this field has been the concomitant development of sophisticated computer programmes for analysing the diffraction data obtained.

In this contribution we intend to restrict our discussion to the conformations of microbial capsular polysaccharides. Even within this class of biopolymers the variety of structures is formidable and so we have chosen to select a few of the polysaccharides from the genus *Klebsiella*. This genus alone contains at least 83 different serotypes [Nimmich, 1968]. In general these polysaccharides are extracellular and surround the bacteria as an additional outer layer. These capsules may act as a viscoelastic barrier against desiccation and phagocytosis. The capsular material exhibits specific antigenic properties, and the classification of *Klebsiella* is based on serological antigen reactions of the capsule (K-antigen) [Kauffmann, 1966]. In all known cases the isolated capsular material consists of heteropolysaccharides with a regular repeating unit of up to eight residues, often including a mono- or disaccharide as a side appendage. The presence of a uronic acid, pyruvate or both seems to be a characteristic of these glycans [Nimmich, 1968].

Methodology

The major tool for the elucidation of the three dimensional molecular structure at atomic level is the technique of X-ray diffraction. Maximum information is available from this technique if the material occurs naturally as, or can be induced to form, single crystals. Polysaccharide single crystals rarely occur naturally and it has not been found possible to produce them synthetically. A few polysaccharides, for example cellulose and chitin, occur naturally in a semi-crystalline form exhibiting fibre symmetry and these are the materials which featured in early work on polysaccharide conformations. However, in common with many other linear synthetic and biological polymers, polysaccharides should in general be capable of orientation and crystallization. Usually the best that can be expected is a sample containing small crystallites with their chain axes aligned in a unique direction, but with random azimuthal orientations, i.e. with fibre symmetry. This gives rise to a fibre type diffraction pattern, examples of which are shown in Figures 2, 5, 8, 12, 15, 18, 21 and 23, which are somewhat similar to a single crystal rotation photograph. Clearly an essential step in producing such samples was the development of techniques for orientation and crystallization and a breakthrough occurred in the early seventies using stress fields and annealing (sometimes at elevated temperatures) in a manner analogous to those that had previously been used on synthetic polymer samples [Atkins and Mackie, 1972].

Analysis of fibre diffraction patterns can be taken to various stages of refinement, depending on the quality of the data available. Such patterns exhibit diffracted intensity confined to approximately horizontal layer lines which reflects the periodic nature along the chain direction. Thus measurement of the layer line spacing provides the pitch of the helix. Further, for a helix with n units in a single turn, that is per pitch p, diffracted intensity on the meridian only occurs on layer lines whose integer index l is a multiple of n. The quotient p/n yields the axial projection h which generally corresponds to the chemical repeat. (There are exceptions to these simple rules, such as intertwining helices and structures in which not all chemical repeats are related by symmetry, but these can usually be deduced by the experienced diffractionist.) This information combined with the known chemical repeating sequence is sufficient for molecular models to be built that conform with the observed helical parameters. (Clearly at this stage information on standard bond length, bond angles and torsion angles must be incorporated to limit the number of possible structures.)

If the fibre pattern contains diffraction spectra to sufficient resolution, then the next stage of refinement involves a complete structural determination of the unit cell contents using a full set of intensity information. Current procedures for such a determination usually generate molecular conformations using a linked atom description similar to that reported by Arnott and Wonacott [1966] in which atomic positions are defined in terms of bond lengths, bond angles and torsion angles. This allows the simple incorporation of standard values for bond lengths and angles and a rigid structure for the standard saccharide ring, derived from an averaged set [Arnott and Scott, 1972]. Concurrent with refinement of the model to fit the intensity data in a least squares manner, non-bonded steric interactions may be minimised following the method developed by Williams [1969] for low molecular weight hydrocarbons and now adopted for applications to polysaccharides. In the most favourable cases positioning of counterions and water molecules may be achieved, generally using Fourier methods.

Serotypes

Nimmich [1968] has classified 83 different serotypes of *Klebsiella*. At the time of preparing this manuscript the chemical structures of 39 serotypes are known and the majority of the remainder are under investigation. X-ray diffraction patterns, of a quality sufficient to undertake model building, have been obtained for 22 serotypes although some of these are of serotypes whose chemical structure is not yet elucidated. In this contribution we present a progress report on eight *Klebsiella* serotypes offering stereochemically satisfactory models consistent with the X-ray parameters. These eight serotypes: K5, K8, K9, K16, K25, K54, K57 and K63, provide us with the opportunity to visualize the molecular shapes of these complex polysaccharides and to examine the geometry, and effect on polymer shape, of novel glycosidic linkages.

Serotype K5

The repeating sequence of *Klebsiella* K-type 5 polysaccharide has been established by Dutton and Mo-Tai Yang [1973]. It consists of a linear trisaccharide repeat as illustrated in Figure 1. The repeat contains a 1,3-linked β-D-mannose, a 1,4-linked β-D-glucuronic acid and a 1,4-linked β-D-glucose residue. It is anticipated that all the sugar residues are in the normal $^4C_1$ chair conformation, resulting in a backbone linkage geometry of Man-(1 eq - 4 eq)- GlcUA - (1 eq - 4 eq) - Glc - (1 eq - 3 eq) - Man (eq represents equatorially oriented and ax represents axially oriented bonds. See Figure 1). In addition to the

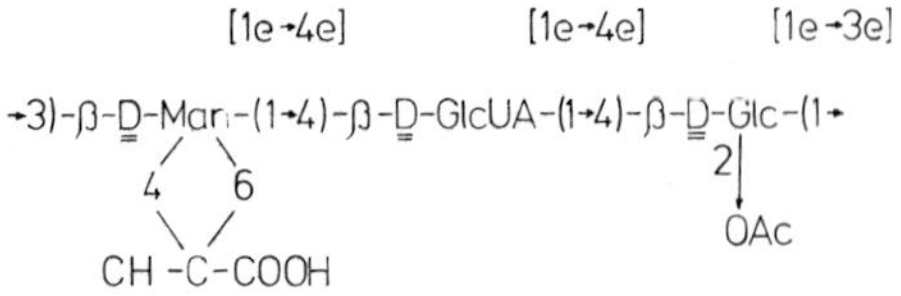

```
               [1e→4e]          [1e→4e]        [1e→3e]
→3)-β-D-Man-(1→4)-β-D-GlcUA-(1→4)-β-D-Glc-(1→
        /  \                          2|
       4    6                          ↓
        \  /                          OAc
     CH -C-COOH
```

Fig. 1 Repeating chemical structure for *Klebsiella* serotype K5.

glucuronic acid residue the molecule contains a charged group in the form of 4,6 ketal pyruvate attached to the mannose unit. A further appendage worth noting is the *O*-acetate at the 2-position of the glucopyranose ring.

The X-ray diffraction pattern obtained from an oriented film of the sodium salt form of the K5 polysaccharide is shown in Figure 2. The pattern has a layer line spacing of 2.70 nm with meridional reflections occurring on even layer lines only. Meridional reflections are those which lie on the vertical bisector of the X-ray diffraction pattern and represent axially projected repeats within the structure. The straightforward interpretation of these observations is that the molecule adopts a two-fold helical conformation with an axially projected repeat of 1.35 nm. This value is comparable with that expected for the K5 trisaccharide repeating sequence. A maximum theoretical extension of 1.56 nm would be expected if all the vectors joining successive glycosidic oxygen atoms were to line up pre-

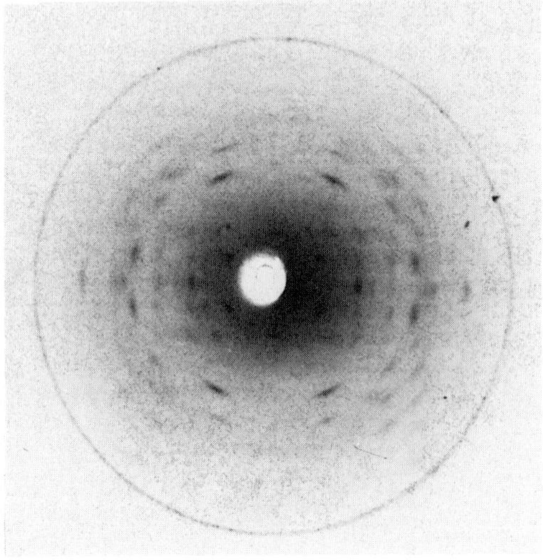

Fig. 2 X-ray fibre diffraction pattern obtained from an oriented film of the sodium salt of K5. The layer line spacing is 2.70 nm and meridional reflections occur on even layer lines indicating a two-fold helix.

cisely. The observed value of 1.35 nm is some 14% less than the theoretical maximum, a degree of extension similar to that found in a variety of plant and animal polyuronides (4% - 18%) [Atkins *et al.*, 1974].

Computer model building procedures have been used to construct molecular chain conformations which are consistent with the observed helical parameters, as deduced from the X-ray pattern, while simultaneously attempting to form a maximum number of stabilizing hydrogen bonds. It was found possible to form three hydrogen bonds, one across each glycosidic linkage, and still meet the helical constraints. Computer drawn projections are shown in Figure 3. The structure incorporates an O(5)...H-O(3) hydrogen bond at both the β-D-Man-(1,4)-β-D-GlcUA and the β-D-GlcUA - (1,4) - β-D-Glc glycosidic linkages. This hydrogen bond is present in many homopolysaccharides such as cellulose [Gardner and Blackwell, 1974] and the connective tissue polydisaccharide structures [Guss *et al.*, 1975]. It is interesting to find it also appearing in this more complex structure. At the third glycosidic linkage (β-D-Glc-(1,3) -β-D-Man) there is an O(2)...O(2) hydrogen bond.

The torsional angles at the 1,3 linkage dispose the charged pyruvate group on the radial periphery of the molecule and in close proximity to the *O*-acetyl group. It has been suggested that non-carbohydrate constituents such as *O*-acetyl and pyruvate groups are antigenic determinants and the close association of these appendages may represent a determinant site.

Serotype K8

The covalent chemical repeat of K8 has been established [Sutherland, 1970a] and is given in Figure 4. It is a tetrasaccharide consisting of three sugar residues in the backbone and one residue in the side chain. The backbone contains three commonly occurring neutral sugars: a 1,3-linked β-D-galactose followed by a 1,3-linked α-D-galactose and finally a 1,3-linked β-D-glucose, whilst the side chain is an α-D-glucuronic acid residue attached at the 4-position to β-D-galactose. All four saccharide units are expected to exist in the normal $^4C_1$ chair conformation giving rise to the linkage geometry shown in Figure 4. Thus the backbone contains two 1 eq - 3 eq glycosidic linkages and one 1 ax - 3 eq linkage. If the vectors between successive glycosidic oxygen atoms were to align precisely the maximum theoretical extension per covalent repeat would be 1.38 nm.

The X-ray diffraction pattern for the sodium salt of the K8 polysaccharide is shown in Figure 5. The material is highly oriented and crystalline and from the systematic absences of odd order meridional reflections it can be seen that the molecule is a 2_1 helix. However, the layer line

2.70 nm
Glc
GlcUA
carboxyl
acetate
Pyr
carboxyl

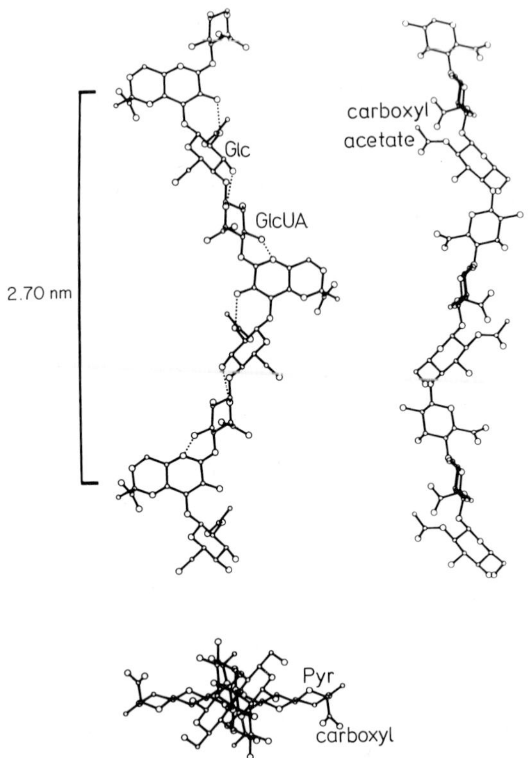

Fig. 3 Computer drawn projections of the K5 two-fold helix. The projections at the top correspond to views perpendicular to the helix axis and are perpendicular to each other. The projection at the bottom represents a view looking along the helix axis. The dotted lines represent hydrogen bonds. Note that the pyruvate groups attached to the mannose residue project well away from the helix axis. Also the carboxyl group and acetate group are in close proximity (top right hand projection).

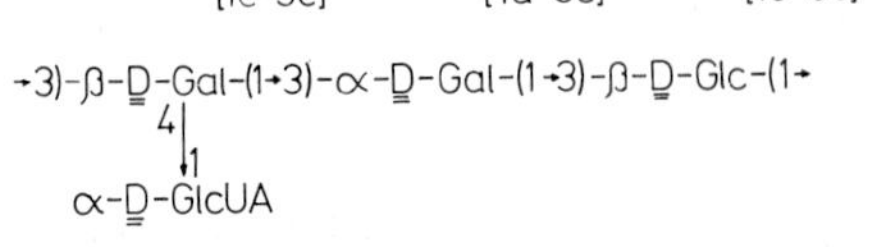

[1e→3e]
[1a→3e]
[1e→3e]
→3)-β-D-Gal-(1→3)-α-D-Gal-(1→3)-β-D-Glc-(1→
4
1
α-D-GlcUA

Fig. 4 Repeating chemical structure for *Klebsiella* serotype K8.

spacing of 5.078 nm is far too large for a repeat of two asymmetric units. In fact it is very close to the theoretical maximum extension for four complete covalent repeats, i.e. 4 x 1.38 = 5.52 nm. Thus the observed repeat is only

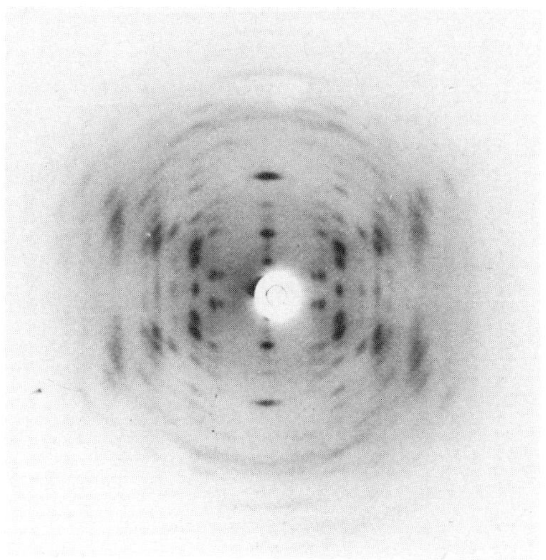

Fig. 5 X-ray fibre diffraction obtained from an oriented film of the sodium salt of K8. The layer line spacing is 5.078 nm with meridional reflections on even layer lines only. Such a spacing is compatible with four covalent repeats and we may interpret this as a slightly perturbed four-fold helix. By altering the relative humidity meridionals only occur on layer lines with l = 4n confirming this interpretation.

some 10% less than the maximum permissible extension.

For preliminary model building it appears reasonable to assume that the structure of the isolated model is a perfect four-fold helix with an axial advance per covalent repeat (h) = 1.27 nm. Perturbations from an idealized four-fold helix would be expected to result in a lower symmetry and consequently the packing of the molecular chains in an orthorhombic rather than a tetragonal unit cell. The phenomenon has also been observed in hyaluronic acid [Guss *et al.*, 1975].

Stereochemically feasible models were constructed to fit a four-fold helical symmetry and incorporating the maximum number of intra-molecular hydrogen bonds. It was found impossible to construct a model which incorporated hydrogen bonds at all three backbone linkages. Only a left-handed helix allowed the formation of two hydrogen bonds in the backbone. Projections of this structure are shown in Figure 6. The conformation contains hydrogen bonds O(5)... O(4) and O(5)...O(2) at the β-D-Gal-(1,3) - α-D-Gal and α-D-Gal-(1,3) -β-D-Glc glycosidic linkages respectively. It was found not possible to form a four-fold helix (with the observed extension) which contained a hydrogen bond at the β-D-Glc-(1,3) - β-D-Gal glycosidic linkage. This linkage is adjacent to the attachment site of the uronic acid side chain.

Lacking any information to define the position of the side chain in the isolated molecule, other than that it must be a stereochemically allowed position, it was necessary to use X-ray intensity information to determine the orientation of the pendant residue [K.H. Gardner and E.D.T. Atkins, unpublished work].

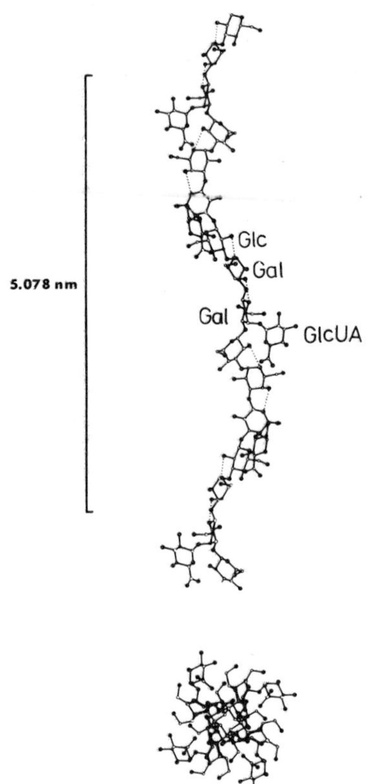

Fig. 6 Computer generated projections of a left-handed four-fold helical structure for serotype K8; note the glucuronic acid side chain.

Serotype K9

The primary structure of serotype K9 polysaccharide has been established by Lindberg *et al.* [1972]. The regularly repeating sequence is a pentasaccharide consisting of four sugar residues in the backbone, three α-L-rhamnose residues and a single α-D-galactose, and a β-D-glucuronic acid side chain (Figure 7). It is interesting to consider the conformations of the individual residues. Starting with α-L-rhamnose, the X-ray crystal structure refinement of the

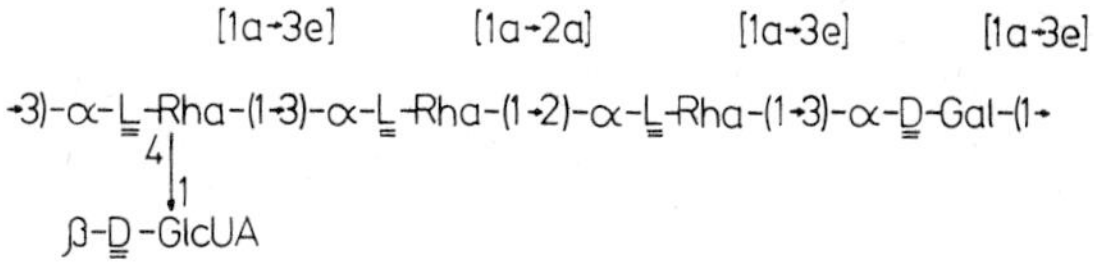

[1a→3e] [1a→2a] [1a→3e] [1a→3e]

→3)-α-L-Rha-(1→3)-α-L-Rha-(1→2)-α-L-Rha-(1→3)-α-D-Gal-(1→
4
↑
1
β-D-GlcUA

Fig. 7 Repeating chemical structure for *Klebsiella* serotype K9.

monohydrate showed that the residue exists in the $^1C_4$ chair [McGeachin and Beevers, 1957]. It may also be noted that α-D-mannose in the $^4C_1$ chair is the mirror image (except for the substituent at C(5)) of α-L-rhamnose in the $^1C_4$ chair. Since both n.m.r. [Angyal, 1969] and theoretical energy calculations [Angyal, 1968] favour the $^4C_1$ chair conformation for α-D-mannose this is additional indirect evidence predicting the $^1C_4$ chair for α-L-rhamnose. In this chair conformation the O(1) and O(2) atoms are axially disposed whilst the O(3), O(4) and C(6) are positioned equatorially [see Figure 7]. For α-D-galactose the $^4C_1$ chair is favoured by X-ray single crystal structure refinement [Sheldrick, 1976], n.m.r. [Angyal, 1969] and theoretical energy calculations [Angyal, 1968]. In this conformation all substituent groups are equatorial except for O(1) and O(4) which are axially disposed. It is assumed that the β-D-glucuronic acid will exist in the normal $^4C_1$ chair conformation with all its substituent groups equatorially positioned.

Thus following the covalent repeating sequence in Figure 7 the backbone linkage geometry is Rha - (1 ax - 3 eq) - Rha - (1 ax - 2 ax) - Rha - (1 ax - 3 eq) - Gal - (1 ax - 3 eq) - Rha. The D-glucuronic acid side chain is attached (1 eq - 4 eq) to a rhamnose residue.

The X-ray fibre diffraction pattern obtained from a stretched film of the sodium salt form of *Klebsiella* serotype K9 polysaccharide is shown in Figure 8. The photograph has a layer line spacing of 4.13 nm with meridional reflections occurring only on layer lines $l = 3n$. This is most simply interpreted as resulting from a three fold helical structure with an axially projected chemical repeat of 1.377 nm. This observed value is some 15% less than the maximum theoretical extension of 1.63 nm predicted for one chemical repeating sequence with all the vectors joining successive glycosidic oxygen atoms lined up precisely.

Trial structures for both left and right handed helices were generated, on the basis of stereochemical acceptability and potential hydrogen bond formation. The most favourable backbone conformation found was a left handed helix containing three intrachain hydrogen bonds. These hydrogen bonds are O(5)...H-O(4) across the Rha-(1,3) - Rha linkage, O(5)...H-O(4) across the Rha-(1,3) - Gal linkage and O(2)-H...H-O(2) across the Gal-(1,3) - Rha linkage. It was also found possible to incorporate an

O(5)...H-O(3) hydrogen bond across the GlcUA-(1,4)-Rha linkage to the side chain (Figure 9).

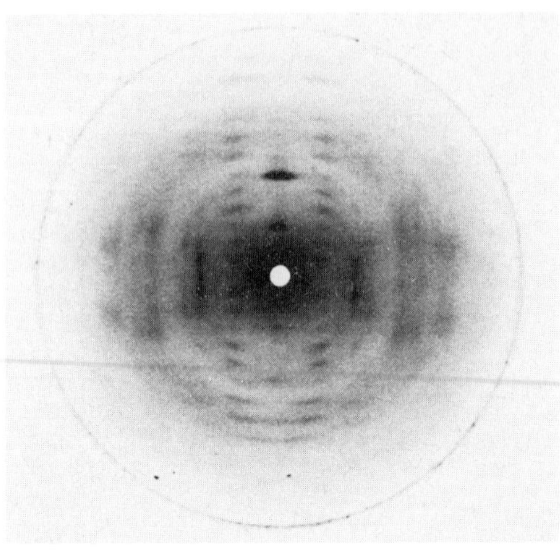

Fig. 8 X-ray diffraction pattern obtained from an oriented film of the sodium salt form of K9. The layer line spacing is 4.13 nm with meridional reflections occurring on layer lines with l = 3 n indicating a three-fold extended helical conformation.

This is the first instance to our knowledge that a polymer containing α-L-rhamnose residues has been crystallized. Rhamnose is not a commonly occurring sugar in polysaccharides and so its presence in the *Klebsiella* K9 serotype deserves some attention. The existence of this residue in the $^1C_4$ chair conformation is consistent with our experimental results. The repeating sequence of K9 serotype polysaccharide contains three rhamnose residues, one linked α 1,2 and two linked α 1,3. The former has been considered (together with β 1,2-linked rhamnose) to have a kinking function in pectic substances [Rees and Wight, 1971] where it occasionally interrupts a sequence of α 1,4-linked D-galacturonic acid. However, such a kinking effect is not evident in the condensed phase of K9 polysaccharide, which exhibits a fairly extended conformation in common with many other plant and animal polyuronides [Atkins *et al.*, 1974] and bacterial polysaccharides. Indeed in our proposed model, the virtual bond across the 1,2-linked α-L-rhamnose residue lies close to and nearly parallel with the helix axis as shown in Figure 10.

The only charged appendage is the carboxyl group of the glucuronic acid side chain. Examination of the molecular model, illustrated in Figure 9, shows that this charged group lies well out (~0.9 nm) from the helix axis, almost on the radial periphery of the molecule. This side chain

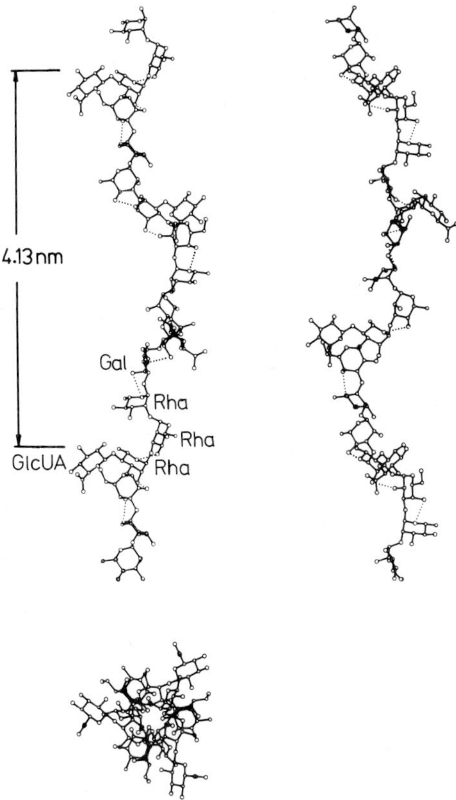

Fig. 9 Computer generated projections of a left-handed three-fold helix for K9.

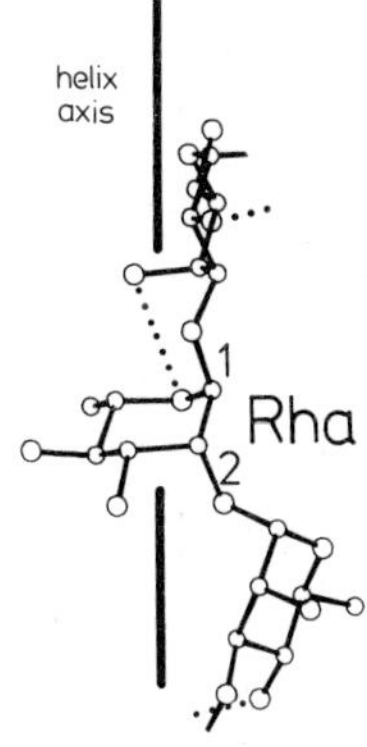

Fig. 10 An enlargement of part of the conformation shown in Figure 9 showing the 1,2-linked rhamnose residue. Note that such a glycosidic linkage does not kink the backbone similar to that postulated for rhamnose in pectin [Rees and Wight, 1971].

positioning is found when the Rha-O(3)...H-O(2)- GlcUA hydrogen bond is incorporated across the glycosidic linkage to the side chain. If this stabilizing hydrogen bond is removed it is possible for the side chain to rotate and change the vertical position of the carboxyl group but still disposing it near the radial periphery. Even if we were to release the backbone stabilizing intra-chain hydrogen bonds we would not expect substantial variation in this side group behaviour, since the extended nature of the chain gives rise to this feature. The choice between different side chain conformations will have to await further detailed packing calculations, but at this stage we favour a model incorporating the stabilizing hydrogen bond.

Serotype K16

The chemical repeating sequence in *Klebsiella* K16 polysaccharide has been determined by Chakraborty and Niemann [S. Stirm, personal communication], and the details are shown in Figure 11. It is a polytetrasaccharide with three residues in the backbone, including the charged glucuronic acid, and a single galactose residue as a side chain. The glucose and galactose residues are likely to be in the

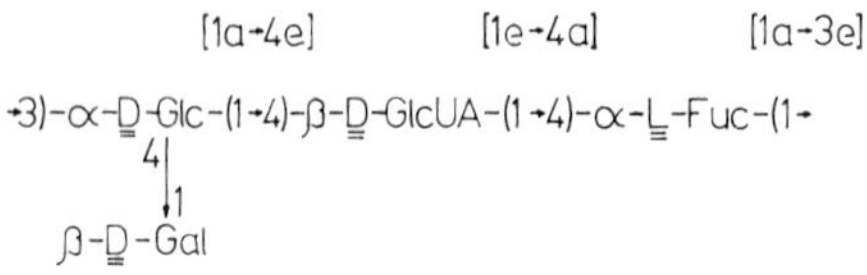

[1a→4e] [1e→4a] [1a→3e]

→3)-α-D-Glc-(1→4)-β-D-GlcUA-(1→4)-α-L-Fuc-(1→
4
↑
1
β-D-Gal

Fig. 11 Repeating chemical structure for *Klebsiella* serotype K16.

normal $^4C_1$ conformation but the α-L-fucose residue we expect to exist in the alternative $^1C_4$ chair, as predicted by the single crystal refinement of Longchambon *et al.* [1975]. This gives rise to the linkage geometry in Figure 11. Thus the backbone consists of α-L-Fuc linked through 1 ax, 4 ax, β-D-GlcUA through 1 eq, 4 eq and α-D-Glc through 1 ax, 3 eq with the β-D-Gal attached to it via a 1 eq, 4 eq, linkage.

Figure 12 shows the X-ray fibre diffraction pattern obtained from a sample of K16 polysaccharide after stretching a film of the material [Elloway, 1977]. It shows a strong meridional on the sixth layer line and a weaker meridional on the third layer line at a spacing of 1.292 nm. This is a contraction of 10% on the maximum theoretical extension of 1.414 nm and so the simple interpretation of the pattern is a three fold helical conformation with a projected chemical repeat of 1.291 nm.

Attempts have been made to computer generate sterically acceptable left and right handed helices with the observed pitch and containing stabilizing intramolecular hydrogen bonds. It was found impossible to build right handed models without any unacceptable short contacts across the linkages. Thus the most favourable model is a left-handed conformation as illustrated in Figure 13. This contains two hydrogen bonds in the backbone, namely an O(5)...H-O(3) bond across the β-D-GlcUA(1,4)-α-L-Fuc linkage and an O(2)-H...H-O(2) bond across the α-L-Fuc-(1,3)-α-D-Glc linkage. No simple stabilizing hydrogen bond could be incorporated simultaneously across the other backbone linkage, α-D-Glc-(1,4)-β-D-GlcUA, although there is a possibility of an O(6)-H...H-O(6) attraction. A hydrogen bond O(5)... H-O(3) across the β-D-Gal-(1,4)-α-D-Glc linkage to the side chain may also be incorporated.

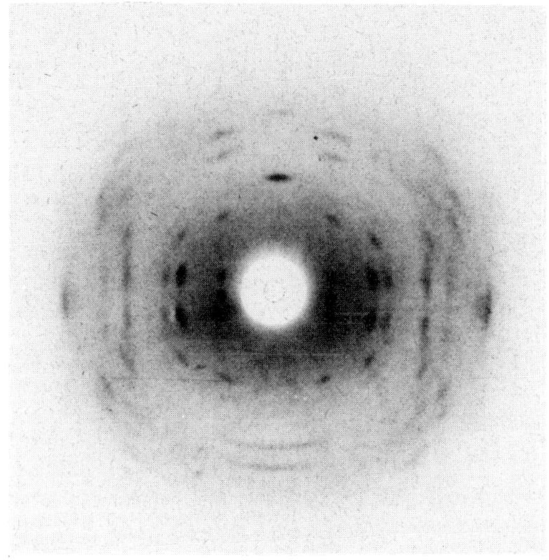

Fig. 12 X-ray fibre diffraction pattern obtained from an oriented film of the sodium salt of K16. The meridional reflections occur on layer lines with 1 = 3 n indicating a three-fold helical conformation for this serotype.

Serotype K25

Klebsiella serotype K25 polysaccharide has a tetrasaccharide repeating sequence [Niemann *et al.*, 1977] consisting of two β-D-Glucose residues, one β-D-Galacturonic acid and one β-D-Galactose residue (see Figure 14). All these residues are expected to exist in the normal $^4C_1$ chair conformation. The backbone has a disaccharide repeat unit (one glucose residue and one galactose residue) and there is a side chain containing the other two residues of the repeat. Thus the backbone linkage geometry is Gal-(1 eq - 4 eq)-Glc-(1 eq - 3 eq)-Gal with the side chain attached to the axial O(4) atom of the galactose residue, giving

3.87 nm
Fuc
GlcUA
Glc
Gal
1nm

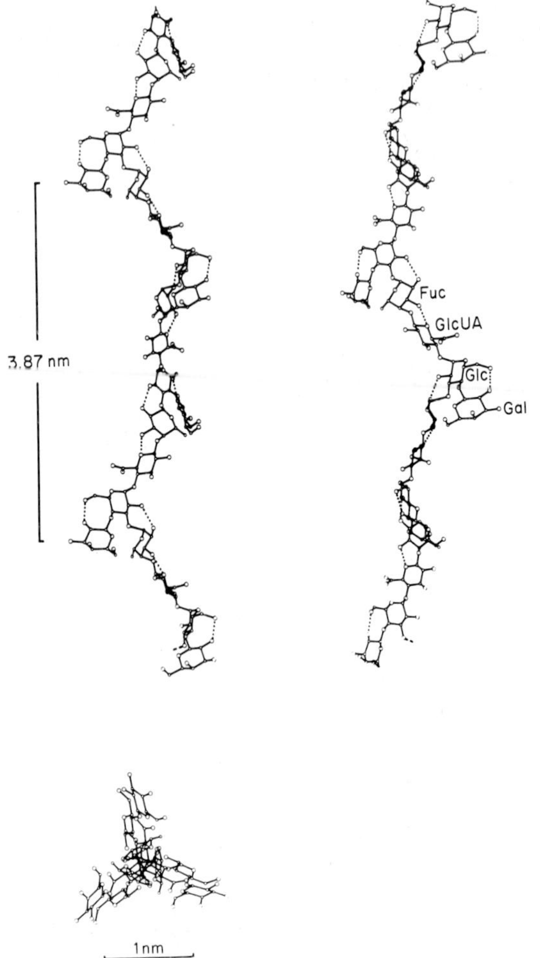

Fig. 13 Computer generated projections of a left-handed three-fold helical conformation for K16. The side appendage is a galactose residue.

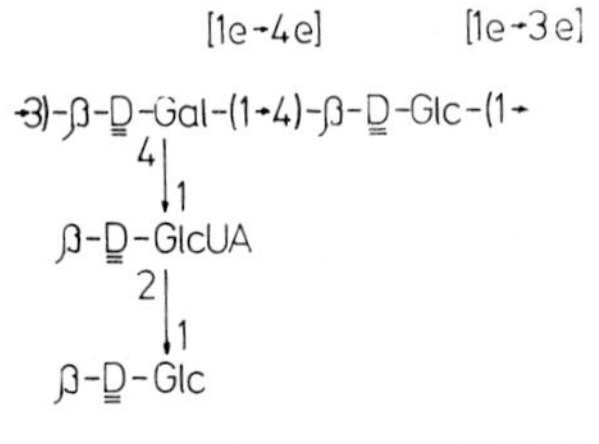

[1e→4e]
[1e→3e]
→3)-β-D-Gal-(1→4)-β-D-Glc-(1→
4
1
β-D-GlcUA
2
1
β-D-Glc

Fig. 14 Repeating chemical structure for *Klebsiella* serotype K25.

side chain linkage geometry of Glc-(1 eq - 2 eq) - GlcUA - (1 eq - 4 ax)-Gal. It is interesting to note that the backbone linkage geometry is the same as that found in the connective tissue polysaccharides hyaluronic acid, chondroitin sulphate and dermatan sulphate.

The X-ray fibre diffraction pattern (Figure 15) obtained from a stretched film of the sodium salt form of K25 polysaccharides has a layer line spacing of 2.91 nm with meridional reflections occurring only on those layer lines (l) with l = 3 n. This is most simply interpreted as resulting from a three fold helical structure with an axially projected chemical repeat of 0.97 nm. This is some 5% less than the maximum theoretical extension of 1.02 nm predicted for one chemical repeating sequence when the vectors joining successive glycosidic oxygen atoms line up.

Attempts were made to build trial helices meeting the observed helical parameters and incorporating hydrogen bonds across the backbone linkages. Both left and right handed helices were generated and it was found that left handed helices were more favourable, as had previously been

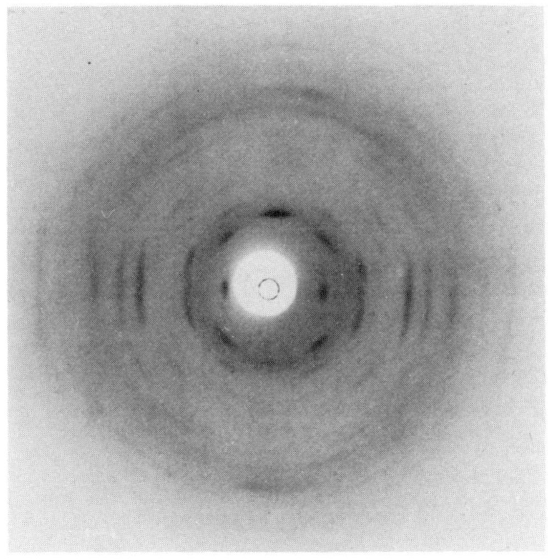

Fig. 15 X-ray fibre diffraction pattern obtained from oriented film of the sodium salt of K25. Meridional reflections occur on layer lines with l = 3 n indicating a three-fold helical conformation.

observed in the similar backbone conformations of connective tissue polysaccharides [Atkins *et al.*, 1974]. This left handed conformation (Figure 16) allowed the incorporation of stabilizing hydrogen bonds across both backbone linkages, namely an O(5)...H-O(2) hydrogen bond across the β 1,4 linkage and an O(2)-H...H-O(2) hydrogen bond across the β 1,3 link. Initial interpretation of the X-ray diffraction data gives no information regarding the conformational angles of the side chain. These torsion angles were therefore fixed for our trial helical model on the basis of hydrogen bonding and steric restrictions only. Thus the linkage between the two residues of the side chain was fixed so that the hydrogen bond Glc-O(2)..H..O(3)-GlcUA was formed. The conformation at the linkage joining the side chain to the backbone was not so obviously defined since there are no potential hydrogen bonds. These torsional angles were eventually fixed in such a position that Gal-O(6) was 0.28 nm from GlcUA-O(5) giving the potential of forming such a hydrogen bond.

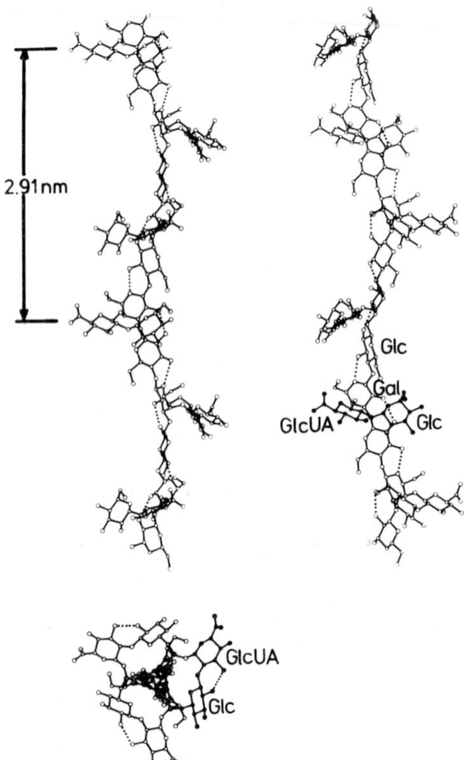

Fig. 16 Computer generated projections of a left-handed three-fold helical conformation for K25. The backbone geometry is similar to the connective tissue polysaccharides hyaluronic acid, chondroitin sulphates, dermatan sulphate and keratan sulphate. The atoms of the disaccharide side chain are shown as filled circles to help clarify the structure details. Note how the side chain wraps itself around the backbone (bottom left) and exposes the carboxyl group of the glucuronic acid residue.

Comparison of this backbone conformation with those seen in the connective tissue polysaccharides is worthwhile since the backbone linkage geometry is the same. In particular a left handed three fold structure of chondroitin-4-sulphate of similar pitch has recently been refined by Winter *et al.* [1977] and is comparable since this also contains a glucose and a galactose residue in the backbone. This structure was also found to incorporate the intramolecular hydrogen bond O(3)-H...O(5) across the β 1,4 linkage. However, since chondroitin sulphate contains an *N*-acetyl group in the 2 position of the galactose residue this affects the possibility of stabilizing the β 1,3 linkage with a hydrogen bond and no such bond was incorporated. It is also instructive to compare the K25 polysaccharide trial structure with the refined structures of hyaluronic acid. Three types of conformation have been refined recently, a four fold contracted single helix [Guss *et al.*, 1975], a three fold extended single helix [Winter *et al.*, 1975] and a contracted four fold double stranded helix [Sheehan *et al.*, 1977]. The double stranded helix contains no intramolecular hydrogen bond and probably relies on interchain interactions for its stability. In both the single helix structures the β 1,3 linkage is stabilized by a O(4)-H...O(5) hydrogen bond, but the comparison of this linkage is not so justifiable since the O(4) in hyaluronic acid is equatorial as opposed to its axial position in K25. Also in K25 this is a glycosidic linkage to the side chain and so there is no hydrogen free to denote. Instead in K25 an O(2)..H..O(2) hydrogen bond is postulated. At the β 1,4 link a O(3)-H...O(5) hydrogen bond is incorporated in our trial model of K25. This is also seen in the extended 3-fold structure of hyaluronate but not the contracted four fold structures. In the four fold single helix an N(2)..O(6) hydrogen bond is introduced across this linkage. For comparison native cellulose contains an O(3)-H..O(5) hydrogen bond and an O(2)...O(6) hydrogen bond across the β 1,4 linkage [Gardner and Blackwell, 1974]. We were therefore interested to look at the positions of the hydroxymethyl oxygen atoms of each residue to see whether they could form hydrogen bonds. At the β 1,4 backbone linkage we anticipate that O(2)...O(6) could occur in addition to O(3)...O(5) since this can take place in cellulose, although we found that two 0.28 nm bonds could not be formed simultaneously. The hydroxymethyl of the galactose residue could hydrogen bond to O(5) of the glucuronic acid residue. The carboxyl of the glucuronic acid residue is disposed on the periphery of the molecule and so is not in a position to form an intra-chain hydrogen bond although it is likely to be involved with inter-chain interactions and interactions with counterions and water. The oxygen atom of the hydroxymethyl group of the glucose side residue is at about 0.28 nm from the backbone Glc O(2) and Gal O(2), so there exists the possibility of three hydrogen bonds between these three atoms. It may be seen that the side chain does not extend far away from the helix axis but seems to lie close to the backbone (Figure 16).

Serotype K54

The chemical repeating structure of *Klebsiella* serotype K54 polysaccharide was established by Conrad *et al.* [1966] and is shown in Figure 17. The presence of both acetyl groups [Sutherland and Wilkinson, 1968] and formate [Sutherland, 1970b] have also been detected although both the extent of acetylation and formylation and positions of substitution remain unknown. Four possible positions of

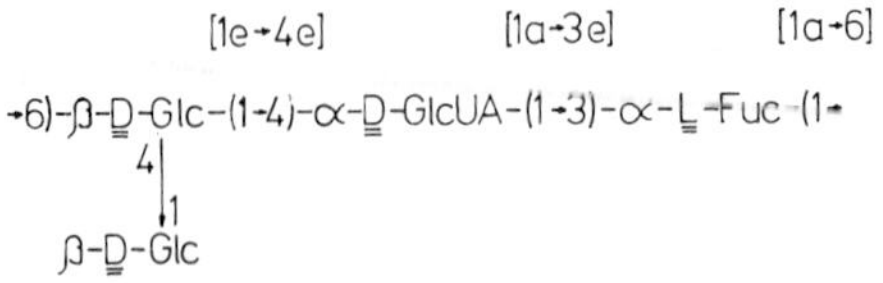

```
            [1e→4e]          [1a→3e]          [1a→6]
→6)-β-D-Glc-(1→4)-α-D-GlcUA-(1→3)-α-L-Fuc-(1→
        4|
         |1
    β-D-Glc
```

Fig. 17 Repeating chemical structure for *Klebsiella* serotype K54.

acetylation have been suggested: the 2 or 3 positions of the glucuronic acid residue or the 2 or 4 positions of the fucose residues. The two glucose and the glucuronic acid residues are likely to exist in the normal $^4C_1$ chair conformation whereas the α-L-Fucose residue, in common with the known single crystal structure [Longchambon *et al.*, 1975], is likely to be in the alternative $^1C_4$ chair conformation. This gives rise to the linkage geometry shown in Figure 17. The presence of a 1,6 linkage is a novel feature which introduces a third variable torsional angle to this linkage.

Samples of the sodium salt form of K54 after stretching gave an X-ray fibre diffraction pattern as illustrated in Figure 18 [Elloway, 1977]. This shows a strong meridional on the sixth layer line at a spacing of 0.618 nm and on tilting a weaker meridional reflection is observed on the third layer line at 1.236 nm. The maximum theoretical extension of the chemical repeating unit is about 1.45 nm, some slight uncertainty in this figure arising in assessment of the virtual bond length across the 1,6-linked residue. Thus the simplest interpretation of the diffraction pattern is a molecule with a three fold conformation and a projected chemical repeat of 1.236 nm, a contraction of about 17% from the maximum permissible extension.

The possibility of building both left and right handed stereochemically acceptable three fold helices with the observed pitch was investigated. The additional flexibility of the 1,6 linkage in the backbone was reduced by restricting the O(6) conformation to one of the three acceptable positions, namely gauche-gauche, trans-gauche and gauche-trans [Sundaralingham, 1968]. Left handed helices with any of these three O(6) conformations involved a marginally acceptable contact between C(4) of α-D-Glc and O(5) of β-D-Glc which did not occur in right handed models. Right handed helices could be built only if O(6) was placed in the gauche-gauche conformation (in practice it was found necessary to rotate O(6) by 2° from the gauche-gauche position to relieve one slight short contact). This conforma-

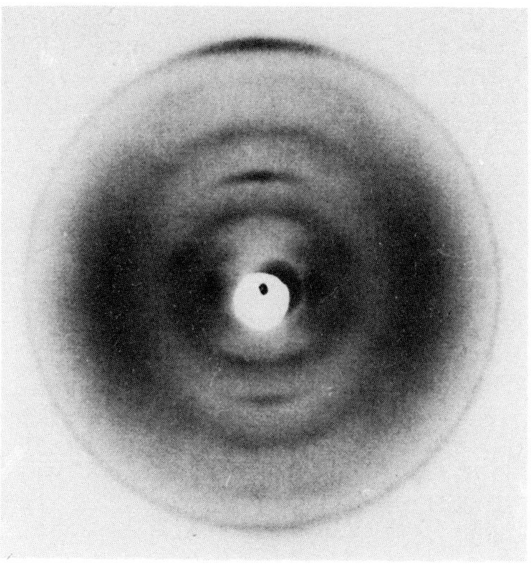

Fig. 18 X-ray fibre diffraction pattern obtained from an oriented film of sodium salt of K54. The reflections indicate a three-fold helical structure.

tion for O(6) is in fact similar to that found in the single crystal structure of the 1,6-linked disaccharide α-melibiose [Kanters *et al.*, 1976] and the trisaccharide raffinose [Berman, 1970]. The most favourable overall model for K54 is shown in Figure 19. This backbone conformation incorporates two hydrogen bonds, namely O(5)...H-O(3) across the β-D-Glc-(1,4)-α-D-GlcUA linkage, and O(2)-H...H-O(2) across the α-D-GlcUA-(1,3)-α-L-Fuc linkage. No intramolecular hydrogen bond could be formed to stabilize the 1,6 linkage. A hydrogen bond across the linkage to the side chain may also be incorporated: this is an O(5)...H-O(3) bond across the β-D-Glc-(1,4)-β-D-Glc link. A further feature of the model is the positioning of the carboxyl group close to the periphery of the helix, where it is most likely to be able to interact with water and counterions. The various suggested positions for the acetyl group were investigated briefly to see if any could be eliminated on stereochemical grounds, but it was found that none of the possible sites could be either eliminated or favoured [Elloway, 1977].

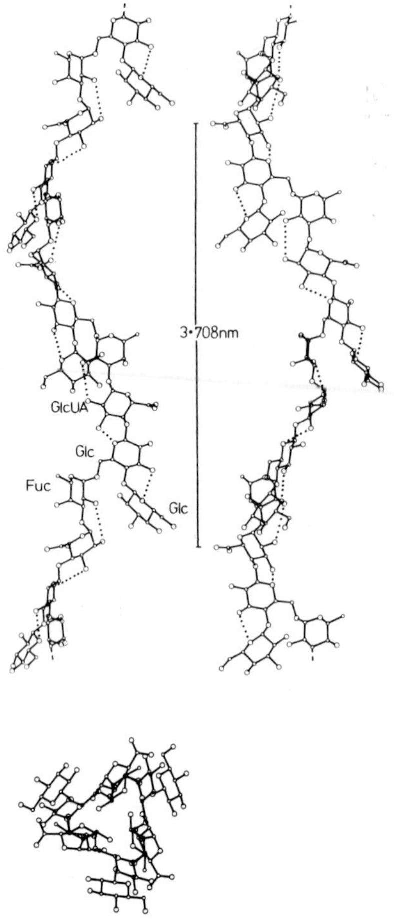

Fig. 19 Computer generated projections of a right-handed three-fold helical conformation for K54. An interesting feature of this serotype is a 1,6 glycosidic linkage within the backbone.

Serotype K57

K57 is a polytetrasaccharide consisting of three sugar residues in the backbone and with one single residue in the side chain. The detailed chemical covalent repeat has been established by Kamerling *et al.* [1975] and is given in Figure 20. The backbone consists of two neutral sugars and a uronic acid residue. This 1,3-linked α-D-galacturonic acid residue is attached to a 1,2-linked α-D-mannose residue followed by a 1,3-linked β-D-galactose residue. The side group (α-D-mannose) is attached to the uronic acid

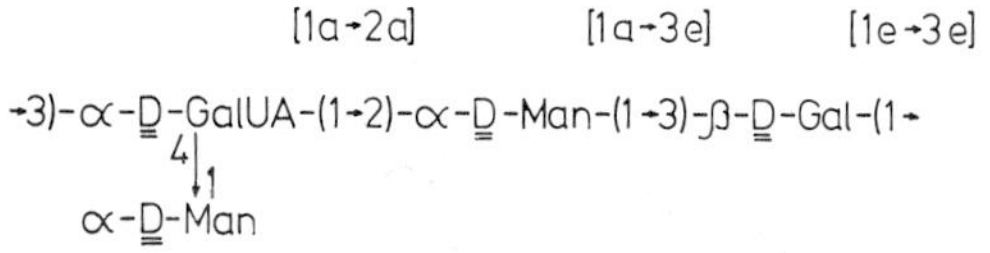

[1a→2a] [1a→3e] [1e→3e]

→3)-α-D-GalUA-(1→2)-α-D-Man-(1→3)-β-D-Gal-(1→
4↑1
α-D-Man

Fig. 20 Repeating chemical structure for *Klebsiella* serotype K57.

residue. It is anticipated that all the sugar residues exist in the normal $^4C_1$ chair conformation resulting in one 1,3-diequatorial glycosidic linkage, a 1,2 diaxial linkage and one 1 ax - 3 eq linkage (see Figure 20). In addition the mannopyranose residue is 1,4 diaxially attached. *A priori* this structure presents us with some novel glycosidic linkage geometries to examine and with the added complication of a small side chain. The maximum theoretical extension for the chemical repeat, following the method described earlier, is 1.27 nm.

The X-ray diffraction pattern from the K57 polysaccharide is shown in Figure 21. The layer line spacing was measured to be 3.429 nm with meridional reflections present only on layer lines 1 = 3 n. The simplest interpretation of this pattern is that the polysaccharide backbone forms a three-fold helix with a projected axial repeat of 1.143 nm. This value, 10% less than the maximum permissible, correlates with a single covalent repeat and suggests a fairly extended structure.

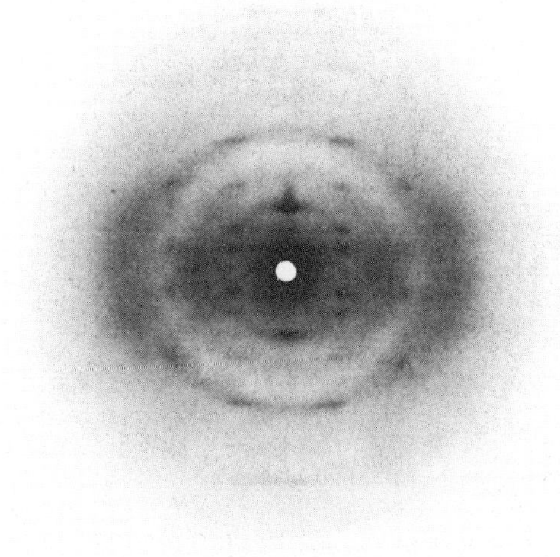

Fig. 21 X-ray fibre diffraction pattern obtained from the sodium salt form of K57. The basic features of the pattern indicate a three-fold helical conformation.

Trial models have been generated conforming to the helical symmetry and dimensions. Both left-handed and right-handed models have been generated using the techniques and criteria outlined above. Attempts were made to form the maximum number of intra-chain hydrogen bonds. It was found that no model could be constructed that included hydrogen bonds across all three backbone glycosidic linkages. Only a left-handed helix allowed the formation of two intra-chain hydrogen bonds in the backbone: α-D-GalUA-O(2)...O(3)-α-D-Man and α-D-Man-O(5)...H-O(2)-β-D-Gal. Further calculations incorporating energy minimization of the structure have confirmed that both right-handed and left-handed helical conformations (Figure 22) are stable structures. Cylindrically averaged Fourier transform calculations are unable to distinguish between the two models [Isaac *et al.*, 1978]. The left-handed model has the advantage of the two intra-residue hydrogen bonds.

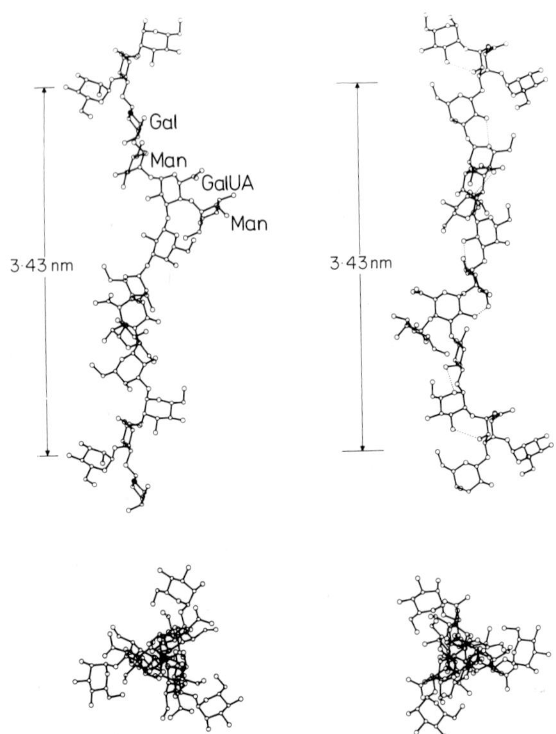

Fig. 22 Projections of molecular models for *Klebsiella* K57 with three-fold helical structures repeating in 3.43 nm: (a) the best right-handed model; and (b) the best left-handed model. The dotted lines show the hydrogen bonds that may be incorporated in this model (see text).

Serotype K63

As we shall show later in this section, the X-ray fibre diffraction studies of *Klebsiella* K63 polysaccharide were important in helping to establish the correct chemical repeating sequence. Thus it is instructive to consider some of these diffraction results before looking at the details of the covalent structure. In fact this is also the chronological order of the investigations since the X-ray patterns were obtained before the chemical structure was known [Elloway, 1977].

The X-ray fibre diffraction photograph obtained from K63 polysaccharide (in collaboration with H. Chanzy, Grenoble) is shown in Figure 23. This pattern has been indexed on a monoclinic unit cell with a = 1.025 nm, b = 1.210 nm, c = 2.37 nm and γ = 108.82°. Meridional reflections are observed on even order layer lines only, the first one occurring at 1.185 nm.

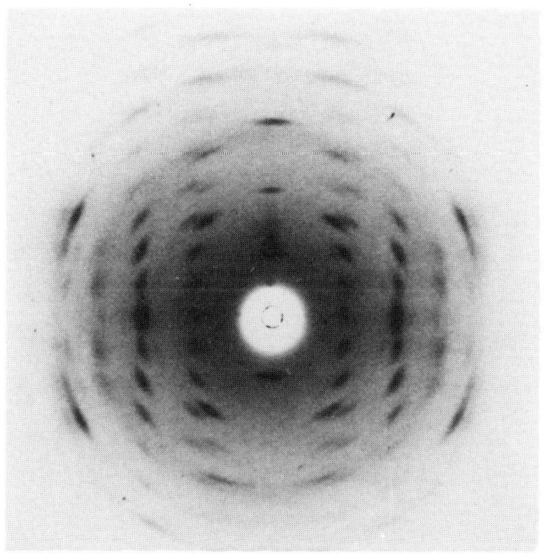

Fig. 23 X-ray fibre diffraction pattern obtained (in conjunction with Dr. H. Chanzy) for oriented film of the sodium salt of K63. The meridional reflections occur on even layer lines only indicating a two-fold helical conformation.

Preliminary investigations on the chemical structure of *Klebsiella* serotype K63 by J-P. Joseleau, Grenoble (personal communication) strongly favoured a covalent repeating sequence of all α 1,3 residues in a linear backbone. The analyses however were not complete and two possibilities needed to be considered. A linear pentasaccharide as shown in Figure 24 and a linear trisaccharide shown in Figure 26. In both cases only the three saccharides:

$$\rightarrow 3\,\mathrm{Gal}\,1 \xrightarrow{\alpha} 3\,\mathrm{GalUA}\,1 \xrightarrow{\alpha} 3\,\mathrm{Fuc}\,1 \xrightarrow{\alpha} 3\,\mathrm{Fuc}\,1 \xrightarrow{\alpha} 3\,\mathrm{Gal}\,1 \xrightarrow{\alpha}$$

Fig. 24 Provisional repeating pentasaccharide structure for *Klebsiella* serotype K63 when the X-ray diffraction pattern was obtained.

galactose, fucose and galacturonic acid were present.

The maximum theoretical extension for the pentasaccharide repeat is 2.115 nm and so to correlate the chemical repeat with the observed 1.185 nm projected repeat required an unusually great contraction of about 44%. This did not appear to fit in with the generally observed contractions of between 10% and 20%. However, models were

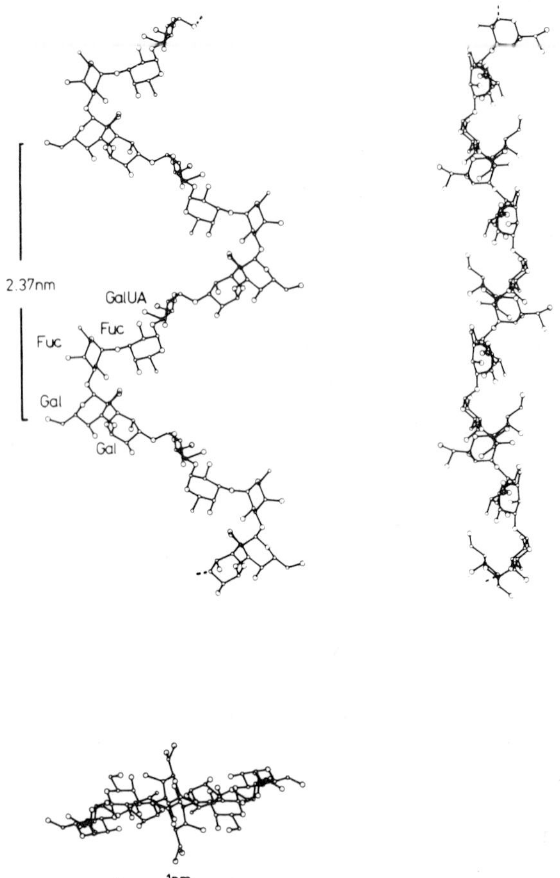

Fig. 25 Projections of a model for *Klebsiella* serotype K63 based on the pentasaccharide repeat (Figure 24). Such a model is unfavourable stereochemically and can be ruled out when all the X-ray diffraction information is taken into account.

built using this pentasaccharide repeat and fitting the observed helical parameters. Such models were found to be not very favourable stereochemically and generally ungainly as may be seen from a typical example in Figure 25. Indeed with a shape in cross section approximating to a rectangle of dimensions greater than 2 nm by 1 nm, it is difficult to believe that such a structure would pack into the observed unit cell. Further, density considerations predict that the unit cell could contain only one chain plus about 45 water molecules, another a suspicious feature.

[1a→3e] [1a→3e] [1a→3e]

→3)-α-D-Gal-(1→3)-α-D-GalUA-(1→3)-α-L-Fuc-(1→

Fig. 26 Correct repeating trisaccharide structure for *Klebsiella* serotype K63.

The maximum theoretical extension of the trisaccharide repeat (Figure 26) is 1.269 nm. Such a structure would give rise to an extended molecule with only a 6% contraction, and in the absence of side chains two such chain segments could pass through the unit cell, which would also contain about thirty water molecules. Model building was undertaken using the trisaccharide chemical repeat compatible with the observed helical parameters. Of those generated the most favourable stereochemically is shown in Figure 27. This contains hydrogen bonds across two of the linkages but it was not found possible to simultaneously incorporate one across the α-D-GalUA-(1,3)-α-L-Fuc linkage. These hydrogen bonds are O(2)-H...H-O(2) across the α-L-Fuc(1,3)-α-D-Gal linkage and O(5)...H-O(2) across the α-D-Gal-(1,3)-α-D-GalUA link, with a further possibility at this linkage of an O(6)-H...O(2) hydrogen bond. It may be seen that the carboxyl groups are situated on the periphery of the helix where they are likely to play a dominant role in interactions between chains and where they are readily accessible to water and counterions.

It is clear from the above considerations that the trisaccharide model is in agreement with the X-ray results. More recently the complete chemical analysis of *Klebsiella* serotype K63 has also confirmed a trisaccharide repeating sequence [J-P. Joseleau and M-F. Marais, personal communication]. This is an interesting example of how the X-ray diffraction methods have been useful in guiding interpretation of the chemical structure.

Discussion

We have presented examples of model building on a small selection of the large variety of microbial polysaccharides. This has recently become possible with the crystallization

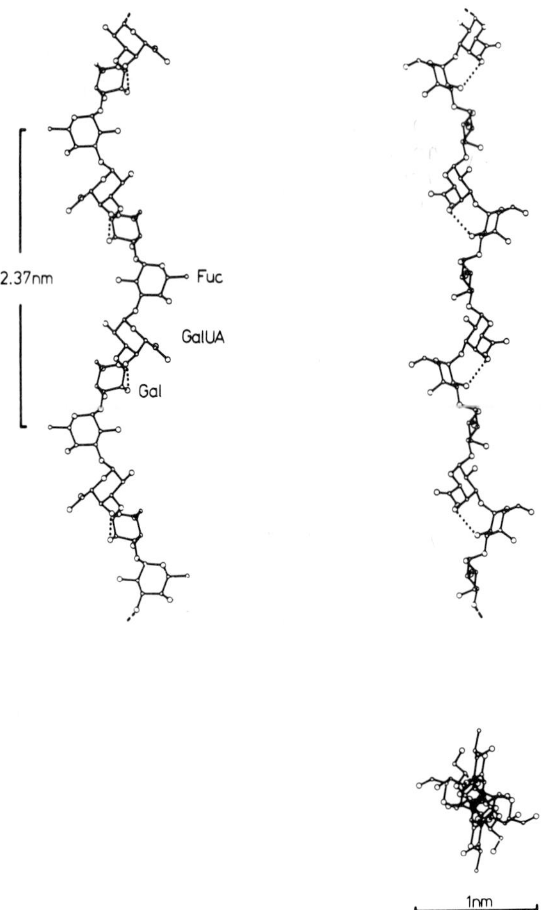

Fig. 27 Projections of a two-fold helix based on the trisaccharide repeat (Figure 26). Such a model is quite feasible stereochemically and is compatible with the X-ray diffraction information.

of these molecules in a form suitable for X-ray diffraction studies. It is only through these X-ray studies that the helical parameters necessary for meaningful model building can be obtained. Even though the examples we have chosen have substantially different primary structures, certain salient points are apparent. First, all the structures presented exist in extended conformations. This feature appears to be independent of the helical symmetry exhibited by the molecule.

In the preliminary stages of this model building exercise the criteria used for selecting the best conformation able to meet the observed helical parameters were the stereochemical constraints of conformational analysis coupled with the incorporation of a maximum number of stabilizing intramolecular hydrogen bonds. This approach has been justified [Isaac *et al.*, 1978] in the case of K57 polysaccharide which was found to favour a similar conformation when more sophisticated interaction terms were incorporated into the calculations. It is thus likely that these procedures are generally applicable, particularly to those polysaccharides in which the interchain packing is not well defined and hence whose conformations are most dependent on the intrachain interactions. Indeed it may be argued that model building of isolated chains is likely to produce helices which are more closely related to the conformations occurring in solution and *in vivo* conditions.

We are currently systematically investigating the conformation and structure of the other *Klebsiella* polysaccharides which will hopefully enable certain aspects of polysaccharide conformation to be anticipated from the primary structure. The crystallization and X-ray diffraction analysis of such a large array of complex polysaccharides offers the polysaccharides field a unique opportunity to examine the behaviour of particular glycosidic linkage geometries and a selection of saccharide units. From an immunological view-point it is known that certain *Klebsiella* bacteriophage depolymerize specific *Klebsiella* serotypes. For example *Klebsiella* bacteriophage No. 13 depolymerizes *Klebsiella* serotypes K2, K13, K22 and K37 [Niemann *et al.*, 1978]. It will be useful, when all these four serotypes are crystallized, to examine their detailed three-dimensional structure and ascertain the spatial distribution of particular groups, such as carboxyl, in relation to the glycosidic linkage being cleaved.

Acknowledgements

We wish to thank Professor G.G.S. Dutton, Dr. J-P. Josleau, Professor S. Stirm, Dr. I. Sutherland for supplying us with *Klebsiella* serotypes, and Dr. K.H. Gardner for his help and advice with computing the various conformations. We also thank the Science Research Council for financial support for D.H.I. and H.F.E.

References

Angyal, S.J. (1968). Conformational Analysis in Carbohydrate Chemistry. I. Conformational free energies. The conformations and α:β ratios of Aldopyranoses in aqueous solution. *Australian Journal of Chemistry* **21**, 2737-2746.

Angyal, S.J. (1969). The Composition and Conformation of Sugars in Solution. *Angewandte Chemie International Edition, England* **8**, 157-166.

Arnott, S. and Scott, W.E. (1972). Accurate X-ray Diffraction Analysis of Fibrous Polysaccharides containing Pyranose Rings. Part I. The linked-atom approach. *Journal of the Chemical Society Perkins*

Transactions II, 324-335.

Arnott, S. and Wonacott, A.J. (1966). The refinement of the crystal and molecular structures of polymers using X-ray data and stereochemical constraints. *Polymer* **7**, 157-166.

Atkins, E.D.T., Isaac, D.H., Nieduszynski, I.A., Phelps, C.F. and Sheehan, J.K. (1974). The polyuronides: their molecular architecture. *Polymer* **15**, 263-271.

Atkins, E.D.T. and Mackie, W. (1972). Criteria for the crystallisation of polysaccharides. *Biopolymers* **11**, 1685-1691.

Atkins, E.D.T. and Sheehan, J.K. (1972). Structure of hyaluronic acid. *Nature New Biology* **235**, 253-254.

Berman, H.M. (1970). The Crystal Structure of a Trisaccharide, Raffinose Pentahydrate. *Acta Crystallographica* **B26**, 290-299.

Conrad, H.E., Bamburg, J.R., Epley, J.D. and Kindt, T.J. (1966). The Structure of the *Aerobacter aerogenes* A3(S1) Polysaccharide. II. Sequence Analysis and Hydrolysis Studies. *Biochemistry* **5**, 2808-2817.

Dutton, G.G.S. and Yang, M-T. (1973). The Structure of the Capsular Polysaccharide of *Klebsiella* K-type 5. *Canadian Journal of Chemistry* **51**, 1826-1832.

Elloway, H.F. (1977). X-ray Fibre Diffraction Studies of a Variety of Animal, Plant and Bacterial Polysaccharides. Ph.D. Thesis, University of Bristol.

Gardner, K.H. and Blackwell, J. (1974). The Structure of Native Cellulose. *Biopolymers* **13**, 1975-2001.

Guss, J.M., Hukins, D.W.L., Smith, P.J.C., Winter, W.T., Arnott, S., Moorhouse, R. and Rees, D.A. (1975). Hyaluronic Acid: Molecular conformations and interactions in two sodium salts. *Journal of Molecular Biology* **95**, 359-384.

Isaac, D.H., Gardner, K.H., Atkins, E.D.T., Elsasser Beile, U. and Stirm, S. (1978). Molecular structures for microbial polysaccharides. X-ray diffraction results for *Klebsiella* serotype K57 capsular polysaccharide. *Carbohydrate Research* **66**, 43-52.

Kamerling, J.P., Lindberg, B., Lönngren, J. and Nimmich, W. (1975). Structural studies of the *Klebsiella* Type 57 capsular polysaccharide. *Acta Chemica Scandinavica* **B29**, 593-598.

Kanters, J.A., Roelofsen, G., Doesburg, H.M. and Koops, T. (1976). The crystal structure of a disaccharide, α-melibiose monohydrate (O-α-D-galactopyranosyl - (1-6)-α-D-glucopyranoside). *Acta Crystallographica* **B32**, 2830-2837.

Kauffmann, F. (1966). *The Bacteriology of Enterobacteriaceas*. Copenhagen: Munksgoord.

Lindberg, B., Lönngren, J., Thompson, J.L. and Nimmich, W. (1972). Structural Studies of the *Klebsiella* Type 9 Capsular Polysaccharide. *Carbohydrate Research* **25**, 49-57.

Longchambon, P.F., Ohannessian, J., Avenel, D. and Neuman, A. (1975). Structure cristalline du β-D-galactose et de l' α-L-fucose. *Acta Crystallographica* **B31**, 2623-2627.

McGeachin, H. McD. and Beevers, C.A. (1957). The Crystal Structure of α-Rhamnose Monohydrate. *Acta Crystallographica* **10**, 227-232.

Niemann, H., Beilherz, H. and Stirm, S. (1978). Kinetics and Substrate Specificity of the Glycanase Activity associated with Particles of *Klebsiella* Bacteriophage No. 13. *Carbohydrate*

Research **60**, 353-60.

Niemann, H., Kwiatkowski, B., Westphal, U. and Stirm, S. (1977). *Klebsiella* Serotype K25 Capsular Polysaccharide: primary structure and depolymerisation by a bacteriophage-borne glycanase. *Journal of Bacteriology* **130**, 366-374.

Nimmich, W. (1968). Zur Isolierung und qualitativen Bausteinanalyse der K-Antigene von Klebsiellen. *Zeitschrift für Medizinische Mikrobiologie und Immunologie* **154**, 117-131.

Rees, D.A. and Wight, A.W. (1971). Polysaccharide Conformation. Part VII. Model building computations for α 1,4 galacturonan and the kinking function of L-rhamnose residues in pectin substances. *Journal of the Chemical Society* **B**, 1366.

Sheehan, J.K., Gardner, K.H. and Atkins, E.D.T. (1977). Hyaluronic Acid: a double helical structure in the presence of potassium at low pH and found also in the presence of cations ammonium, rubidium and caesium. *Journal of Molecular Biology* **117**, 113-135.

Sheldrick, B. (1976). The Crystal Structures of the α- and β-Anomers of D-Galactose. *Acta Crystallographica* **B32**, 1016-1020.

Sundaralingham, M. (1968). Some aspects of stereochemistry and hydrogen bonding of carbohydrates related to polysaccharide conformations. *Biopolymers* **6**, 189-213.

Sutherland, I.W. (1970a). Structure of *Klebsiella aerogenes* Type 8 Polysaccharide. *Biochemistry* **9**, 2180-2185.

Sutherland, I.W. (1970b). Formate, a New Component of Bacterial Exopolysaccharides. *Nature* **228**, 280.

Sutherland, I.W. and Wilkinson, J.F. (1968). The Exopolysaccharide of *Klebsiella aerogenes* A3(S1) (Type 54). The isolation of the *O*-acetylated octasaccharide, tetrasaccharide and trisaccharide. *Biochemical Journal* **110**, 749-750.

Williams, D.E. (1969). A Method of Calculating Molecular Crystal Structures. *Acta Crystallographica* **A25**, 464-470.

Winter, W.T., Smith, P.J.C. and Arnott, S. (1975). Hyaluronic Acid: Structure of a fully extended 3-fold helical sodium salt and comparison with the less extended 4-fold helical forms. *Journal of Molecular Biology* **99**, 219-235.

Winter, W.T., Arnott, S., Isaac, D.H. and Atkins, E.D.T. (1978). Chondroitin 4-sulfate: the structure of a sulfated glycosaminoglycan. *Journal of Molecular Biology* 125, 1-19.

Chapter 8

ECONOMIC VALUE OF BIOPOLYMERS AND THEIR USE IN ENHANCED OIL RECOVERY

A. GABRIEL

Shell International Petroleum Company Ltd., London, UK

Introduction

When an exploratory well has been drilled, in the swamps of Borneo or in the North Sea, and oil discovered on a commercial scale, the assumed reaction is that the problems of the oil companies are over. This conveniently overlooks the hard and increasingly expensive efforts to delineate the field, to build and install production platforms and to transport the oil and gas to where it is wanted. This is especially true of localities like the North Sea where the initial investment costs are currently between $6,000 and $10,000 per daily barrel.

During the first period of actual production, the rate of oil flow normally increases rapidly as a result of new wells coming in and usually reaches a peak a few years after completion of the drainage well pattern. From then on the flow rate starts to fall, mainly because the reservoir pressure is declining, but also depending on the depletion rate and the geological aspects of the particular structure. Gradually less and less natural energy is available to push the oil in the reservoir towards the well system and thereafter through the well-pipes up to the surface.

The first remedy to keep up the production is to assist nature in lifting of the oil to the surface, by the installation of pumps, for example. However, even after applying artificial lift, as this is called, the pressure in the reservoir continues to decline and the rate of production may well reach a stage where it will no longer be economic to continue natural flow production of the well system. This is normally the end of the primary production phase.

Historically the next step was to assist nature in pushing the oil to the well by injecting water or gas. The additional oil which could be produced in this way is called Secondary Oil. Nowadays it is common practice not to wait with the injection of water or gas until the end of the primary production stage, but to start far earlier and to maintain the pressure at a higher level. In this way higher overall production levels can be maintained and this has the advantage that the production life of the field can be shortened, without loss in ultimate cumulative production.

However, in the case of water injection, a time comes when the water breaks through the oil bank and more and more of it appears in the producing well until considerably more water is pumped up than oil. The producing structure is then rapidly approaching the end of the secondary production phase and often, unfortunately, the end of its commercial life. At this stage, it is important to ask: How much oil is still left in the ground? The astonishing answer is that on average for the whole world it is about two thirds. In other words, for every barrel of oil actually produced, another two remain in the ground, in a sense lost and effectively locked up in a near impregnable safe. For the larger North Sea fields, the situation is more favourable but even here some 40 - 50% is still left underground.

A small improvement in the overall extraction level would be sufficient to improve the world's reserves significantly whilst the viability of some marginal new ventures would also be enhanced. The use of polymers in Enhanced Oil Recovery (E.O.R.) Systems is just one of the potential keys to partially unlocking some of the many oil safes around the world.

Terminology

The following terms are customarily used [see Mayer-Gurr, 1976]. Primary production is the amount of oil which can be produced by using only the natural energy available in the reservoir to push the oil in the reservoir towards the well. Secondary production is the additional amount of oil which can be produced above the primary displacement by using fluid injection as a means of supplying the additional energy in the reservoir for pushing the oil towards the well. A widely employed method, water flooding, involves the injection of water to maintain or increase reservoir pressure and to displace the crude oil. This has now been in use for several decades. Another is gas injection in which natural gas is reinjected to maintain, or increase reservoir pressure and to displace the crude. This is increasingly used where no pipelines are available for the associated gas. By the application of pumps or gaslift, lower flowing pressures can be realised at the bottom of the well and consequently higher production rates can be achieved at the same reservoir pressure than without application of artificial lift.

There remains the possibility of third stage, frequently referred to as Tertiary Recovery, where more elaborate methods are used to improve the recovery of oil which would otherwise be left in place. Currently both secondary and tertiary stages are simply linked together in the phrase Enhanced Recovery (Figure 1).

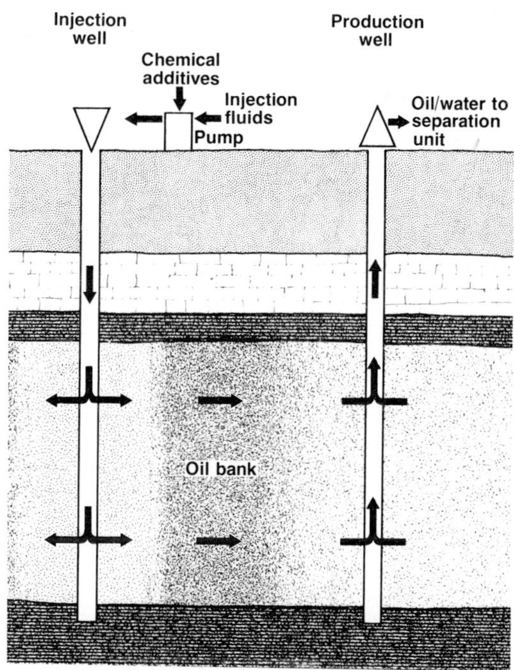

Fig. 1 A schematic illustration of different enhanced recovery methods.

Types of Enhanced Recovery Systems

The various types of E.O.R. systems [Herbeck *et al.*, 1976a, b; Mayer-Gurr, 1976] may be summarized as follows:

i) *Thermal methods* - involve heating by steam (commercial) or burning (developmental) of a portion of the crude *in situ* to lower the viscosity of the crude in the reservoir.

ii) *Miscible drive* - it is possible to inject natural gas or CO_2 under pressures that result in miscible or near miscible conditions. This method reduces the oil normally left behind in the reservoir after primary depletion or ordinary gas injection.

iii) *Caustic flooding* - a developmental method of injecting an alkaline solution to react with the crude and form "*in situ* surfactants" which in turn free the oil.

iv) *Polymer flooding* - a developmental technique of injecting a water thickening polymer into the field to improve the sweep efficiency.

v) *Microemulsion or micellar systems* - an experimental method of injecting a water/brine/surfactant/co-surfactant mix to free residual oil, followed by injection of polymer-thickened water to push out the expanded oil emulsion bank formed.

Polymer flooding

This method, which is largely the subject of this paper, is the use, in one form or another, of thickened water drive or rather a thickened water front [Jennings *et al.*, 1971; Herbeck *et al.*, 1976a, b]. The problem with using water is that it often leads to break-through or fingering in the formation. The efficiency of the water injection can be improved by making the water more viscous which in turn gives a more stable displacement front to push out the oil.

This is a crude and simplified explanation of what in reality is a very complex process. Of course the oil is not in a convenient pool but dispersed throughout a porous sandstone or limestone rock. To give some idea of the complexities: one cubic foot of typical sandstone reservoir rock may have less than a gallon of oil, plus several pints of water, all distributed through a maze of microscopic pores having a total surface area of more than an acre.

A Polymer Flood is designed primarily to lower the mobility ratio of the displacing water so as to improve the sweep efficiency and avoid water fingering. This simple statement unfortunately introduces a series of complex ideas which may be expanded as follows:

The mobility of a given fluid is determined by the permeability of the rock to that fluid divided by the viscosity of the fluid. The mobility ratio is the mobility of the injected water (at residual oil saturation), divided by the mobility of the oil (at connate water saturation) and this is usually very high at 10:1, 50:1 and even higher ratios. A mobility ratio approaching unity (or less) is much more favourable to oil displacement and a reduction of a high, unfavourable ratio, as in a simple water flood, considerably improves the sweep efficiency thus increasing the total displacement of oil into the collecting well system [Jennings *et al.*, 1971; Herbeck *et al.*, 1976a, b; Mayer-Gurr, 1976].

Examination of the factors contributing to a favourable mobility ratio indicates that a Water Flood can be made more favourable with improved sweep efficiency by any of the following:

i) decrease the effective permeability of the reservoir rock to water;

ii) increase the effective permeability of the oil;
iii) decrease the oil viscosity (usually by heat);
iv) increase the viscosity of the injected water.

Items (i) and (ii) are largely theoretical and little can be done to improve the flow characteristics of the oil in most reservoirs except by thermal recovery systems. However, the use of polymeric thickening agents in the water flood can have a marked improvement on the mobility ratio at the oil/water front.

Microemulsion or micellar systems

A very similar situation exists for the polymer drive in the case of microemulsion or micellar flooding. The micro-emulsion/micellar slug in a sense scours out the residual oil, but the principal drive still has to be water and the micellar slug has to be protected by a mobility buffer which consists of a thickened water front immediately behind the micellar slug. In practice the viscosity of the poly-mer bank is graded from high viscosity next to the micellar slug or oil bank, down to a low value next to the main water drive. This grading is accomplished by varying the polymer concentration in the solution, say from 500 ppm down to 100 ppm or less. Such a graduated bank not only achieves a better mobility ratio at the interfaces but is also less costly in use [MacWilliams *et al.*, 1973; Doscher, 1977; Grist and Buckley, 1977; Hesselink, 1977; Stewart, 1977].

Requirements of a Thickening Agent or Viscosifier

The technical requirements of a thickening agent for E.O.R. applications are five-fold [MacWilliams *et al.*, 1972; Arnold, 1975; Kelco, 1977]:

i) the thickener must produce a substantial increase in viscosity at very low concentrations, but it should not produce a viscous gel which would simply clog up the pores of the oil formation (Figure 2);
ii) ideally, the thickening agent should be pseudoplastic. In other words a small amount of shear force will produce a marked increase in flow and corresponding decreases in viscosity, but once the shearing forces are removed the thickening effect reasserts itself. It is fairly obvious that this characteristic is most desirable in the oil world, otherwise the products would be very difficult to handle through pumps and filters - as well as in the narrow, oil-filled capillaries of the producing structure itself, without requiring impractical pressures (Figure 3);
iii) the product must be chemically and thermally stable and effective over a period of years. The physical conditions in an oil well are not congenial, with temperatures approaching 100°C or more, together with

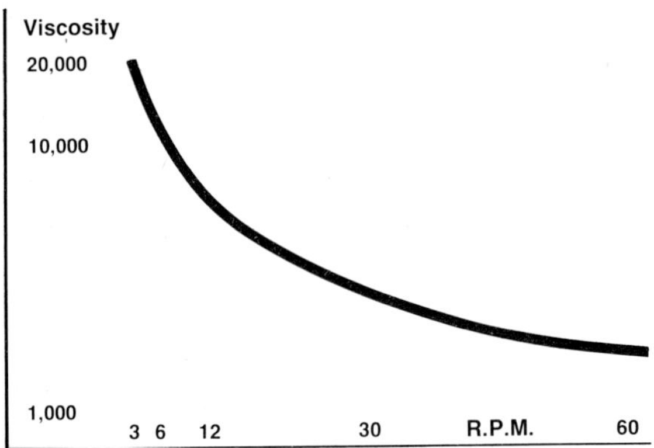

Fig. 2 Viscosity curve of a typical biopolymer at 1% (w/v) solution.

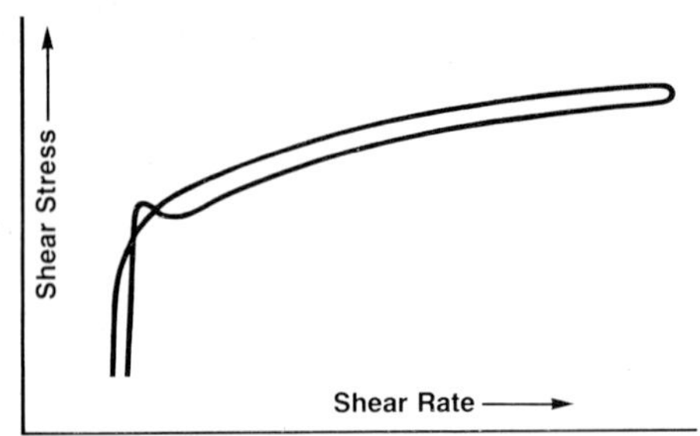

Fig. 3 Pseudoplastic behaviour of a biopolymer used in enhanced oil recovery.

very high salinity levels from surrounding water bearing formations;

iv) the product must also be consistent in quality. It is often said that oilmen will put anything down their wells, if they think they can get more oil out. However, this old adage should not be taken too literally as modern oilmen are quite fussy about any product which is introduced into their well systems;

v) it must be cost effective - measured by the value of the recovered oil over the life of the project.

It is obvious from the above list of primary requirements that not many products are likely to survive intact for several years in this rigorous context.

Major Types of Thickening Agents

Two major types of product have been used, at least experimentally, in enhanced recovery systems. These are the chemical polymers such as the polyacrylamides [Jennings *et al.*, 1971; MacWilliams *et al.*, 1973] and the biopolymers such as the polysaccharides [Doscher, 1977; MacWilliams *et al.*, 1973].

Polyacrylamides

These are chemically produced from acrylates and acrylamides and are essentially high molecular weight polymers either partially hydrolysed or copolymerized to give high molecular weights with different polar end groupings. They are relatively low cost and can be synthesized and modified in a whole variety of ways. There are literally hundreds of so-called polyacrylamides available and most of these have been used or offered in the oil world for some time. Their main limitation is that they are shear sensitive and have a limited stability over a long period of time in actual reservoir conditions. Being normally ionic in character, they are also salt sensitive and highly polar so that they can be adsorbed out of the system thus lowering their effectiveness at the front. However, their mode of action does not depend on simple viscosity considerations since they reduce mobility primarily by decreasing the permeability of the reservoir rock to water.

Polysaccharides

These are a more recent development and offer considerable promise in the field of enhanced oil recovery systems. The products are very high molecular weight hydrated natural polymeric agents with a sugar backbone structure. The most important types are produced as extra-cellular slimes by certain aerobic bacteria, the best known example being xanthan gum produced from the bacterium *Xanthomonas campestris* [see Sutherland, Chapter 1; Evans, Chapter 3 and Powell, Chapter 6].

Properties of xanthan gum Xanthan polymer has remarkable properties [MacWilliams *et al.*, 1973; Kelco, 1977] apart from its very high thickening power per unit concentration. Its rheology demonstrates a marked pseudoplastic effect. The solutions are therefore sensitive to shear and can thus be pumped and handled quite readily, but the thickening power rapidly reasserts itself when the shear forces are removed. It is also reasonably stable with respect to temperature - up to almost 90°C. For all practical purposes, solutions of xanthan gum are largely unaffected by the ionic environment and the products are relatively stable in highly saline solutions (unlike the polyacrylamides). Solutions of xanthan polymer can now be made in a low cell debris form with a uniform and consistent quality with reproducible rheological properties.

The major snag with this material is that by the nature of its production method, it is expensive and the current general market price is in excess of US$4 per pound.

Uses of and prospects for polysaccharides Xanthan gum is only one of several polysaccharide type materials which are of potential interest in enhanced oil recovery [Doscher, 1977; Arnold, 1977]. However, considerable experimental work now in hand both at various exploration and production laboratories and in limited field trials on xanthan type polymers, gives increasing confidence to oilmen as to its longer term performance when it comes to designing larger scale production trials.

Xanthan is by far the polysaccharide in largest scale production at present. Nevertheless, the current world production capacity does not exceed 15,000 tons. This is probably sufficient to supply the main current outlet, namely the food market and also most industrial applications, including drilling muds, that is all uses except enhanced oil recovery. However, this scale of production would be quite inadequate to supply the oil industry if enhanced oil recovery became widely established with no better product available than xanthan gum or current type chemical polymers.

Determining Factors in Enhanced Oil Recovery

Until recently, two types of polymer flooding enhanced oil recovery systems had been tried out experimentally [Arnold, 1977]: Solo Polymer systems, where only water and a polymer agent were used and Detergent Polymer systems, in which a detergent front is pushed through the oil bearing strata by a slug of polymer thickened water.

Both of these systems had been tried out on a limited scale in the U.S.A. even before 1973, but were largely academic exercises because in most cases the costs involved substantially exceeded the value of the recovered oil at that time.

The most important single limiting factor in enhanced oil recovery is not a technological one, but rather it is more a question of cost/benefit. Most enhanced oil recovery systems are very heavily front-end loaded projects, in other words, one often has to drill extra injection wells and also to provide the water injection and recovery systems compatible with good environmental standards - before one can even start the project.

The polymer itself is also not an insignificant cost item, even in a simple Solo Polymer system. In more sophisticated Chemical slug/Micellar systems, the total cost of the chemical products which have to be injected represents an even larger initial capital outlay. The benefits only accrue over a period of 3 to 7 years subsequent to the initial slug injection, assuming that all goes well. In fact, few field trials have so far fully lived up to their original expectations!

World oil prices have changed radically since the Autumn of 1973 but it took another two and a half years for "tertiary oil" in the U.S.A. to be classified as "new oil" so that the price received could get somewhere near the range where enhanced recovery could be justified on economic grounds. With continued increases in the price of oil, enhanced recovery systems will undoubtedly become more attractive not only in the U.S.A. but also elsewhere. This applies even more where production drilling is very expensive and necessarily restricted, as for example in the North Sea, where it is imperative to get the most out of each and every structure, bearing in mind the huge investment costs involved during the exploration and primary field production stages. Even a few per cent additional overall production could justify substantial enhanced recovery systems at projected future oil prices.

Estimates of Future Production by Enhanced Recovery

In trying to make a rational assessment of the scale and applicability of future activities in this field, it should be recognized that enhanced recovery using polymer or detergent/polymer systems are by no means suitable to all fields and structures.

Before the question of chemical requirements can be approached, estimates have to be made as to the likely future of various enhanced recovery systems. For historical economic and technical reasons, enhanced oil recovery will be more significant in the U.S.A. to start with and indeed, the U.S.A. will remain the dominant user of these systems and their associated chemical products for at least the next two decades. Middle East requirements are likely to be minimal in the immediate future but eventually as existing fields age and pressures fall, both secondary and tertiary recovery systems will be utilized on an increasing scale if production rates are to be maintained in line with world demands.

When discussing Europe, the dominant area is, of course, the North Sea which in production terms is a very young area and therefore still in the first flush of primary production in most fields and structures. Nevertheless, it is no secret that serious consideration is being given to various injection and pressure maintenance systems at a relatively early stage in some North Sea structures. However, the chemical requirements which may materialize will be quite modest by standards in the U.S.A., which remains the most promising market for the next two decades [Gulf Universities Research Consortium, 1973; Federal Energy Administration, 1976; National Petroleum Council, 1976; Noran, 1978].

Requirements for Polymers for Enhanced Oil Recovery

Once the likely pattern of future enhanced recovery systems has been estimated, it is then possible to determine the consequential future . . . . polymer, requirements provided the quantities of chemicals or polymer per barrel of recovered oil can be agreed upon.

The original, well-publicized, estimates by Gulf Universities Research Consortium [1973], Federal Energy Administration [1976], National Petroleum Council [1976] and Noran [1978] were made in the wake of the 1973/74 OPEC price increases but these estimates and most others of that period were grossly exaggerated. Figures of hundreds of thousands of tons of polymers and millions of tons of detergent/solvents were quoted and frequently requoted out of context. Fortunately not everyone believed these original figures and now several official and semi-official bodies have published substantially revised estimates for the future of various enhanced oil recovery systems.

The basic assumptions in most of these systems are now on reasonably common ground. Something like one pound of polymer is estimated to be required for every barrel of extra oil produced by the Solo Polymer method. Likewise in Detergent Polymer systems between 10 and 20 pounds of detergent, plus up to 5 pounds of solvent or co-surfactant may be required per barrel, plus about one pound per barrel of polymer.

Most U.S. estimates of polymer usage have simply assumed that polyacrylamides and biopolymers would be used in roughly equal proportions. However, our present view is that the most probable ratio of biopolymers to chemical polymers for enhanced recovery systems during the next decade will be of the order of 2:1, or even higher.

This makes a significant difference to predictions of future biopolymer usage in enhanced oil recovery systems. These differences have been incorporated into Table 1 which otherwise is based on the U.S. National Petroleum Council data.

Allowing for the change in the biopolymer/polyacrylamide ratio, as well as the significant increase in average price, it has been estimated that there could be a potential cumulative market for suitable biopolymers in the U.S. alone of some $5 billion by the year 2000 with oil at $15/bbl and at least four times that by the end of the century with oil at $25/bbl (calculated in 1976 money) (Figure 4). Translating this to the world outside the U.S.A., the probability is that a cumulative market of similar scale is likely to develop but its timing will be at least a decade later (Figure 5).

However, these figures should not be taken at their face value, since they represent an extremely optimistic point of view. Simple calculation shows that the estimated costs of the chemical components alone in a micellar flooding exercise are very high, being over $11.5 (with oil at $15/bbl) and over $15/bbl (with oil at $25/bbl). It is clear that the incentive for major commitments at present is marginal since the overall cost of the system might well be

TABLE 1

Enhanced oil recovery - U.S.A.

Miscellar and polymer flooding	Assumed cost of:	\$ 1976 Money \$15/bbl	 \$25/bbl
Cumulative oil recovery, 1976-2000 (10^9 bbl)		2.1	8.3
Chemical requirements		(lb per bbl)	
Surfactant		17.5	
Alcohol		5.25	
Polymer		1.0	
Total chemical requirements		(10^9 lb)	
Surfactant		36.8	145.3
Alcohol		11.0	43.6
Polymer		2.1	8.3
	Total	49.9	197.2
Average prices		(\$ per lb)	
Surfactant		0.43	0.59
Alcohol		0.20	0.27
Polymer		3.0	3.53
Cumulative chemical market		(\$$10^9$)	
Surfactant		15.8	85.7
Alcohol		2.2	11.8
Polymer		6.3	29.3
	Total	24.3	126.8

Source: U.S. National Petroleum Council
Modified Shell Estimates (ratio - biopolymer/chemical 2:1)

bbl = barrel

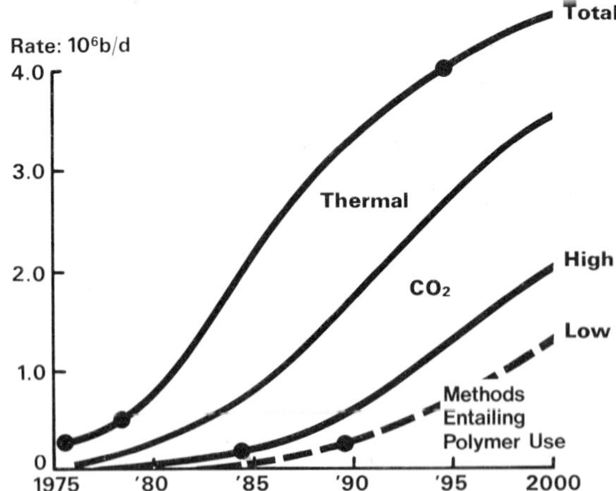

Fig. 4 Prospects for enhanced oil recovery in the U.S.A. as a result of the application of thermal and miscible drive enhanced recovery systems, compared with those entailing the use of polymers showing high and low estimates for the latter.

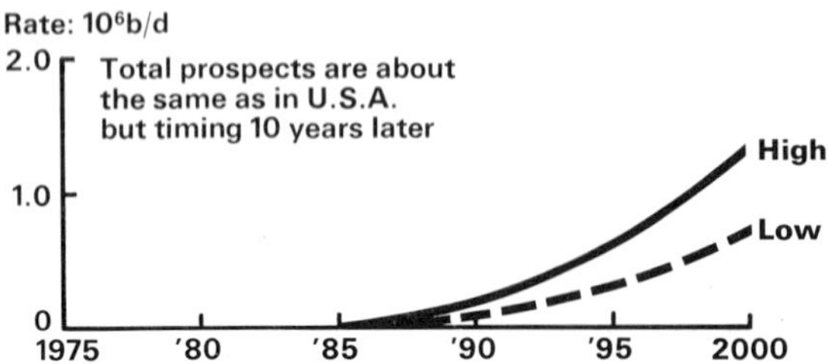

Fig. 5 Prospects for enhanced oil recovery outside North America as a result of the application of enhanced recovery systems entailing the use of polymers.

close to the current value of oil. The risks in such a situation are considerable, especially when so much of the expenditure has to be made at the beginning of the project.

Future of Chemical Polymers in Enhanced Oil Recovery

New and improved grades of polyacrylamides will be produced, and also completely new classes of complex polymers are already being looked at for possible performance in this field. There are some potential candidates, but none of them is likely to be easy or cheap to make and although a comparable performance to current polyacrylamides may be achieved, costs are likely to be of the same order as existing polysaccharides.

Future of Biopolymers in Enhanced Oil Recovery

In realistic terms, it is unlikely that any single polymer type will score the highest marks on all of the five requirements for a polymer for E.O.R. (see p.195) and it is no secret that many firms already in the biopolymer business and several oil and chemical companies are looking for possible performance improvements and product alternatives.

What is safe to predict, however, is that xanthan will continue to be used in a variety of experimental laboratory and field tests and will provide a sound basis of applicational know-how and confidence to the oilman. He may not like the current price, but in practice he tends to prefer an established commercial material rather than a product which is only produced on a pilot scale with an insufficient technical background of full scale field tests.

In looking at the future of biopolymers in enhanced oil recovery systems, one striking feature is their probable price/performance sensitivity. Current biopolymers are by chemical standards very expensive being already, in the U.S.A. $4 plus per pound bracket. At this kind of price level, it can be stated fairly categorically, that very remarkable performance characteristics are not only expected, but most necessary for such products to be taken seriously by oilmen.

There are two approaches to this problem. The first is the screening and selection of new biopolymer candidates with superior or at least different, performance characteristics. This is the natural response of most microbiologists and some interesting potential candidates have been isolated by ourselves and others. However, this is by no means enough as there exists between the laboratory stage and large scale commercial production an enormous technological gap. To produce such new polymers consistently in large volumes and at low prices is yet a further challenge.

Looking at fermentation and associated biological processes in general, it is apparent that, compared to normal chemical processes, most biological processes are low productivity systems. Both the fermentation and the work-up stages involve large and expensive capital equipment with an overall output only a fraction of what would be expected from analogous chemical hardware. The result is inevitably very high capital costs and therefore, by definition, high product costs as well.

So far, the general impression is that more research and development effort has been put into the isolation of new candidates rather than in tackling the more difficult problem of intrinsic low productivity of basic fermentation and extraction systems. It is difficult to be convinced that this balance, and it is a question of balance, is the right one. However, there is little doubt that substantial scope exists for improved bio engineering which has not yet been taken up with sufficient enthusiasm and on an adequate scale for significant commercial results to be achieved in this and associated fields.

References

Arnold, C.W. (1977). Enhanced oil recovery challenges chemists. *Chemtech* **12**, 762-765.

Arnold, C.W. (1975). *Chemicals for Advanced Oil Recovery.* Industrial Marketing Research Association Symposium 28/11. London.

Doscher, T.M. (1977). Tertiary oil recovery and chemistry. *Chemtech* **5**, 232-239.

Federal Energy Administration (1976). *The Potential and Economics of Enhanced Oil Recovery. Contract No. CO-03-50222-000.* Lewin and Associates Inc.: Washington D.C.

Grist, D.M. and Buckley, P.S. (1977). Petroleum displacement in porous media by saline micellar solution. *Journal of Petroleum Technology* **3**, 35-52.

Gulf Universities Research Consortium (1973). *E.O.R. Potential in the USA.* Gulf Universities Research Consortium: Houston, Texas.

Herbeck, E.F., Heintz, R.C. and Hastings, J.R. (1976a). Micellar solution flooding. *Petroleum Engineer* **6**, 44-56.

Herbeck, E.F., Heintz, R.C. and Hastings, J.R. (1976b). Polymer flooding. *Petroleum Engineer* **7**, 48-59.

Hesselink, F. Th. (1977). Effect of surfactant phase behaviour and interfacial activity on the recovery of capillary trapped residual oil. *Journal of Petroleum Technology* **3**, 11-21.

Jennings, R.R., Rogers, J.H. and West, J.T. (1971). Factors influencing mobility control by polymer solutions. *Journal of Petroleum Technology* **3**, 391-401.

Kelco (1977). *Xanthan Gum and Polymer XC.* San Diego: Kelco Company.

MacWilliams, D.C., Rogers, J.H. and West, T.J. (1973). Water soluble polymers in petroleum recovery. *Polymer Science and Technology* **2**, 105-126.

Mayer-Gurr, A. (1976). Petroleum Engineering. *Geology of Petroleum* **3**, 76-107.

National Petroleum Council (1976). *Enhanced Oil Recovery.* National Petroleum Council Workshop Publication: Washington D.C.

Noran, D. (1978). Annual production issue. *Oil and Gas Journal* **3**, 113-140.

Stewart, G. (1977). Review - surfactant processes, disparities between laboratory and pilot studies assessed. In *Enhanced Oil Recovery by Displacement with Saline Solutions*, pp.53-73. London: B.P. Educational Services.

Chapter 9

CHITIN, CHITOSAN AND THEIR DEGRADATIVE ENZYMES

R.C.W. BERKELEY

Department of Bacteriology, University of Bristol, Bristol, UK

Introduction

Two microbial polysaccharides not dealt with elsewhere in this book are chitin and its close relative chitosan. Another chemically related compound, bacterial peptidoglycan, which may be regarded as a chitin ether of lactic acid [Muzzarelli, 1977], is considered in Chapter 10 by Rogers. Both chitin and chitosan are synthesized by some fungi. The former is also produced by certain diatoms and protozoa and is synthesized in massive amounts by the arthropods which are the principal non-microbial source of this polymer.

Chitin and chitosan are of interest not only as substrates for the enzymes which attack them but as materials of commercial potential. This interest stems from the relatively recent realisation that these polysaccharides, unique amongst the naturally occurring types in having basic as opposed to neutral or acidic characteristics, represent a substantial renewable source of material with many possible applications. In addition, chitinous residues from certain industries are of considerable nuisance value because of the problems associated with their disposal. Thus any productive use to which they can be put is doubly useful.

Both polymers are eminently biodegradable substances and hence materials manufactured from them are likely to be subject to biodeterioration. Thus the enzymic degradation of these materials is of interest in this context as well as in those of their decay in the natural environment and of their physiological role.

In this chapter the distribution of chitin and chitosan and their potential uses are surveyed and the activities of chitinolytic and chitosanolytic enzymes in biodeterioration, decay and physiological processes are considered.

Chitin and Chitosan

There has been much discussion in the literature about the proper and precise meaning of these terms. It is as well then to explain at the outset the way in which they are used in this article.

Idealized chitin is a long chain polymer of *N*-acetyl-glucosamine (2-acetamido-2-deoxy-D-glucose) linked by β 1,4 bonds (Figure 1). Some authors have used the name chitan for material which corresponds with this ideal and contains no free amino groups [Falk *et al*., 1966] but the generally

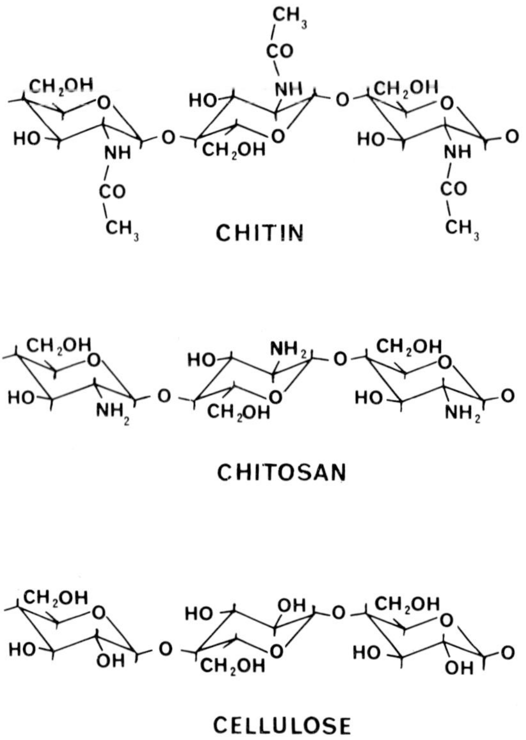

Fig. 1 The chemical structures of the repeating units of idealized chitin, chitosan and cellulose.

accepted usage is that the pure polymer as well as that in which there is a small proportion of free amino groups is referred to as chitin and the term chitosan is used for the long chain polymer of glucosamine (2-amino-2-deoxyglucose) in which there may be a few *N*-acetylated groups. With few exceptions chitin and chitosan occur in nature in close association with other materials and in these forms are usefully designated 'native chitin' [Hackman, 1960] or 'native chitosan'.

Evidence of the distribution of chitin and chitosan synthesizing ability in living organisms has been accumulating since the existence of these polymers was recognized. Unfortunately the quality of the evidence is uneven. Early workers relied heavily on histochemical methods which are not always sufficiently specific when examining small organisms or structures in which there are only low amounts of the polysaccharide. When large objects are being tested X-ray diffraction methods give reliable results but a disadvantage of this method is that it is not applicable to delicate fragments. In addition the results obtained are only qualitative as are those of an alternative physical method, infra-red spectroscopy. A sensitive and specific method for chitin capable of giving quantitative results has been developed by Jeuniaux [1963]. This depends on the digestion of chitin in the organism or structure being examined by purified chitinase and measurement of the amount of *N*-acetylglucosamine released. A similar method considered by Richards [1978] to be of promise involves the use of chitinase conjugated with a fluorescent dye [Benjaminson, 1969]. Certainly if detection is the sole objective this technique has the advantage over that of Jeuniaux in obviating the need to establish the existence of chitinase digestion products. If however quantitative estimates of the amount of chitin present are required then chitinase digestion must be the method of choice. In addition to its use in animals it has been applied to fungi [Cabib and Bowers, 1971]. The relatively recent discovery of chitosanase [Monaghan *et al.*, 1973] opens the way to the use of an analogous enzymic method for quantitative studies of the distribution of chitosan [Fenton *et al.*, 1978].

The vast majority of fungi including members of the Ascomycotina, Basidiomycotina, Deuteromycotina and Mastigomycotina have walls which contain chitin and glucans or mannans whereas those of the Zygomycotina contain both chitin and chitosan [Bartnicki-Garcia, 1968]. The chitin and/or chitosan contents of the mycelium of some fungi are shown in Table 1. Too much reliance should not be placed upon the numerical values quoted or their comparability as several different methods have been used to obtain them. Furthermore, even making allowance for this, it is clear that considerable variation occurs between strains. For example, with *Aspergillus niger* CMI 17454 a value of about 11% chitin is indicated [Johnston, 1965], with strain 1AM 2020 of 24% [Tani *et al.*, 1968] and with strain QM 8601 of 42% [Stagg and Feather, 1973]. In addition, with the same strain, the stage in the life cycle will affect the amount of chitin present as is illustrated for *Mucor rouxii* in Table 1.

The distribution of chitin in the animal kingdom has been thoroughly surveyed by Jeuniaux [1963]. It is widely encountered in the 'lower' forms and is a particularly im-

TABLE 1

Estimates of the chitin and chitosan content of some fungal cell walls

		Chitin	Chitosan	
		(% dry wt.	of wall)	
Ascomycotina	*Aspergillus niger*	22*	0	See text
	Penicillium chrysogenum	35	0	Tani *et al.* [1968]
Basidiomycotina	*Agaricus bisporus*	43	0	Michalenko *et al.* [1976]
	Schizophyllum commune	4	0	Bartnicki-Garcia [1968]
Deuteromycotina	*Rhodotorula glutensis*	10	0	Kreger [1954]
	Sporobolomyces roseus	10	0	Kreger [1954]
Mastigomycotina	*Allomyces macrogynus*	60	0	Aronson and Machlis [1959]
Zygomycotina	*Mucor rouxii*			
	Yeasts	8	28	Bartnicki-Garcia [1968]
	Hyphae	9	33	Bartnicki-Garcia [1968]

*Mean value

portant component of the arthropod exoskeleton where in some instances up to 80% of the cuticle may be chitin. Amongst the microbes a few protozoa have been shown to possess structures in which this polysaccharide occurs in small quantities. For example, the rhizopods *Allogromia oviformis* and *Plagiopxis* sp. contain respectively 0.3% and 1.6% of the dry weight of the whole body as chitin [Jeuniaux, 1963]. Amongst the ciliates, 14 of 22 species examined on a qualitative basis by the highly sensitive enzymatic method of detection were shown to produce chitin [Bussers and Jeuniaux, 1974]. In all reported instances save one, chitin in animals is found in association with protein. The exception is in an anthozoan, *Pocillopera damicornis* [Wainwright, 1962], but there are apparently no reports of either the existence of chitin in other anthozoa or of the synthesis of the idealized polymer by other animals.

In algae in contrast there is ample evidence that in two marine diatoms, *Thallassiosira fluviatilis* and *Cyclotella cryptica*, pure, homogeneously crystalline chitin is found [Falk *et al.*, 1966] and that this comprises 10-15% of the dry weight of the organism [Allan *et al.*, 1978]. The occurrence of chitin in certain green algae has been reported but this is another instance where confirmatory evidence is required [see Tracey, 1955a; Jeuniaux, 1963].

Possible Uses for Chitin and Chitosan

Although chitin and chitosan were named about 150 years [Odier, 1823] and 80 years ago respectively [Hoppe-Seyler, 1894] and their potential use explored sporadically for at least 50 years, it is only during the current decade that there has been any serious attempt to exploit these two compounds, which are unique amongst the biologically synthesized polymers in being basic rather than acidic or neutral in character [Muzzarelli, 1978]. (In the case of chitin this basic nature is due to the existence of a few free $-NH_2$ groups within the generally 2-acetylated compound.)

The impetus for considering chitin and chitosan as raw materials for use by man has derived from a number of sources [Marine Industry Advisory Service, 1976]. It was generally realized that the world supply of petroleum was not inexhaustible. This encouraged a search for alternatives to petrochemicals for manufacturing purposes and also created a general awareness that it would be prudent to make more use of renewable, that is biologically synthesized, resources. Chitin and chitosan attracted attention on both these counts although, given the realities of supply (see below), the contribution which they could make as an alternative to petroleum-based products is extremely limited. In addition, concern about pollution meant that factories processing the increasingly large shellfish catch were no longer able to dispose, in at least some countries, of their waste by dumping it in a convenient natural water and while some processors are able with a minimum amount of treatment to prepare a saleable poultry feed from the shellfish waste, even this processing causes environmental problems [Perceval, 1978]. Thus the attraction of marketing waste at a higher price as a raw material for industrial use is obvious. Lastly, there are now of the order of 100 specific applications of greater or lesser merit for chitin or chitosan which have been described.

A survey of the proposed uses for which one or both of the two polymers have been considered (Table 2) indicates that there are more potential applications for chitosan than for chitin [see also Hattis and Murray, 1977]. Furthermore the applications for which chitin is uniquely required, those in the medical sphere, depending on degradation of the polysaccharide by human lysozyme, are ones which create demand for only low volumes of the polymer.

TABLE 2

Applications areas for chitin and chitosan

Application	Polymer
Adhesive	Chitosan, chitin
Chelating agent	Chitosan, chitin
Coagulant	Chitosan
Drug carrier	Chitin
Healing accelerator	Chitin
Paper and textile additive	Chitosan, chitin
Photographic products	Chitosan
Surgical adjuncts	Chitin
Textile finishers	Chitin, chitosan

Parenthetically it may be remarked that wound healing acceleration has been attributed to both chitin and chitosan. Evidence has however been presented that it is soluble lysozyme digestion products containing *N*-acetyl-D-glucosamine which are responsible for the speeding up of this process [Balassa and Prudden, 1978]. Furthermore it is known that *N*-acetyl groups in this sugar are important for lysozyme activity [Berger and Weiser, 1957; Rupley, 1967] although it is evident that this is more than a simple quantitative effect. Amano and Ito [1978] have shown that the position of deacetylated residues in oligosaccharides derived from lysozyme digests of 70% deacetylated chitin determines where further hydrolysis occurs. Thus until information bearing on the extent of deacetylation of the chitosan preparations used in trials is available and until there is evidence that glucosamine or its oligomers are active in promoting healing, it would be unwise to infer that polyglucosamine is an effective compound.

Thus it appears certain that the quantity of chitosan needed will be greater than that of chitin although the relative rates of future development of the various promising applications may be such as to cause the reverse to be true.

A consideration of the supply of raw materials is appropriate at this point and is given in the next section.

Supply of Raw Materials

It is perhaps difficult to accept the conventional wisdom that chitin is only second to cellulose in terms of abundance in nature since it is estimated that the marine copepods alone synthesize 10^9 tonnes of chitin annually as compared with the total annual crop of 10^{11} tonnes of cellulose [Tracey, 1957]. It is however equally hard to produce figures to show that the other chitin/chitosan synthesizing organisms produce enough of this polysaccharide to eliminate the two orders of magnitude gap between the two estimates given above. The animals and fungi concerned are principally marine or small or microscopic soil types and it is notoriously difficult to obtain reliable data on the animal populations in even a restricted sea area or to determine accurately the chitin-containing biomass in a single soil let alone to extrapolate from these estimates to give global figures in which any real confidence can be placed. Until such information is available it must remain possible that the amount of chitin synthesized annually equals or even exceeds that of cellulose. What there can be no dispute about is that the chitin crop is not as accessible for man's use as that of cellulose and if the anticipated expansion in use of chitin and chitosan takes place it may well be that there will be a shortfall of supply over demand. With this possibility in mind a critical evaluation of the potential sources of chitin and chitosan has recently been undertaken [Allan *et al.*, 1978].

Since it is chitosan rather than chitin for which the greater demand is anticipated [Hattis and Murray, 1977] the situation with regard to supply of this polymer is examined first.

Chitosan

It is possible that as the deacetylated polymer occurs in the members of the Zygomycotina (See Table 1) it could be obtained from a natural source and not as is currently the case by chemical modification of chitin. There are however two possible objections to this approach but the strength of these is difficult to assess because little if any work has been done along these lines. The first difficulty is that in all chitosan-containing fungi so far examined chitin occurs as well. There are in addition anionic polyuronides [Bartnicki-Garcia, 1968] which would probably have to be separated from the chitosan. Given the solubility characteristics of the deacetylated amino-polysaccharide but depending on the tightness of its association with other wall polymers this may not be a difficult task although it is one which would add to the cost of production and thus tends to undermine the whole *raison d'être* of this approach. Jeffries *et al.* [1977] suggest that in *Rhizopus arrhizus* the wall polymers might be in separate layers but the evidence of Datema *et al.* [1977] is that in *Mucor mucedo* homopolymers of glucosamine (chitosan) do not exist and that this sugar occurs in complexes with *N*-acetylglucosamine and linked to protein. Perhaps a more compelling objection is that although *Mucor* and related organisms are used in several food preparation processes, notably in Asia and the Far East [Fenton *et al.*, 1978], there are no substantial concentrations of waste chitosan-containing mycelium available for extraction of the aminopolymer. However, if the extraction is easy and the product has attractive characteristics, a fermentation process perhaps using some waste material as substrate could be set up with chitosanaceous mycelium as the end product. But until more is known about biologically produced chitosan, the practicalities and economics of its extraction and the characteristics of the end product, this polymer will be produced by chemical deacetylation of chitin.

Chitin

Potential sources of chitin may be divided into three categories: it may be derived from a purpose-directed fermentation process, or from the waste of a sophisticated industry or from that of an unsophisticated one. From the point of view of the quality and reproducibility of the raw materials from which the extraction process starts these are in descending order of preference.

At first sight one of the most attractive sources of chitin is the pure material produced by diatoms *Cyclotella cryptica* and *Thallassiosira fluviatilis* because there is neither any need to purify chitin from associated polymers nor subsequently to dispose of this material and because the fibres can be separated from the cells by simple physical processes and obtained in 85-95% yields. The possibility that these diatoms could be cultured artificially as a source of chitin has been investigated [Allan *et al.*, 1978]. In both batch and continuous culture the organisms grew slowly, the cell masses produced were small, although within the range normally obtained in autotrophic algal cultures, and yields of chitin were at best only about 20 mg/litre. Thus the promise of this approach in which the raw material quality could have been rigorously controlled has not been fulfilled.

Less well investigated, and perhaps not entirely serious, suggestions for production of the raw material for chitin extraction include cockroach and waterbeetle farming [see Allan *et al.*, 1978; Hattis and Murray, 1977].

The fermentation industries based on fungi generate, it is estimated, some 790,000 tonnes of waste each year [Allan *et al.*, 1978]. Of this some 41,000 tonnes are produced as a result of citric acid production by *Aspergillus niger*. Taking the amount of chitin in the mycelium as 22% of the total (see Table 1) this represents some 9,000 tonnes of chitin potentially available from a sophisticated industry. If chitin extraction plants were established in association with those for citric acid production, the polysaccharide could be prepared under rigorously controlled conditions from start to finish.

Unfortunately chitin from this source is tightly associated, although probably not covalently linked, with an α 1,3 glucan [Stagg and Feather, 1973] which is not easily separable from the aminopolysaccharide. This apparently does not debar the use of chitosan produced from this chitin source as a strength additive for certain paper [see Allan *et al.*, 1978].

Thus the theoretical advantages of obtaining chitin from well controlled sources are negated by the economics of the process or the untried nature of the product in many applications. However, from what is known of the vagaries in the supply of raw material from an unsophisticated industry (see below) and assuming that the expected expansion in the use of chitin and chitosan occurs, it may be that effort invested in the investigation of the production of high quality chitin from reproducible microbial sources will be rewarded in due course. In the short term though, it is certain that the major, if not sole, source of this aminopolysaccharide will be marine animals.

It would be inappropriate here to deal at length with this source and a thorough consideration of the various aspects of the supply of chitin from shellfish waste in the U.S.A. has been published [Hattis and Murray, 1977] so that it should be possible to assess rapidly the situation in any other country. But since economic considerations provide the principal argument militating against the use of microbial chitin/chitosan it is appropriate to summarize briefly the situation with respect to the supply from marine animals.

Substantial amounts of chitinous waste are produced by the seafood industry but much of it is dispersed in such a way as to make it necessary to collect material for processing. The results of their study of the use of shellfish waste in the U.S.A. led Hattis and Murray [1977] to suggest that an economically viable scheme could be established which would involve collection of shell from areas within a radius of 50 square miles (Figure 2), separation by mechanical means of the waste meat from the shell and transportation of the latter in a dry state to a chitin/chitosan production centre. They estimate, using pessi-

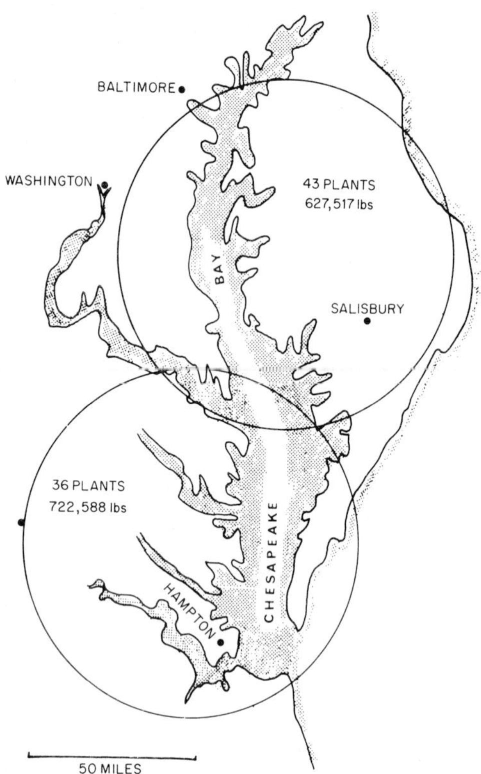

Fig. 2 Potential chitin production from shellfish processing plants located within two 50 mile radius circles in the Chesapeake Bay area, U.S.A. Data modified from Hattis and Murray [1977].

mistic assumptions, that 1 to 4 x 10^6 pounds of chitosan could be produced at a market price of $1.00 - $2.00 per pound and, using optimistic assumptions, that up to 8 x 10^6 pounds could be produced at this price. Perceval, however, calculates that for the Chesapeake Bay area alone 1 x 10^6 pounds a year could be produced to sell at $1.00 to $2.00 per pound for chitin and chitosan respectively. Furthermore, on the basis of West Coast experience Johnson and Peniston [1978] suggest that with a plant producing 1.25 x 10^6 pounds of chitosan per year the market price would be $2.00. Thus the worst assumptions of Hattis and Murray [1977] about the potential amount of chitin which could be available at these prices may lead to an estimate that is unduly pessimistic.

At this cost chitin/chitosan can realistically be considered as an alternative to certain polymers in existing use [Murray and Hattis, 1978]. If the demand for these amino polysaccharides, either because of their unique properties or because of a price advantage, exceeds 1-4 x 10^6 pounds per year, and assuming that the other major sea-food waste generating countries consume their own waste-derived polymers, then sources other than those considered in detail by Hattis and Murray [1977] would have to be tapped. Other sources in the U.S.A. are obvious candidates, in particular those in Alaska. In the world as a whole the waste from Antarctic krill could supply massive amounts of chitin. Assuming a 500 x 10^6 tons annual catch, a chitin content of 3.2% [Anderson *et al.*, 1978] and an 80% water content this would generate about 3 x 10^6 tons of chitin each year.

The most promising of the microbial sources is the fungal mycelium waste of the fermentation industries [Hattis and Murray, 1977], followed by algal material. To emphasize the correctness of this order of choice it is only necessary to look at the mass to be handled for each organism in order to produce 1 x 10^6 pounds of chitin (Table 3). The prospect that raw material from such controlled sources may be used in the future underlines the need incumbent upon the infant shellfish waste chitin/chitosan industry to take account in the development of its products of the fact that some potential applications demand a high quality, reproducible material. It is explicitly recognized [Perceval, 1978; Johnson and Peniston, 1978] that quality control of raw materials is a problem.

TABLE 3

Mass of raw material required to produce 10^6 pounds of chitin from different organisms

Organism	Nature of material	Mass (tons)
Callinectes sapidus	Waste shell	10 x 10^3
Aspergillus niger	Waste mycelium	20 x 10^3
Cyclotella cryptica	Whole culture	22 x 10^6

As well as ageing of the waste prior to processing, Perceval [1978] points out that seasonal biological variation in the animals will be a significant factor to be considered in achieving a standard product and Averbach [1978] points out the different properties of chitosan from different species.

Ageing of the material will, in most instances, reflect microbial decay of the residual meat in the shell but concurrent and/or sequential attack of the chitin and consequential alteration of its properties may also occur.

This leads to a consideration of the biodeterioration of the raw materials, purified polysaccharides and the manufactured products but before this the fabrication of chitin/chitosan artefacts is considered briefly.

Fabrication of Chitin and Chitosan Artefacts

For many of the uses envisaged for chitin/chitosan, purified material may only have to be subjected, prior to use., to minimal physical treatment such as grinding to a more or less coarse powder. For other purposes however a requirement exists for the fabrication of an artefact, usually either a fibre or a membrane. Such manufacture involves solution of the polymer. This presents no problem for chitosan which is soluble in several organic acids [Hayes *et al.*, 1978; Filar and Wirick, 1978]. From solutions, for example in 2% formic or acetic acid, films and fibres may readily be produced [Averbach, 1978]. Chitin in contrast, is insoluble in conventional solvents and this has been a real barrier to the exploitation of this polymer. The discovery [see Rutherford and Austin, 1978] of novel solvent systems such as *N*,*N*-dimethylacetamide and *N*-methyl-2-pyrrolidone both containing 5% lithium chloride, of which the first component acts as a swelling agent allowing solution by the lithium chloride, has removed this obstacle and chitin filaments and sheets can also now be spun and cast.

Biodeterioration

The susceptibility to biodeterioration of cellulose, the polysaccharide most extensively exploited by man and chemically closely related to chitin and chitosan (Figure 1), is well and thoroughly documented [Reese, 1963]. In contrast there is little information available about the spoilage of chitinous materials, which is not to say that the problem has not been recognized.

Perceval [1978] identifies raw material ageing as an important factor to be considered in the maintenance of product quality; Johnson and Peniston [1978], in the context of levelling out fluctuation in the supply of shellfish waste, remark, without specifying the means, that it is possible to prevent bacterial spoilage for up to three weeks and, in assessing the industrial prospects for the utilization of shellfish waste and discussing the location of primary processing sites, Hattis and Murray [1977] make the assumption that collection sites must lie within a circle of fifty miles radius in order that 'excessive' spoilage is avoided.

One instance of degradation of purified material, which was kept for 2-3 days under ambient conditions in Texas in a wet condition, has occurred; the causative organisms being an *Achromobacter* sp. and an unidentified fungus [P.M. Perceval, personal communication].

Only one case is known to the author of biodeterioration of a chitinous artefact. A chitin membrane cast from a solution in dimethylacetamide/lithium chloride (see above) was stored under ambient conditions in a laboratory drawer in Delaware [C.J. Brine, personal communication]. During the summer evidence of an unidentified fungal growth appeared on the surface.

Similarly there is one record of degradation of chitosan in bulk form. During an investigation into the absorption of metals from seawater in Oregon degradation during the summer months of chitosan column packing forced the discontinuation of the use of this adsorbent in favour of another material [C.E. Wicks, personal communication].

Undoubtedly the number of cases of biodeterioration will increase as does the use of chitin/chitosan although this should not be in parallel since in the majority of instances spoilage may be avoided. The procedures open to users of these materials have been discussed in general terms [Berkeley, 1978] but now that a clearer picture has emerged of the few likely principal uses to which chitin and chitosan will be put in the near future, the biodeterioration risk attached to each of them can be considered briefly.

Of the chitosan currently being manufactured, probably the greatest amount is being used for flocculation purposes. In this application, involving the addition to aqueous waste of low concentrations of the polycation, there can be almost absolute certainty that, given proper storage conditions, there will be no spoilage problem prior to use and one of the attractions of chitosan as a coagulant is that it is biodegradable.

As was noted earlier, biological degradation has been encountered in the use of chitosan-packed columns which have been exposed to seawater for protracted periods at relatively high ambient temperatures. It can with some confidence be predicted that under comparable conditions this will recur. The problem can be avoided by the use of thin chitosan membranes whose exchange capacity would be saturated in too short a time to allow microbial attack to occur. Of course, because of the cost of casting such films they will be more expensive to prepare than bulk chitosan. The price comparison of this material with that of conventional exchange resins [Hattis and Murray, 1977] thus becomes less informative but it seems that for certain specialized extraction applications there is still some justification for using chitosan.

Attempts to assess the likely importance of chitin biodeterioration in the medical context have been reported elsewhere [Berkeley and Pepper, 1979]. In short, since the human body provides a moist, warm environment in which nutrients for microbial growth are readily available and since it is known that chitinolytic bacteria of medical provenance exist [Clarke and Tracey, 1956] it might seem that degradation will be inevitable and provide a strong argument against the development of the use of chitin for medical purposes. On the contrary, the limited studies so far carried out indicate that isolates from clinical material with chitinolytic ability do not occur very commonly and that medical bacteria examined for their ability to decrease the strength of chitin fibres do not appear to do so at a rate likely to cause problems [Berkeley and Pepper, 1979].

Biodeterioration is of course, a homocentric view of what for the microorganism is a perfectly normal physiological activity. In the following sections the enzymes responsible for biodeterioration of chitin and chitosan and their role in physiological processes are examined.

Enzymes Degrading Chitin

The modes of action of purified enzymes from microbial chitinase systems have been examined in remarkably few instances. Such studies have been carried out by Jeuniaux [1963] on the enzymes from *Streptomyces antibioticus*, by Berger and Reynolds [1958] using *Streptomyces griseus*, and Ohtakara [1971, 1964] with *Streptomyces griseus* and *Aspergillus niger*. The general pattern to emerge from these studies is of random cleavage of the polysaccharide to form oligosaccharides, principally di-*N*,*N*'-acetylchitobiose, followed by degradation of the oligomers to *N*-acetyl-glucosamine [Ohtakara *et al.*, 1978].

Exceptionally, there is evidence that in an *Arthrobacter* sp. there is a prehydrolytic factor, CH_1, analogous to the C_1 of the cellulase system [see Gooday, Chapter 19], which causes changes in the 'crystalline' chitin allowing hydrolysis by the randomly acting endo-chitinase [Monreal and Reese, 1969]. In one instance, relating to a crude enzyme preparation from *Lycoperdon pyriforme* there is a suggestion that the chitinase acts in an exo-fashion producing *N*-acetyl-glucosamine alone as its end product [Tracey, 1955b]. Unpublished work [Berkeley and Ohtakara] however, shows that in similar enzyme preparations from this organism, there is chitobiase activity and although the possibility that the chitinase and chitobiase activity are due to the same enzyme has not yet been formally excluded, the probability must be that this enzyme system conforms to the pattern of action of the others so far described. Furthermore in *Lycoperdon perlatum* an exo-β-*N*-acetylglucosaminidase has been described [Powning and Irzykiewicz, 1964].

In addition to the polysaccharases described above it is possible that other enzymes attack the chitin chain, producing changes other than the breakage of the glycosidic bond. Zobell and Rittenberg [1938] first recognized the existence of bacteria with such enzymes and coined the name chitinoclasts for them. There are a number of theoretically possible changes that could occur prior to cleavage of the glycosidic link, including opening of the carbon ring, modification of one or more of the hydroxyl groups, removal of the whole acetamido group or deacetylation followed by deamination. There is neither any evidence from Zobell and Rittenberg concerning the first two of these possibilities nor any suggestion that they might occur. These workers suggest that a process involving liberation of ammonia from the polysaccharide chain either directly, following deacetylation, or as a result of the breakdown of liberated acetamide may occur but their evidence is equivocal as the possibility that the ammonia was derived from residual amino acids in the purified chitin can not be excluded.

Similarly their use of acid production in growing cultures as an indication of chitinoclastic activity must be regarded cautiously. Veldkamp [1955] provides more reliable evidence by showing that not only were acetic acid and glucosamine present in the culture fluids of *Pseudomonas chitinovorans* and a *Cytophaga* sp. but that the concentration of these products increased when the cultures were kept under conditions inhibiting metabolism. The demonstration of a deacetylase from *Bacillus cereus* which is inactive against *N*-acetylglucosamine and which will deacetylate glycol chitin [Araki *et al.*, 1971] provides little support for this suggested pathway as the enzyme is much more active against peptidoglycan than chitin and is regarded as a peptidoglycan deacetylase. The same is probably true of some, if not all, of the deacetylases identified in other bacteria by Araki and Ito [1975] although of course a dual role for all these enzymes can not be excluded. Whether deacetylation of chitin as opposed to *N*-acetylglucosamine is a degradative route of any significance in bacteria must remain in doubt.

In the fungi examined by Araki and Ito [1975], the enzyme was present at levels comparable to those found in most instances in bacteria but in *Mucor rouxii* the activity is elevated and believed to be responsible for the formation, from chitin, of the chitosan which occurs in the mycelium of this organism.

Enzymes Degrading Chitosan

Paradoxically although chitosanases have only recently been discovered and, compared to chitinases, few organisms producing them have so far been recognized, there is information available about the mode of action of a number of purified enzymes of this type [Fenton *et al.*, 1978]. All of them have an endo-action on chitosan. In the case of *Streptomyces* sp. No. 6 chitosanase, the products are di- and tri-saccharides [Price and Storck, 1975], and in that of *Penicillium islandicum*, acting upon 30-60% acetylated chitosan, they are oligosaccharides, trisaccharide and some *N*-acetylglucosamine [Fenton *et al.*, 1978]. Furthermore since the Penicillium enzyme yields products with *N*-acetylglucosamine terminal reducing residues it must have a high degree of specificity towards bonds adjacent to C-1 of this sugar. Yet, in common with all other chitosanases so far examined, it has little or no activity towards chitin. The homogeneous chitosanase preparation from myxobacter strain AL-1 also attacks internal bonds, in a random fashion and, perhaps surprisingly, is also active against carboxymethylcellulose [Hedges and Wolfe, 1974]. In both these characteristics it is similar to the chitosanase from two other unidentified bacteria [Fenton *et al.*, 1978].

The underlying molecular mechanisms were they fully understood would surely provide information fascinating in the evolutionary [see Gooday, Chapter 19] as well as the biochemical contexts. In the absence of such understanding and given the overlapping specificities of chitinase, chitosanase and lysozyme, not to mention endo-β-*N*-acetylglucosaminidase [Wadstrom and Hisatsune, 1970], it is clear that there is scope for considerable confusion if studies do not employ a suitably wide range of accurately described substrates.

Such confusion is not likely to be restricted to biochemical studies. Attention has been drawn earlier in this article to the possibility that chitin rather than chitosan was the active compound in one particular application.

Ideally, information on chain length and nature, degree and distribution of substituent groups is needed. Such a requirement is easier to specify than to achieve but as far as chitin and chitosan are concerned, in addition to the chemical methods for establishing the degree of acetylation used by Fenton *et al.* [1978], Price and Storck [1975] and Hayes and Davies [1978], preliminary results of a study aimed at quantitating this using mass spectra have been published [Hayes and Davies, 1978]. Wu and Bough [1978] have described the use of high pressure liquid chromatography for the determination of molecular weight.

The sequence of residues in chitin/chitosan oligomers could be established with the use of the highly specific exo-β-*N*-acetylglucosaminidase such as, for example, that of *Bacillus subtilis* B [Berkeley *et al.*, 1973], in conjunction with the exo-glucosaminidase, complementary to chitosanase, which has yet to be identified and described, but which is presumed to exist.

Physiological Role of Chitinase and Chitosanase

It is possible to consider the role of microbial chitinases and to a very much lesser extent, because of the paucity of information concerning these newly discovered enzymes, chitosanases, in the physiology of the organisms producing them, under three headings: scavenging, heterolytic and autolytic activities. The difference between the first two categories depends on the nature of the substrate; it is impossible to identify with any certainty when a piece of chitinous mycelium in the soil becomes dead and thus whether an enzyme attacking it should be regarded as having scavenging rather than heterolytic activity. Similarly the distinction between an autolysin and a heterolysin may be apparent rather than real.

Scavenging activity

Chitinase or chitosanase degrading non-living residues are active in a scavenging role, playing a part in the provision of an oxidizable substrate, a source of carbon or nitrogen.

The massive amounts of chitin and chitosan which are synthesized annually have been noted earlier. Clearly from the absence of any evidence of the accumulation of these materials in the soil, the sea or bottom deposits, those situations where any tendency to accumulate would occur, equally massive amounts must be degraded each year.

The rate of breakdown of purified chitin strips in three such environments in northern Europe has been studied by Gray *et al.* [1967]. Decomposition was most rapid in sea-water, the strip being impossible to recover after 12 days, in intertidal mud recovery became impossible after 25 days, in forest soil in the A_1 horizon after 84 days and in the C horizon extensive decomposition occurred only after 224 days. In the marine environments bacteria were predominantly associated with the strips and no fungi were observed. Kohlmeyer [1972], however, has reported the existence of marine fungi decomposing chitin. In acidic forest soils Gray *et al.* found that fungi were principally involved with the chitin strips although bacteria and streptomycetes had a role. In dry, non-acidic soils decomposition of chitin is probably substantially due to these prokaryotic organisms. They occur very commonly in such situations [see Gooday, Chapter 19] and of 100 strains examined by Jeuniaux [1955] 98 produced chitinase. Other data although not strictly comparable with those of Gray *et al.* indicate that mineralization of chitin is rapid in all environments except those such as the C horizons of soil in which the microbial population is sparse. Complete disappearance of chitin strips added to a Nigerian soil in 21 days has been observed by Okafor [1978] and particulate chitin is 98% degraded in an estuarine environment in a semi-tropical region in 23 days [Hood and Meyers, 1977].

In the same estuarine environment Hood and Meyers [1978] have demonstrated the crucial importance of exact definition of chitin substrates in ecological studies. Chitin from the shrimp *Penaeus setiferus* prepared in four different ways was examined for rate of breakdown at the water-sediment interface of the estuarine environment. The relative rates of degradation, over a 24 hour period, of native chitin containing 68% chitin, 10% protein and 22% inorganic material, of moulted chitin from the same source containing relatively more chitin, of 2% HCl-treated chitin and of purified chitin treated with acid, alkali and ethanol

[Campbell and Williams, 1951] were 100:86:63:19 respectively.

In natural environments, hexosaminides will usually be insoluble and too large to be transported across the cytoplasmic membrane for metabolism. Microorganisms have evolved two means of circumventing this problem; some *Cytophaga johnsonae* strains, apparently unique in the genus in this respect [see Gooday, Chapter 19], establish close contact between their cells and the chitin particles and degrade the substrate by means of surface located enzymes. In a study of one *Cytophaga* strain which deals with the polysaccharide in this way, and one which liberates an enzyme into the culture medium, Sundarrj and Bhatt [1972] showed that in static cultures the rate of degradation of chitin was very similar for both strains. The speed with which myxobacteria are able to oxidize insoluble substrates by surface contact is emphasized by the demonstration of Stanier [1942] that the oxygen uptake rates of *Cytophaga hutchinsonii* were closely similar in the presence of cellulose or glucose but slightly lower when cellobiose was provided.

The second and most common method used by microorganisms to deal with the problem of metabolizing an insoluble substrate is to liberate chitinase into the environment. In only relatively few instances is there unequivocal evidence that this is done by an export mechanism compatible with the continuation of the normal physiological processes of the cells producing the enzyme, as opposed to production by autolysis [Pollock, 1962; Glenn, 1976; Priest, 1977]. From the point of view of the population as a whole however, it may matter little if a few cells die in order to produce the soluble enzyme necessary to hydrolyse an insoluble material to products utilizable by their mothers, sisters or daughters [Rogers, 1971].

However they arise, the amino-sugar products of the chitinase system are metabolized in many, if not all, environments by many more inhabitants than have the capability to degrade the polymers from which they are derived (Table 4). This being the case, it might be attractive to suppose that those organisms which degrade polymers by a surface contact mechanism have an advantage over those which use an extracellular enzyme system, since in the former instance soluble products will not become available for utilization by other members of the microflora. Such use has the effect of negating the catabolite control mechanism operating on the producer of the extracellular hydrolytic enzymes thus causing it to continue to produce enzyme yet deriving little or no benefit from the products. The demonstration that there is *N*-acetylglucosamine in the culture filtrates of *Cytophaga johnsonae* C 35, the strain which degrades chitin by surface contact, undermines this idea at first sight. But since it was also shown that a substantial amount of this sugar occurs in the supernatant

TABLE 4

Percentage of the population of heterotrophic bacteria from different environments capable of decomposing chitin and utilizing N-acetylglucosamine or glucosamine

	Oak leaf litter	Ash leaf litter	Sea	Bottom mud
Chitin degraders	0	2	1	10
Amino sugar utilizers	3	36	100	100

Hissett and Gray [1974]; Okutani [1978].

fluid of cultures grown in the presence of glucosamine the *N*-acetylated sugar may arise by other than hydrolytic means.

The final stage of the scavenging activity, where benefit is reaped, is the utilization of the amino sugars. In the case of *N*-acetylglucosamine it is normally assumed, although there is only sparse and fragmentory evidence to substantiate this, that after deacetylation follows uptake of the glucosamine and then intracellular deamination prior to oxidation [Hood, 1973]. Transport of the intact *N*-acetylated sugar occurs in *Bacillus subtilis* [Bates and Pasternak, 1965]. In *Cytophaga johnsonae* a route involving deacetylation of *N*-acetylglucosamine, oxidation of glucosamine to glucosaminic acid and deamination to yield gluconic acid, which can be detected in the culture filtrates, appears to be followed [Sundarrj and Bhatt, 1972].

The ultimate fate of the products of aminopolysaccharase action is the same be the enzymes scavenging or heterolytic.

Heterolytic activity

It is well known that addition of chitin to the soil results in increased microbial activity. In cultivated soils a very high proportion of the organisms developing following chitin amendment are actinomycetes. Streptomycetes have long been implicated in the fungitoxity of the soil [Lockwood, 1959] and since a high proportion of these organisms produce chitinase [Jeuniaux, 1959] and many plant pathogens contain chitin in their walls it is tempting to conclude that the control, by addition of chitin to the soil, of such organisms as *Fusarium solani* f. *phaseoli*, which causes root rot of beans [Mitchell and Alexander, 1961] is due to chitinolytic activity. *Pythium debaryanum* which is not suppressed in this way does not contain chitin [Mitchell and Alexander, 1963]. However, just as the mechanism by which bacteria inoculated into the soil in certain specialized environments controls diseases is unknown [Brown, 1975], the real reason for the suppression of pathogenic activity by added chitin is not clear. It has been suggested that instead of being due to heterolytic action it may be due to autolysis caused by starvation [Lloyd and Lockwood, 1966; Ko and Lockwood, 1970] or be the result of inhibition by non-specific lipid compounds of microbial origin [Sneh and Henis, 1970]. Van Eck [1978] argues that the lysis of *Fusarium solani* f. sp. *cucurbitae* chlamydospores is not due to other organisms because there is no visible sign of penetration or disruption of the wall and because no microorganisms were observed within the lysed spore walls.

If this conclusion was to be generally applicable it would argue that the contents of the above paragraph should appear under another heading but as doubt must still exist it serves to underline the difficulties in observing interactions between microorganisms in natural environments. The size of the microhabitats militates against success; mere attempts at observation necessarily often change conditions to such an extent that the very conditions under which one is trying to observe behaviour are radically altered. Even if it were possible to monitor an unchanged environment the interaction between microbes may be mediated by low levels of soluble molecules quite undetectable by currently available methods. The subtleties of antagonism and competition are such that the kind of varying interpretations noted above are always likely.

One instance in which the type of interaction seems clear is that between giant soil amoebae and fungal spores. It has been shown that these animals perforate the conidia of *Cochliobolus sativus*, for example, by the excision of a circular disc from the wall. It is presumed, since this spore is known to contain chitin, that chitinase is involved. Following penetration the amoeba does not digest this disc but retains it within its body or egests it and no further lysis of the spore is observed. Thus it is the spore contents which are used as nutrients rather than the spore coat [Old and Darbyshire, 1978]. Here then the heterolytic attack on chitin, if that it be, is on a very restricted scale. In the case of attack of crayfish by the causative organism of crayfish plague, *Aphanomyces astaci*, it is thought that the protease secreted by the fungus is not for the purpose of allowing penetration of the hyphal tip through the arthropod cuticle but to render accessible the chitin for effective chitinase action [Söderhäll and Unestam, 1975]. Here is an example where it seems that chitinase, in heterolytic attack, is acting in a nutritional capacity.

Both scavenging and heterolytic activity are, however, in essence nutritional in character. In the case of autolysis morphogenetic as well as nutritional aspects must be considered.

Autolytic activity

Statements such as: 'In the puff-ball (*Lycoperdon*) an autolytic process is responsible for preparing the spores in the spore body for release. Among the enzymes involved in the liquefaction phase is the chitinase system which helps to break down the chitinous membranes within the spore body. Acetylglucosaminidase at this time assists in the breakdown process to yield acetylglucosamine which may be resorbed into the mycelium' [Powning and Irzykiewicz, 1964], indicate a generally held view which was based on no more than the very reasonable assumption that this is indeed the case but whether in microorganisms products of wall autolysis are, in addition to being incorporated into new polymers, used as respirable substrates is a moot point. The conventional view is that the wall polymers of bacteria are not implicated as substrates for endogenous metabolism [Dawes and Ribbons, 1964]. It is however considered that in fungi they may be [Trinci and Thurston, 1976].

In bacteria it is difficult to envisage how it can be unequivocally shown in cultures that the products of wall autolysis are or are not used for endogenous respiration. The wall is located outside the cytoplasmic membrane and so it is likely that autolysis products of the wall will diffuse into the culture medium and be used by other cells in the culture as well as by that which is autolysing. Mutual exchange of material in this way excludes, by definition [Dawes and Ribbons, 1964], endogenous metabolism and implies cryptic growth [see Postgate, 1976]. From the point of view of the selfish genome, to adapt from Dawkins [1976], the distinction between cannibalism and endogenous metabolism at the expense of a wall polymer as a means of survival is not a very real one. This is, of course, a development of the concept of the sacrifice of a few individuals for the benefit of a population of bacteria, mentioned already in the context of externalization of digestive enzymes.

A study of single cells should enable clarification of this question but until this has been achieved, are there any lines of evidence bearing on whether wall peptidoglycan is used for endogenous metabolism?

It is perhaps suggestive that there exist autolytic endo-β-*N*-acetylglucosaminidases which produce the 'wrong' end group for further biosynthetic addition by currently known pathways, to the glycan of the wall peptidoglycan [Rogers, Chapter 10]. In addition, in *Bacillus subtilis* there are found an exo-β-*N*-acetylglucosaminidase [Berkeley *et al.*, 1973] and an exo-β-*N*-acetylmuramidase [Del Rio and Berkeley, 1976]. These enzymes are produced late in batch cultures, are substantially and tightly associated with the cell wall, do not have any activity on intact glycan chains from peptidoglycan, are highly specific for the two major alternative disaccharide digestion products from this polymer, and are possibly coordinately controlled; all of which suggests a role in cell wall metabolism. These three enzymes, acting together, could produce amino sugars for endogenous metabolism. Whether they do or not remains to be seen but the possibility that they are involved in sporogenesis in the way suggested by Sudo and Dworkin [1973] can be ruled out as a mutant lacking both the exo-enzymes apparently sporulates normally.

In fungi the situation, with respect to the use of wall polymers for endogenous metabolism, is further complicated by the existence in one individual of hyphal compartments and the lack of information as to whether one particular compartment can use its own wall polymer for this purpose [Trinci and Thurston, 1976].

The autolytic degeneration of the tip of a condiophore of *Alternaria brassicicola* is clearly visible in Figure 3. It can also be seen that the pores in the cross-septum are blocked, thus material from cell wall digestion, at least in the later stages of breakdown, must reach the other hyphal compartments, if at all, by an external route. This raises, as with bacterial wall-glycan digestion products, the possibility of cross feeding.

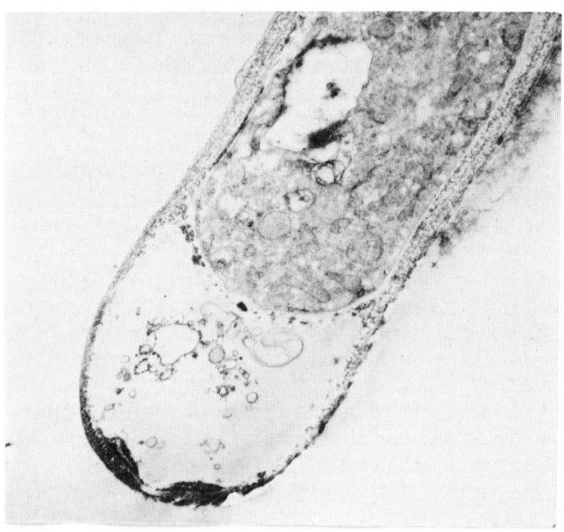

Fig. 3 The autolytic degeneration of the tip of an *Alternaria brassicicola* conidiophore showing blocked septal pores. Reproduced from Campbell [1970] with the permission of the Trustees of the New Phytologist.

Some doubt has been cast on the role of chitinase in autolysis. Although in *Aspergillus nidulans* it is observed that, correlated with glucose depletion, there is an increase in α 1,3 glucanase activity and a breakdown of the cell wall α 1,3 glucan [Zonneveld, 1972], nevertheless in *Aspergillus flavus* it is suggested that the relative constancy of the glucosamine analysis figures during autolysis over a period of 27 days indicates that chitin is little affected [Lahoz and Gonzalez Ibeas, 1968]. Other interpretations of these data are possible and more recent studies with *Aspergillus nidulans* show that for the first 3-5 hours of autolysis wall polymer is degraded at a linear rate, releasing glucose, mannose, galactose and *N*-acetylglucosamine and soluble oligosaccharides, accounting for 3% of the wall polymer. Thereafter the rate is substantially slower but *N*-acetylglucosamine is still released [Polacheck and Rosenberger, 1975]. Thus although the doubt about the involvement of chitinase in autolysis seems unfounded there must remain considerable uncertainty as to how the products of this enzyme are used under different conditions.

As to its involvement in morphogenetic processes; in filamentous organisms 'it seems obvious that the controlled enzymic lysis must occur during morphogenesis of fungal hyphae, although it is difficult to obtain direct evidence for it' [Gooday, 1973]. In hyphae the major sites of chitin synthesis, as demonstrated by incorporation of radioactively labelled *N*-acetylglucosamine, are the apex and branch points [Gooday, 1971; Bartnicki-Garcia, 1973]. Thus if, as is envisaged, a balance between hydrolysis and synthesis exists, which is essential for growth and morphogenesis [Bartnicki-Garcia, 1973], the apex and the branch points must be the principal site for chitinase activity too. The addition of lysozyme, which has chitinase activity, leads to an imbalance causing rupture of the hyphal tip of *Mucor rouxii* [Bartnicki-Garcia, 1973]. Also, the application, at low concentrations, of snail digestive juice, known to contain chitinase, amongst other hydrolytic enzymes, caused *Neurospora crassa* to grow in a highly branched colonial fashion. Hydrolysates of cell walls of mutants grown in the same way contained more glucosamine than did wild type [De Terra and Tatum, 1963]. Thus one can speculate, since chitin plays a quantitatively greater part in the cell wall and is not protected to such a great extent by the other polymers in which it is embedded in the wild type [Mahadevan and Tatum, 1967], that the same amount of enzyme leads to the formation of a multiplicity of branch points, so leading to the colonial form. This implies that the copolymers of chitin are important in the control of morphogenesis.

Indeed it is suggested [Mahadevan and Mahadakar, 1970] that it is protease and glucanase which participate in branching of a *Neurospora crassa* spreading colonial mutant *spco*-1, rather than chitinase which is active in autolysis only at a low level. Supporting this conclusion is the evidence obtained using immunological techniques that β 1,3 glucanase and protease occur in connection with the cross septa and branching tips of hyphae and that the mutant *spco*-1, with greater frequency of branching, possessed higher levels of these enzymes [Sukumaran and Mahadevan, 1975]. This and other work [Mehta and Mahadevan, 1975], shows the role that protease activity has in 'softening' the wall of *Neurospora crassa* prior to branching but it does not exclude the involvement of chitinase. Indeed continuing with the assumption that balanced degradation and synthesis are necessary for morphogenetic changes, the demonstration that potential for chitin synthetase activity occurs along the entire length of *Aspergillus niger* hyphae [Katz and Rosenberger, 1971] provides one further reason for not yet discounting the involvement of chitinase. Another is that chitin synthetase is stimulated by *N*-acetylglucosamine. Here however, it must be noted that this sugar seems to mimic UDP-*N*-acetylglucosamine and is not incorporated into chitin *in vivo* [Bartnicki-Garcia *et al.*, 1978].

The same is true in the stipe of the basidiomycete *Coprinus cinereus* [Gooday and De Rousett-Hall, 1975] thus although *N*-acetylglucosamine and its disaccharide may be active *in vitro* they are not themselves necessary *in vivo* [Gooday, 1977]. Conceivably they would have some sparing action on the UDP-derivative if present as seems certain to be the case in *Coprinus* since chitinase activity has been detected. Furthermore the demonstration that net chitin synthesis occurs at the expense of other cell constituents in excised elongating stipes and that exogenous *N*-acetylglucosamine was incorporated into chitin is a further indication that the products of chitinase action are used for growth and that there is a relationship between breakdown and synthesis [Gooday, 1975]. Nevertheless, some doubt must remain over the role of chitinase in branching and apical growth in mycelial forms and in chitin synthetase activation.

One instance where there is no question about the part chitinase plays is in the autolysis of the mature fruiting bodies of *Coprinus lagopus*. Approximately one hour before the release of spores from the gills, newly synthesized chitinase can be detected. This reaches its maximum level some six hours later when the gills are completely lysed and the spores released [Iten and Matile, 1970]. Whether the *N*-acetylglucosamine produced is reused in the subtending mycelium can not be determined from the evidence presented.

Conclusions

In the first instance any commercial exploitation of chitin/chitosan will depend on chitin from macroscopic animals rather than from microbes. The theoretical advantages of microbial sources of these aminopolysaccharides are not yet sufficient to outweigh their cost disadvantage and performance uncertainty.

With the projected increase in the use of these polymers in commercial applications it seems certain that given the known wide distribution of chitinolytic and chitosanolytic ability amongst the microflora, biodeterioration problems will occur. To what extent these are avoided will depend on the previous thought which is given to the use of these materials.

In the applied use of these compounds as well as in biochemical and ecological studies it is clear that it is essential to characterize the substrate properly. It seems possible that newly described methods will make this a much less intractable problem than hitherto.

The role of chitinases in scavenging chitin containing organic residues are fairly well understood in several environments. The consideration given to chitosanases in this respect is not surprisingly scant in view of their recent discovery. The distinction between scavenging activity and heterolysis is far from clear and given the difficulties in performing suitably critical studies in this field is likely to remain so.

The autolytic role of these enzymes is twofold. In both bacterial and fungal nutrition, although in the former instance not in the context of chitinases but of related enzymes, it is difficult to determine whether the concept of endogenous metabolism has any significance when the wall is being considered as a substrate.

The morphogenetic role of chitinase in filamentous organisms is currently in some doubt but in fruiting types this enzyme certainly seems to play a part at very least in the maturation of the cap leading to release of spores.

Acknowledgements

I am grateful to Mr. P.M. Perceval, Dr. G.W. Gooday and Dr. R.E. Campbell for valuable discussion during the preparation of this paper.

References

Allan, G.G., Fox, J.R. and Kong, N. (1978). A critical evaluation of potential sources of chitin and chitosan. In *Proceedings of the First International Conference on Chitin/Chitosan*, pp.64-78. Edited by R.A.A. Muzzarelli and E.R. Pariser. MIT Sea Grant Report MITSG 78-7.

Amano, K-I. and Ito, E. (1978). The action of lysozyme on partially deacetylated chitin. *European Journal of Biochemistry* **85**, 97-104.

Anderson, C.G., De Pablo, N. and Romo, C.R. (1978). Antarctic krill (*Euphausia superba*) as a source of chitin and chitosan. In *Proceedings of the First International Conference on Chitin and Chitosan*, pp.54-63. Edited by R.A.A. Muzzarelli and E.R. Pariser. MIT Sea Grant Report MITSG 78-7.

Araki, Y., Fukuoka, S., Oba, S. and Ito, E. (1971). Enzymatic deacetylation of N-acetylglucosamine residues in peptidoglycan from *Bacillus cereus* cell walls. *Biochemical and Biophysical Research Communications* **45**, 751-758.

Araki, Y. and Ito, E. (1975). A pathway of chitosan formation in *Mucor rouxii*. *European Journal of Biochemistry* **55**, 71-78.

Aronson, J.M. and Machlis, L. (1959). The chemical composition of the hyphal walls of the fungus *Allomyces*. *American Journal of Botany* **46**, 292-300.

Averbach, B.L. (1978). Film forming capacity of chitosan. In *Proceedings of the First International Conference on Chitin/Chitosan*, pp.199-209. Edited by R.A.A. Muzzarelli and E.R. Pariser. MIT Sea Grant Report MITSG 78-7.

Balassa, L.L. and Prudden, J.F. (1978). Applications of chitin and chitosan in wound healing acceleration. In *Proceedings of the First International Conference on Chitin/Chitosan*, pp.296-305. Edited by R.A.A. Muzzarelli and E.R. Pariser. MIT Sea Grant Report MITSG 78-7.

Bartnicki-Garcia, S. (1968). Cell wall chemistry, morphogenesis and taxonomy of fungi. *Annual Review of Microbiology* **22**, 87-108.

Bartnicki-Garcia, S. (1973). Fundamental aspects of hyphal morphogenesis. In *Microbial Differentiation*, pp.245-267. Edited by J.M. Ashworth and J.E. Smith. Cambridge: University Press.

Bartnicki-Garcia, S., Bracker, C.E. and Ruiz-Herrera, J. (1978). Synthesis of chitin microfibrils *in vitro* by chitin synthetase particles, chitosomes, isolated from *Mucor rouxii*. In *Proceedings of the First International Conference on Chitin/Chitosan*, pp.450-463. Edited by R.A.A. Muzzarelli and E.R. Pariser. MIT Sea Grant Report MITSG 78-7.

Bates, C.J. and Pasternak, C.A. (1965). The incorporation of labelled amino sugars by *Bacillus subtilis*. *Biochemical Journal* **96**, 155-158.

Benjaminson, M.A. (1969). Conjugates of chitinase with fluorescein isothiocyanate or lissamine rhodamine as specific stains for chitin *in situ*. *Stain Technology* **44**, 27-31.

Berger, L.R. and Reynolds, D.M. (1958). The chitinase system of a strain of *Streptomyces griseus*. *Biochimica et Biophysica Acta* **29**, 522-534.

Berger, L.R. and Weiser, R.S. (1957). The β-glucosaminidase activity of egg-white lysozyme. *Biochimica et Biophysica Acta* **26**, 517-521.

Berkeley, R.C.W. (1978). Chitinolytic and chitosanolytic microorganisms and the potential biodeterioration problem in the commercial application of chitin and its derivatives. In *Proceedings of the First International Conference on Chitin/Chitosan*, pp.570-577. Edited by R.A.A. Muzzarelli and E.R. Pariser. MIT Sea Grant Report MITSG 78-7.

Berkeley, R.C.W., Brewer, S.J., Ortiz, J.M. and Gillespie, J.B. (1973). An exo-β-N-acetylglucosaminidase from *Bacillus subtilis* B; characterization. *Biochimica et Biophysica Acta* **309**, 157-168.

Berkeley, R.C.W. and Pepper, E.A. (1979). Biodeterioration of chitin in the context of its medical applications. *Proceedings of the 4th International Biodeterioration Symposium*, in press.

Brown, M.E. (1975). Rhizosphere Microorganisms - opportunists, bandits or benefactors. In *Soil Microbiology*, pp.21-38. Edited by N. Walker. London: Butterworths.

Bussers, J.C. and Jeuniaux, Ch. (1974). Recherche de la chitine dans les productions métaplastique de quelques ciliés. *Protistologica* **10**, 43-46.

Cabib, E. and Bowers, B. (1971). Chitin and yeast budding. Localization of chitin in yeast bud scars. *Journal of Biological Chemistry* **246**, 152-159.

Campbell, L.L. and Williams, O.B. (1951). A study of chitin-decomposing microorganisms of marine origin. *Journal of General Microbiology* **5**, 894-905.

Campbell, R. (1970). An electron microscope study of exogenously dormant spores, spore germination, hyphae and conidiophores of *Alternaria brassicicola*. *New Phytologist* **69**, 287-293.

Clarke, P.H. and Tracey, M.V. (1956). The occurrence of chitinase in some bacteria. *Journal of General Microbiology* **14**, 188-196.

Datema, R., Wessels, J.G.H. and Van Den Ende, H. (1977). The hyphal wall of *Mucor mucedo*. 2. Hexosamine containing polymers. *European Journal of Biochemistry* **80**, 621-626.

Dawes, E.A. and Ribbons, D.W. (1964). Some aspects of the endogenous metabolism of bacteria. *Bacteriological Reviews* **28**, 126-149.

Dawkins, R. (1976). *The Selfish Gene*. Oxford: University Press.

Van Eck, W.H. (1978). Autolysis of chlamydospores of *Fusarium solari* f. sp. *cucurbitae* in chitin and laminarin amended soils. *Soil Biology and Biochemistry* **10**, 89-92.

Falk, M., Smith, D.G., McLachlan, J. and McInnes, A.G. (1966). Studies on chitan (β-(1→4)-linked 2-acetamido-2-deoxy-D-glucan) fibres of the diatom *Thallassiosira fluviatilis*. II. proton magnetic presonance, infrared, and X-ray studies. *Canadian Journal of Chemistry* **44**, 2269-2281.

Fenton, D., Davis, B., Rotgers, C. and Eveleigh, D.E. (1978). Enzymatic hydrolysis of chitosan. In *Proceedings of the First International Conference on Chitin/Chitosan*, pp.525-534. Edited by R.A.A. Muzzarelli and E.R. Pariser. MIT Sea Grant Report MITSG 78-7.

Filar, L.J. and Wirick, M.G. (1978). Bulk and solution properties of chitosan. In *Proceedings of the First International Conference on Chitin/Chitosan*, pp.169-176. Edited by R.A.A. Muzzarelli and E.R. Pariser. MIT Sea Grant Report MITSG 78-7.

Glenn, A.R. (1976). Production of extracellular proteins by bacteria. *Annual Review of Microbiology* **30**, 41-62.

Gooday, G.W. (1971). An autoradiographic study of hyphal growth of some fungi. *Journal of General Microbiology* **67**, 125-133.

Gooday, G.W. (1973). Differentiation in the Mucorales. In *Microbial Differentiation*, pp.269-294. Edited by J.M. Ashworth and J.E. Smith. Cambridge: University Press.

Gooday, G.W. (1975). The control of differentiation in fruit bodies of *Coprinus cinereus*. *Report of the Tottori Mycological Institute* **12**, 151-160.

Gooday, G.W. (1977). Biosynthesis of the fungal wall - mechanisms and implications. *Journal of General Microbiology* **99**, 1-11.

Gooday, G.W. and De Rousett-Hall, A. (1975). Properties of chitin synthetase from *Coprinus cinereus*. *Journal of General Microbiology* **89**, 137-145.

Gray, T.R.G., Baxby, P., Hill, I.R. and Goodfellow, M. (1967). Direct observation of bacteria in soil. In *The Ecology of Soil Bacteria*, pp.171-192. Edited by T.R.G. Gray and D. Parkinson. Liverpool: University Press.

Hackman, R.H. (1960). Studies on Chitin. IV. The occurrence of complexes in which chitin and protein are covalently linked. *Australian Journal of Biological Sciences* **13**, 568-577.

Hattis, D. and Murray, A.E. (1977). *Industrial Prospects for Chitin and Protein from Shellfish Wastes*. MIT Sea Grant Report MITSG 77-3.

Hayes, E.R. and Davies, D.H. (1978). Characterization of chitosan. II. The determination of the degree of acetylation of chitosan and chitin. In *Proceedings of the First International Conference on Chitin/Chitosan*, pp.406-414. Edited by R.A.A. Muzzarelli and E.R. Pariser. MIT Sea Grant Report MITSG 78-7.

Hayes, E.R., Davies, D.H. and Munroe, V.G. (1978). Organic acid solvent systems for chitosan. In *Proceedings of the First International Conference on Chitin/Chitosan*, pp.103-106. Edited by R.A.A. Muzzarelli and E.R. Pariser, MIT Sea Grant Report MITSG 78-7.

Hedges, A. and Wolfe, R.S. (1974). Extracellular enzyme from myxobacter AL-1 that exhibits both β 1,4 glucanase and chitosanase activities. *Journal of Bacteriology* **120**, 844-853.

Hissett, R. and Gray, T.R.G. (1974). Bacterial populations of litter and soil in a deciduous woodland. I. Qualitative studies. *Revue d'Ecologie et Biologie de Sol* **4**, 495-508.

Hood, M.A. (1973). Chitin Degradation in the Salt Marsh Environment. Ph.D. Thesis. Louisiana State University.

Hood, M.A. and Meyers, S.P. (1977). Rates of degradation in an estuarine environment. *Journal of the Oceanographical Society of Japan* **33**, 328-334.

Hood, M.A. and Meyers, S.P. (1978). Chitin degradation in estuarine environments and implications in crustacean biology. In *Proceedings of the First International Conference on Chitin/Chitosan*, pp.563-569. Edited by R.A.A. Muzzarelli and E.R. Pariser. MIT Sea Grant Report MITSG 78-7.

Hoppe-Seyler, F. (1894). Ueber Chitin und Cellulose. *Berichte der Deutschen Chemischen Gesellschaft* **27**, 3329-3331.

Iten, W. and Matile, P. (1970). Role of chitinase and other lysosomal enzymes of *Coprinus lagopus* in the autolysis of fruiting bodies. *Journal of General Microbiology* **61**, 301-309.

Jefferies, T.W., Eveleigh, D.E., MacMillan, J.D., Parrish, F.W. and Reese, E.T. (1977). Enzymatic hydrolysis of the cell walls of yeast cells and germinated fungal spores. *Biochimica et Biophysica Acta* **499**, 10-23.

Jeuniaux, C. (1955). Production d'exochitinase par des *Streptomyces*. *Comptes Rendus des Seances de la Société de Biologie, Paris* **149**, 1307-1308.

Jeuniaux, C. (1963). *Chitine et Chitinolyse*. Paris: Masson et Cie.

Johnson, E.L. and Peniston, Q.P. (1978). The production of chitin and chitosan. In *Proceedings of the First International Conference on Chitin/Chitosan*, pp.80-87. Edited by R.A.A. Muzzarelli and E.R. Pariser. MIT Sea Grant Report MITSG 78-7.

Johnston, I.R. (1965). The composition of the cell wall of *Aspergillus niger*. *Biochemical Journal* **96**, 651-664.

Katz, D. and Rosenberger, R.F. (1971). Hyphal wall synthesis in *Aspergillus nidulans*: effect of protein synthesis inhibition and osmotic stabilizer shock on chitin insertion and morphogenesis. *Journal of Bacteriology* **108**, 184-190.

Ko, W-H. and Lockwood, J-L. (1970). Mechanism of lysis of fungal mycelium in soil. *Phytopathology* **60**, 148-157.

Kohlmeyer, J. (1972). Marine fungi deteriorating hydrozoa and keratin-like annelid tubes. *Marine Biology* **12**, 277-284.

Kreger, D.R. (1954). Observations on cell walls of yeasts and some other fungi by X-ray diffraction and solubility tests. *Biochimica et Biophysica Acta* **13**, 1-9.

Lahoz, R. and Gonzalez Ibeas, J. (1968). The autolysis of *Aspergillus flavus* in an alkaline medium. *Journal of General Microbiology* **53**, 101-108.

Lloyd, A.B. and Lockwood, J.L. (1966). Lysis of fungal hyphae in the soil and its possible relation to autolysis. *Phytopathology* **56**, 595-602.

Lockwood, J.L. (1959). *Streptomyces* spp. as a natural cause of fungi-toxicity in soils. *Phytopathology* **49**, 327-331.

Mahadevan, P.R. and Mahadakar, V.R. (1970). Role of enzymes in growth and morphology of *Neurospora crassa*: cell-wall-bound enzymes and their possible role in branching. *Journal of Bacteriology* **101**, 941-947.

Mahadevan, P.R. and Tatum, E.L. (1967). Localization of structural polymers in the cell wall of *Neurospora crassa*. *Journal of Cell Biology* **35**, 295-302.

Marine Industry Advisory Service (1976). *Chitin and Chitin Derivatives*. MIT Sea Grant Report MITSG 76-5.

Mehta, N.M. and Mahadevan, P.R. (1975). Proteases of *Neurospora crassa*. Their role in morphology. *Indian Journal of Experimental Biology* **13**, 131-134.

Michalenko, G.O., Hohl, H.R. and Rast, D. (1976). Chemistry and architecture of the mycelial wall of *Agaricus bisporus*. *Journal of General Microbiology* **92**, 251-262.

Mitchell, R. and Alexander, M. (1961). Chitin and the biological control of Fusarium disease. *Plant Disease Reporter* **45**, 487-489.

Mitchell, R. and Alexander, M. (1963). Lysis of soil fungi by bacteria. *Canadian Journal of Microbiology* **9**, 169-177.

Monaghan, R.L., Eveleigh, D.E., Tewari, R.M. and Reese, E.T. (1973). Chitosanase, a novel enzyme. *Nature New Biology* **245**, 78-80

Monreal, J. and Reese, E.T. (1969). The chitinase of *Serratia marcescens*. *Canadian Journal of Microbiology* **15**, 689-696.

Murray, A.E. and Hattis, D. (1978). Approaches to a practical assessment of supply and demand for chitin products in the United States. In *Proceedings of the First International Conference on Chitin/Chitosan*, pp.30-44. Edited by R.A.A. Muzzarelli and E.R. Pariser. MIT Sea Grant Report MITSG 78-7.

Muzzarelli, R.A.A. (1977). *Chitin*. Oxford, New York, Toronto, Sydney, Paris, Frankfurt: Pergamon Press.

Muzzarelli, R.A.A. (1978). Chitin, an important natural polymer. In *Proceedings of the First International Conference on Chitin/Chitosan*, pp.1-3. Edited by R.A.A. Muzzarelli and E.R. Pariser. MIT Sea Grant Report MITSG 78-7.

Odier, A. (1823). Mémoire sur la composition chimique des parties cornées des insects. *Memoires Société Histoire Naturelle (Paris)*

1, 29-42.

Ohtakara, A. (1964). Studies on the chitinolytic enzymes of black-koji mould. Part VII. Degradation of glycol chitin and chitin by the chitinase system of *Aspergillus niger*. *Agricultural and Biological Chemistry Japan* **28**, 811-818.

Ohtakara, A. (1971). The chitinase system of a strain of *Streptomyces griseus*. *Bulletin of Hiroshima Womens University* **6**, 1-12.

Ohtakara, A., Uchida, Y. and Mitsutomi, M. (1978). Chitinase systems in microorganisms and the commercial use of chitin. In *Proceedings of the First International Conference on Chitin/Chitosan*, pp.587-605. Edited by R.A.A. Muzzarelli and E.R. Pariser. MIT Sea Grant Report MITSG 78-7.

Okafor, N. (1978). Chitin decomposition by microorganisms in soil. In *Proceedings of the First International Conference on Chitin/Chitosan*, pp.578-581. Edited by R.A.A. Muzzarelli and E.R. Pariser. MIT Sea Grant Report MITSG 78-7.

Okutani, K. (1978). Chitin- and *N*-acetylglucosamine-decomposing bacteria in the sea. In *Proceedings of the First International Conference on Chitin/Chitosan*, pp.582-586. Edited by R.A.A. Muzzarelli and E.R. Pariser. MIT Sea Grant Report MITSG 78-7.

Old, K.M. and Darbyshire, J.F. (1978). Soil fungi as food for giant amoebae. *Soil Biology and Biochemistry* **10**, 93-100.

Perceval, P.M. (1978). The economics of chitin recovery and production. In *Proceedings of the First International Conference on Chitin/Chitosan*, pp.45-53. Edited by R.A.A. Muzzarelli and E.R. Pariser. MIT Sea Grant Report MITSG 78-7.

Polacheck, Y. and Rosenberger, R.F. (1967). Autolytic enzymes in hyphae of *Aspergillus nidulans*: their action on old and newly formed walls. *Journal of Bacteriology* **121**, 332-337.

Pollock, M.R. (1962). Exoenzymes. In *The Bacteria*, Volume IV, pp. 121-178. Edited by I.C. Gunsalus and R.Y. Stanier. New York and London: Academic Press.

Postgate, J.R. (1976). Death in Macrobes and Microbes. In *The Survival of Vegetative Microbes*, pp.1-18. Edited by T.R.G. Gray and J.R. Postgate. Cambridge: University Press.

Powning, R.F. and Irzykiewicz, H. (1964). β-acetylglucosaminidase in the cockroach (*Periplaneta americana*) and in the puff-ball (*Lycoperdon perlatum*). *Comparative Biochemistry and Physiology* **12**, 405-515.

Price, J.S. and Storck, R. (1975). Production, purification and characterization of an extracellular chitosanase from *Streptomyces*. *Journal of Bacteriology* **124**, 1574-1585.

Priest, F.G. (1977). Extracellular enzyme synthesis in the genus *Bacillus*. *Bacteriological Reviews* **41**, 711-753.

Reese, E.T. (1963). *Advances in Enzymic Hydrolysis of Cellulose and Related Materials*. Oxford: Pergamon Press.

Richards, A.G. (1978). The detection and estimation of chitin in insect organs. In *Proceedings of the First International Conference on Chitin/Chitosan*, pp.22-25. Edited by R.A.A. Muzzarelli and E.R. Pariser. MIT Sea Grant Report MITSG 78-7.

Del Rio, L.A. and Berkeley, R.C.W. (1976). Exo-β-*N*-acetylmuramidase - a novel hexosaminidase. Production by *Bacillus subtilis* B, purification and characterization. *European Journal of Biochemistry*

65, 3-12.

Rogers, H.J. (1961). The dissimilation of high molecular weight substances. In *The Bacteria*, Volume II, pp.257-318. Edited by I.C. Gunsalus and R.Y. Stanier. New York and London: Academic Press.

Roseman, S. (1957). Glucosamine metabolism. I. N-acetylglucosamine deacetylase. *Journal of Biological Chemistry* **226**, 115-124.

Rupley, J.A. (1967). The binding and cleavage by lysozyme of N-acetylglucosamine oligosaccharides. *Proceedings of the Royal Society Series B* **167**, 416-428.

Rutherford, F.A. and Austin, P.R. (1978). Marine chitin properties and solvents. In *Proceedings of the First International Conference on Chitin/Chitosan*, pp.182-192. Edited by R.A.A. Muzzarelli and E.R. Pariser. MIT Sea Grant Report MITSG 78-7.

Sneh, B. and Henis, Y. (1970). Production of antifungal substances active against *Rhizoctonia solani* in chitin amended-soil. *Phytopathology* **62**, 595-600.

Söderhäll, K. and Unestam, T. (1975). Properties of extracellular enzymes from *Aphanomyces astaci* and their relevance in the penetration process of crayfish cuticle. *Physiologia Plantarum* **35**, 140-146.

Stagg, C.M. and Feather, M.S. (1973). The characterization of a chitin associated D-glucan from the cell walls of *Aspergillus niger*. *Biochimica et Biophysica Acta* **320**, 64-72.

Stanier, R.Y. (1942). The cytophaga group: a contribution to the biology of myxobacteria. *Bacteriological Reviews* **6**, 143-196.

Sudo, S.Z. and Dworkin, M. (1973). Comparative biology of prokaryotic resting cells. *Advances in Microbial Physiology* **9**, 153-224.

Sukumaran, C.P. and Mahadevan, P.R. (1975). Localization of enzymes in cell walls of *Neurospora crassa*. *Indian Journal of Experimental Biology* **13**, 127-130.

Sundarrj, N. and Bhat, J.V. (1972). Breakdown of chitin by *Cytophaga johnsonii*. *Archiv für Mikrobiologie* **85**, 159-167.

Tani, I., Kii, K. and Mori, M. (1968). Studies of the surface structure of fungi. I. Chitin contents of the cell wall of several fungi. *Japanese Journal of Bacteriology* **23**, 191-199.

De Terra, and Tatum, E.C. (1963). A relationship between cell wall structure and colonial growth in *Neurospora crassa*. *American Journal of Botany* **50**, 669-677.

Tracey, M.V. (1955a). Chitin. In *Modern Methods in Plant Analysis*, Volume II, pp.264-274. Edited by K. Paech and M.V. Tracey. Berlin: Springer Verlag.

Tracey, M.V. (1955b). Chitinase in some Basidiomycetes. *Biochemical Journal* **61**, 579-586.

Tracey, M.V. (1957). Chitin. *Reviews of Pure and Applied Chemistry* **7**, 1-14.

Trinci, A.P.J. and Thurston, C.F. (1976). Transition to the non-growing state in eukaryotic microorganisms. In *The Survival of Vegetative Microbes*, pp.55-79. Edited by T.R.G. Gray and J.R. Postgate. Cambridge: University Press.

Veldkamp, H. (1955). A study of the aerobic decomposition of chitin by microorganisms. Thesis. University of Leiden.

Wadstrom, T. and Hisatsune, K. (1970). Bacteriolytic enzymes from *Staphylococcus aureus*. Specificity of the action of end-β-*N*-

acetylglucosaminidase. *Biochemical Journal* **120**, 735-744.

Wainwright, S.A. (1962). An anthozoan chitin. *Experientia* **18**, 1-3.

Wu, A.C.M. and Bough, W.A. (1978). A study of the variables in the chitosan manufacturing process in relation to molecular-weight distribution, chemical characteristics and waste treatment effectiveness. In *Proceedings of the First International Conference on Chitin/Chitosan*, pp.88-102. Edited by R.A.A. Muzzarelli and E.R. Pariser. MIT Sea Grant Report MITSG 78-7.

Zobell, C.E. and Rittenberg, S.C. (1938). The occurrence and characteristics of chitinoclastic bacteria in the sea. *Journal of Bacteriology* **35**, 275-287.

Zonneveld, B.J.M. (1972). Morphogenesis in *Aspergillus nidulans*. The significance of α 1,3 glucan of the cell wall and α 1,3 glucanase for cleistothecium development. *Biochimica et Biophysica Acta* **273**, 174-187.

Chapter 10

THE FUNCTION OF BACTERIAL AUTOLYSINS

H.J. ROGERS

Division of Microbiology, National Institute for Medical Research, London, UK

Introduction

Among the earliest knowledge about microbes is that some organisms disintegrate when suspended and incubated under unfavourable conditions for growth and anabolism. Indeed autolysates of yeasts and other cells were some of the earliest sources of soluble enzymes, and did trojan work in the heroic days of the investigations of intermediary carbohydrate metabolism. With the development of the electron microscope and consequent ultrastructural awareness, cell walls were recognized, isolated and their chemistry studied. Autolysis could then be seen to be due to dissolution of the walls. As the chemical structure of the polymers making up the walls was elucidated, it became possible to designate the bonds specifically hydrolysed by various autolytic enzymes and thus to account for the solubilization of the walls and the disintegration of the cells. Since the supportive polymers in microbial cell walls are often complex, for example, the peptidoglycans of bacteria, or alternatively consist of a number of polymers interacting to provide mechanical support for the cell as in some microfungi, autolysins with a variety of bond specificities are often to be found in the same cell. Although these are not all "polysaccharases" they will be involved frequently in the discussion that follows because in considering the more sophisticated ideas about function, it is not always possible to distinguish which autolysin is responsible for the particular effect, or to exclude the necessity for the combined action of more than one enzyme.

At the outset, the purely degradative aspect of autolysins was emphasized and one of their functions is presumably to remove what might otherwise be an embarrassing accumulation of insoluble microbial matter. Recently, however, the possibility of more positive functions has been widely discussed and tested. One of the paradoxes inherent in microbes with cell walls is that they have a surface completely covered with insoluble polymers, but they must be able to expand during growth, and modify the surface so that division into two new individuals can take place. Obvious candidates that might modify surfaces are the autolysins but unambiguous evidence that they are necessarily involved during growth has been more difficult to obtain. Work with micro-fungi and with bacteria has been actively proceeding for some ten years to test such hypotheses and with the former hopeful correlative evidence has been obtained. In bacteria evidence has so far been rather negative but certainly not unambiguously so. It is to these latter aspects of the functions of autolysins that this article will pay particular attention.

The Substrates

The nature of the substrates

In view of the known specificity of the autolytic enzymes described later, it is justified for present purposes to accept, without too much questioning, that the preparations usually isolated as cell walls are what they are claimed to be. These are made from broken cells by depositing the insoluble material and repeatedly washing it to remove membraneous, proteinaeous or other cytoplasmic material by the most expeditious means. This approach has been particularly successful with bacteria, partly because some of the polymers in the walls thus obtained are unique and contain unique constituents. From Gram-positive species, the isolation procedure yields thick structures similar to those seen in the bacteria. From Gram-negative bacteria it yields thin strong layers or sacculi which are now considered [see Rogers, 1979] to provide the mechanical strength of the walls; at all events, it is towards the polymers in this layer that the specificity of the autolytic enzymes is directed. The walls of micro-fungi, on the other hand, although containing unique polymers, often contain constituents in common with the cytoplasmic contents. For example wall proteins appear to be either involved in maintaining wall integrity or in regulating access of hydrolases to their polysaccharide substrates. Even the wall polysaccharides, such as chitin, glucans and mannans do not always have unique constituents whereby to monitor the "cleanliness" of the wall preparations.

Bacterial peptidoglycans

The structures of peptidoglycans have been repeatedly reviewed [Ghuysen, 1968; Rogers and Perkins, 1968; Schleifer and Kandler, 1972; Rogers, 1974] and only a generalized picture will be given. They all consist of long glycan strands of β 1,4-linked amino sugars which are joined together by short peptides. The amino sugar residues are usually *N*-acetylglucosamine and *N*-acetylmuramic acid, although other acyl groups may sometimes be present instead of acetyl. In most if not all Gram-positive species other polymers are covalently linked to the glycan chains probably through the 6-OH groups of the muramyl residues. In some species, such as staphylococci, $-OCH_3$ groups also occur. The peptides in the material from any one species of organism contain only a very limited number of types of amino acid, not usually exceeding four or five and both L and D isomers are present alternating in sequence. A D-alanine residue is always present. These peptides, which are attached to the -COOH groups of the *N*-acyl muramyl residues, are linked together in three or four basic manners. Two of the sorts of linkage are shown in Figure 1: the one a direct linkage which is present in peptidoglycan from many bacilli and from all the Gram-negative species examined; the other having a bridge between the -COOH terminal D-alanyl residue of one peptide and the amino group of the diamino acid in a neighbouring peptide.

The Enzymes

Bond specificity and distribution

The bond specificities of the four common types of autolytic enzyme are also shown in Figure 1. They are: 1) a lysozyme-like enzyme (a muramidase) that hydrolyses the *N*-acetylmuramyl β 1,4 *N*-acetylglucosamine bonds in the glycans to liberate free-reducing groups of *N*-acetylmuramic acid. 2) A β *N*-acetylglucosaminidase that liberates the free-reducing groups of *N*-acetylglucosamine. Both 1) and 2) are endo-enzymes and the corresponding exo-enzymes, which are also known [Berkeley *et al.*, 1973; Del Rio and Berkeley, 1976], do not seem to act as autolysins, being unable to hydrolyse bonds in whole peptidoglycans. 3) An *N*-acetylmuramyl-L-alanine amidase (amidase) that hydrolyses the bond between the glycan chains and the peptide. 4) Peptidases that can hydrolyse some of the main peptides and the bridge peptides when they occur between the D-alanyl terminus and the amino group of a contiguous peptide chain (see Figure 1). Also, commonly occurring in many organisms are D-alanine carboxypeptidases which can cleave off any carboxy terminal D-alanyl residues of the peptidoglycan. These enzymes, however, are not known to act as autolysins and therefore will not be considered. A further most interesting autolytic enzyme is formed by *Escherichia coli*. This not only hydrolyses the β 1,4 bond between *N*-acetylmuramic acid and *N*-acetylglucosamine but carries out a dehydration reaction so that the disaccharide formed from peptidoglycan is non-reducing because a 1,6 anhydro-*N*-acetylmuramic acid residue is present. This product was first unambiguously identified from the action of an enzyme in phage λ and V_1II endolysin lysates of *Escherichia coli*

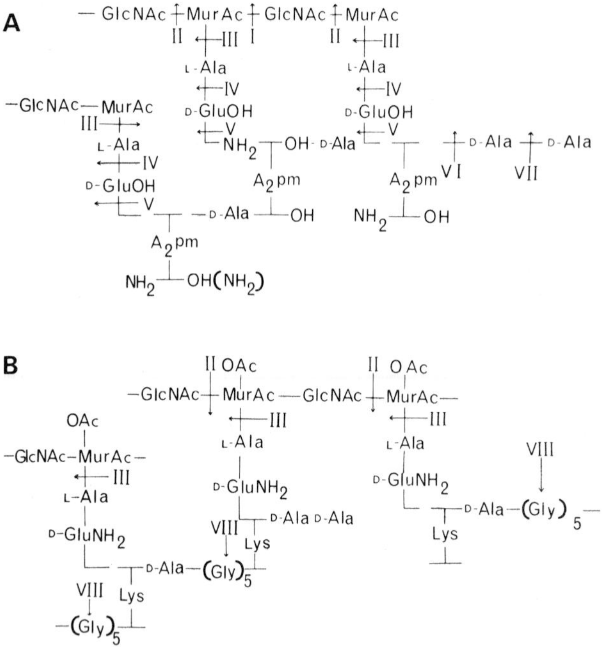

Fig. 1 (a) The structure of the peptidoglycans present in many bacilli and in all the Gram-negative species examined. The arrows indicate the bonds hydrolysed by the known autolytic enzymes. These enzymes are i) endomuramidase, (ii) endo-β *N*-acetylglucosaminidase, (iii) *N*-acetylmuramyl-L-alanine amidase (amidase), (iv) and (v) peptidases associated with spores or sporulating cultures, (vi) and (vii) D-alanine carboxypeptidases present in many organisms.
(b) The structure of a peptidoglycan with a 'bridge' peptide. The one illustrated is present in *Staphylococcus aureus*. The notation for autolytic enzymes is as in Figure 1 together with (viii) which is a peptidase present in lysostaphin and produced by other staphylococci.

[Taylor *et al.*, 1975]. Subsequently it was isolated and purified from disintegrated cells of *Escherichia coli* strain W7. It completely degraded peptidoglycan sacculi from the organism into this non-reducing disaccharide [Holtje *et al.*, 1975]. The distribution of this enzyme is not known.

Table 1 shows the distribution of the four autolytic enzymes. The commonest ones are the amidase and the endo-β *N*-acetylglucosaminidase which most frequently occur together in the same organisms. The muramidases appear, from existing evidence, more often to exist as the sole or predominant autolytic enzyme of the bacteria concerned, but relatively few organisms have been examined in sufficient detail to make this certain. Among them are *Streptococcus faecalis* [Shockman *et al.*, 1967] and *Lactobacillus acidophilus* [Coyette and Ghuysen, 1970]. Some enzymes such as those formed by *Streptomyces*, have been used extensively to investigate the structures of bacterial peptidoglycans but it is not entirely clear whether all of them are involved in autolysis of the organisms that produce them. They have at present, therefore, to be regarded as bacteriolytic enzymes rather than as autolytic ones.

Purification

Rather few autolysins have been purified to yield homogeneous proteins when examined by techniques such as polyacrylamide electrophoresis. Undoubtedly one of the reasons for this is the difficulty that has frequently been met in isolating them from wall autolysates where the enzymes are often firmly bound to wall constituents, such as the teichoic acids [Brown *et al.*, 1970; Brown and Young, 1970]. Success has been more readily met in purifying autolysins that are exported from the cells into the culture fluid. For example, each of the three types of staphylococcal enzyme has been obtained as a homogeneous protein [Wadström and Hisatsune, 1970a; Wadström and Vesterberg, 1971] as have the endo-muramidases of the strains of *Bacillus subtilis* [Okada and Kitahara, 1973; Takahara *et al.*, 1974]. Among the autolysins of Gram-negative bacteria the amidase of *Escherichia coli* strain K12 has been purified from the supernatant fluid of spheroplasts [van Heijenoort *et al.*, 1975]. The bacteriolytic amidase, peptidase and two glycanases have been purified from the culture supernatants from cultures of streptomyces species [Ghuysen *et al.*, 1968; Ward and Perkins, 1968].

Nevertheless some successes have been obtained by direct fractionation of cell wall hydrolysates. For example, at an early stage a streptococcal endo-β-*N*-acetylglucosaminidase was purified by some 5,300-fold from phage lysates with a yield of 36% [Barkulis *et al.*, 1964]. In many other examples, however, what was probably only partial purification could be achieved. The difficulties of separating the autolytic enzymes from wall polymers has been overcome in recent times by applying [Fan, 1970; Brown, 1972] the observation first made by Pooley *et al.* [1970]. These workers found that the autolytic muramidase could be removed from cell walls of *Streptococcus faecalis* by extraction with strong salt solutions. Solutions of CsCl, NH_4Cl and LiCl at concentrations of up to 10 molar, were used and both the latent and active forms of the enzyme were removed by solutions of 4.0 and 10.0 molar respectively. LiCl was chosen for further study because the relatively low density of strong solutions facilitated deposition of wall preparations during centrifuging. Application of this method has allowed the purification of the autolytic amidase from the walls of *Bacillus megaterium* and *Bacillus subtilis* [Chan and Glaser, 1972; Herbold and Glaser, 1975a].

Properties of the isolated autolysins

The molecular weights of the isolated enzymes show considerable differences, ranging from 20,000 for the amidase of *Bacillus megaterium* [Chan and Glaser, 1972] to 100,000 for the endo-β-*N*-acetylglucosaminidase of streptococci [Barkulis *et al.*, 1964] and 90,000 for the muramidase from one strain of *Bacillus subtilis* [Okada and Kitahata, 1973]. Some indication has been found for differences in molecular weight for enzymes of the same specificity isolated from different strains or closely related species. For example, the staphylococcal endo-β-*N*-acetylglucosaminidase isolated from culture filtrates of strain M 18 had a molecular weight of 70,000 whereas that in lysostaphin was 55,000. Even more impressive, endo-muramidases isolated from two strains of *Bacillus subtilis* had molecular weights of 13,000 [Takahara *et al.*, 1974] and 90,000 [Okada and Kitahata, 1973].

The glucosaminidases at least from some sources appear to depend upon -SH groups for their activity. The inactivation of the enzyme from staphylococci during storage at 4°C was prevented by the presence of cysteine, or dithioerythritol, and enzyme activity was reduced to 50% by 1 mM-*N*-ethylmaleimide, 0.1 mM p-hydroxymercuribenzoate or 0.1 mM iodoacetate [Wadström, 1970]. Likewise the enzyme from haemolytic streptococci was inhibited by iodoacetamide [Barkulis *et al.*, 1964]. On the other hand the activity of muramidase of *Streptococcus faecalis* in partially purified preparations did not depend on reduced -SH groups [Shockman *et al.*, 1967] and neither did that of this enzyme from one strain of *Bacillus subtilis* [Takahara *et al.*, 1974]. The amidase from *Bacillus subtilis* [Herbold and Glaser, 1975a], the streptococcal β-*N*-acetylglucosaminidase [Barkulis *et al.*, 1964] and the muramidase of *Bacillus subtilis* [Takahara *et al.*, 1974] are critically influenced in activity by their ionic environment at low concentrations of salt, whereas the β-*N*-acetylglucosaminidase from staphylococci is not so, except at high concentrations (0.2-0.4 molar) when it is inactivated [Wadström, 1970]. The latter enzyme, which like lysozyme has a high iso-electric point of pH 9.5 [Wadström and Hisatsune, 1970b], is very heat stable at pH values from 3 to 4, resisting 100°C for 30 min, but becomes inactivated at pH 9.5 in 5 min at the same temperature. Despite the heat stability at low pH it is 50% inactivated by keeping for 2 h at 4°C in a pH of 3.2 and all activity had disappeared in 24 h [Wadström, 1970]. The muramidase of *Bacillus subtilis* on the contrary was of intermediate heat stability being inactivated by 1 h at 90°C at pH 6.2. Thus the autolysins seem to be a rather widely assorted set of enzyme proteins which, irrespective of bond specificity, show a variety of characteristics which differ not only according to the species of microorganism from which they come but even, in some examples, according to the strain of the same species.

Specificity

Each of the four groups of enzymes seems to be quite specific for a particular bond in the peptidoglycan, except possibly for some of the peptidases which have not been investigated in detail. The ability of autolysins to act on walls from different species is probably as much related to their interaction with the polymers attached to the peptidoglycan as upon the structure of the polymer itself. For example, the presence and type of teichoic acid attached to peptidoglycan in strains of *Bacillus subtilis* was shown to be important to the action of the isolated amidase and even more to its avidity for the walls [Herbold and Glaser, 1975b]. In general the strength with which the autolytic amidases bind to walls, however, does not seem to be a reliable guide to their enzymic effectiveness. For example, the amidase from *Bacillus megaterium* walls does not bind significantly to homologous wall preparations [Chan and Glaser, 1972] whereas that from *Bacillus subtilis* binds strongly with a specificity dependent upon the teichoic acid present [Herbold and Glaser, 1975b]; both enzymes rapidly lysed their homologous walls.

The staphylococcal endo-β-*N*-acetylglucosaminidase [Wadström and Hisatsune, 1970b] is rather catholic in its taste for walls of different species of bacteria, lysing cells of eleven species at a significant rate - a wider choice than egg-white lysozyme. It could also hydrolyse chito-dextrins at a slow rate [Wadström, 1971] with an optimum pH of 4.1 compared with a value of about 6.0 when acting on bacterial cell walls. The product was di-*N*,*N*'-acetylchitobiose and not free *N*-acetylglucosamine. Some muramidases are highly specific, for example, the enzyme from *Lactobacillus acidophilus* is able to hydrolyse walls from only one other species of bacteria out of nine tested [Coyette and Ghuysen, 1970]. Unlike egg-white lysozyme, the bacterial muramidases appear usually to be unable to hydrolyse chitin and its dextrins.

The amidases may be rather more specific than the β-glucosaminidases but less so than the muramidases. For example, the enzyme from *Bacillus subtilis* will hydrolyse only walls of *Bacillus megaterium* as rapidly as homologous walls, but those from *Bacillus licheniformis* are hydrolysed at about 30% of the rate, and surprisingly those from one strain but not another of *Streptococcus lactis* are hydrolysed slightly faster (50%) than those from *Bacillus licheniformis* [Chan and Glaser, 1972]. Neither of the isolated enzymes from bacilli would hydrolyse peptidoglycan in which glycoside bonds had been extensively hydrolysed

[Herbold and Glaser, 1975a; Chan and Glaser, 1972]. The enzyme from staphylococci appeared to be able to do this [Wadström and Vesterberg, 1971] but very slowly. In contrast one of the best substrates for the amidase from *Escherichia coli* was the *N*-acetylmuramyl-pentapeptide [van Heijenoort *et al.*, 1975]. Peptidoglycans freed from their associated polymers such as the teichoic or teichuronic acids are frequently not such a good substrate for autolytic enzymes as are whole cell walls. The combined result of this observation and the acute specificity demands for the correct teichoic acid to be attached to the peptidoglycan, already exemplified, has led to suggestions of a necessary specific interaction between the teichoic acids and the autolysins. Alternatively it has been suggested [Robson and Baddiley, 1977] that teichuronic acid and not teichoic acid is essential for rapid autolysis of walls of *Bacillus licheniformis* strain ATCC 9945. A chain forming, novobiocin-resistant mutant was almost completely deficient in cell wall teichuronic acid and selective removal of teichoic acid from walls of the wild-type had little effect on sensitivity to added lysin whereas removal of teichuronic acid made them as resistant as those of the mutant strain.

Location

To be able to lyse the walls around bacteria without damage to the underlying protoplast membrane, at least some of one or more of the autolytic enzymes produced must be associated with the walls, even if the steady state level during growth is low. Intact functioning protoplasts, for example, have been produced from *Streptococcus faecalis* by incubating suspensions of cells in buffered sucrose and allowing the wall-associated autolysin to remove the walls - they were referred to as autoplasts [Rosenthal and Shockman, 1975; Rosenthal *et al.*, 1975]. It is technically very difficult nevertheless, to design methods which unambiguously define the level during growth because as soon as the cells are disrupted, redistribution of the enzymes can take place. Alternatively attempts to estimate enzyme activity associated with walls in growing cells from the known or suspected functions of autolysins, such as wall turnover, are bedevilled by insufficient knowledge about the processes themselves, and about the importance of the chemical topology of the surface substrate which may regulate autolytic activity. In *Streptococcus faecalis* a rather high proportion of the total cellular enzyme is in a latent form that can be activated by trypsin treatment of the walls. Both this latent form and the active enzyme have very high affinities for wall preparations with the result that most of the total autolytic activity appears as wall-associated in a disintegrated cell preparation [Shockman *et al.*, 1967]. Later work [Pooley and Shockman,

1969] suggested nevertheless, that a high proportion of the latent enzyme might be associated with the cytoplasm in the living organism. In L-forms of *Bacillus licheniformis* strain 6346 on the other hand 93% of the autolytic enzymes was associated with the membrane in disrupted cell preparations, from which it could be removed by incubation with isolated walls [Forsberg and Ward, 1972]. In the vegetative cells, significant amounts of the enzyme were also found in the membranes. In *Lactobacillus acidophilus* about 25% of the muramidase remained associated with the cytoplasm in disrupted preparations and this was thought to be due to the limited number of binding sites in the walls [Coyette and Shockman, 1973]. In *Escherichia coli* [Pelzer, 1963] and in some strains of *Staphylococcus aureus* [Singer *et al.*, 1972] very high proportions of the autolytic activities remained in the soluble cytoplasmic fractions. More recently [Wolf-Watz and Normark, 1976] a rather high proportion of the amidase in the former organism was found in the outer membrane of the bacteria. Presumably these observations may reflect the relative affinities of the enzymes for the different cell fractions rather than their true locations *in vivo*. At first sight, the most surprising feature of autolysins is that in a number of strains of Gram-positive species such as *Staphylococcus aureus* and some bacilli [Takahara *et al.*, 1974; Okada and Kitahata, 1973; Priest, 1977] a large proportion of the activity is found soluble in the supernatant fluid of the cultures. Examination of the distribution of lytic factors in cultures of *Staphylococcus aureus* strain Oeding [Huff and Silverman, 1969] for example, showed the presence of 4 times as much activity, soluble in the supernatant fluid as there was associated with the cells. Subsequent to the demonstration [Tipper, 1969] that three autolysins were present in staphylococci, all three were shown to be present in the supernatant fluids of another strain of *Staphylococcus aureus* M18 [Wadström and Vesterberg, 1971]. Lysostaphin indeed is a soluble bacteriolytic preparation produced from cultures of a colony of cocci occurring on staphylococcal plates and causing bacteriolysis of the surrounding staphylococci [Schindler and Schuhardt, 1964]. It contains all three autolysins [Browder *et al.*, 1965; Wadström and Vesterberg, 1971]. A soluble extra cellular glycanase attacking the walls of *Micrococcus luteus* was found to be produced by a strain of *Bacillus subtilis* [Richmond, 1959a, b] and extra-cellular endo-muramidases were recognized in two other strains of the organism [Takahara *et al.*, 1974; Okada and Kitahara, 1973]. These latter observations point to the great strain variation in this respect since other common laboratory strains of *Bacillus subtilis* such as 168 and W23, do not form soluble autolysins. Likewise a strain of a closely related species, *Bacillus licheniformis*, with extremely active autolysins when compared with some strains of *Bacillus subtilis*, also did not liberate the enzymes into the supernatant fluid. A mutant of the *Bacillus licheniformis* strain however, could be isolated, called 'superlytic', that did so during the exponential phase of growth. Examination of cultures of this mutant strain for a known cytoplasmic enzyme, α-glucosidase, showed that an exactly equivalent proportion was present in the supernatant fluids. Thus it seems probable that in this instance some of the cells were lysing even during the exponential phase of growth and leakage rather than export of both enzymes was occurring [Forsberg and Rogers, 1974]. It nevertheless seems unlikely that such an explanation applies to other organisms where very high proportions of the total enzymes are extra-cellular. Despite the apparent absence of soluble autolytic enzymes in exponential phase cultures of some strains of bacilli, they can be transferred between cells in cultures of *Bacillus subtilis*. If wild-type cells are grown in the same culture as an autolytic-deficient mutant, aspects of the phenotype such as cell separation and wall turnover of the latter are restored towards those of the wild-type [Fein and Rogers, 1976; Rogers *et al.*, 1974]. The two organisms must be grown together in mixed culture; no soluble factor is present and no permanent change of the mutant cell is involved. Thus the autolytic enzymes are apparently so superficially located in the wild-type that they can be transferred to the surfaces of the mutant cells. Indeed two quite different approaches [Pooley *et al.*, 1972; Higgins *et al.*, 1973] have suggested that effective autolytic enzymes might always be exported and then act from the outside of the cell, rather than acting within the wall or on its inner face as might be expected.

Another approach attempting to answer questions about the location of autolytic enzymes is to examine sections of cells that are undergoing autolysis. In this way it was shown that in a *Streptococcus faecalis* cell the leading edge of the forming septum that will eventually divide the cells is the first site of action of the autolytic muramidase [Higgins *et al.*, 1970]. Only after the septum has been extensively damaged is the peripheral wall in the vicinity removed. This technique certainly gives a clear impression of localization of enzyme action in cells placed under conditions that are suitable for autolysis (i.e. non-growing suspensions incubated in buffer), but it cannot, of course, be extrapolated with certainty to the situation in rapidly growing bacteria. The results obtained may also depend on the local susceptibility of wall polymers as well as on localization of the enzyme. Other evidence [Shockman *et al.*, 1967] supports the presence of autolytic enzyme in the region of the septum. When exponentially growing cultures are first labelled with ^{14}C-labelled L-lysine for 0.1-0.4 of a generation, and the walls isolated from disrupted cells and allowed to autolyse, radioactivity due to the L-lysine in the peptidoglycan is solubilized more rapidly than the walls as a whole. In other words, the most recently synthesized wall is most rapidly released as might be expected since quite different evidence [see Rogers *et al.*, 1978, for summary] shows that the walls of streptococci grow from the septal region, near to where autolysin is seen to be acting most vigorously in non-growing cells. Thus despite the known relocalization of autolysin during cell disruption, autolysis still occurs near the same place as it does in whole cells which may be evidence for differences in local susceptibility to autolytic enzymes in the walls, and raises further questions about the interpretation of the localization of autolysin seen in whole cells by the sectioning technique.

Functions

Autolysins and the growth of microbial cells

If indeed walls consist of polymers covalently linked together into 'huge bag-shaped' molecules [Weidel and Pelzer, 1964], or even if they contain polymers like cellulose covalently linked together in one plane but with the fibrils held together by secondary valency bonds in two other planes, enzymic nicking would be an expeditious method of allowing the introduction of new units to expand the network during growth. If newly biosynthesized units are to be added to the nicked polymers, the appropriate acceptor groups must be made available by the hydrolases present. This is particularly relevant of course, when complex wall polymers such as the peptidoglycans in bacteria are involved. Either the reducing group of the *N*-acetylmuramyl residues, a suitable amino group of a diamino acid, or -COOH groups of D-alanine in the peptide, must be available for the biosynthetic enzymes to function. In fungal cell walls, the requirements are less stringent since most of the autolytic hydrolysases break bonds in the main backbone of homopolysaccharides, even when branching is present. They thus inevitably liberate suitable functional groups for the addition of new material.

As far as bacteria are concerned, the first organism to be seriously considered in the context of a possible essential role for autolytic enzymes in growth was *Streptococcus faecalis* [Toennies and Shockman, 1958; Shockman *et al.*, 1967]. In some ways this was fortunate since it has a muramidase as the sole autolytic enzyme (see Table 1) and the specificity of this enzyme is suitable for revealing potentially effective functional *N*-acetyl muramyl reducing groups as acceptors. Secondly, evidence already mentioned showed that autolytic activity is concentrated in just the region of the cells in which new wall material is being deposited during growth and division. Thus it seemed, and still seems, a wholly reasonable hypothesis to be applied to this organism. Unfortunately as can be seen from Table 1

TABLE 1

The distribution of common bacterial autolysins

Microorganism	*N*-acetylmuramyl-L-alanine amidase (amidase)	β-*N*-acetyl-glucosaminidase	Muramidase	Peptidase	References
Arthrobacter crystallopoietes	0	0	+	0	Krulwich and Ensign [1968]
Staphylococcus aureus	+ (E)	+ (E)	0	+ (E)	Tipper [1969]
Bacillus subtilis 168	+	+	0	0	Young [1966]
Bacillus subtilis W23	+	+	0	0	Brown and Young [1970]
Bacillus subtilis YT25	?	?	+ (E)	?	Takahara *et al.* [1974]
Bacillus subtilis K77	?	?	+ (E)	?	Okada and Kitahata [1973]
Bacillus cereus	+	+	0	0	Hughes [1971]
Bacillus licheniformis	+	+	0	0	J.S. Thompson and H.J. Rogers, unpublished observation
Bacillus thuringensis var. *thuringensis* (sporulating)	+	0	+	+	Kingan and Ensign [1968]
Bacillus stearothermophilus	+	0	0	0	Grant and Wicken [1970]
Clostridium welchii	0	+	+	0	Martin and Kemper [1970]; J.B. Ward and R. Williamson (unpublished)
Clostridium botulinum type A	+	+	0	0	Takumi *et al.* [1971]
Streptococcus faecalis	0	0	+	0	Shockman *et al.* [1967]
Streptococcus pyogenes[b]	0	+	0	0	Barkulis *et al.* [1964]
Diplococcus pneumoniae	+				

Lactobacillus acidophilus (AM Gasser)	0	0	+	0	Coyette and Ghuysen [1970]
Mycobacterium smegmatis	+	0	0	+	Kilburn and Best [1977]
Escherichia coli	+	+	0	+	Pelzer [1963]
Proteus vulgaris	+	?	?	?	van Heijenoort *et al.* [1975]
Brucella abortus	+	?	?	?	van Heijenoort *et al.* [1975]
Listerella monocytogenes	+	(+)[a]	(+)[a]	0	Tinnelli [1968]
Streptomyces griseus	?	0	+ (E)	?	Ward and Perkins [1968]
Streptomyces sp.	+	+	+ (E)		Munoz *et al.* [1966]
Aeromonas hydrophila				+ E	Coles *et al.* [1969]
Myxobacter sp.	+			+ E	Ensign and Wolfe [1965]; Hungerer *et al.* [1969]

[a]Glycanase present but bond specificity not determined; [b]Bacteriophage infected; E = Extracellular.

many other bacteria have autolytic enzymes of unsuitable specificity which could not easily liberate suitable functional groups, and the location of autolytic activity has not been so easy to demonstrate in other organisms as it was with *Streptococcus faecalis*. Where some localization has been demonstrated in the septal regions as for example in *Lactobacillus acidophilus* [Higgins *et al.*, 1973] it has appeared rather to be acting from the outside of the cells and to be part of cell separation events, rather than of cell growth. Suggestive negative evidence against an essential role for autolysins in growth has also now accumulated from the study of defective mutants. It should immediately be said, however, that this evidence does not yet exclude unambiguously a role for a very small proportion of the wild-type autolytic activity, strategically positioned in the cells. Successful isolation of autolytic-deficient mutants has been made from strains of *Streptococcus faecalis* [Pooley *et al.*, 1972; Cornett *et al.*, 1978], *Streptococcus pneumoniae* [Lacks, 1970], *Staphylococcus aureus* [Chatterjee *et al.*, 1976; Koyama *et al.*, 1977], *Bacillus licheniformis* [Forsberg and Rogers, 1971, 1974] and *Bacillus subtilis* [Fan and Beckman, 1971; Ayusawa *et al.*, 1975; Yoneda and Maruo, 1975; Fein and Rogers, 1976]. Conditions have also been found [Tomasz, 1968] whereby, due to a change in the wall teichoic acid of *Streptococcus pneumoniae*, the autolytic activity is grossly reduced. The activities of the strains in these various experiments were reduced by 50-95% but only in one instance [Fan and Beckman, 1971] was a significant effect on either growth rate or individual cell morphology recorded; this experiment will be discussed later. In the work of Fein and Rogers [1976] the mutants of *Bacillus subtilis* were 90-95% deficient in both of the autolytic enzyme activities present (i.e. an endo-β-*N*-acetylglucosaminidase and an amidase). Despite an effect on both activities, all the genetic evidence supports the presence of only a single lesion. The growth rates of these mutants in five different media, in which the doubling times varied from 19.3 to 65.7 min, were identical within 15% to those of the isogenic wild-type. The morphology of the individual cells of the mutant appeared normal under the light microscope but detailed ultrastructural work was not done. It is against this evidence that the observation [Fan and Beckman, 1971] of a positive role for autolysins in the growth of *Bacillus subtilis* has to be assessed. A mutant strain was shown to grow better at 51°C in a broth medium when either lysozyme or an extract containing autolysin from the wild-type was added to the medium and the distorted morphology of the individual cells was corrected. Although very interesting, the meaning of this work has to be queried. The mutant was not nearly as deficient in autolysin as many others but it is difficult to be precise because the kinetics of lysis of the mutant and wild-type walls differed considerably.

At best a figure of 50-70% deficiency for the mutant compared with the wild-type activity would be reasonable. The organism had been mutagenized with nitrosoguanidine on two separate occasions during isolation, and then mutants were severely selected for survival to 0.01M KCN at 51°C, and for ability to have improved growth in whole wall autolysate prepared from the original parent organisms. The properties of the mutant could not be transferred by transformation and it seems likely that many different lesions were present. That its growth rate was improved and its morphology partially corrected when grown with lytic factors is certainly interesting but must await further work before being taken as proof of an essential role for autolysins in bacterial growth, particularly in view of the normal behaviour in these respects, of mutants that were very much more deficient.

One final piece of evidence shows that the bulk of the expressed autolytic activity of bacteria is not necessary for morphological changes. Conditional mutants of several species of bacteria have been described which can grow either as rods or cocci. Examination of one such mutant of *Bacillus subtilis* has shown that the coccus, like *Streptococcus faecalis* forms its peripheral wall from the septal region, unlike the rod which extends within the cylindrical region of the cell [I.D.J. Burdett and H.J. Rogers, unpublished work]. One might expect that the coccus would show some disability in growth if deprived of autolysin since the presence of autolytic activity at the base of the septum, to allow the 'peeling' apart of the peripheral walls as they are formed, is required for the model of Higgins and Shockman [1976]. However, insertion of the *lyt* gene from one of the mutants described by Fein and Rogers [1976] into strains already carrying *rod* genes had little or no effect on either the morphological change or the growth of the coccal forms [Rogers and Thurman, 1978; H.J. Rogers, P.J. Piggot, C. Taylor and P.F. Thurman, unpublished work].

A clear weakness, on the other hand, of the negative evidence is that none of the mutants so far described is entirely deficient in autolysins and there is no yardstick to measure how much is enough. An indication of our ignorance in this respect is illustrated by the finding [J.B. Ward and C. Taylor, unpublished work], that in one of the mutants isolated by Fein and Rogers [1976], the glycan chains in the peptidoglycan had been hydrolysed to yield free *N*-acetylglucosamine-reducing groups to nearly the same extent as those in the walls of the isogenic wild-type (Table 2), yet this mutant appeared to have only 3-4% of the wild-type endo-*N*-acetylglucosaminidase activity in LiCl extracts made from it. The exact interpretation of the results for 'biosynthetic' and *in vivo* chain lengths [Ward, 1973] is possibly open to some debate, but the appearance of an undiminished amount of free-reducing groups of *N*-acetylglucosamine in the walls is strongly suggestive of continued enzyme action despite the low enzymic activity as measured in extracts acting on exogenous substrates.

TABLE 2

Average chain lengths of glycans in peptidoglycans from an autolytic-deficient mutant (lyt⁻) and its isogenic wild-type of Bacillus subtilis
[*J.B. Ward and C. Taylor, unpublished work*]

	Average chain lengths				Ratios: Biosynthetic / *in vivo*	
	Biosynthetic[a]		*in vivo*[a]			
Mutant	487.6	313.4	192.9	158.3	2.53	1.98
Wild-type	604.1		209.3		2.89	

[a]These values for two separate batches of cell walls are calculated from results obtained by the method of Ward [1973]. The biosynthetic value indicates the proportion of free-reducing groups of *N*-acetylmuramyl residues to the total whilst the length *in vivo* indicates the proportion of *N*-acetylglucosamine reducing groups.

Cell separation and motility

Study of the deficient mutants has demonstrated clearly that depriving bacterial cells of a large proportion of their autolytic activity leads to their failure to separate from each other. Mutants of rods or streptococci form long chains of cells [Pooley *et al.*, 1972; Forsberg and Rogers, 1971, 1974; Fein and Rogers, 1976], as wild-type strains of pneumococci do when phenotypic wall changes lead to gross reduction in autolytic activity [Tomasz, 1968]. With mutants of organisms such as staphylococci which usually grow as small irregular groups of cells, large regular packets are formed [Koyama *et al.*, 1977] from some strains though possibly not others [Chatterjee *et al.*, 1976], as would be expected if cells of these mutants dividing successively in planes at right angles to each other can not separate. When normal amounts of autolysin are formed, it must be supposed that the cells can move relative to one another after septation and division. Another mutant possibly not forming regular packets was described as growing in large clumps of cells [Chatterjee *et al.*, 1976].

Most of the mutants so far examined for the specificity of their remaining enzyme complement have less of all of those present in the wild-type and they must be supposed to have lesions in regulatory rather than structural genes. For this reason, it is not possible to know which enzyme is responsible for separation or whether all must be present. A recently examined mutant of *Bacillus subtilis* isolated by

M.J. Tilby [personal communication] however, has only 2% of endo-β-*N*-acetylglucosaminidase activity but 60% of the amidase and separates normally to grow as single motile cells [H.J. Rogers and C. Taylor, unpublished work]. This may suggest that the amidase alone can effect cell separation. No mutant depressed only in the amidase has been found.

Another characteristic of the autolytic-deficient mutants of *Bacillus subtilis* and *Bacillus licheniformis* is that the bacteria have lost their flagella and are therefore non-motile [Ayusawa *et al.*, 1975; Yoneda and Maruo, 1975; Fein and Rogers, 1976; J.E. Fein, unpublished work]. Some of these [Yoneda and Maruo, 1975] have been shown to have a pool of flagellin in the cytoplasm, suggesting that the reduced autolytic activity in some way prevents the extrusion and/or organization of the flagella from this protein. Motile revertants are all fully autolytic. Again the single mutant of *Bacillus subtilis* so far available with a grossly depressed β-*N*-acetylglucosaminidase but a reasonably high amidase activity is motile [H.J. Rogers and C. Taylor, unpublished work].

Since the lytic-deficient mutants of *Bacillus subtilis* have complex phenotypes and are most likely to have a lesion in some regulatory gene, it is possible to argue that the lack of flagella is one more aspect of the phenotype not directly connected in a causative manner with the absence of active autolytic enzymes. Examination of autolytic-deficient phosphoglucomutase-deficient mutants of *Bacillus licheniformis* and *Bacillus subtilis* [Forsberg and Rogers, 1971; 1974] showed that these too did not separate and were non-flagellated. This observation alone would not help the argument since their phenotype is equally complicated, but it is possible to circumvent the effect of the phosphoglucomutase lesion in a biochemical way. When grown in media containing galactose together with an assimilable carbon source such as glycerol, the phenotype of these mutants is partially reversed. Polymers dependent upon the formation of UDP-glucose appear in the walls and the autolytic enzyme activity is partially restored [Forsberg *et al.*, 1973]. More important in the present context, flagella are formed and the bacteria are once more motile [J.E. Fein, unpublished observations].

Evidence has also been obtained [Wolf-Watz and Normark, 1976] that cell separation and the splitting of the peptidoglycan septum [Burdett and Murray, 1974a, b] is related to the autolytic amidase activity in Gram-negative species such as *Escherichia coli*. The specific activity of amidase in the *env*A mutant, which forms chains of unseparated cells is some five times lower than in the *env*A$^+$ strain at high growth rates but only 30% lower at slow rates. Correspondingly the chains were longer at high growth rates and the specific activity of the amidase varied as would be expected if it were related to separation. Addition of

N-acetylmuramyl-L-ala-D-glu-meso DAP-stopped cell separation altogether suggesting competitive inhibition of the amidase by a good enzyme substrate [van Heijenoort *et al.*, 1975].

Turnover of bacterial wall polymers

Examination [Chaloupka, 1967; Chaloupka *et al.*, 1962; Boothby *et al.*, 1973; Mauk *et al.*, 1971; Pooley, 1976a, b] of the fate of pulses of radioactive compounds labelling peptidoglycans, or of cells already so labelled but growing in non-radioactive medium, has shown that the walls of many organisms such as bacilli or lactobacilli are in a state of flux. In other species such as *Streptococcus faecalis* this is not so [Boothby *et al.*, 1973] and the walls do not show turnover, which is the term usually applied to this process of loss and renewal of wall polymers. It is of course, not strictly turnover, as applied to small molecules, or even to proteins, since it probably represents the continued partial breakdown of macromolecules and either the further growth of their remaining parts or the initiation of new chains. The negative side of such a process involves the formation of soluble material from the insoluble wall and it is, therefore, reasonable to suppose that autolytic enzymes have to act for it to occur. One report [Glaser, 1973] very disturbing to this hypothesis, however, is that of a strain of *Bacillus subtilis* W23 with a normal complement of autolysins and normal walls but with a wall turnover reduced by 90%. Although little other exacting work has been done to test the idea such results as are available suggest a role for autolysins in turnover. Examination of an autolytic-deficient, phosphoglucomutase-negative strain of *Bacillus licheniformis* [Rogers *et al.*, 1974] and the chain forming strain Ni15 of *Bacillus subtilis* [Pooley, 1976a] by first labelling the walls with [^{14}C] *N*-acetylglucosamine and then growing them in medium containing the non-radioactively labelled compound, showed that both had greatly reduced turnover compared with the wild-types from which they were derived. When small volumes of a crude autolysate from the wild-type *Bacillus subtilis* were added to cultures of Ni15, turnover was increased in relation to the volume of autolysate added [Pooley, 1976a]. Examination of the turnover of the walls of the autolytic-deficient mutant of *Bacillus subtilis*, FJ6 isolated by Fein and Rogers [1976] showed that none could be detected [R.S. Buxton and J.E. Fein, unpublished work]. This was again done by the complete labelling technique. Work with another of the latter mutants (FJ3) by treatment with short pulses of ^{14}C *N*-acetylglucosamine followed by a chase, appears, however [J. Mandelstam, personal communication], to demonstrate that turnover can occur under some circumstances. On the other hand similar pulse experiments with Ni15 did not allow a different conclusion from the experiment with completely labelled walls already described [H.M. Pooley, personal communication], since greatly reduced turnover was found. Again it is not possible from these experiments to say whether one or both of the autolytic enzymes in the organisms are likely to be necessary for turnover. It could be argued that some activity of both would be expected to be necessary since both glycan and peptide bonds hold the peptidoglycan together. However, only one enzyme, a muramidase, is present in *Lactobacillus acidophilus* [Coyette and Ghuysen, 1970] yet the peptidoglycan of this organism actively turns over [Boothby *et al.*, 1973; Daneo-Moore *et al.*, 1975]. This raises interesting problems about the geometry of the process. Two possibilities can be envisaged: 1) two contiguous glycan strands may be hydrolysed that are cross-linked by peptides but are not thus joined to other glycans; 2) one strand may be hydrolysed but only in regions where peptide cross-linking is not complete. It is particularly interesting that whereas the peptidoglycan of *Streptococcus faecalis* does not turnover, that of *Lactobacillus acidophilus* does, despite the presence in both organisms of only muramidase as an autolytic enzyme. It may be relevant to note that a minimum fraction equivalent to 10-20% of the peptidoglycan in the latter organism was immune from the process [Daneo-Moore *et al.*, 1975]. This proportion corresponds well with the figure of 13% for the conserved radioactivity in the poles of bacilli [H.M. Pooley and D. Karamata, personal communication], and with 15% for the measured area of the pole formed as a proportion of the whole surface [Burdett and Higgins, 1978]. These facts might with profit be related to those suggesting that autolysins act from the outside of the cell both in promoting turnover [Rogers *et al.*, 1974; Pooley, 1976a, b] and in cell separation [Higgins *et al.*, 1973]. The poles of bacilli but not the cylindrical region are partially resistant to the action of autolysin, at least under some conditions [Fan *et al.*, 1972], and exponentially growing whole cells of *Streptococcus faecalis* do not fix autolysin and are resistant to high concentrations of lysozyme [Toennies *et al.*, 1961]. Supernatant fluids from cultures of this latter organism, however, dechain the streptococci [Lominski *et al.*, 1958], suggesting the presence of autolysin. One might see in these various observations a reason for the failure of the walls of *Streptococcus faecalis* to show turnover whereas bacilli, for example, do so: the walls of growing streptococci cannot be attacked from the outside and this is necessary before turnover can occur. Exponentially growing lactobacilli on the other hand are highly sensitive to lytic enzyme along the cylinder of the rod-shaped organism. Nevertheless, the possibility must still be kept in mind that the difference between *Lactobacillus acidophilus* and *Streptococcus faecalis* lies rather in the organization of the wall in such a way that muramidase is able to liberate soluble material from the former but not the latter microbe. On grounds of resistance to external lytic enzyme, one might suppose that the fraction of the wall of *Lactobacillus acidophilus* resistant to turnover is the polar region of the cells. However, a claim has been made that the polar regions of *Bacillus subtilis* show turnover at nearly the same rate as the wall of the cylindrical part [Fan *et al.*, 1974].

Autolysins in transformation

Another attractive role for the autolysins of bacteria is to allow entry to the cell through the cell wall of macromolecules such as DNA. Such a role in transformation has been suggested over a period of many years in a number of different systems [Akrigg *et al.*, 1967; Akrigg and Ayad, 1970; Akrigg *et al.*, 1969; Ephrussi-Taylor and Freed, 1964; Ranhand *et al.*, 1971; Ranhand, 1973; Tichy and Landman, 1969; Young *et al.*, 1964; Seto and Tomasz, 1975]. A variety of evidence supporting the idea has been obtained. For example, correlations have been found between competence and the rates of autolysis of cells or walls isolated from them in *Bacillus subtilis* [Young and Spizizen, 1963] and in Group H streptococci [Ranhand *et al.*, 1971; Ranhand, 1973]. No doubt the extremely complicated nature of the competent state is, however, responsible for difficulties met in providing unambiguous evidence that autolysis of walls is an essential part of it. When competence factor (CF) is added to streptococci, the rate of lysis increases for both Group A streptococci [Ranhand, 1973] and pneumococci [Seto and Tomasz, 1975]. In the former this also corresponds with increasing appearance of extra cellular lytic activity. Competence reaches a peak at the maximum of autolytic rate and extra cellular lytic activity. Seto and Tomasz [1975] propose that the CF causes some form of membrane change allowing exit of autolytic enzymes and access to the walls. That autolytic activity has a role to play in the transformation of streptococci is further suggested by the finding [Ranhand, 1974] that reagents such as *N*-ethylmaleimide or mercuric chloride inhibit both autolysis and the development of competence. The inhibition of both is reversed by 2-mercaptoethanol. When placed in hypertonic media, pneumococci form spheroplasts rather than protoplasts since they retain about 13-14% of the wall muramic acid. Competence and the rate of spheroplast formation are usually correlated [Lacks and Neuberger, 1975]. Once again in this work, as in studies with other organisms [Young and Wilson, 1972], this correlation is not complete, although undoubtedly suggesting involvement of autolytic activity in some way in transformation. For example, the mutation *ntr* causes pneumococci to be genetically incompetent and whilst *ntr*-2 mutants form spheroplasts very slowly indicating reduced autolytic activity, other *ntr* mutations such as *ntr*-9 have no such effect. The autolytic-deficient mutant of CW-1 is fully transformable [Lacks, 1970] as were the *lyt* mutants of *Bacillus subtilis* [Fein and Rogers, 1976]. Complete removal of the wall renders some Gram-positive species unable to take up DNA, but its partial removal in *Bacillus subtilis* has been claimed [Tichy and Landman, 1969] to increase DNA penetration and hence competence. Fractionation of water extracts from competent cells of *Bacillus subtilis* yields a material showing both amidase and competence increasing activities [Akrigg and Ayad, 1970] but it also contained a nuclease, forming single stranded DNA [Ayad and Shimmin, 1974]. Thus the evidence for a role of autolysins in transformation is teasingly suggestive without being, as yet, compelling.

Autolysins and the Action of Antibiotics

As we have seen, during growth of many species wall synthesis and degradation are balanced and so-called turnover of the wall ensues. If wall synthesis is inhibited in such situations, without a corresponding inhibition of either the formation or action of the autolysins, cell lysis might be expected to happen and indeed occurs. Wall synthesis may be inhibited either by temperature-sensitive lesions affecting it [Matsuzawa *et al.*, 1969; Lugtenberg *et al.*, 1972; Chatterjee and Young, 1972; Good and Tipper, 1972; Buxton, 1978], removal of specifically essential amino acids [Rhuland, 1957; Toennies and Gallant, 1949], or by adding antibiotics or other substances that are more or less specifically inhibitory [Rogers, 1967; Prestidge and Pardee, 1957].

Antibiotics that inhibit wall synthesis are bactericidal, unlike most of those inhibiting protein synthesis which are bacteriostatic. The question posed many years ago [Rogers, 1964] but not frequently mentioned, is the following: is simple inhibition of wall synthesis sufficient alone to kill microorganisms, and when the cells lyse is this because their walls can no longer expand so that the cell bursts [McQuillen, 1958], or is autolytic action an essential part of the killing of sensitive bacteria? One possible way to answer this question would seem to be to stop the formation of autolysins by blocking protein synthesis. Combinations of inhibitors of protein synthesis, such as chloramphenicol, with penicillin are not bactericidal for *Escherichia coli* [Jawetz *et al.*, 1951; Prestidge and Pardee, 1957] or for staphylococci [Rogers, 1967]. Lysis of staphylococci could also be stopped [Rogers, 1967] as could that of *Bacillus subtilis* [Rogers and Forsberg, 1971] and *Streptococcus faecalis* [Shockman *et al.*, 1965] by the inhibition of protein synthesis.

Another more specific approach to the relationship between autolysins of bacteria and the bactericidal action of wall inhibitors is to study mutants or circumstances making the bacteria deficient in autolytic activity. Mutants of *Bacillus licheniformis* [Rogers and Forsberg, 1971] and wild-type *Diplococcus pneumoniae* with modified walls [Tomasz *et al.*, 1970], died very much less rapidly than the unmodified wild-type strains when treated with a variety of antibiotics inhibiting wall synthesis. There are, however, certain points to be noted about both of these studies. Both organisms had walls that were modified so that they were resistant to the action of the cells' own autolytic enzymes as well as being deficient in the autolytic enzymes themselves. The *Bacillus licheniformis* mutants were lacking phosphoglucomutase [Forsberg *et al.*, 1973] and therefore could not glycosylate the teichoic acid in their walls, neither could they form teichuronic acid. The non-autolytic *Diplococcus pneumoniae* cells were produced by growing the wild-type strain in medium containing ethanolanine instead of choline. Nevertheless in this latter study, a mutant of the organism apparently with normal walls but with low autolytic ability was also resistant to the killing action of several wall inhibitors. The above work is all consistent with the hypothesis that autolysis is necessary for cell death when wall synthesis is inhibited but a note of caution is introduced by a study of autolytic-deficient mutants of *Bacillus subtilis* [Fein and Rogers, 1976]. These organisms had walls of normal composition and susceptibility to autolysins. Their autolysis in the presence of three different wall inhibiting antibiotics was undoubtedly very much slower but only when rather low concentrations were used. With larger amounts (5 $\mu g\ ml^{-1}$) only small barely significant differences between wild-type and mutant could be shown. A study of *Staphylococcus aureus* H (Str) [Chatterjee *et al.*, 1976] superficially led to still more disturbing conclusions. The addition of sufficient benzyl-penicillin to the wild-type of this strain to stop growth only began to lead to slow lysis four or five generations later. This might be explained however, by the fact that the penicillin was added to the cultures towards the end of the exponential phase of growth. Lysis might be expected to be, and indeed is, more rapid during the true exponential growth phase when rapid protein synthesis is occurring; despite the negative conclusions drawn by the authors, clear differences were still apparent between the autolytic-deficient mutant and the wild-type, even under their unfavourable conditions for autolysis of the latter strain [see Figure 5 of Chatterjee *et al.*, 1976].

A problem possibly raised by the results with the *lyt* strains of *Bacillus subtilis*, mentioned above, is that of just how far the presence of penicillin itself modifies autolytic enzyme formation and action. Clearly it is possible that penicillins may induce enzyme activity or alternatively some results could be explained by inhibition of the formation of autolytic enzymes by higher concentrations of some penicillins [Rogers, 1967]. No evidence on these matters has yet appeared so that they must remain speculations but a different explanation for increased activity of autolysins during inhibition of wall synthesis has been suggested [Horne *et al.*, 1977]. It is known from other work [Holtje and Tomasz, 1975; Cleveland *et al.*, 1975] that amphipathic substances such as lipoteichoic acids and the Forsman antigen from *Diplococcus pneumoniae* can strongly inhibit the action of some autolysins. It has now been found [Horne *et al.*, 1977] that during inhibition of peptidoglycan synthesis by growing cultures, lipid-containing materials are released from the bacteria without lag. This was done by first incorporating [^{3}H]-acetate and measuring the proportion of soluble radioactivity appearing in the cultures. In streptococci 80-90% of the radioactivity in the cells had been incorporated into lipid and representatives of all the membrane phospholipids could be recognized in the culture supernatants after the addition of wall inhibiting antibiotics. Clearly dilution of substances such as lipoteichoic acids in the culture supernatants might lead to greater activities of cellular autolysins, that is, if lipoteichoic acids or lipids regulate the action of the enzymes *in vivo*.

Summary

Autolysins are enzymes hydrolysing specific bonds in peptidoglycan and are weapons of potential suicide. They are retained in isolated wall preparations by weak, possibly ionic bonds. Their locations and functions in the living cells have still to be decided by unambiguous experimental work. In growing organisms, their formation and action is so regulated that, in many but not all Gram-positive bacteria, wall is lost due to their action but renewed by biosynthesis so that the total amount remains constant. How this process is regulated is unknown but many potential regulatory systems have been demonstrated. Wall turnover allows more rapid change in microbial surfaces when nutritional circumstances, such as phosphate-limitation, trigger it. Many autolysins are formed in sufficient excess to appear in quantity in the supernatant fluid of cultures during the latter parts of the exponential phase of growth in batch cultures. If production is severely restricted, even in less bountiful organisms, by introducing *lyt* genes, the cells fail to separate from each other and motile cells become non-flagelled and non-motile. Cell-separation as a part of the total cell-division process is essential for bacterial dissemination in the environment, as is motility. Thus autolytic enzymes are important survival parameters for the microorganisms.

We cannot yet afford to be dogmatic about the roles, or lack of roles, of bacterial autolysins in the growth of the individual cells. Enzymic activity can nevertheless be reduced by about twenty-fold without harmful effects. Many cells have enzymes that hydrolyse only inappropriate bonds for the purpose of adding new units to the peptidoglycan. The process of growth of the surface structures is so little understood at the molecular level that even the 5% or so of the wild-type complement of enzyme remaining in *lyt* mutants may be enough to play some vital role in cell surface growth. Despite this, the function of the known autolytic enzymes cannot be that of opening the peptidoglycan structure to receive newly biosynthesized units into covalent bondage, except in those few organisms known to produce muramidases. For this to happen in other bacteria, new and different biosynthetic pathways or small amounts of different autolytic enzymes would have to exist. Apart from their roles in the growth of bacteria, there is suggestive work that the autolysins play parts in transformation and in the bactericidal nature of some antibiotics. Again none of this evidence is so strong as to be unassailable.

The solution of many problems awaits the isolation of structural gene mutants, conditional or otherwise, affected in the formation of individual autolytic enzymes. So far a number of years of searching for such by many different selection techniques, in *Bacillus licheniformis* and *Bacillus subtilis* has failed [C.W. Forsberg, J.E. Fein, R. Williamson, J.B. Ward and H.J. Rogers, unpublished work] and only presumed regulatory mutants have been found.

Areas in which autolytic enzymes are almost certainly involved in a necessary way, but not discussed, are in sporulation and spore germination. Although much interesting work already exists on these subjects, including the recognition of a membrane bound peptidase with a unique specificity [Guinand *et al.*, 1974], too little has really been done to say how far generally the enzymes differ from those in the normal vegetative cells, or exactly how and when they function in the cell differentiation process.

References

Akrigg, A. and Ayad, S.R. (1970). Studies on the competence-inducing factor of *Bacillus subtilis*. *Biochemical Journal* **117**, 397-403.

Akrigg, A., Ayad, S.R. and Barker, G.R. (1967). The nature of a competence-inducing factor in *Bacillus subtilis*. *Biochemical and Biophysical Research Communications* **284**, 1062-1067.

Akrigg, A., Ayad, S.R. and Blamire, J. (1969). Uptake of DNA by competent bacteria a possible mechanism. *Journal of Theoretical Biology* **24**, 266-272.

Ayad, S.R. and Shimmin, E.R.A. (1974). Properties of the competence-inducing factor of *Bacillus subtilis* 168I$^-$. *Biochemical Genetics* **11**, 455-474.

Ayusawa, D., Yoneda, Y., Yamane, K. and Maruo, B. (1975). Pleiotropic phenomena in autolytic enzyme(s) content, flagellation and simultaneous hyperproduction of extracellular α-amylase and protease in a *Bacillus subtilis* mutant. *Journal of Bacteriology* **124**, 459-469.

Barkulis, S.S., Smith, C., Boltralik, J.J. and Heymann, H. (1964). Structure of streptococcal cell walls. IV. Purification and properties of streptococcal phage muralysin. *Journal of Biological Chemistry* **239**, 4027.

Berkeley, R.C.W., Brewer, S.J., Ortiz, J.M. and Gillespie, J.B. (1973). An exo-β-N-acetylglucosaminidase from *Bacillus subtilis* B; characterization. *Biochimica et Biophysica Acta* **309**, 157-168.

Boothby, D.L., Daneo-Moore, L., Higgins, M.L., Coyette, J. and Shockman, G.D. (1973). Turnover of bacterial cell wall peptidoglycans. *Journal of Biological Chemistry* **248**, 2161-2169.

Browder, H.P., Zygmunt, W.A., Young, J.R. and Tavormina, P.A. (1965). Lysostaphin: enzymatic mode of action. *Biochemical and Biophysical Research Communications* **19**, 383-389.

Brown, W.C. (1972). Binding and release from cell walls: a unique approach to the purification of autolysins. *Biochemical and Biophysical Research Communications* **47**, 993-996.

Brown, W.C., Fraser, D.K. and Young, F.E. (1970). Problems in purification of a *Bacillus subtilis* autolytic enzyme caused by association with teichoic acid. *Biochimica et Biophysica Acta* **198**, 308-315.

Brown, W.C. and Young, F.E. (1970). Dynamic interactions between cell wall polymers extracellular proteases and autolytic enzymes. *Biochemical and Biophysical Research Communications* **38**, 564-568.

Burdett, I.D.J. and Higgins, M.L. (1978). Study of pole assembly in *Bacillus subtilis* by computor reconstruction of septal growth zones seen in central longitudinal twin sections of cells. *Journal of Bacteriology* **133**, 959-971.

Burdett, I.D.J. and Murray, R.G.E. (1974a). Septum formation in *Escherichia coli*: characterization of septal structure and the effects of antibiotics in cell division. *Journal of Bacteriology* **119**, 303-324.

Burdett, I.D.J. and Murray, R.G.E. (1974b). Electron microscope study of septum formation in *Escherichia coli* strains B and B/r during synchronous growth. *Journal of Bacteriology* **119**, 1039-1053.

Buxton, R.S. (1978). A heat-sensitive lysis mutant of *Bacillus subtilis* 168 with a low activity of pyruvate carboxylase. *Journal of General Microbiology* **105**, 175-185.

Chaloupka, J. (1967). Synthesis and degradation of surface structures by growing and non-growing *Bacillus megaterium*. *Folia Microbiologica* **12**, 264-273.

Chaloupka, J., Rihova, L. and Krekova, P. (1962). The mucopeptide turnover in the cell walls of growing cultures of *Bacillus megaterium* K.M. *Experentia (Basel)* **18**, 362-363.

Chan, L. and Glaser, L. (1972). Purification of *N*-acetylmuramic acid-L-alanine amidase from *Bacillus megaterium*. *Journal of Biological Chemistry* **247**, 5391-5397.

Chatterjee, A.N., Wong, W., Young, F.E. and Gilpin, R.W. (1976). Isolation and characterisation of a mutant of *Staphylococcus aureus* deficient in autolytic activity. *Journal of Bacteriology* **125**, 961-967.

Chatterjee, A.N. and Young, F.E. (1972). Regulation of the bacterial cell wall: isolation and characterisation of peptidoglycan mutants of *Staphylococcus aureus*. *Journal of Bacteriology* **111**, 220-230.

Cleveland, R.F., Holtje, J-V., Wicken, A.J., Tomasz, A., Daneo-Moore, L. and Shockman, G.D. (1975). Inhibition of bacterial wall lysins by lipoteichoic acids and related compounds. *Biochemical and Biophysical Research Communications* **67**, 1128-1135.

Coles, N.W., Gibbo, C.M. and Broad, A.J. (1969). Purification, properties and mechanism of action of a staphylolytic enzyme produced by *Aeromonas hydrophila*. *Biochemical Journal* **111**, 7-15.

Cornett, J.B., Redman, B.E. and Shockman, G.D. (1978). Autolytic-deficient mutant of *Streptococcus faecalis*. *Journal of Bacteriology* **133**, 631-640.

Coyette, J. and Ghuysen, J-M. (1970). Wall autolysin of *Lactobacillus acidophilus* strain 63 AM Gasser. *Biochemistry* **9**, 2952-2956.

Coyette, J. and Shockman, G.D. (1973). Some properties of the autolytic *N*-acetylmuramidase of *Lactobacillus acidophilus*. *Journal of Bacteriology* **114**, 34-41.

Daneo-Moore, L., Coyette, J., Sayare, M., Boothby, D. and Shockman, G.D. (1975). Turnover of the cell wall peptidoglycan of *Lactobacillus acidophilus*: The presence of a fraction immune to turnover. *Journal of Biological Chemistry* **250**, 1348-1353.

Del Rio, L.A. and Berkeley, R.C.W. (1976). Exo-β-*N*-acetylmuramidase - a novel hexosaminidase. *European Journal of Biochemistry* **65**, 3-12.

Ensign, J.C. and Wolfe, R.S. (1965). Lysis of bacterial cell walls by an enzyme isolated from a *Myxobacter*. *Journal of Bacteriology* **90**, 395-402.

Ephrussi-Taylor, H. and Freed, B.A. (1964). Incorporation of thymidine and amino acids into deoxyribonucleic acid and acid insoluble cell structures in pneumococcal cultures synchronised for competence to transform. *Journal of Bacteriology* **87**, 1211-1215.

Fan, D.P. (1970). Cell wall binding properties of the *Bacillus subtilis* autolysin(s). *Journal of Bacteriology* **103**, 488-493.

Fan, D.P. and Beckman, M.M. (1971). Mutant of *Bacillus subtilis* demonstrating requirement of lysis for growth. *Journal of Bacteriology* **105**, 629-636.

Fan, D.P., Beckman, B.E. and Beckman, M.M. (1974). Cell wall turnover at the hemispherical caps of *Bacillus subtilis*. *Journal of Bacteriology* **117**, 1330-1334.

Fan, D.P., Pelvit, M.C. and Cunningham, W.P. (1972). Structural differences between walls from ends and sides of the rod-shaped bacterium *Bacillus subtilis*. *Journal of Bacteriology* **109**, 1266-1272.

Fein, J.E. and Rogers, H.J. (1976). Autolytic enzyme-deficient mutants of *Bacillus subtilis* 168. *Journal of Bacteriology* **127**, 1427-1442.

Forsberg, C.W. and Rogers, H.J. (1971). Autolytic enzymes in growth of bacteria. *Nature, London* **229**, 272.

Forsberg, C.W. and Rogers, H.J. (1974). Characterisation of *Bacillus licheniformis* 6346: mutants which have altered lytic enzyme activities. *Journal of Bacteriology* **118**, 358-368.

Forsberg, C.W. and Ward, J.B. (1972). *N*-acetylmuramyl-L-alanine amidase of *Bacillus licheniformis* and its L-form. *Journal of Bacteriology* **110**, 878-888.

Forsberg, C.W., Ward, J.B., Wyrick, P.B. and Rogers, H.J. (1973).

Effect of phosphate limitation on the morphology and wall composition of *Bacillus licheniformis* and its phosphoglucomutase-deficient mutants. *Journal of Bacteriology* **113**, 969-984.

Ghuysen, J-M. (1968). The use of bacteriolytic enzymes in the determination of wall structure and their role in cell metabolism. *Bacteriological Reviews* **32**, 425-464.

Glaser, L. (1973). Bacterial cell surface polysaccharides. *Annual Review of Biochemistry* **42**, 91-112.

Good, C.M. and Tipper, D.J. (1972). Conditional mutants of *Staphylococcus aureus* defective in cell wall precursor synthesis. *Journal of Bacteriology* **111**, 231-241.

Grant, W.D. and Wicken, A.J. (1970). Autolysis of cell walls of *Bacillus stearothermophilus* B65 and the chemical structure of the peptidoglycans. *Biochemical Journal* **118**, 859-868.

Guinand, M., Michel, G. and Tipper, D.J. (1974). Appearance of a Y-D-glutamyl-(L)-meso-diaminopimelate peptidoglycan hydrolase during sporulation in *Bacillus sphaericus*. *Journal of Bacteriology* **120**, 173-184.

Herbold, D.R. and Glaser, L. (1975a). *Bacillus subtilis N*-acetylmuramic acid L-alanine amidase. *Journal of Biological Chemistry* **250**, 1676-1682.

Herbold, D.R. and Glaser, L. (1975b). Interaction of *N*-acetylmuramic acid L-alanine amidase with cell wall polymers. *Journal of Biological Chemistry* **250**, 7231-7238.

Higgins, M.L., Coyette, J. and Shockman, G.D. (1973). Sites of cellular autolysis in *Lactobacillus acidophilus*. *Journal of Bacteriology* **116**, 1375-1382.

Higgins, M.L., Pooley, H.M. and Shockman, G.D. (1970). Site of initiation of cellular autolysis in *Streptococcus faecalis* as seen by electron microscopy. *Journal of Bacteriology* **103**, 504-512.

Higgins, M.L. and Shockman, G.D. (1976). Study of a cycle of cell wall assembly in *Streptococcus faecalis* by three dimensional reconstruction of thin sections of cells. *Journal of Bacteriology* **127**, 1346-1358.

Holtje, J-V., Mirelman, D., Sharon, N. and Schwarz, U. (1975). Novel type of murein transglycosylase in *Escherichia coli*. *Journal of Bacteriology* **124**, 1067-1076.

Holtje, J-V. and Tomasz, A. (1975). Lipoteichoic acid: a specific inhibitor of autolysin activity in pneumococcus. *Proceedings of the National Academy of Sciences, U.S.A.* **72**, 1690-1694.

Horne, D., Hakenbeck, R. and Tomasz, A. (1977). Secretion of lipids induced by inhibition of peptidoglycan synthesis in streptococci. *Journal of Bacteriology* **132**, 704-717.

Huff, E. and Silverman, C.S. (1968). Lysis of staphylococcal walls by a soluble staphylococcal enzyme. *Journal of Bacteriology* **95**, 99-106.

Hungerer, K.D., Fleck, J. and Tipper, D.J. (1969). Structure of the cell wall peptidoglycan of *Lactobacillus casei* RO94. *Biochemistry* **8**, 3567-3573.

Hughes, R.C. (1971). Autolysis of *Bacillus cereus* cell walls and isolation of structural components. *Biochemical Journal* **121**, 791-802.

Jawetz, E., Gunnison, J.B., Speck, R.C. and Coleman, V.R. (1951). Studies on antibiotic synergism and antagonism; the interference of chloramphenicol with action of penicillin. *Archives of Internal Medicine* **87**, 349-359.

Kilburn, J.O. and Best, G.K. (1977). Characterisation of autolysins from *Mycobacterium smegmatis*. *Journal of Bacteriology* **129**, 750-755.

Kingan, S.L. and Ensign, J.C. (1968). Isolation and characterisation of three autolytic enzymes associated with sporulation of *Bacillus thuringensis* var. *thuringensis*. *Journal of Bacteriology* **96**, 629-638.

Koyama, T., Yamada, M. and Matsuhashi, M. (1977). Formation of regular packets of *Staphylococcus aureus* cells. *Journal of Bacteriology* **129**, 1518-1523.

Krulwich, T.A. and Ensign, J.C. (1968). Activity of an autolytic N-acetylmuramidase during sphere-rod morphogenesis in *Arthrobacter crystallopoietes*. *Journal of Bacteriology* **96**, 857-859.

Lacks, S. (1970). Mutants of *Diplococcus pneumoniae* that lack deoxyribonucleases and other activities possibly pertinent to genetic transformation. *Journal of Bacteriology* **101**, 373-383.

Lacks, A. and Neuberger, M. (1975). Membrane location of deoxyribonuclease implicated in the genetic transformation of *Diplococcus pneumoniae*. *Journal of Bacteriology* **124**, 1321-1329.

Lominski, I., Cameron, J. and Wylie, G. (1958). Chaining and unchaining *Streptomyces faecalis* - a hypothesis of the mechanism of bacterial cell separation. *Nature, London* **181**, 1477.

Lugtenberg, E.J.J., De Haas-Menger and Ruyters, W.H.M. (1972). Murein synthesis and identification of cell wall precursors of temperature-sensitive mutants of *Escherichia coli*. *Journal of Bacteriology* **109**, 326-335.

McQuillen, K. (1958). Lysis resulting from metabolic disturbance. *Journal of General Microbiology* **18**, 498-512.

Martin, H.H. and Kemper, S. (1970). Endo-*N*-Acetylglucosaminidase from *Clostridium perfringens* lytic for cell wall murein of Gram-negative bacteria. *Journal of Bacteriology* **102**, 347-350.

Matsuzawa, H., Matsuhashi, M., Oka, A. and Sugino, Y. (1969). Genetic and biochemical studies on cell wall peptidoglycan synthesis in *Escherichia coli* K12. *Biochemistry and Biophysical Research Communications* **36**, 682-689.

Mauk, J., Chan, L. and Glaser, L. (1971). Turnover of the cell wall of Gram-positive bacteria. *Journal of Biological Chemistry* **246**, 1820-1821.

Munoz, E., Ghuysen, J-M., Leyh-Bouille, M., Petit, J-F. and Tinelli, R. (1966). Structural variations in bacterial cell wall peptidoglycans studied with *Streptomyces* F_1 endo-*N*-acetylmuramidase. *Biochemistry* **5**, 3091-3098.

Mosser, J.L. and Tomasz, A. (1970). Choline-containing teichoic acid as a structural component of pneumococcal cell wall and its role in sensitivity to lysis by an autolytic enzyme. *Journal of Biological Chemistry* **245**, 287.

Okada, S. and Kitahata, S. (1973). Purification and properties of bacterial lysozyme. *Journal of Fermentation Technology* **51**, 705-712.

Pelzer, H. (1963). Mucopeptidhydrolases in *Escherichia coli*. 1. Nachweis und Wirkungsspezifität. *Zeitschrift für Naturforschung* **18b**, 950-956.

Pooley, H.M. (1976a). Turnover and spreading of old wall during sur-

face growth of *Bacillus subtilis*. *Journal of Bacteriology* **125**, 1127-1138.
Pooley, H.M. (1976b). Layered distribution, according to age within the cell wall of *Bacillus subtilis*. *Journal of Bacteriology* **125**, 1139-1147.
Pooley, H.M., Porres-Juan, J.M. and Shockman, G.D. (1970). Dissociation of an autolytic enzyme cell-wall complex by treatment with unusually high concentration of salt. *Biochemical and Biophysical Research Communications* **38**, 1134-1140.
Pooley, H.M. and Shockman, G.D. (1969). Relationship between the latent form and the active form of the autolytic enzyme of *Streptococcus faecalis*. *Journal of Bacteriology* **100**, 617-624.
Pooley, H.M., Shockman, G.D., Higgins, M.L. and Porres-Juan, J. (1972). Some properties of two autolytic-defective mutants of *Streptococcus faecalis* ATCC 9790. *Journal of Bacteriology* **109**, 423-431.
Prestidge, L.S. and Pardee, B. (1957). Induction of bacterial lysis by penicillin. *Journal of Bacteriology* **74**, 48-59.
Priest, F.G. (1977). Extracellular enzyme synthesis in the genus *Bacillus*. *Bacteriological Reviews* **41**, 711-753.
Ranhand, J.M. (1973). Autolytic activity and its association with the development of competence in Group H streptococci. *Journal of Bacteriology* **115**, 607-614.
Ranhand, J.M. (1974). Inhibition of the development of competence in *Streptococcus sanguis* (Wicky) by reagents that interact with sulfhydryl groups: discernment of the competence process. *Journal of Bacteriology* **118**, 1041-1050.
Ranhand, J.M., Leonard, C.G. and Cole, R.M. (1971). Autolytic activity associated with competent group H streptococci. *Journal of Bacteriology* **106**, 251-268.
Rhuland, L.E. (1957). Role of α,ε-diaminopimelic acid in the cellular integrity of *Escherichia coli*. *Journal of Bacteriology* **73**, 778-783.
Richmond, M.H. (1959a). Formation of a lytic enzyme by a strain of *Bacillus subtilis*. *Biochimica et Biophysica Acta* **33**, 78-91.
Richmond, M.H. (1959b). Properties of a lytic enzyme produced by a strain of *Bacillus subtilis*. *Biochimica et Biophysica Acta* **33**, 92-101.
Robson, R.L. and Baddiley, J. (1977). Role of teichuronic acid in *Bacillus licheniformis*: defective autolysis due to deficiency of teichuronic acid in a novobiocin-resistant mutant. *Journal of Bacteriology* **129**, 1051-1058.
Rogers, H.J. (1964). The mode of action of some antibacterial substances. In *Experimental Chemotheraphy*, vol. II, part 1, pp.37-76. Edited by R.J. Schnitzer and F. Hawking. New York and London: Academic Press.
Rogers, H.J. (1967). Killing of staphylococci by penicillins. *Nature, London* **213**, 31-33.
Rogers, H.J. (1974). Peptidoglycans (mucopeptides) structure, function and variations. *Annals New York Academy of Science* **235**, 29-51.
Rogers, H.J. (1979). Biogenesis of the wall in bacterial morphogenesis. *Advances in Microbial Physiology* **20**, in press.
Rogers, H.J. and Forsberg, C.W. (1971). Role of autolysins in the killing of bacteria by some bactericidal antibiotics. *Journal of*

Bacteriology **108**, 1235-1243.
Rogers, H.J. and Perkins, H.R. (1968). The Mucopeptides. In *Cell Walls and Membranes*, pp.231-258. Edited by C. Long. London: E. and F.N. Spon Limited.
Rogers, H.J., Pooley, H.M., Thurman, P.F. and Taylor, C. (1974). Wall and membrane growth in bacilli and their mutants. *Annales de Microbiologie (Institut Pasteur)* **125b**, 135-147.
Rogers, H.J. and Thurman, P.F. (1978). Temperature sensitive nature of the *rod B* mutation in *Bacillus subtilis*. *Journal of Bacteriology* **133**, 298-305.
Rogers, H.J., Ward, J.B. and Burdett, I.D.J. (1978). Structure and growth of the walls of Gram-positive bacteria. *Society for General Microbiology Symposium* **28**, 139-175.
Rosenthal, R.S., Jungkind, D., Daneo-Moore, L. and Shockman, G.D. (1975). Evidence for the synthesis of soluble peptidoglycan fragments by protoplasts of *Streptococcus faecalis*. *Journal of Bacteriology* **124**, 398-409.
Rosenthal, R.S. and Shockman, G.D. (1975). Synthesis of peptidoglycan in the form of soluble glycan chains by growing protoplasts (autoplasts) of *Streptococcus faecalis*. *Journal of Bacteriology* **124**, 419-423.
Schindler, C.A. and Schuhardt, V.T. (1964). Lysostaphin: a new bacteriolytic agent for the staphylcoccus. *Proceedings of the National Academy of Sciences, U.S.A.* **51**, 4-14.
Schleifer, K.H. and Kandler, O. (1972). Peptidoglycan types of bacterial walls and their taxonomic implications. *Bacteriological Reviews* **36**, 407-477.
Seto, H. and Tomasz, A. (1975). Protoplast formation and leakage of intra-membrane cell components: induction by the competence activator substance of pneumococci. *Journal of Bacteriology* **121**, 344-353.
Shockman, G.D., Thompson, J.S. and Conover, M.J. (1965). Replacement of lysine by hydroxylysine and its effects on cell lysis in *Streptococcus faecalis*. *Journal of Bacteriology* **90**, 575-588.
Shockman, G.D., Thompson, J.S. and Conover, M.J. (1967). The autolytic enzyme system of *Streptococcus faecalis*. II. Partial characterisation of the autolysin and its substrate. *Biochemistry* **6**, 1054-1065.
Singer, H.J., Wise, E.M. and Park, J.T. (1972). Properties and purification of *N*-acetylmuramyl-L-alanine amidase from *Staphylococcus aureus* H. *Journal of Bacteriology* **112**, 932-939.
Takahara, Y., Machigaki, E. and Maruo, S. (1974). General properties of endo-*N*-acetylmuramidase of *Bacillus subtilis* YT-25. *Agricultural Biological Chemistry* **38**, 2357-2365.
Takumi, K., Kawata, T. and Hisatsune, K. (1971). Autolytic enzyme system of *Clostridium botulinum*. II. Mode of action of autolytic enzymes in *Clostridium botulinum* Type A. *Japanese Journal of Microbiology* **15**, 131-141.
Taylor, A., Das, B.C. and van Heijenoort, J.U. (1975). Bacterial cell wall peptidoglycan fragments produced by phage λ or V_1 II endolysin and containing 1-6 anhydro *N*-acetylmuramic acid. *European Journal of Biochemistry* **53**, 47-54.
Tichy, P. and Landman, O.E. (1969). Transformation in quasi-sphero-

plasts of *Bacillus subtilis*. *Journal of Bacteriology* **97**, 42-51.

Tinelli, R. (1968). Structure de la paroi de *Listeria monocytogenes*. 1. Fractionnement et identification partielle des produits obtenus par autolyse du glycopeptide: nature des enzymes autolytiques pariétaux. *Bulletin de la Société de Chimie Biologique* **51**, 283-297.

Tipper, D.J. (1969). Mechanism of autolysis of isolated cell walls of *Staphylococcus aureus*. *Journal of Bacteriology* **97**, 837-847.

Toennies, G. and Gallant, D.L. (1949). Bacterial turbimetric studies. II. The role of lysine in bacterial maintenance. *Journal of Biological Chemistry* **177**, 831-839.

Toennies, G., Izard, L., Rogers, N.B. and Shockman, G.D. (1961). Cell multiplication studied with an electronic particle counter. *Journal of Bacteriology* **82**, 1054-1065.

Toennies, G. and Shockman, G.D. (1958). Growth Chemistry of *Streptococcus faecalis*. *Proceedings of the 4th International Congress of Biochemistry* **13**, 365-394.

Tomasz, A. (1968). Biological consequences of the replacement of choline by ethanolamine in the cell wall of pneumococcus: chain formation loss of transformability and loss of autolysis. *Proceedings of the National Academy of Sciences, U.S.A.* **59**, 86-93.

Tomasz, A., Albino, A. and Zaneti, E. (1970). Multiple antibiotic resistance in a bacterium with suppressed autolytic system. *Nature, London* **227**, 138-140.

van Heijenoort, J., Parquet, C., Flouret, B. and van Heijenoort, Y. (1975). Envelope-bound-*N*-acetylmuramyl-L-alanine amidase of *Escherichia coli* K12. *European Journal of Biochemistry* **58**, 611-619.

Wadström, T. (1970). Bacteriolytic enzymes from *Staphylococcus aureus*. Properties of the endo-β-*N*-acetylglucosaminidase. *Biochemical Journal* **120**, 745-752.

Wadström, T. (1971). Chitinase activity and substrate specificity of endo-β-*N*-acetylglucosaminidase of *Staphylococcus aureus*. *Acta Chemica Scandinavica* **25**, 1807-1812.

Wadström, T. and Hisatsune, K. (1970a). Bacteriolytic enzymes from *Staphylococcus aureus*: purification of an endo-β-*N*-acetylglucosaminidase. *Biochemical Journal* **120**, 725-734.

Wadström, T. and Hisatsune, K. (1970b). Bacteriolytic enzymes from *Staphylococcus aureus*: specificity of action of endo-β-*N*-acetylglucosaminidase. *Biochemical Journal* **120**, 735-744.

Wadström, T. and Vesterberg, O. (1971). Studies on endo-β-*N*-acetylglucosaminidase, staphylolytic peptidase and *N*-acetylmuramyl-L-alanine amidase in lysostaphin and from *Staphylococcus aureus*. *Acta Pathologica Microbiologica Scandinavica, Section B* **79**, 248-264.

Weidel, W. and Pelzer, H. (1964). Bag-shaped macromolecular - a new outlook in bacterial cell walls. *Advances in Enzymology* **26**, 193.

Ward, J.B. (1973). The chain length of glycans in bacterial walls. *Biochemical Journal* **133**, 395-398.

Ward, J.B. and Perkins, H.R. (1968). The purification and properties of two staphylolytic enzymes from *Streptomyces griseus*. *Biochemical Journal* **106**, 69.

Weidel, W. and Pelzer, H. (1964). Bag-shaped macromolecular - a new outlook in bacterial cell walls. *Advances in Enzymology* **26**, 193.

hyperproduction of α-amylase and protease and its synergistic effect. *Journal of Bacteriology* **124**, 48-54.

Young, F.E. (1966). Autolytic enzyme associated with cell walls of *Bacillus subtilis*. *Journal of Biological Chemistry* **241**, 3462-3467.

Young, F.E. and Spizizen, J. (1963). Biochemical aspects of competence in the *Bacillus subtilis* transformation system. II. Autolytic enzyme activity of cell walls. *Journal of Biological Chemistry* **238**, 3126-3130.

Young, F.E., Tipper, D.J. and Strominger, J.L. (1964). Autolysis of cell walls of *Bacillus subtilis*. *Journal of Biological Chemistry* **239**, 2600-2602.

Young, F.E. and Wilson, G.A. (1972). Genetics of *Bacillus subtilis* and other Gram-positive sporulating *Bacilli*. In *Spores*, vol. V, pp.77-106. Edited by H.O. Halvorsen, R. Hanson and L.L. Campell. Washington D.C.: American Society for Microbiology.

Chapter 11

FACTORS LIMITING THE ACTION OF POLYSACCHARIDE DEGRADING ENZYMES

J.S.D. BACON

Rowett Research Institute, Aberdeen, UK

Introduction

Plants are the major makers of polysaccharides, animals are not particularly well equipped to digest them, and so microorganisms bear the major responsibility for their degradation in the biosphere. The chief exceptions to this rule are reserve polysaccharides, which are made by all classes of organisms and degraded by them as the need arises. These polysaccharides, which are made to be re-used, are often the objects of choice by predators and parasites. Structural polysaccharides are not made with any thought for their dismantling, and in fact the maker may have good reason to introduce all kinds of complexities into the structure in order to protect itself against invasion and digestion.

Faced with these protective devices it is not surprising that microorganisms have produced a great variety of enzymes capable of polysaccharide degradation, and that more than one mode of degradation has been applied to the same polysaccharide.

Our knowledge of the structure of polysaccharides (none of which can really be called simple) has developed hand in hand with our knowledge of the specificities of the enzymes that hydrolyse them, and in neither case can we yet say that knowledge is complete. Sometimes we know a great deal, and might be tempted to think that we know almost everything, about the structure of a polysaccharide and its enzymic degradation (starch, for example), but too often we have only a sketchy idea about what is going on.

This paper is not a comprehensive survey, but an attempt to outline achievements so far, and to point out the fields in which future research will need to be concentrated if we are to understand in enzymic terms the complex processes that are described in other chapters in this book.
So far as possible references are made to recent publications from which the history of a particular topic can be

traced. (Many details of the enzymes mentioned can be found in the various volumes of the third edition of "The Enzymes", edited by P.D. Boyer, and in the two previous editions.)

The Glycosidic Bond

Let us begin by considering the glycosidic bond (which is the main basis of polysaccharide structure) as it relates to a single sugar residue. Compounds in which such a residue is attached to a non-carbohydrate molecule (the "aglycone") are called glycosides, and many such compounds occur naturally in living organisms. The organic chemist is able to make similar compounds and it was the pioneer researches of Emil Fischer that led to the discovery that at least two different glycosides of the same aglycone could be made from a single sugar. Enzyme preparations were found that would hydrolyse one form, but not the other, and this study of the simple α- and β-glucosidases led to the lock-and-key concept of interaction between enzyme and substrate and to a recognition of the importance of stereochemistry for enzymic activity. It is somewhat ironic that we now can see that the fit between the glucosidases and their substrates is not as precise as it is in many other classes of enzymes, and that attempts at the classification of carbohydrases are among the least successful efforts of the Commission on Enzyme Nomenclature; more of this later.

Although in the hydrolysis of glycosides the enzymes are evidently designed to deal with the configuration of the carbon atom which participates in ring-closure and the glycosidic bond, they may differ in their reaction to changes in the rest of the sugar molecule, or in the aglycone. Generally they do not tolerate much variation in the former, but it has long been known that inversion of the configuration at C4, converting the sugar residue from glucose to galactose, does not completely abolish the action of the β-glucosidase in sweet-almond emulsin [Heyworth and Walker, 1962]; the reverse is true for β-galactosidase [Wallenfels and Weil, 1972]. Barnett *et al.* [1967] discovered that if C1 carries a fluorine atom, instead of the *O*-aglycone structure, hydrolysis is still catalysed by α- and β-glucosidases, α-galactosidase and α-mannosidase. The aglycone can be altered in many respects without abolishing the enzymic activity, but with considerable effects on the affinity and on the rate of hydrolysis [e.g. Nath and Rydon, 1954]. We may expect that when we turn to the enzymes attacking polysaccharides the same considerations will apply.

The Non-reducing Ends of Polysaccharide Chains - Exo-polysaccharases

When we consider attack upon the residue at the non-reducing end of a polysaccharide chain some interesting phenomena are revealed. Some of the simple glycosidases will not tolerate a sugar residue in the place of the aglycone, so these are completely excluded from action on polysaccharides. Others show activity towards short chains of sugar residues such as di- or tri-saccharides, but have little action on longer chains; these are generally classed with the former as glycosidases. A third class of enzyme will act on short chains but prefers long ones; these are genuine polysaccharases. Because they act upon the outer regions of the polysaccharide molecule they are called exo-enzymes to distinguish them from the endo-polysaccharases which break bonds in the interior of the polysaccharide structure. These exo-enzymes have a peculiar feature: they release the sugar residue with the configuration at C1 inverted.

This was first detected by observations of changes in optical rotation in solutions of the substrate undergoing hydrolysis. More recently two other methods have been applied to the reaction mixture: g.l.c. of the sugars released [Parrish and Reese, 1967] and n.m.r. of the products [Eveleigh and Perlin, 1969]. All these observations seem to confirm a rule, first proposed by Reese *et al.* [1968], that exo-glucanases may be distinguished from glucosidases because they bring about inversion of configuration at the residue released; they studied these relationships in a series of fungal enzymes which released glucose from aryl glucosides, glucose oligosaccharides or glucans. Similar results have been found for enzymes attacking mannose derivatives [Gorin *et al.*, 1969].

There has been speculation about the significance of the inversion, which does not occur with endo-polysaccharases. Koshland [1953] suggested that catalysis involving a single step could produce inversion, whereas a two-step process would regenerate the original configuration, but this attractive idea has not been supported by subsequent research. Present ideas of α-amylase action lean towards a mechanism in which the ring conformation of the sugar residue involved in the glycosidic linkage is strained by being bound at the active site, configuration at C1 is retained in the enzyme-bound intermediate, and the configuration of the product is determined by the direction from which the water molecule is permitted to approach the active site.

It is not easy to see why there should be a sharp distinction in this respect between exo- and endo-polysaccharases. The fact that the former are aided in their catalytic action by the lengthening of the polysaccharide chain suggests that some of the residues internal to the point of hydrolysis are bound to the enzyme; this has been supported by experiments on amyloglucosidase [Hiromi *et al.*, 1973]. The endo-polysaccharases also form associations with several sugar residues (see below). The difference may lie in the fact that the exo-polysaccharase is attacking a "free" end of the polysaccharide structure, and so its mode of binding may differ from that between an endo-polysaccharase and the internal regions of the molecule. As we shall presently see, in these regions there are likely to be forces at work holding the chain in close proximity to other parts of the molecular structure, and correspondingly limited opportunities for enzyme substrate interactions. When this much has been said, though, it must be recognized that the stereochemistry of the binding at the active site ultimately determines the course of catalysis. Amyloglucosidase, an exo-enzyme removing the terminal glucose unit from the end of an α 1,4 glucan chain, releases β-glucose whether it is acting upon polysaccharide or upon the much simpler α-glucosyl fluoride [Barnett, 1971].

The classical example of inversion of configuration is β-amylase, which removes two residues at a time (constituting a molecule of β-maltose) from the non-reducing end of the α 1,4 glucan chains in amylose and amylopectin. This prompts the question: how far from the non-reducing end of the chain can a polysaccharase act and still be an exo-enzyme?

A Special Case of Chain-end Modification

A very remarkable enzyme that is an exo-enzyme in the strict sense, but does not bring about inversion of configuration, is the Schardinger enzyme, which produces cyclic dextrins in high yield from the starch polysaccharides. Its action is not hydrolytic and so it cannot be detected by the release of reducing groups; it was discovered because the products crystallize from aqueous solution. After many years of confusion [described by French, 1957], it was finally realized that the products were rings of six, seven or more α 1,4-linked glucose residues, the two former predominating. The action of the enzyme is a transglucosylation from an internal linkage to the non-reducing end of the amylose chain, and as is general with transglucosylations the residue involved keeps its configuration when the ring is joined.

At first sight it is difficult to imagine why the enzyme should favour the sixth or seventh residue from the end of the chain, except by the possession of a very large and unusual combining site with appropriate specificity. The most likely explanation, first pointed out by Freudenberg, is that it is the structure of the polysaccharide that offers the opportunity of cyclization. The α 1,4 glucan chain has long been suspected to assume a helical form.

The crystal structures in native starches are now believed to be based on a double helix [Wu and Sarko, 1978a, b], and the characteristic iodine complex probably upon a single helix. In each case six glucose residues constitute one turn of the helix. It seems likely that the same form is present in solution [Cael *et al.*, 1973]. This means that the residues to be combined by transglycosylation may lie quite close to one another.

No other polysaccharides are known to be degraded in this way, but it must be remembered that the methods generally used to follow their degradation might not reveal it.

Another Unusual Enzyme

An enzyme occurs in fungi and bacteria that produces large oligosaccharides, none smaller than a pentasaccharide, as the products of hydrolysis of β 1,3 glucans. First noticed in marine fungi by Chesters and Bull [1963], it was later found to be an actively lytic component of a mixture of myxobacterial β-glucanases acting upon the yeast cell wall [Bacon *et al.*, 1970]. Doi *et al.* [1973] have studied the action pattern of a similar enzyme in an *Arthrobacter* species, with results suggesting that it is not an exo-enzyme.

The crystalline form of β 1,3 glucans is probably a triple helix (with water molecules associated) in which six residues occupy one turn [Jelsma and Kreger, 1975]. This prompts the speculation that this enzyme also has a specificity directed towards the helical form of its substrate, and must recognize two complete turns of the helix before it catalyses hydrolysis. The alternative explanation would be that at least ten residues of the glucan chain must be bound to the enzyme, perhaps in an extended form.

Multiple Binding Sites on Polysaccharases

This last suggestion raises the question of "sub-sites", i.e. sites on the enzyme, other than the catalytic site, which play a part in the binding of the substrate and in determining specificity. It has already been pointed out above that exo-polysaccharases act more readily as the chain length of oligosaccharide substrates is increased, suggesting that several residues need to be bound to the enzyme in order to maximise the catalytic effect. In some polysaccharides all the residues in the chain are the same, so that a series of sub-sites could fit with equal ease almost anywhere on the chain except at the ends or at branch-points. Amylase action on starch provides a good example, and has been studied in recent years with the help of linear dextrins of known chain length. The results indicate that sub-sites situated on both sides of the active centre play a part in substrate binding. This evidence for the existence of multiple binding sites antedated the discovery that the tertiary structure of lysozyme provides a channel into which several sugar residues of the substrate will fit, some on either side of the catalytic site [Blake *et al.*, 1967].

Where the polysaccharide does not have a homogeneous structure, as for example in the mixed β 1,3-, β 1,4-linked glucans from cereal seeds, more complex specificities are shown. The mixed linkage β-glucans may be attacked by β 1,3 glucanases (laminarinases) or β 1,4 glucanases ("cellulases") with hydrolysis of the corresponding linkages, but another class of enzymes exists, specific for the mixed-linkage glucans, and these hydrolyse a β 1,4 linkage on the "reducing side" of a β 1,3-linked residue, and have no action on homogeneous β 1,3 or β 1,4 glucans [Manners and Wilson, 1976]. A fungal β 1,3 glucanase has been extensively purified and shown still to attack β 1,4 linkages in mixed-linkage glucans [Clark *et al.*, 1978].

Polysaccharide Solubility

Up to this point we have been visualizing polysaccharides in solution, their chains of sugar residues relatively free to take up the conformations demanded by the binding sites on enzymes. This is unfortunately a grossly over-simplified view, because most polysaccharides, despite their abundance of hydroxyl groups, are not very soluble in water; regions of the same or neighbouring molecules associate with one another through hydrogen bonding and in the extreme case, exemplified by cellulose, all water is squeezed out of the crystalline structure. With other polysaccharides, as recent work on electron diffraction by hydrated polysaccharides has shown, water molecules form an essential part of the ordered structure. Between the conditions of substances like cellulose and of freely soluble polysaccharides there lie many intermediate states, including gels and slimes, which must be taken into account in considering polysaccharase action.

Although the two starch polysaccharides are usually thought of as soluble in water, they are laid down by the plant in an insoluble form, and, despite intensive study, uncertainties still exist about the significance of the differences in crystal structure shown by native starches. Some amylases show a much greater ability than others to attack the intact starch granule [Walker and Hope, 1963]. The cooking of starchy foods greatly increases their digestibility in animals; the polysaccharides do not dissolve, but the granules swell to form a gelatinous mass. A typical meal, in fact, is likely to contain examples of the many states of order in which polysaccharides can exist, ranging from starch in solution in the soup to the fibrous elements in the cell walls of vegetables and fruit.

A further complication in the case of starch is that it has long been known that starches "retrograde" from aqueous solution, forming precipitates that do not stain with iodine and are not attacked by amylase [Watson, 1964]. This retrogradation introduces difficulties into the study of enzyme action on solutions of the starch polysaccharides, especially amylose, and has to be overcome by adjustments of temperature, enzyme concentration, and so on.

Cell Wall Polysaccharides

Concepts of secondary and tertiary structure and of subunit binding are familiar topics in protein biochemistry. They are equally real problems in our attempt to understand the enzymology of polysaccharide degradation. They arise especially when one considers the degradation of cell walls, an important part of the processes described elsewhere in this book (Chapter 18).

The cell walls of fungi and higher plants usually contain fibrillar elements embedded in an "amorphous" matrix [Monro *et al.*, 1976]. Although it was once thought that fibre-forming polysaccharides could be recognized by their basic chemical structure it is now clear that many different sugars and linkage types occur in microfibrils. It is also clear that the degree of order within fibrillar elements may vary greatly, cellulose being very highly-ordered in the cotton fibre, much less so in some plant cell walls [French, 1978].

Similarly, the "amorphous" matrix must include within itself certain elements of order. A theory of polysaccharide gel formation, perhaps not universally applicable, put forward by Rees [1972], envisages a network of intermolecular associations formed by helical regions of the polysaccharide chains, with disordered regions lying between them. In these disordered regions the water content will be high. Rises in temperature, by increasing the extent of disorder, will usually weaken the gel structure, but not all gels respond in this way. Curdlan, a linear β 1,3 glucan produced by a mutant of *Salmonella alcaligenes (Alcaligenes faecalis)*, forms firm gels when the powdered polysaccharide is heated in aqueous suspension, and these are stable at 100°C [Harada, 1974]. Marchessault *et al.* [1977] found that heating the gel at 160°C, a procedure that "anneals" polysaccharide crystals, caused the gel to shrink, and suggested that this was due to an increase in order and consequent expulsion of water from the structure. (This would be analogous to the familiar "drying" of wet sand when one stands on it.) It is thus difficult to generalize about the effects of physical treatment on polysaccharide gels, and consequently about their effects on cell walls, but we must be prepared to find that heating and drying will alter resistance to polysaccharases [cf. Scherrer *et al.*, 1977]. Such considerations have led some workers to use substrates in a "never-dried" state.

A gel could be sufficiently porous to allow enzyme molecules to pass into and through it, thus permitting its own destruction and that of fibrillar elements embedded in it, but one might guess that the porosity of cell walls will be such that proteins are excluded. A few measurements of cell wall porosity have been made using solutions of polymers such as polyethylene glycol of known molecular size range. The validity of such measurements depends very much on knowing the size distribution of the test polymer samples [Scherrer and Gerhardt, 1971].

The action of polysaccharases on relatively homogeneous insoluble materials with different degrees of order may be illustrated by two examples.

The sclerotial cell walls of the fungus *Sclerotinia sclerotiorum* contain a β 1,3 glucan in which some of the residues are substituted at C6 by a β-glucose residue. (This substituent can probably be accommodated in the triple helical structure of the glucan.) The lysis of these cell walls, and solubilization of gels formed by the isolated polysaccharide, is brought about by exo- and endo-β 1,3 glucanases. Acting singly each enzyme is halted before the process is complete, presumably by barriers within the primary, secondary or tertiary structure of the polysaccharides. However, when acting together the two enzymes bring about rapid and almost complete solubilization [Bacon, 1973; Jones *et al.*, 1974]. It seems likely that organisms generally will elaborate more than one type of polysaccharase when dealing with complex insoluble substrates.

This is well illustrated by the mixture of enzymes that brings about the conversion of cellulose to glucose [see Eriksson, Chapter 12; Eveleigh, Chapter 14]. Certain filamentous fungi when grown on substrates containing cellulose release into the culture medium an assembly of enzymes capable of digesting the most highly ordered forms of cellulose (exemplified by the cotton hair cell wall). Reese *et al.* [1950] showed that two thermolabile factors had to be present for extensive solubilization of cotton to occur. One was an endo-β 1,4 glucanase capable of hydrolysing soluble cellulose derivatives (such as carboxymethyl cellulose) and disordered forms of cellulose; the other they called C_1, and suggested that it somehow produced disorder in the crystalline regions of the cellulose substrate, and so made it possible for the other to act. Intensive studies of the group of enzymes involved [see, for example, Wood and McCrae, 1977], in which they were purified by a variety of chromatographic and electrophoretic procedures and tested on a range of substrates, revealed that the role allotted to the C_1 factor could be filled by an exo-polysaccharase, a cellobiohydrolase, which was present in all the culture filtrates that solubilized cotton cellulose. Wood believes, very reasonably, that this cellobiohydrolase is the C_1 factor, and that it is not necessary to postulate the existence of a further disordering enzyme. He envisages the initial attack on the surface of the crystalline cellulose is the scission of a β 1,4 linkage by the endo-glucanase, and that the exo-glucanase, forming a loose complex with the endo-enzyme, then removes a cellobiose unit from non-reducing chain-end exposed by the initial breaking of this glycosidic bond, and so prevents the re-formation of the bond by the powerful forces that hold the cellulose molecules in the crystalline array [c.f. Rowland, 1975].

Such hypotheses encourage the highly speculative view that in the case of organisms (e.g. *Cytophaga* and *Sporocytophaga*; see Gooday, Chapter 19) that can degrade cotton cellulose but do not release all the necessary enzymes into the medium a complex of two or more enzymes may be present in their outer surface structure.

Substituent Groups

So far we have thought of polysaccharides as consisting wholly of sugar residues, but many have non-carbohydrate substituents attached to them, there being always one or more hydroxyl groups available for substitution on each residue. Ester linkages to the polysaccharides are formed by a variety of acids [see Sutherland, Chapter 1]. Of these acetic acid is the best known, but several di-carboxylic acids also occur, so linked that free acidic groups are exposed. (Pyruvic acid linked through a ketal structure also provides a carboxyl function on some polysaccharides.) Esters of phenolic acids related to cinnamic acid are found in the cell walls of the Monocotyledons and Chenopodiaceae [Harris and Hartley, 1976], and it seems possible that subsequent polymerization through such phenolic substituents could create covalent linkages between polysaccharides and lignin.

The presence of quite small ester groups, such as acetyl, interferes with the intermolecular bonding of the polysaccharide molecules and increases their solubility in water. To this extent they are likely to be more susceptible to enzyme attack, but the substituents may also interfere with enzyme-substrate binding and protect regions of the polysaccharide chain from attack. The kind of effects that might be produced are known from experiments with chemically modified polysaccharides [Weill *et al.*, 1975]. Substitution on half the residues can have a large effect, and degrees of substitution of this order are not unknown. We have suggested that acetylation is one of the factors affecting rumen digestibility of forages [Morris and Bacon, 1977].

A special case of substitution is the occurrence of branch-points in the polysaccharide structure, where another chain of residues is attached glycosidically to a hydroxyl group not occupied in chain formation. A well-known example is the occurrence of branch-points in glycogen and amylopectin, where the main chain is linked C1 to C4 and the branch-points are at C6. Like other substituents, chains of sugar residues inserted at branch-points interfere with intermolecular bonding, and polysaccharides with a branched structure are usually soluble in water [Whistler, 1973].

The action of exo-enzymes (e.g. β-amylase) is often halted at or near a branch-point, but not always. An exoglucanase that attacks β 1,3 glucans will degrade sclerotan, a β 1,3 glucan in which on average every third residue carries a β-glucose substituent at C6. The enzyme releases glucose from the open parts of the chain and gentiobiose from the substituted residues.

Endo-enzymes may work close to a branch-point, with the result that small oligosaccharides containing the branch-point accumulate in the reaction mixture, e.g. the α-limit dextrins produced from starch. These oligosaccharides are usually susceptible to the action of glycosidases, which show varying degrees of linkage specificity.

Specific de-branching enzymes also occur [Manners, 1971] so that the purely carbohydrate substituents may be removed very early or very late in polysaccharide digestion. A debranching enzyme acting alone will tend to reduce the solubility of its substrate.

Whereas much attention has been paid to branch-points during the elucidation of polysaccharide structure, relatively little attention has been given to the non-carbohydrate substituents. Unless a specific enzyme to remove them is present they might be expected to protect some of the polysaccharide from complete degradation, even when digestive activity is intense [Gaillard and Richards, 1975].

Polysaccharide Interactions

Finally, we must consider how one polysaccharide may prevent another from being digested, when they are incorporated together into the structure of the cell wall. After what we have already discussed about the associations between polysaccharide chains, it is evident that one polysaccharide could block the degradation of another. Indications that this is the case are given by the order of appearance of polysaccharases when plant pathogenic fungi are grown with higher plant cell walls as carbon source [Bateman, 1976]. The first enzymes to appear are directed towards the pectic substances, the next towards hemicelluloses, and the last towards cellulose. This order conforms in a general way to expectation, because cellulose is present in microfibrils embedded in a matrix of hemicellulose, while the pectic substances occur most abundantly in the primary wall, the first polysaccharide-containing barrier to invasion of the cell.

In this context it is important to notice that the

effects seen may not be due to a permeability barrier preventing access of enzyme molecules to macroscopic regions of the cell wall, but to intermolecular associations such as those which have been discussed at length above. The idea of cell wall digestion being prevented by encrustation of the fibrils with lignin seems a very crude concept when one sees the subtle way in which barriers to enzyme action can be created at the molecular level. Nor should we think of inhibition of digestion being an all-or-none process; in some contexts, such as rumen digestion, reduced rates of digestion are as important as complete inhibition, and would be better explained by a slow dissolution of intermolecular associations than by an impermeable barrier.

Some Practical Considerations

It follows from all the discussion above that it is dangerous to discuss the polysaccharide-degrading abilities of biological systems in terms of the activities of cellulases, hemicellulases, pectinases, etc., established by the use of polysaccharides isolated from the natural substrate. The Commission on Enzyme Nomenclature in 1961 called cellulase "β-1,4-glucan 4-glucanohydrolase: hydrolyses β-1,4-glucan links in cellulose". Such descriptions cannot say much about the state of the substrate and yet this may be crucial. Therefore measurement of degradation of the polysaccharide as it occurs in the system under study is usually to be preferred to the use of purified substrates.

A further point to notice is the danger of measuring initial rates of degradation. The substrate, although pure in the sense of its primary structure, may be heterogeneous in terms of secondary and tertiary structure. Thus the measurements of a small conversion of cellulose to reducing sugar and expression of the results as "cellulase activity" may conceal the fact that the major part of the substrate is completely resistant to attack by the enzyme preparation.

A more serious problem is the question of the purity of the enzyme preparation being used. Where the structure of a substrate is completely known one can usually visualize and eliminate other ways in which the substrate could be being transformed. Where the structure of the substrate is only partly known enzymic activities may be at work that one might not expect. For example, the high molecular weight β-glucans in barley endosperm cell walls were recently found by Forrest and Wainwright [1977] to be depolymerized by the action of a proteolytic enzyme, or by hydrazinolysis. The product was still of high molecular weight (0.6×10^6) suggesting that the peptide linkages involved constitute a very small part of the wall glucan structure.

Some General Remarks and Conclusions

The discussion above has been concentrated upon the less complex polysaccharides, and particularly upon glucans, because so much research has been devoted to them. Highly-branched structures in which more than one sugar is present are now the subject of intensive investigations, because they form the carbohydrate portions of glycoproteins. Glycosidases have been successfully employed in elucidating the structure of some of these materials [c.f. Conchie and Strachan, 1978], since they will usually remove sugar residues situated at chain ends. These residues play an important role in immune reactions and hence probably in cell recognition phenomena [Hughes, 1975].

These substances are not strictly polysaccharides, but their susceptibility to enzyme attack indicates the manner in which complex plant polysaccharides, the hemicelluloses and gums, are degraded. Highly-branched structures, especially those with uronic acid residues in them, are likely to be more hydrated than the homopolymers discussed above, and hence more accessible to enzymic attack. It does not seem likely that any new principles will have to be invoked in describing their hydrolysis.

This review has been restricted deliberately to hydrolytic reactions because these appear to be dominant in the degradation of polysaccharides. It must be remembered though that hydrolysis of the glycosidic linkage is a special case of glycosylase activity, and that transfer reactions may have a place in polysaccharide degradation. Hehre has made a powerful case for a new classification of carbohydrases along these lines [Hehre *et al.*, 1973]. At present the classification lays emphasis on the distinctions between "glycanohydrolases" and transferases. He and his colleagues have shown that the action of amyloglucosidase is reversible with, as expected, β-glucose as the substrate. More importantly, they have shown that amylases, hitherto thought of as purely hydrolytic in action, will synthesize dextrins from α-maltosylfluoride.

Phosphorolysis, though important in the degradation of α 1,4 glucans, does not seem to be a very common mechanism of polysaccharide degradation.

Polyuronic acids are often de-polymerized by eliminases. Here the glycosidic bond is split in such a way that the glycone side is released as a reducing sugar, while the aglycone side suffers dehydration and an unsaturated sugar acid results; as a consequence the reaction may be followed spectrophotometrically. Pectic substances and alginates may both be degraded in this way.

I have not attempted to describe here the effects on polysaccharase action of pH or temperature. One may safely assume that the enzymes will respond in the expected way, but there may be effects upon the substrate which will need to be interpreted in terms of the stability of its secondary and tertiary structure. When uronic acids are present pH changes can affect the state of ionisation, but this effect is likely to occur at pH values rather far removed from the optimum pH of most polysaccharases.

Any study of the effects of substrate concentration raises difficult problems where insoluble substances are concerned, and even more difficult problems where the enzymes themselves are bound to the surface of the cell that produces them. Whereas it may be possible to state the molecular weight of a protein precisely and so express its concentration in terms of molarity and to know the numbers of particular monomer residues present, there is no reason to believe that this will ever be possible with the majority of polysaccharides.

For these reasons we again face problems in making the description of polysaccharase activity conform to the accepted rules for enzymes acting upon simple soluble substrates. Activities usually have to be expressed in terms of weight solubilized, "reducing power" liberated, etc. This is not to say, though, that studies of the relationships between quantity of substrate and quantity of enzyme cannot be of great importance in this field.

The overall impression that one gets from reviewing our present knowledge of polysaccharases is that the interaction between knowledge of chemical structure and knowledge of enzyme specificity which has been the dominating feature of so much research in the past will continue to dominate. As our picture of the primary structure of polysaccharides becomes more complete we shall have to give more attention to the physical state of the molecules. Susceptibility to enzyme attack will continue to be a valuable indication of structural differences. Here the complete purification and physical characterization of the enzymes concerned will play an important part, because eventually we shall need to assure ourselves that our knowledge of substrate structure, of enzyme structure, and of the binding of susceptible regions at the active site, form a coherent three-dimensional picture.

Physical methods of investigation of chemical structure are already being used widely in the study of polysaccharides, and it would only be stating the obvious to say that research must develop further along these lines. What is more important at present is to consider ways in which the biologist, and more specifically the microbiologist, can contribute. It seems to me that the greatest contribution must be to explore the diversity of polysaccharase action still further, and to identify those organisms that are especially suitable for large-scale production and purification of this diversity of enzymes. In some cases there is already a commercial or technological stimulus to these developments, but the subject has a fascination that makes it certain that it will continue to attract workers in many disciplines for a long time to come.

References

Bacon, J.S.D. (1973). The contribution of β-glucanases to the lysis of fungal cell walls. In *Yeast, Mould and Plant Protoplasts*, pp.61-74. Edited by J.R. Villanueva, I. García-Acha, S. Gascón and F. Uruburu. London and New York: Academic Press.

Bacon, J.S.D., Gordon, A.H., Jones, D., Taylor, I.F. and Webley, D.M. (1970). The separation of β-glucanases produced by *Cytophaga johnsonii* and their role in the lysis of yeast cell walls. *Biochemical Journal* **120**, 67-78.

Barnett, J.E.G. (1971). The hydrolysis of glycosyl fluorides by glycosidases. Determination of the anomeric configuration of the products of glycosidase action. *Biochemical Journal* **123**, 607-611.

Barnett, J.E.G., Jarvis, W.T.S. and Munday, K.A. (1967). The hydrolysis of glycosyl fluorides by glycosidases. *Biochemical Journal* **105**, 669-672.

Blake, C.C.F., Johnson, L.N., Mair, G.A., North, A.C.T., Phillips, D.C. and Sarma, V.R. (1967). Crystallographic studies of the activity of hen egg-white lysozyme. *Proceedings of the Royal Society, Series B* **167**, 378-388.

Bateman, D.F. (1976). Plant cell wall hydrolysis by pathogens. In *Biochemical Aspects of Plant-Parasite Relationships*, pp.79-103. Edited by J. Friend and D.R. Threlfall. London, New York and San Francisco: Academic Press.

Cael, J.J., Koenig, J.L. and Blackwell, J. (1973). Infrared and Raman spectroscopy of carbohydrates. Part III. Raman spectra of the polymorphic forms of amylose. *Carbohydrate Research* **29**, 123-134.

Chesters, C.G.C. and Bull, A.T. (1963). The enzymic degradation of laminarin. 2. The multicomponent nature of fungal laminarinases. *Biochemical Journal* **86**, 31-38.

Clark, D.R., Johnson, J., Chung, K.H. and Kirkwood, S. (1978). Purification, characterisation and action pattern studies on the endo-(1→3)-β-D-glucanase from *Rhizopus arrhizus* QM 1032. *Carbohydrate Research* **61**, 457-477.

Conchie, J. and Strachan, I. (1978). The carbohydrate units of ovalbumin: complete structures of three glycopeptides. *Carbohydrate Research* **63**, 193-213.

Doi, K., Doi, A., Ozaki, T. and Fukui, T. (1973). Lytic β 1,3 glucanase from *Arthrobacter*: pattern of action. *Agricultural and Biological Chemistry* **37**, 1629-1633.

Eveleigh, D.E. and Perlin, A.S. (1969). A proton magnetic resonance study of the anomeric species produced by D-glucosidases. *Carbohydrate Research* **10**, 87-95.

Forrest, I.S. and Wainwright, T. (1977). The mode of binding of β-glucans and pentosans in barley endosperm cell walls. *Journal of the Institute of Brewing* **83**, 279-286.

French, A.D. (1978). The crystal structure of native ramie cellulose. *Carbohydrate Research* **61**, 67-80.

French, D. (1957). The Schardinger dextrins. *Advances in Carbohydrate Chemistry* **12**, 189-260.

Gaillard, B.D.E. and Richards, G.N. (1975). Presence of soluble lignin-carbohydrate complexes in the bovine rumen. *Carbohydrate Research* **42**, 135-145.

Gorin, P.A.J., Spencer, J.F.T. and Eveleigh, D.E. (1969). Enzymic degradation of the yeast cell-wall mannans and galactomannans to polymeric fragments containing α-(1→6)-linked D-mannopyranose residues. *Carbohydrate Research* **11**, 387-398.
Harada, T. (1974). Succinoglucan and gel-forming beta-1,3-glucan. *Process Biochemistry* **9**, 21-25.
Harris, P.J. and Hartley, R.D. (1976). Detection of bound ferulic acid in cell walls of the Gramineae by ultraviolet fluorescence microscopy. *Nature, London* **259**, 508-510.
Hehre, E.J., Okada, G. and Genghof, D.S. (1973). Glycosylation as the paradigm of carbohydrase action. In *Carbohydrates in Solution (Advances in Chemistry Series)*, pp.309-333. Edited by R.F. Gould. Washington D.C.: American Chemical Society.
Heyworth, R. and Walker, P.G. (1962). Almond-emulsin β-D-glucosidase and β-D-galactosidase. *Biochemical Journal* **83**, 331-335.
Hiromi, K., Nitta, Y., Numata, C. and Ono, S. (1973). Subsite affinities of glycoamylase: examination of the validity of the subsite theory. *Biochimica et Biophysica Acta* **302**, 362-375.
Hughes, R.C. (1975). The complex carbohydrates of mammalian cell surfaces and their biological roles. *Essays in Biochemistry* **11**, 1-36.
Jelsma, J. and Kreger, D.R. (1975). Ultrastructural observations on (1→3)-β-D-glucan from fungal cell-walls. *Carbohydrate Research* **43**, 200-203.
Jones, D., Gordon, A.H. and Bacon, J.S.D. (1974). Co-operative action by endo- and exo-β-(1→3)-glucanases from parasitic fungi in the degradation of cell-wall glucans of *Sclerotinia sclerotiorum* (Lib.) de Bary. *Biochemical Journal* **140**, 47-55.
Koshland, D.E. (1953). Stereochemistry and the mechanism of enzymic reactions. *Biological Reviews of the Cambridge Philosophical Society* **28**, 416-436.
Manners, D.J. (1971). Specificity of debranching enzymes. *Nature, London* **234**, 150-151.
Manners, D.J. and Wilson, G. (1976). Purification of malted-barley endo-β-D-glucanases by ion-exchange chromatography: some properties of an endo-barley-β-D-glucanase. *Carbohydrate Research* **48**, 255-264.
Marchessault, R.H., Deslandes, Y., Okawa, K. and Sundararajan, P.R. (1977). X-ray diffraction data for β-(1→3)-D-glucan. *Canadian Journal of Chemistry* **55**, 300-303.
Monro, J.A., Penny, D. and Bailey, R.W. (1976). The organisation and growth of primary cell walls of lupin hypocotyl. *Phytochemistry* **15**, 1193-1198.
Morris, E.J. and Bacon, J.S.D. (1977). The fate of acetyl groups and sugar components during the digestion of grass cell walls in sheep. *Journal of Agricultural Science, Cambridge* **89**, 327-340.
Nath, R.L. and Rydon, H.N. (1954). The influence of structure on the hydrolysis of substituted phenyl β-D-glucosides by emulsin. *Biochemical Journal* **57**, 1-10.
Parrish, F.W. and Reese, E.T. (1967). Anomeric form of D-glucose produced during enzymolysis. *Carbohydrate Research* **3**, 424-429.
Rees, D.A. (1972). Shapely polysaccharides: the Eighth Colworth Medal Lecture. *Biochemical Journal* **126**, 257-273.

Reese, E.T., Maguire, A.H. and Parrish, F.W. (1968). Glucosidases and exo-glucanases. *Canadian Journal of Biochemistry* **46**, 25-34.

Reese, E.T., Siu, R.G.H. and Levinson, H.S. (1950). The biological degradation of soluble cellulose derivatives and its relationship to the mechanism of cellulose hydrolysis. *Journal of Bacteriology* **59**, 485-497.

Rowland, S.P. (1975). Selected aspects of the structure and accessibility of cellulose as they relate to hydrolysis. In *Cellulose as a Chemical and Energy Resource*, pp.183-191. Edited by C.R. Wilke. New York: John Wiley & Sons.

Scherrer, R. and Gerhardt, P. (1971). Molecular sieving by the *Bacillus megaterium* cell-wall and protoplast. *Journal of Bacteriology* **107**, 718-735.

Scherrer, R., Berlin, E. and Gerhardt, P. (1977). Density, porosity and structure of dried cell walls isolated from *Bacillus megaterium* and *Saccharomyces cerevisiae*. *Journal of Bacteriology* **129**, 1162-1164.

Walker, G.J. and Hope, P.M. (1963). The action of some α-amylases on starch granules. *Biochemical Journal* **86**, 452-462.

Wallenfels, K. and Weil, R. (1972). β-Galactosidase. In *The Enzymes*, 3rd edn., vol. 7, pp.617-663. Edited by P.D. Boyer. New York: Academic Press.

Watson, S.A. (1964). Determination of the rate of starch retrogradation. In *Methods in Carbohydrate Chemistry*, vol. 4, pp.150-152. Edited by R.L. Whistler. New York: Academic Press.

Weill, C.E., Nickel, J.B. and Guerrara, J. (1975). The action of amylase on 6-amino-6-deoxyamyloses. *Carbohydrate Research* **40**, 396-401.

Whistler, R.L. (1973). Solubility of polysaccharides and their behaviour in solution. In *Carbohydrates in Solution (Advances in Chemistry Series 117)*, pp.242-255. Edited by R.F. Gould. Washington D.C.: American Chemical Society.

Wood, T.M. and McCrae, S.I. (1977). Cellulase from *Fusarium solani*: purification and properties of the C_1 component. *Carbohydrate Research* **57**, 117-133.

Wu, H-C.H. and Sarko, A. (1978a). The double-helical structure of crystalline β-amylose. *Carbohydrate Research* **61**, 7-25.

Wu, H-C.H. and Sarko, A. (1978b). The double-helical structure of crystalline α-amylose. *Carbohydrate Research* **61**, 27-40.

Chapter 12

BIOSYNTHESIS OF POLYSACCHARASES

K-E. ERIKSSON

Swedish Forest Products Research Laboratory, Stockholm, Sweden

Introduction

The polysaccharases to be dealt with here are the enzymes involved in degradation of cellulose by the white-rot fungus *Sporotrichum pulverulentum*. These enzymes have been extensively studied in our laboratory. The present knowledge is presented in Figure 1.

The hydrolytic degradation of cellulose takes place through the concerted action of (1) five endo-β 1,4 glucanases attacking at random the β 1,4 glucosidic linkages along the cellulose chain, (2) one exo-β 1,4 glucanase, splitting off cellobiose or glucose units from the non-reducing end of the cellulose, (3) two β 1,4 glucosidases hydrolysing cellobiose to glucose and cellobionic acid to glucose and gluconolactone (Figure 2) [Eriksson, 1978; Eriksson and Pettersson, 1975a, b; Almin *et al.*, 1975; Streamer *et al.*, 1975; Deshpande *et al.*, 1978].

It seems now to be generally accepted that essentially the same picture is true also for cellulose hydrolysis by *Trichoderma viride* even if a few differences exist, such as the number of the various enzymes [Emert *et al.*, 1974].

Until recently polysaccharases would exclusively have meant hydrolytic enzymes. In addition to the hydrolytic enzymes, however, an oxidative enzyme has now also been demonstrated to be of importance for degradation *in vitro* of cotton cellulose. It was shown that with a cell-free culture solution from *Sporotrichum pulverulentum* cotton degradation was twice as fast in an atmosphere of oxygen than in an atmosphere of nitrogen [Eriksson *et al.*, 1975]. An oxidative enzyme has recently been purified from *Sporotrichum pulverulentum* culture solution and been found to be a cellobiose oxidase [Ayres *et al.*, 1978]. The molecular weight of this enzyme is around 100,000. The enzyme has been found to be a haem-protein and also to contain a flavin component. In addition to cellobiose, cellotriose, cellotetraose, cellopentaose and cellohexaose are also

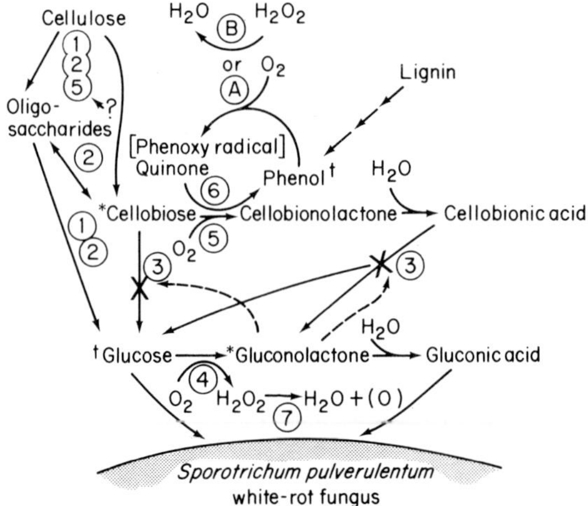

Fig. 1 Enzyme mechanisms for cellulose degradation and their extracellular regulation in *Sporotrichum pulverulentum*.
Enzymes involved in cellulose degradation: 1) endo-β 1,4 glucanases, 2) exo-β 1,4 glucanase, 3) β-glucosidases, 4) glucose oxidase, 5) cellobiose oxidase, 6) cellobiose-quinone:oxidoreductase, 7) catalase.
Enzymes involved in lignin degradation: a) laccase, b) peroxidase.
*Products regulating enzyme activity: gluconolactone inhibits (3). cellobiose increases transglycosylations.
#Products regulating enzyme synthesis: glucose, gluconic acid - catabolite repression, phenols - repression of glucanases.

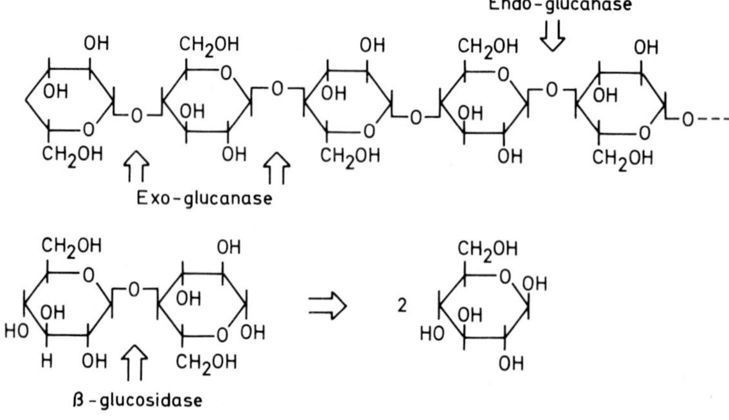

Fig. 2 Degradation of cellulose by three different types of hydrolytic enzymes.

oxidized at approximately the same rate. Figure 3 illustrates the ultraviolet and visible spectra of the oxidized and reduced form of the purified enzyme. The reduction is obtained in the presence of substrate. Whether or not the enzyme also oxidizes the reducing end group formed in cellulose once a β 1,4 glucosidic bond is split by the endo-enzymes has not yet been elucidated. However, combination experiments will be carried out with purified endo-glucanases, exo-glucanases, β-glucosidase and cellobiose oxidase.

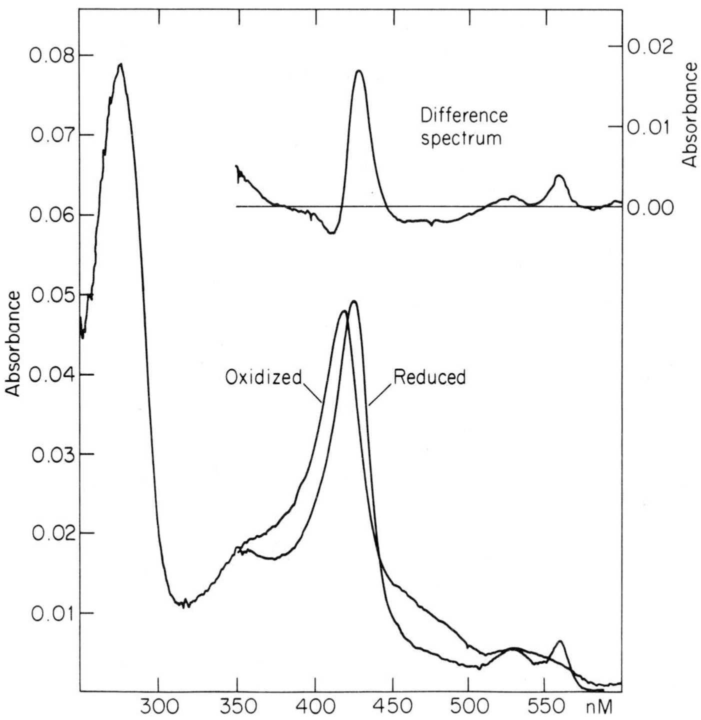

Fig. 3 Ultraviolet and visible spectra of purified cellobiose oxidase. The reduced form of the enzyme was obtained in the presence of substrate. From Ayers *et al.* [1978].

If the cellobiose oxidase does not participate in the initial attack on the cellulose another possible function can be that it withdraws cellobiose from transglycosylation reactions and from inhibitory reactions.

It seems likely that the oxidation of cellobiose to cellobionic acid is important for the fungus, since it has two metabolic routes for this oxidation. Route one is oxidation of cellobiose by the cellobiose oxidase as described above. Route two is dependent upon lignin. Cello-

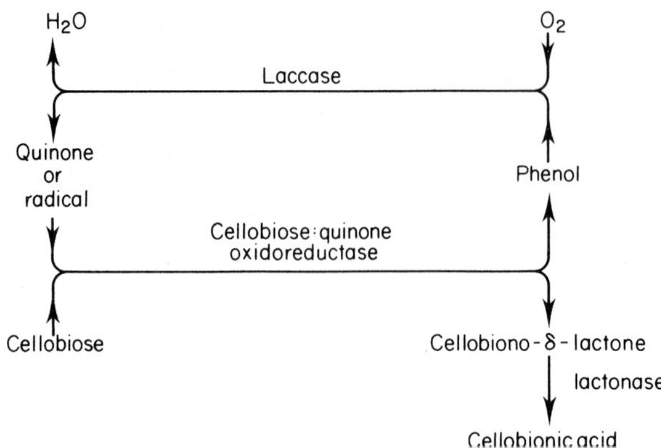

Fig. 4 Reaction mechanism of the enzyme cellobiose: quinone oxidoreductase of importance for the degradation of both cellulose and lignin. From Westermark and Eriksson [1974a,b].

biose is then oxidized concomitantly with the reduction of quinones and/or phenoxy radicals by the enzyme cellobiose: quinone oxidoreductase (Figure 4) [Westermark and Eriksson, 1974a, b; 1975]. Phenoxy radicals and quinones have been formed from lignin phenols by their oxidation through mechanisms A or B (Figure 1).

Cellobiose:quinone oxidoreductase is a flavoprotein with FAD as a prosthetic group and it produces cellobiono-δ-lactone as a product of cellobiose oxidation. Cellopentaose is also oxidized but no oxidation of cellulose has been detected. The enzyme oxidizes lactose and β 4-glucosyl mannose but not β 4-mannosyl glucose, which implies that the C2 hydroxyl of the non-reducing end of the disaccharide is important for substrate specificity. However, the quinone requirement is less specific and the enzyme is able to reduce both *ortho*- and *para*-quinones.

Induction and Catabolite Repression of the Endo-glucanases

Much of our understanding of enzyme regulation is derived from studies on prokaryotes, in particular the *lac* operon of *Escherichia coli*. Regulatory phenomena of the biosynthesis of polysaccharases from fungi have not been studied to the same extent. We have found a deeper understanding of these questions important since it is probable that the fungus, *Sporotrichum pulverulentum*, will be used in technical processes for enzyme production, protein production and water purification [Ek and Eriksson, 1978].

The two main controlling elements of the synthesis of polysaccharide degrading enzymes are the induction and catabolite repression phenomena. Catabolic enzymes which attack polymeric substrates are, generally speaking, induced by water soluble degradation compounds derived from the substrate. Cellobiose is thus known to trigger the mechanism of cellulase production. It seems likely that, in rot fungi, cellobiose causes induction of all the enzymes necessary for cellulose degradation and probably also for hemicellulose degradation [Eriksson and Goodell, 1974].

One group of the basic types of enzymes involved in cellulose degradation are the endo-glucanases (Figure 2). These enzymes were chosen for studies of the induction and repression phenomena. A very sensitive viscometric method based upon the viscosity-lowering effect of the endo-glucanases on solutions of carboxymethyl-cellulose was therefore worked out [Eriksson and Hamp, 1978]. Mycelial pellets were produced by shaking a spore suspension for 18 h in a glucose solution. After washing, the pellets were aseptically transferred to a test solution containing either carboxymethyl-cellulose alone or in combination with the studied inducer or repressor sugars in known concentrations. With this method the effect of inducers and repressors could be determined whether the enzymes were localized on the cell wall surfaces or released into the surrounding solution.

Cellobiose and sophorose have long been known as inducers of endo-glucanase activity [Mandels and Reese, 1960; Mandels *et al.*, 1962]. To study the influence of these compounds upon induction of endo-glucanase, activity was studied as a function of different concentrations of cellobiose and sophorose. The cellobiose results are given in Figure 5. The inducing effect of this disaccharide was demonstrated clearly at an initial cellobiose concentration of only 1 mg/l. The inducing effect remained up to 50 mg/l. Still higher cellobiose concentrations influenced the induction time only slightly. The shortest induction time obtained in these experiments (c.f. Figure 5) was around 45 min for cellobiose concentrations of about 10 mg/l.

Sophorose had a similar effect. The induction time was, however, longer for sophorose than for cellobiose at similar mycelium concentrations and similar concentrations of sophorose and cellobiose. In *Sporotrichum pulverulentum* cellobiose is thus a better inducer of endo-glucanase activity than sophorose.

To study at which concentrations glucose and other monosugars have a catabolite repression effect upon endo-glucanase production, additions of glucose and other sugars were made to the carboxymethyl cellulose solution. These solutions were then inoculated with mycelial pellets. The results of different D-glucose concentrations upon the induction of endo-glucanase are shown in Figure 6. Where no glucose or less than 50 mg/l of glucose was present, endo-glucanase activity was obtained after 3.2 h (low mycelial concentration). In cultures with higher glucose

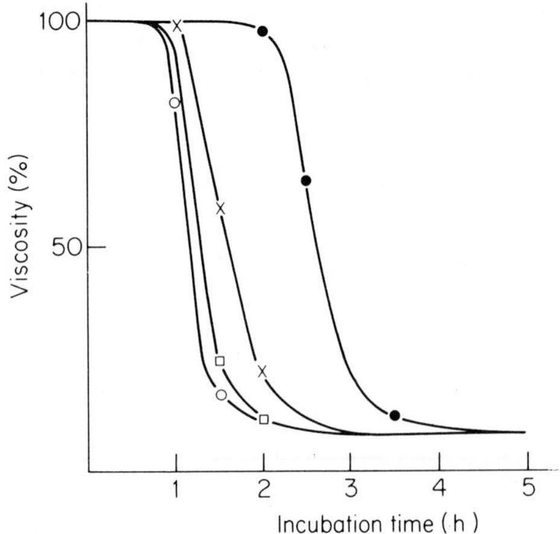

Fig. 5 Influence of cellobiose upon the induction time of endo-β 1,4 glucanase in *Sporotrichum pulverulentum* in the CM-cellulose medium. Mycelium concentration 0.47 g/l. Initial concentration of cellobiose: —●— 0 mg/l; —×— 1 mg/l; —□—10 mg/l; —○— 50 mg/l. From Eriksson and Hamp [1978].

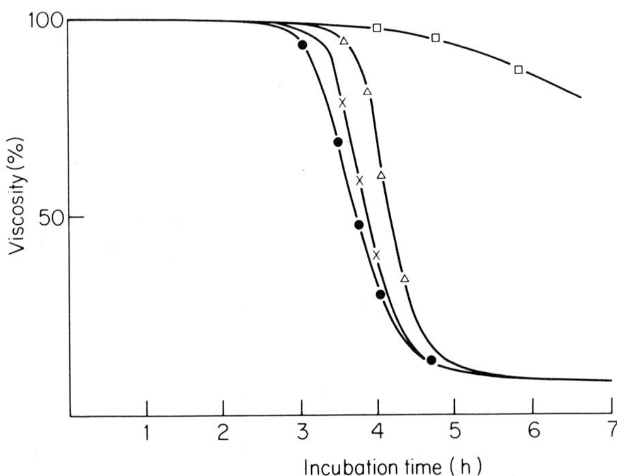

Fig. 6 Influence of glucose concentration upon the induction of endo-β 1,4 glucanase in *Sporotrichum pulverulentum* in the CM-cellulose medium. Mycelium concentration 0.09 g/l. Initial concentration of glucose: —●— 0 mg/l; —×— 50 mg/l; —△—100 mg/l; —□—1000 mg/l. From Eriksson and Hamp [1978].

concentrations repression conditions were obtained and there was a delay in the enzyme production. At an initial glucose concentration of 1000 mg/l there was little decrease in viscosity. Similar results were obtained with D-mannose.

In earlier regulatory studies with *Trichoderma viride* other methods (not as sensitive as the viscometric one developed by us) were used [Mandels and Reese, 1960; Mandels *et al.*, 1962; Nisizawa *et al.*, 1971, 1972; Loewenberg and Chapman, 1977]. It was therefore of considerable interest to study induction and catabolite repression of endo-glucanase activity in *Trichoderma viride* using our method. Since cellobiose, under our conditions, did not cause endo-glucanase induction in this organism sophorose was used as an inducer. Results obtained with sophorose and carboxymethyl-cellulose alone are given in Figure 7. It can be seen that the carboxymethyl-cellulose solution alone did not induce any detectable endo-glucanase formation in *Trichoderma viride* during the run of the experiment (23 h).

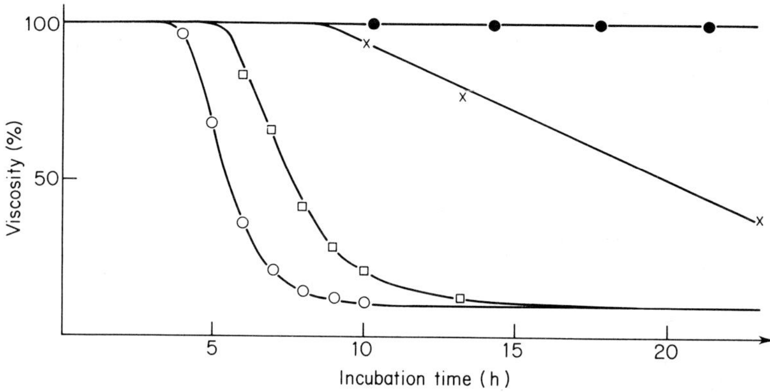

Fig. 7 Influence of sophorose upon the induction time of endo-β 1,4 glucanase in *Trichoderma viride* in the CM-cellulose medium. Mycelium concentration 0.17 g/l. Initial concentration of sophorose: —●— 0 mg/l; —x— 1 mg/l; —□— 10 mg/l; —○— 50 mg/l. From Eriksson and Hamp [1978].

Induction by sophorose, however, could be demonstrated even at concentrations of 1 mg/l. The enzyme production then started after 8.5 h. The induction time was shortened with increasing sophorose concentrations but even at concentrations as high as 50 mg/l the induction time was around 4 h.

It is obvious from these results that striking differences exist in regulation of endo-glucanases in *Sporotrichum pulverulentum* and *Trichoderma viride*. In *Trichoderma viride* carboxymethyl-cellulose alone cannot induce endo-glucanase production. Inducible enzymes are usually formed to a small extent by cells even in the absence of inducer. This phenomenon is known as basal synthesis. Since the polymeric molecule of carboxymethyl-cellulose in itself cannot act as an inducer, the basal synthesis of endo-glucanases seems to be higher in *Sporotrichum pulverulentum* than in *Trichoderma viride*.

For the *lac* operon, basal synthesis by wild-type *Escherichia coli* results in up to 5 active molecules of β-galactosidase per cell. In the presence of galactosidase inducers the level rises to 5000 molecules per cell. The degree of inducibility, known as the induction ratio, can vary from approximately 1000 for β-galactosidase to 10 for penicillin β-lactamases of Gram-negative bacteria [Smith, 1963]. At the induction ratios, the distinction between induced and constitutive synthesis is, of course, marginal. For different purposes it is desirable to have access to an organism that constitutively synthesizes enzymes. This can be achieved by turning the wild-type organism into a mutant which is not dependent upon inducers and also is not catabolite repressed by high sugar concentrations. Since the repressor acts at the operator site, mutations in this region will prevent binding of the repressor and allow constitutive synthesis from the adjacent structural gene. Constitutive mutants can also arise from mutations in the regulator gene.

Repression of Endo-glucanases by Phenols

It was demonstrated by Varadi [1972] that a wide variety of phenols repress the production of cellulases and xylanases in the fungi *Schizophyllum commune* (Figure 8) and *Chaetomium globosum*. At concentrations of less than 1 mM, vanillic acid, vanillic alcohol and vanillin considerably repressed production of these enzymes. In a recent study in our laboratory it was shown that in a phenoloxidaseless mutant (*Phe* 3) of *Sporotrichum pulverulentum* the production of endo-glucanases in the presence of kraft lignin or phenols at a concentration of 10^{-3} M was drastically repressed (Table 1) [Ander and Eriksson, 1976]. However, both the wild-type (WT) and a phenoloxidase-positive revertant (*Rev* 9) produced the enzymes without significant repression. Addition of kraft lignin even stimulated endo-glucanase production in these strains. If a highly purified laccase preparation was added to the growth medium of *Phe* 3 in the presence of phenols the endoglucanase production increased to normal. These results indicate that kraft lignin and phenols decrease endoglucanase synthesis in *Phe* 3 due to the absence of phenol oxidase production. Phenol oxidases may thus function in regulating the production of both lignin- and polysaccharide-degrading enzymes by oxidation of lignin and lignin-related phenols, acting as repressors of endo-glucanase production, when *Sporotrichum pulverulentum* is growing on wood.

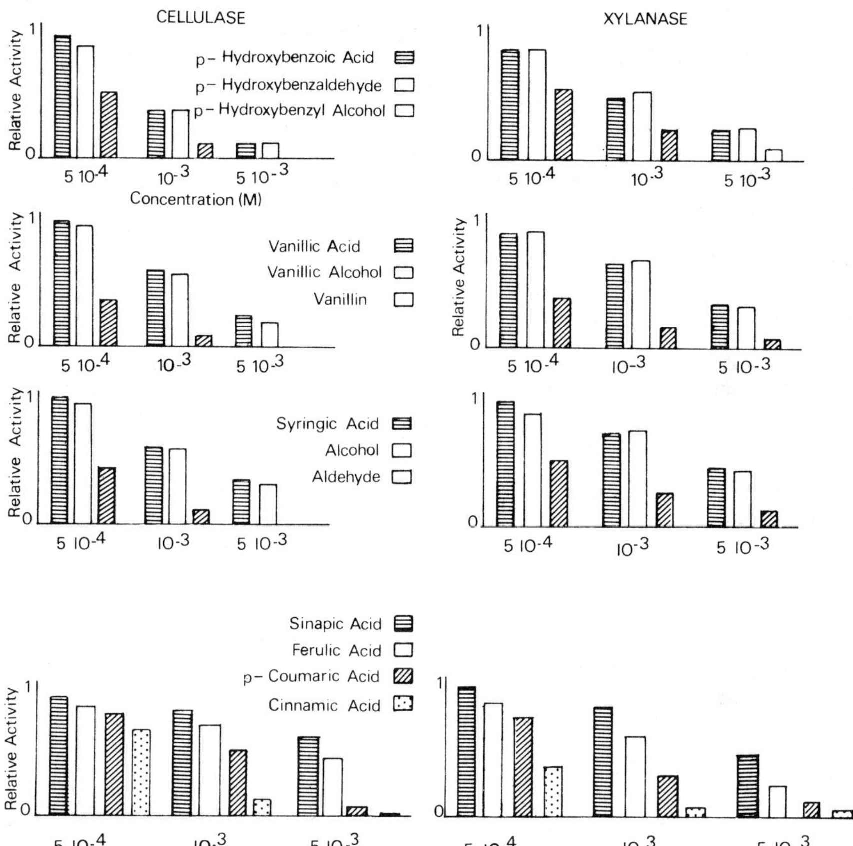

Fig. 8 Effect of phenolic substances on the extracellular enzyme production by *Schizophyllum commune*. From Varadi [1972].

Appearance of β-Glucosidase Activity and its Regulation by Gluconolactone

The activity of cellulose-degrading enzymes in culture filtrates of *Sporotrichum pulverulentum* is dependent not only on mechanisms regulating their biosynthesis but also on the presence of specific regulatory inhibitors of the enzymes themselves. One such inhibitor is gluconolactone. It is produced by oxidation of glucose by the enzyme glucose oxidase but also by hydrolytic cleavage of cellobionolactone (Figure 1). The importance of gluconolactone for the regulation of β-glucosidases from *Sporotrichum*

TABLE 1

Endo-β 1,4 glucanase production by cultures of Sporotrichum pulverulentum

Addition	Enzyme activity*		
	WT	*Phe* 3	*Rev* 9
Ethanol and/or boiled laccase	0.47	0.92	0.56
Vanillic acid	0.22	0.06	0.64
Vanillic acid plus 2 μg laccase	NT	0.14	NT
Vanillic acid plus 8 μg laccase	NT	0.44	NT
p-Hydroxybenzoic acid	0.18	0.05	0.52
Ferulic acid	0.03	0.02	0.08
Kraft lignin	1.08	0.02	1.30
Kraft lignin plus 2 μg laccase	NT	0.03	NT
Kraft lignin plus 8 μg laccase	NT	0.12	NT
Kraft lignin plus 16 μg laccase	NT	0.20	NT

*Endo-β 1,4 β glucanase production (units/ml) by WT, *Phe* 3 and *Rev* 9 in cellulose shake flasks with or without 0.25% kraft lignin and 10^{-3}M phenols. The phenols were added one day after inoculation, whereas kraft lignin was present during sterilization. Laccase was added 6 h after addition of vanillic acid, the addition of laccase to the kraft lignin flasks being carried out at the same time. NT = not tested; WT = wild-type; *Phe* 3 = phenoloxidase-less mutant; *Rev* 9 = phenol-oxidase-positive revertant.

TABLE 2

Kinetic constants for two substrates of β-glucosidases from Sporotrichum pulverulentum

Enzyme	p-nitrophenyl-β-D-glucoside		cellobiose
	K_m (m)	K_i (m)	K_m (m)
Enzyme A	1.5×10^{-4}	3.5×10^{-7}	4.5×10^{-3}
Enzyme B	2.1×10^{-4}	15×10^{-7}	3.7×10^{-3}
Cellbound enzyme	2×10^{-3}	1.2×10^{-4}	-

pulverulentum has recently been studied in our laboratory [Deshpande *et al.*, 1978]. The extracellular β-glucosidase activity could be split into two main peaks. The K_m/K_i of the free enzymes for gluconolactone is approximately 13,000 (Enzyme A) and 2,500 (Enzyme B) with cellobiose as substrate (c.f. Table 2). As can be seen in this table the K_m/K_i ratio for the cellbound β-glucosidase activity is considerably lower. Why the cellbound β-glucosidase is less sensitive to competitive inhibition by gluconolactone is not known but it may be an important factor in the regulation of cellulose degradation.

References

Almin, K.E., Eriksson, K-E. and Pettersson, B. (1975). Extracellular enzyme system utilized by the fungus *Sporotrichum pulverulentum (Chrysosporium lignorum)* for the breakdown of cellulose. 2. Activities of the five endo-1,4-β-glucanases towards carboxymethylcellulose. *European Journal of Biochemistry* **51**, 207-211.

Ander, P. and Eriksson, K-E. (1976). The importance of phenol oxidase activity in lignin degradation by the white-rot fungus *Sporotrichum pulverulentum. Archives of Microbiology* **109**, 1-8.

Ayers, A.R., Ayers, S.B. and Eriksson, K-E. (1978). Cellobiose oxidase, purification and partial characterization of a hemoprotein from *Sporotrichum pulverulentum. European Journal of Biochemistry* **90**, 171-181.

Deshpande, V., Eriksson, K-E. and Pettersson, B. (1978). Production, purification and partial characterization of 1,4-β-glucosidase enzymes from *Sporotrichum pulverulentum. European Journal of Biochemistry* **90**, 191-198.

Emert, G.H., Gum, E.K., Lang, A.A., Lin, T.H. and Brown, R.D. (1974). Cellulases. *Advances in Chemistry Series* **136**, 79-100.

Ek, M. and Eriksson, K-E. (1978). Conversion of waste fibres into protein. In *Proceedings of Symposium on Bioconversion of Cellulosic Substances into Energy, Chemicals and Microbial Protein*, pp.449-454. Edited by T.K. Ghose. New Delhi and Zurich: Indian Institute of Technology and Swiss Federal Institute of Technology.

Emert, G.H., Gum, E.K., Lang, A.A., Lin, T.H. and Brown, R.D. (1974). Cellulases. *Advances in Chemistry Series* **136**, 79-100. *Biotechnology and Bioengineering* **20**, 317-332.

Eriksson, K-E. and Goodell, E.W. (1974). Pleiotropic mutants of the wood-rotting fungus *Polyporus adustus* lacking cellulase, mannanase and xylanase. *Canadian Journal of Microbiology* **20**, 371-378.

Eriksson, K-E. and Hamp, S. (1978). The regulation of endo-1,4-β-glucanase production in *Sporotrichum pulverulentum. European Journal of Biochemistry* **90**, 183-190.

Eriksson, K-E. and Pettersson, B. (1975a). Extracellular enzyme system utilized by the fungus *Sporotrichum pulverulentum (Chrysosporium lignorum)* for the breakdown of cellulose. 1. Separation, purification and physico-chemical characterization of five endo-1,4-β-glucanases. *European Journal of Biochemistry* **51**, 193-206.

Eriksson, K-E. and Pettersson, B. (1975b). Extracellular enzyme system utilized by the fungus *Sporotrichum pulverulentum (Chrysosporium lignorum)* for the breakdown of cellulose. 3. Purification and physico-chemical characterization of an exo-1,4-β-glucanase. *European Journal of Biochemistry* **51**, 213-218.

Eriksson, K-E., Pettersson, B. and Westermark, U. (1975). Oxidation: an important enzyme reaction in fungal degradation of cellulose. *FEBS Letters* **49**, 282-285.

Loewenberg, J.R. and Chapman, C.M. (1977). Sophorose metabolism and cellulase induction in *Trichoderma. Archives of Microbiology* **113**,

61-64.
Mandels, M., Parrish, F.W. and Reese, E.T. (1962). Sophorose as an inducer of cellulase in *Trichoderma viride*. *Journal of Bacteriology* **83**, 400-408.
Mandels, M. and Reese, E.T. (1960). The induction of cellulase in fungi by cellobiose. *Journal of Bacteriology* **79**, 816-826.
Nisizawa, T., Suzuki, H., Nakayama, M. and Nisizawa, K. (1971). Inductive formation of cellulase by sophorose in *Trichoderma viride*. *Journal of Biochemistry (Tokyo)* **70**, 375-385.
Nisizawa, T., Suzuki, H. and Nisizawa, K. (1972). Catabolite repression of cellulase formation in *Trichoderma viride*. *Journal of Biochemistry (Tokyo)* **71**, 999-1007.
Smith, J.T. (1963). Penicillinase and ampicillin resistance in a strain of *Escherichia coli*. *Journal of General Microbiology* **30**, 299-305.
Streamer, M., Eriksson, K-E. and Pettersson, B. (1975). Extracellular enzyme system utilized by the fungus *Sporotrichum pulverulentum (Chrysosporium lignorum)* for the breakdown of cellulose. Functional characterization of five endo-1,4-β-glucanases and one exo-1,4-β-glucanase. *European Journal of Biochemistry* **59**, 607-613.
Váradi, J. (1972). The effect of aromatic compounds on cellulase and xylanase production of fungi *Schizophyllum commune* and *Chaetomium globosum*. In *Biodeterioration of Materials*, Vol. 2, pp.129-135. Edited by A.H. Walters and E.H. Hueck-van der Plas. London: Applied Science Publishers.
Westermark, U. and Eriksson, K-E. (1974a). Carbohydrate-dependent enzymic quinone reduction during lignin degradation. *Acta Chemica Scandinavica B* **28**, 204-208.
Westermark, U. and Eriksson, K-E. (1974b). Quinone oxidoreductase, a new wood-degrading enzyme from white-rot fungi. *Acta Chemica Scandinavica B* **28**, 209-214.
Westermark, U. and Eriksson, K-E. (1975). Purification and properties of cellobiose: quinone oxidoreductase from *Sporotrichum pulverulentum*. *Acta Chemica Scandinavica B* **29**, 419-424.

Chapter 13

TECHNOLOGY OF MICROBIAL POLYSACCHARASE PRODUCTION

D.E. BROWN

Department of Chemical Engineering, UMIST, Manchester, UK

Introduction

The design of a fermentation process for the manufacture of a product such as industrially useful polysaccharases involves the consideration of a wide range of interacting technologies. Although much of the existing expertise has been acquired in an empirical manner, the growing knowledge in both biological and engineering sciences is resulting in improved procedures for process design and control.

The production of enzymes on an industrial scale is usually restricted to products that have some commercial significance. Thus polysaccharase manufacture is in general associated with applications in starch-based industries. A growing knowledge of the mechanisms of control of protein biosynthesis can be extended to explain microbial polysaccharase production. In particular, it has enabled technologists to establish a range of techniques for process development. Some degree of understanding of the kinetics of enzyme formation has also been achieved which will help in future design and development strategies.

The design of the fermenter to be used for the culture of microorganisms for the production of polysaccharases has to take account of many interacting factors. Process requirements for medium pH and sterilization procedures influence materials of construction. Rheological properties of the culture and its oxygen requirements dictate aspects of mixing, heat transfer and aerator design.

The information now available makes it possible to consider the combination of microbiological, biochemical and engineering technologies in the formulation of basic process design procedures.

Polysaccharides of Commercial Importance

Although there are many polysaccharides that can be identified and characterised [Kulp, 1975] there are a limited number that are of current industrial significance. The most important polysaccharides occur as starch or starch degradation products in the brewing and food industries. Starch is composed of the α 1,4-linked glucose dimer, maltose, in amylose and with α 1,6 branches in amylopectin. Pullulan, a linear polymer of α 1,4- and α 1,6-linked glucose molecules also occurs in some starch materials. Different degrees of hydrolysis of the native starch molecules can be implemented depending on the product requirement. This treatment can vary from minor modifications of flour prior to dough-making operations [Barrett, 1975] to a complete conversion of the starch into glucose in syrup manufacturing [Banks *et al.*, 1967; MacAllister *et al.*, 1975].

Another important polysaccharide in the cellular structure of most plants is cellulose. However, not until the recent general concern arose over the continuing supply of energy has much interest been directed at the planned degradation of celluloses. The cellulose molecule is of β 1,4-linked glucose molecules in a linear chain. The repeating dimer is cellobiose [Norkrans, 1967]. Cellulose is commonly associated with lignin, a polymeric material of phenyl propane units [Forss and Fremer, 1975] and hemicellulose acting as a binder. Hemicellulose is a variously bonded mixture of xylose, glucose, galactose, mannose and arabinose with uronic acid [Schurz, 1978].

Biosynthesis of Enzymes

A detailed discussion of the biosynthesis of enzymes will not be undertaken, but some general principles will be outlined because they effect the overall process design strategy.

Enzyme protein synthesis is controlled by mechanisms which operate at the initial transcription level. One important aspect of the various control mechanisms is the fact that the messenger RNA which assists in the assembly of the amino-acids into protein molecules is relatively short-lived. Reports indicate mRNA life-times of 60 to 80 min [Shinmyo *et al.*, 1969; Both *et al.*, 1972]. Thus in order that a microorganism can continue to produce a given enzyme, it must continue to produce the appropriate mRNA. Production of a given mRNA can be limited by mechanisms in operation on the DNA molecule [Demain, 1972].

Within the information contained by the prokaryotic chromosome is a specific section associated with the production of the particular mRNA. In a controlled situation, the production of mRNA is prevented by a repressor molecule produced by the adjacent regulator gene. This repressor molecule associates with the operator gene and prevents transcription. The production of mRNA and thus the enzyme molecule which will be formed can be induced by a molecule which combines with the repressor molecule more readily than the extent to which the repressor molecule binds to the operator gene. The inducer molecule is often the substrate [Smith and Dean, 1972] or sometimes a substrate analogue [Herzenberg, 1959] of the enzyme that will ultimately be made by the mRNA-ribosome complex.

An alternative arrangement is that the repressor gene produces an incomplete repressor, the aporepressor, that requires a co-repressor with which to combine in order to produce the repressor molecule that can combine with the operator gene. Thus, mRNA and then enzyme is produced initially and in this situation the product of the enzyme reaction takes the role of the co-repressor and results in feed-back repression of the system.

Overriding these mechanisms of repression and induction there is also the phenomenon known variously as the glucose effect, catabolic or metabolic repression [Pastan and Perlman, 1970]. Since the observation that a sharp decrease in cyclic adenosine 3',5'-monophosphate (cAMP) occurs in the presence of glucose [Makman and Sutherland, 1965], attention has been directed towards the determination of the role of this compound in enzyme synthesis. Most of the studies have been carried out on *Escherichia coli* and β-galactosidase synthesis and they reveal [Emmer *et al.*, 1970] that cAMP is involved with a further receptor protein (crp) at the promotor gene or in the transcription of the mRNA. The reason for the reduction of cAMP in the cell in the presence of easily metabolizable carbon compounds or as a result of their rapid rates of metabolism has not been established. One possibility is that cAMP diffuses from the cell [Makman and Sutherland, 1965]. Although the role of cAMP and crp in enzyme synthesis and its reduction during metabolic repression have been investigated mainly for β-galactosidase synthesis in *Escherichia coli*, the mechanism might be generally applicable to enzyme systems that show metabolic repression.

Enzyme Process Development

The techniques employed by fermentation technologists in the early days of enzyme production were mainly of an empirical nature and have given the business an artistic rather than scientific image. However, as a result of the greater understanding of the mechanisms and controls which are involved in enzyme biosynthesis, it is possible to itemize certain established procedures [Demain, 1972, 1973] for the design and development of polysaccharase fermentations.

Choice of a suitable microorganism

The isolation of a strain of microorganism which can produce the desired enzyme is a first step. Choices have to be made with regard to the type of medium, the nature and scale of culture technique and the method of assay on which the selection is to be made. A commercially useful enzyme product may need to contain small but necessary quantities of various other enzymes besides the one which is monitored by the assay. Some degree of automation is now being implemented for the selection of new strains and mutants [Calam, 1969].

A second condition which must be applied to microorganisms being selected to produce industrial polysaccharases is that they must not be pathogenic nor produce any toxic or allergenic materials. This is particularly important because most of the applications of microbial polysaccharases are in the human food processing industry. It can be advantageous to consider at this stage the various aspects of fermentation and separation process operations as they might be affected by the different morphologies, metabolic patterns, degrees of autolysis etc. of the different possible species and strains of organism.

Genetic selection of the microorganism

The exposure of a culture of microorganisms to some mutagenic agent can result in the death of most of the cells. Those that survive can be cultured and screened using a variety of selection procedures [Demain, 1973]. The aim is to find mutants which have modified repressor or operator genes such that induction is not necessary. These mutants are designated constitutive for the production of the desired enzyme. Other constitutive characteristics that may arise are the ability to resist feedback control and metabolic repression.

Another useful procedure involving changes at the genetic level is to increase the number of copies of the structural gene associated with the production of the desired mRNA [Pardee, 1969].

Addition of specific chemicals

For the case in which the enzyme requires induction, identification of the most appropriate inducer or perhaps inducer analogue and their optimal concentrations are advantageous. It is possible to block metabolic pathways so that co-repressors are not formed and thus feedback control is unable to occur [Demain, 1971].

Reduction of metabolic repression

From the increased understanding of the mechanisms of

metabolic repression and the accumulated experience of many years of fermentation technology have arisen a range of techniques designed to slow down the overall rate of metabolism.

Choice of carbohydrate materials Since the early days of penicillin production, it has been understood by fermentation technologists that simple carbohydrates such as glucose and sucrose, being easily and rapidly metabolized, caused repression of product formation [Demain, 1971]. The use of lactose, because of its relatively slow rate of catabolism [Demain, 1971; Underkofler, 1966], was a successful development but has been replaced in recent years by the controlled slow feed of glucose [Hockenhull and Mackenzie, 1968]. Many of the traditional recipes for media for enzyme production contain relatively low priced complex carbohydrate materials such as starch, soya bean meal, barley syrup and malt extracts [Burbidge and Collier, 1968]. Their usefulness lies partly in their prices, partly in the various trace elements, amino-acids and unidentified growth factors which they may contain, but particularly in the fact that the carbohydrates become available to the microorganisms at a slow rate [Hernandez and Pirt, 1975], rather like a glucose feed, as the polymers are gradually hydrolysed.

Environmental factors The general effects that temperature and pH have on the rate of growth of microorganisms are well known [Pirt, 1975]. Thus the choice of a particular temperature and pH can be made in order to assist in the adjustment of the overall rate of metabolism so that repression is minimized. A single optimum temperature for a batch fermentation process obtained by empirical methods reflects an averaging effect on growth and enzyme producing phases. However, the temperature optimum for enzyme production is usually less than that for growth [Nyiri, 1971] because the metabolic processes slow down and assist the de-repression at lower temperatures. Temperature profiling of batch fermentation processes to maximize the production of enzymes could be a future development similar to that described for the penicillin process [Constantinides *et al.*, 1970].

The rate of metabolism of the microorganisms is affected by the pH of the medium. However, in an uncontrolled fermentation, the instantaneous pH is the result of metabolism in relation to the starting pH and the nature of the medium. Thus control of the pH at a particular value does not necessarily improve the overall pattern of the fermentation and result in improved enzyme production. The advent of reliable pH electrodes has resulted in an increase of interest in the effects of this environmental factor [Brown and Halsted, 1975] and its interaction with enzyme production rates [Brown *et al.*, 1975; Andreotti *et al.*, 1978]. It is possible to design the medium and starting pH of a batch culture in such a way that the complete utilization of the carbohydrate occurs at the same moment as the pH reaches a value at which metabolism is almost completely stopped. In Figure 1 a pH plateau at approximately 3.0 occurred from the point at which glucose ran out and during which the cellulase was synthesized [Brown *et al.*, 1975].

The pH of the culture medium effects the overall production rate of an enzyme in another way. Most enzymes are only stable within narrow limits of pH. Thus although a low pH may be advantageous in reducing the rate of metabolism and thereby increasing the enzyme production rate, it may also cause rapid decomposition of the enzyme. A further complication may occur in that a particular pH may be found to be the optimum value for the activity of a protease which degrades the desired enzyme [Melling and Phillips, 1975].

Polysaccharases of Commercial Importance

The most important commercial polysaccharase is α-amylase (EC 3.2.1.1). The action of this enzyme is to promote the hydrolysis of endo-α 1,4 links in starch in a random manner to give a mixture of dextrins and oligosaccharides. α-Amylase is obtained from some *Bacillus* species [Fogarty *et al.*, 1974] and some *Aspergillus* species [Banks *et al.*, 1967; Blain, 1975]. Details of the different properties of these enzymes have been collated [Gould, 1975]. Different temperature and pH optima [Marconi, 1974] allow flexibility in their application. Commonly the bacterial α-amylase is produced by *Bacillus subtilis* strains [Burbidge and Collier, 1968] although more thermostable products are now available [see Norman, Chapter 15]. Industrial processes for bacterial α-amylase production are based on the submerged culture of the bacteria in a medium which often includes several types of starch-containing materials. Traditional impressions were that α-amylase was induced by one of the degradation products of starch such as maltose. However, Coleman [1967] investigated the effect of various defined and complex media and concluded that the α-amylase was probably constitutive. A limitation of nucleic acid precursors during growth was proposed as the cause of the very slow enzyme production rate during that period. More recently, Meers [1972] found that the formation of α-amylase in *Bacillus licheniformis* when grown in continuous culture on simple sugars did not require any inducer and suggested that the system is catabolite repressed. A similar constitutive explanation for the production of α-amylase by the fungus *Aspergillus oryzae* has been made by Meyrath and Volavsek [1975], as the enzyme production rate is unaffected by different carbon sources during the growth phase but varies considerably during the stationary phase. Thus, the evidence would support a de-repression towards the end of the batch culture as metabolism is slowed down.

A second important microbial polysaccharase is amyloglucosidase or glucoamylase (EC 3.2.1.3). Its action is to remove glucose units from the non-reducing end of dextrins and oligosaccharides. The 1,6 link can also be hydrolysed with the assistance of this enzyme which makes it very useful in the production of high dextrose syrups. Amyloglucosidase is produced by fungi such as *Aspergillus niger* and *Aspergillus oryzae* [Blain, 1975] and commonly is present in fermentation processes designed to produce fungal α-amylases.

Of growing importance is the enzyme system referred to as cellulase (EC 3.2.1.4). Recent researches have shown [Wood and McCrae, 1978] that several enzymes are present in a cellulase system which is capable of degrading crystalline cellulose [see Bacon, Chapter 11; Eriksson, Chapter 12]. Figure 2 is a diagrammatic representation [Brown and Fitzpatrick, 1976] of the possible enzyme reactions involved. In a recent publication [Enari and Markkanen, 1977] the endo-β 1,4 glucanase is designated EC 3.2.1.4. and thus a further code is required for the exo-β 1,4 glucanase (EC 3.2.1.-) which is present in the mixture. The presence of β 1,4 glucosidase (EC 3.2.1.21) is also evident in preparations which produce high yields of glucose during a cellulose hydrolysis. Although the glucanase enzymes are identified as single enzymes, they are composed of several enzyme proteins operating synergistically. Many cellulase preparations are composed of only the glucanase enzymes with perhaps some glucosidase. The key component which makes a preparation of general use is that enzyme originally referred to as the C_1 component [Reese *et al.*, 1950]. Extensive studies of the mode of action involved in the degradation of cellulose by *Sporotrichum pulverulentum* [Eriksson, 1978; Chapter 12] suggests that the initial step involves oxygen and an oxidative enzyme. However, Wood and McCrae [1978] working with *Trichoderma koningii* cellulase did not find a high dependence on the oxidative reaction.

Cellulose-degrading enzymes are produced by a large number of bacteria and fungi. However, it has not been possible to isolate preparations from many, as they appear to be cell bound. Commercial products are based on a relatively small number of fungi of species of *Trichoderma* and *Aspergillus*. In particular, *Trichoderma viride* and *Trichoderma koningii* produce commercial quantities of enzyme systems suitable for the complete conversion of crystalline cellulose into glucose. Early work on the biosynthesis of the complete cellulase system favoured the growth of the microorganism on a medium containing cellulose and the highest yields are still obtained on media of that type [Mandels and Sternberg, 1976]. It was originally

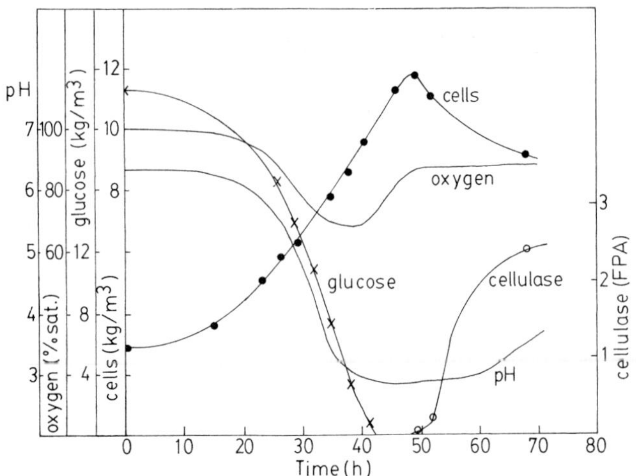

Fig. 1 Cellulase production by *Trichoderma viride* QM9123 in batch culture [Brown *et al.*, 1975].

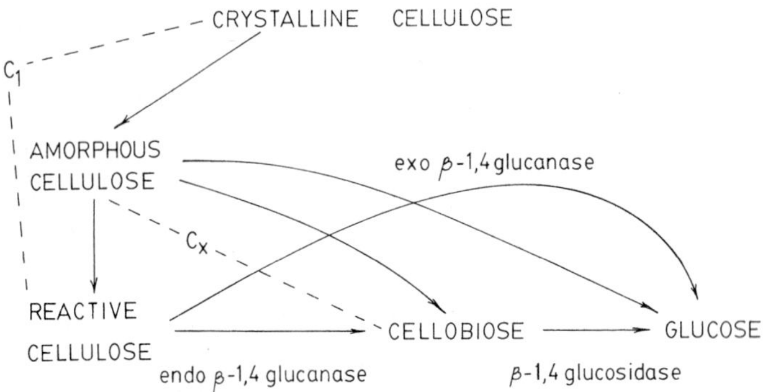

Fig. 2 Possible reaction mechanisms in the enzymic hydrolysis of cellulose to glucose [Brown and Fitzpatrick, 1976].

believed that the cellulase system was induced by cellulose or perhaps by cellobiose. The involvement of the dimer sophorose (2-*O*-β-D-glucose) was reported by Mandels *et al.* [1962] and found by Nisizawa *et al.* [1971] to stimulate the production of cellulase in *Trichoderma viride*. Cellulase biosynthesis in *Myrothecium verrucaria* was claimed [Hulme and Stranks, 1971] to be controlled only by metabolic repression which could be reduced by feeding glucose at a low and controlled rate. Although metabolic repression is clearly a control mechanism [Brown *et al.*, 1975] the presence of trace quantities of sophorose in some commercial glucose products confuses the picture [Brown, 1976].

Another polysaccharase of growing commercial interest is pullulanase (EC 3.2.1.41) which is obtained from the bacterium *Enterobacter* (syn. *Klebsiella, Aerobacter) aerogenes* and has the action of hydrolysing the α 1,6 linkage in pullulan and 1,6-branched dextrins. Thus, in collaboration with α-amylase and amyloglucosidase it can assist in the production of very high glucose syrups. It would seem [Hope and Dean, 1974] that pullulanase biosynthesis is repressed by the metabolism of simple sugars and that cAMP addition can counteract the effect. The requirement for an inducer such as pullulan, or even maltose or maltotriose was indicated but it is not certain whether such materials were only slowing down the rate of metabolism. Pullulanase activity is commonly cell bound although Hope and Dean [1975] found that increasing the cell concentration caused a release of some of the enzyme into the medium. This effect was considered to be associated with the supply of oxygen to the culture but the mechanism was not elucidated.

Kinetics of Polysaccharase Biosynthesis

The general pattern of a batch culture for extracellular polysaccharases is shown in Figure 1. Although this figure shows the results of cellulase production by a fungus, data of this type can be obtained for most systems. The main characteristic is that the cell population grows without any significant enzyme production. Then the curve of enzyme concentration starts at some stage in the process when repression mechanisms have been removed. Different processes will have different lags between cell growth and enzyme formation and bacterial processes will be faster than fungal processes.

Considerable difficulty has been encountered in developing satisfactory unstructured models to described the product formation kinetics of non-growth associated systems. An early approach by Luedeking and Piret [1959], based on a study of the lactic acid fermentation, proposed that the rate of product formation was a function of the growth rate and also of the cell concentration:

$$\frac{dP}{dt} = \alpha\frac{dx}{dt} + \beta x^* \qquad (1)$$

A suitable choice of values for α and β can account for differing time lags between cell growth and product formation.

An approach that introduced a form of structure [Blanch and Rogers, 1971; Brown and Vass, 1973] suggested that the lag between growth and product formation could be explained in terms of the culture containing "immature" and "mature" cells. The mature cells are those that are capable of producing the product of particular interest. The rate of product formation is postulated to be propor-

---

*See page 319 for nomenclature.

tional to the rate of formation of mature cells:

$$\frac{dP}{dt} = k_p \frac{dx_p}{dt} \tag{2}$$

The rate at which mature cells are produced is the same rate at which new cells were being generated t_m hours earlier:

$$\left(\frac{dx_p}{dt}\right)_t = \left(\frac{dx}{dt}\right)_{t-t_m} \tag{3}$$

The values of k_p and t_m are obtained by a trial-and-error plot of $(P)_t$ versus $(x)_{t-\theta}$ for a range of values of θ [Brown and Vass, 1973]. A report of $k_p = 0.37 \times 10^6$ FPA/ kg cells and $t_m = 100$ hours was made for the production of filter paper activity (FPA) cellulase in *Trichoderma viride* QM9123 grown in continuous culture at pH 3.0 and a temperature of 30°C [Brown and Zainudeen, 1977]. 100 hours seems an excessively long time for a portion of cell material to mature to be capable of producing cellulase. However, a controlled low pH resulted in a long lag before cellulase production occurred in batch culture [Brown *et al.*, 1975].

Progress in the development of kinetic models has been made in recent years through a consideration of the microbiological and biochemical mechanisms of enzyme synthesis described above. Terui *et al.* [1967] proposed that the specific rate of enzyme synthesis (ε) is proportional to the quantity of mRNA per unit mass of cells. The application of this principle requires that some relationship can be assumed between the rate of production of mRNA and the growth rate of the cells. In the most general case this relationship was given by:

$$\frac{dr}{dt} = a'\mu - b'\left(\frac{d\mu}{dt}\right) - k'r \tag{4}$$

The first term on the right hand side of equation (4) ($a'\mu$) deals with the fact that mRNA content of the cells increases as growth proceeds. The third term ($-k'r$) represents the decay of mRNA. The second term ($-b'(d\mu/dt)$) is included for cases in which growth-associated repression occurs [Terui, 1972] and which is observed as a negative correlation of ε with μ.

Thus for the case of α-amylase production by *Bacillus subtilis* [Kinoshita *et al.*, 1967] and amyloglucosidase by *Aspergillus niger* [Okazaki and Terui, 1967] and summarized elsewhere [Shinmyo *et al.*, 1969], enzyme production during the growth phase was found to be described very accurately by:

$$\frac{d\varepsilon}{dt} = a\mu - b\frac{d\mu}{dt} - k\varepsilon \tag{5}$$

derived from equation (4).

Values of a, b and k were obtained for the α-amylase production by *Bacillus subtilis* [Kinoshita *et al.*, 1967] and were found to be substantially constant for changing values of initial starch concentration. Experiments carried out at different temperatures between 45°C and 27°C resulted in changing values of the mRNA decay constant k according to Arrhenius' relationship. The values of the constants showed little growth-associated enzyme production ('a' small), some extent of growth associated repression ('b' of significant magnitude) and a relatively stable mRNA ('k' small, approximately = 0.1 h^{-1}).

The continuation of amyloglucosidase production by *Aspergillus niger* after the growth phase had stopped [Okazaki and Terui, 1967] was associated with lower values of k than those observed for *Bacillus subtilis*. A further elaboration of the model was successfully attempted in which it was stated that the mRNA concentration would be composed of a) mRNA remaining from the growth phase and b) new mRNA synthesized from and controlled by the turnover of RNA.

A theoretical analysis of enzyme biosynthesis [van Dedem and Moo-Young, 1973] has considered the genetic control of mRNA production to be as a result of the chemical equilibrium between quantities of repressor molecules (R), operator sites (O) and the inducer or the co-repressor molecules (I):

$$O + R \overset{k_1}{\rightleftarrows} OR \tag{6}$$

$$R + nI \overset{k_2}{\rightleftarrows} RI_n \tag{7}$$

For the case of induction, a value of n = 2 was assumed so that the fraction of available operator gene Q was given by:

$$Q = \frac{[O]}{[O_t]} = \frac{1 + k_2[I]^2}{1 + k_2[I]^2 + k_1[R_t]} \tag{8}$$

The rate of enzyme production was assumed to be proportional to the quantity of mRNA, after Terui *et al.* [1967] and the decay of enzyme was taken to be a first order rate so that:

$$\frac{d[E]}{dt} = k_3[r] - k_4[E] \tag{9}$$

Also the rate of synthesis of mRNA was stated to be proportional to the specific growth rate and to the fraction of available operator (given by equation (8)) and its decay given by the same term as that in equation (4):

$$\frac{d[r]}{dt} = k_5 \mu Q x - k[r] \qquad (10)$$

A combination of equations (8), (9) and (10) can be used to predict the enzyme concentration in a fermentation provided that additional equations are available to describe the growth process on the limiting substrate.

The treatment of metabolic repression by these workers is however not as satisfactory as their model for the induction process. It is proposed that the substrate or some metabolic breakdown product of the substrate binds with an incomplete apo-repressor molecule in the manner in which end-product feed-back control is reputed to work [Demain, 1972]. Thus:

$$R + nI \rightleftarrows RI_n \qquad (11)$$

and

$$O + RI_n \rightleftarrows ORI_n \qquad (12)$$

Similar equations to equations (8), (9) and (10) can be derived using equations (11) and (12) to describe this type of system.

Toda [1976] examined the biosynthesis of invertase in *Saccharomyces carlsbergensis* and concluded that the enzyme was both induced and metabolically repressed. To model this system he proposed that two different operator genes might control the transcription of the appropriate mRNA. This dual control mechanism was modelled by combining the two analyses of induction and metabolic repression described by van Dedem and Moo-Young [1973]. The fraction of the total available operator gene Q was determined as the product of Q_1 for induction and Q_2 for repression. Both the induction and the metabolic repression were assumed to be functions of the limiting substrate and linearized forms of equation (8) were used to obtain the parameter values from continuous culture data.

In a most recent consideration of the dual control of induction and metabolic repression [Imanaka and Aiba, 1977] it was shown that two operator genes were not necessary in modelling the system. Drawing on the role of cAMP and its receptor protein crp [Emmer *et al.*, 1970], they suggested that the available operator gene Q is composed of the product of Q_1 for induction and Q_2 for the probability of crp or a cAMP/crp complex coming in contact with the promotor gene. The quantity of cAMP was set to be inversely proportional to the concentration of the carbon source (glucose) in the medium. Similar analytical expressions to those of Toda [1976] were obtained although the meanings of some of the terms and parameters are different. This scheme [Imanaka and Aiba, 1977] based on the current knowledge of the biosynthesis of enzymes seems to offer the closest kinetic description of this complex dual interaction of induction with metabolic repression. Its general applicability has yet to be confirmed but it forms a useful framework for future investigations.

Fermenter Configuration

The culturing of microorganisms can be carried out in almost any conceivable shape of container. However, each microbial process has individual requirements which sets certain constraints for the designer. Although such designs as bubble column and tower fermenters [Shore and Royston, 1968; Schügerl *et al.*, 1978] attract much interest, the stirred fermenter continues to be widely used in the industrial manufacture of polysaccharases. Several factors unassociated with the fermentation process requirements could result in the permanent retention of this configuration.

The dispersion and suspension of complex medium components such as soya bean flour and corn steep liquor in the make-up water invariably requires some kind of stirrer device. Stirring is usually required for such media throughout the sterilization cycle, otherwise starch particles will gel and agglomerate.

Sterilization is usually carried out using both steam on the jacket of the vessel and also sparged directly into the medium. Thus, both the vessel and its jacket or welded limpet coils must be treated as pressure vessels, and tested according to the relevant code of practice. Pressure relief valves have to be fitted to these systems. The cylindrical shell and dished ends of most stirred vessels are ideally suited for operation at elevated pressures.

Most fermentation factories operate the enzyme production as a series of batch or fed-batch runs. Batch process techniques are preferred to continuous operation for a variety of reasons. For many products the market is intermittant and cannot be anticipated too rigidly. The shelf-life of many commercial enzyme preparations is relatively short. Only in a few cases has there been need for large quantities of a given enzyme on a continuous basis. Processes which have been developed for enzyme production usually include complex substrates which have had little investigation in continuous culture [Hernandez and Pirt, 1975] and rely on appropriate pH profiles for optimum derepression. Batch process operations require an assembly of materials at the start and a container for the system at the end. In some instances such as pullulanase production [Smith and Dean, 1972] there may be the need for some treatment stage to free the enzyme from the cells at the end of the fermentation. All these aspects can be handled very satisfactorily in the fermenter if it is a stirred vessel.

Although some polysaccharase fermentations have pH patterns which do not vary greatly from neutral, the low values of most fungal fermentations are usually in the range pH 2.0 to 3.0. Such pH values require a good stainless steel as the material for construction and this is particularly suitable for ease of cleaning.

Gas-liquid contact requires a high energy dissipation per unit mass of fluid and ideally this should be homogeneous throughout the system. In contradiction, heat transfer problems associated with the removal of metabolic heat require a high fluid velocity at the cooling surface. Adjustments in the design of the impellers can be made to obtain the best compromise between aeration, heat transfer and mixing.

The stirred cylindrical vessel has been found to satisfy most of the requirements needed of a fermenter and is widely used at all scales from 0.001 m^3 to 150 m^3.

Process Design and Aeration Capacity

In the discussion of the biochemistry of enzyme synthesis it was clear that to produce a given polysaccharase requires that the cells must be induced and not repressed by a high rate of metabolism. Variations in the fine details of medium components, temperature and pH can result in variations of the quantity of enzyme produced. However, overriding these complex control aspects in the determination of maximum enzyme production is the importance of the actual quantity of cells grown in the fermentation. The more cells that can be grown and then appropriately de-repressed, the more enzyme will be produced. The quantity of cells that can be produced in a given fermenter will depend on the aeration capacity of the unit. An interaction occurs between the aeration capacity, measured as the absorption coefficient (k_La) and the cell concentration. Figure 3 shows the effect of *Trichoderma viride* mycelial concentration on k_La [Brown and Zainudeen, 1978]. This reduction in k_La is typical of the influence of fungal cells on the aeration capacity [Deindoerfer and Gaden, 1955]. The value of k_La in the presence of the culture might be as low as 20% of the k_La which the agitation and aeration would produce in cell-free nutrient solution.

Figure 4 shows the full cycle of interactions which can confuse the selection of a starting point for fermenter and process design. If the value of k_La in the presence of the microorganism can be decided then it is possible to calculate the maximum cell concentration (x_m) which will result, provided that a given minimum dissolved oxygen tension is maintained. A total volume of culture (V) with

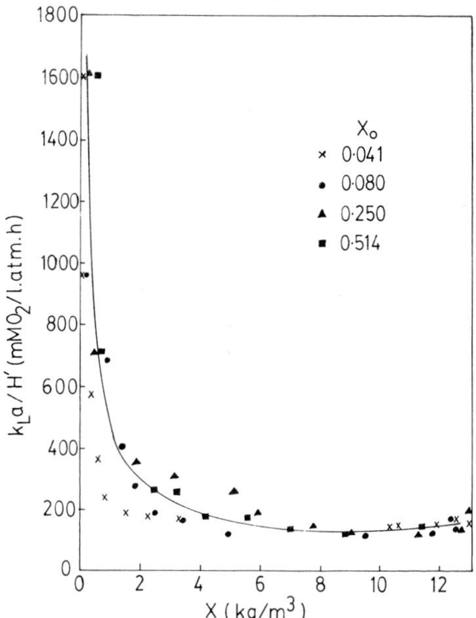

Fig. 3 Effect of cell concentration on the absorption coefficient [Brown and Zainudeen, 1978].

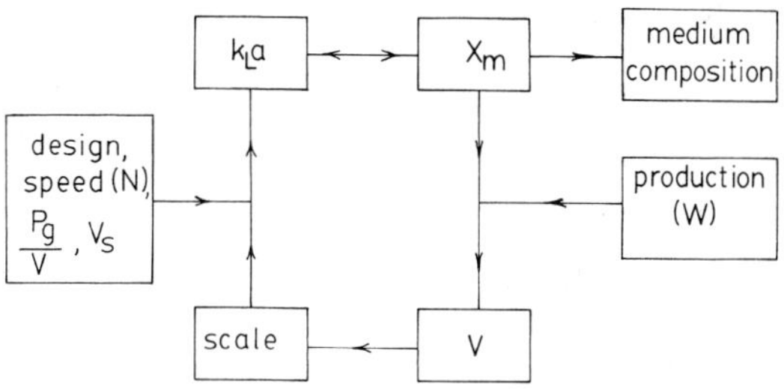

Fig. 4 Interacting factors in fermenter and process design.

a cell concentration of x_m can then be determined if the total production rate is known. The result of this calculation may be a very large volume for a single fermenter and some decision about the scale of operation of a number of smaller fermenters will be necessary. The final problem for the design engineer is to select a suitable impeller and aeration system at the chosen scale which will create the appropriate k_La that was used to calculate x_m.

Determination of the maximum cell concentration For any predetermined value of k_La, it is possible to calculate the maximum cell concentration (x_m) that can be supported and result in a given minimum level of dissolved oxygen tension (T_m). At any moment in time there will be no change observed in the dissolved oxygen tension (T) so that the molar rate of transfer of oxygen will be the same as the molar consumption:

$$\frac{N}{M} = \frac{k_L a}{M} (C^* - C) = Qx \tag{13}$$

For the case of a well stirred fermenter the value of $C^* = p_o/H'$ and $C = T/H'$ [Brown and Zainudeen, 1978] so that for a given minimum oxygen tension, the maximum cell concentration is given by:

$$x_m = \frac{k_L a}{MH'Q} = (p_o - T_m) \tag{14}$$

A value for p_o can be obtained from the oxygen balance over the fermenter:

$$Qx = \frac{G_M}{VP} (p_i - p_o) \tag{15}$$

Thus $$p_o = p_i - \frac{QxV}{G_M} \tag{16}$$

So that $$x_m = \frac{p_i - T_m}{(\frac{QH'M}{k_L a}) + (\frac{QVP}{G_M})} \tag{17}$$

Determination of the culture volume The required production of the plant will have to be decided in consultation with the marketing and sales information and interpreted in terms of cell mass. The volume of the required culture (V) can be calculated by dividing the production rate (W) by the productivity:

$$V = W/PROD \tag{18}$$

The productivity is given by:

$$PROD = x_m/\theta \tag{19}$$

where θ is the total batch time, composed of both the run time and the down time.

Choice of fermenter size Many mechanical engineering problems arise in the consideration of fermenter size. The upper limit to the vessel volume has tended to be about 210 m^3, resulting in dimensions of a diameter of 4.5 m and height of 14 m. More commonly the large vessels will be about 4 m in diameter and 10 m in height with a volume of approximately 125 m^3. Size is often chosen in relation to existing fermenters because this simplifies production planning. Also, many units of separation equipment have been sized in relation to existing fermenter volumes. Thus, new units must not be incompatible with other plant activities. An upper limit in size might be specified in connection with foundation loadings.

An upper limit on the diameter might be set by the distance between beams in existing factory structures. If the vessel is constructed off the site then highway regulations for wide loads may limit the diameter. Of particular importance is the fact that as the diameter is increased, the area-to-volume ratio decreases so that the surface area available for heat transfer becomes limiting. Even with coils inside the vessel as well as on the outside, the limit will be reached at about 4.5 m. Greater diameters will need some form of external cooling which involves pumps and heat exchangers and introduces additional devices which can give sterilization difficulties.

Increase in the height of the vessel usually results in the need for two or more impellers on the shaft. In some cases, an additional shaft bearing will be needed to prevent vibration. As the liquid height increases, a greater pressure drop results from which a greater amount of energy is transferred to the liquid [Pollard and Shearer, 1977]. However, an increased air supply pressure is required with possible changes in the design of compressor used to supply the energy. Increased energy requirement from the stirrer unit is evident in the size and weight of the drive motor and gearbox.

Correlations for the absorption coefficient No single general purpose correlation is available to relate the absorption coefficient (k_La) to all the variables which effect it. The mass transfer coefficient (k_L) in stirred vessels does not change greatly [Calderbank and Moo-Young, 1961a]. Values in the range 0.6 to 3.4 m/h have been reported by Miura [1976] for the air-water system and Robinson and Wilke [1974] show a similar effect in aqueous electrolyte solutions. However, large changes in the value of k_La resulting from changes in stirrer speed and/or air flow rate are predominantly due to changes in the specific interfacial area 'a'. In general terms:

$$k_La = f\text{ (distributions of bubble size and gas hold-up)} \quad (20)$$

A very empirical approach has been based on the assumption that the bubble size distribution and the gas hold-up are determined by the total energy input to the system and its interaction with the air flow rate [Cooper *et al.*, 1944]:

$$k_La = K\left(\frac{P_g}{\rho V}\right)^y (V_s)^z \quad (21)$$

For a given fermenter and fluid properties, an equation of this form has been found satisfactory although some differences are observed in the values of K, y and z from system to system [Sideman *et al.*, 1966]. It is not surprising that these constants will have different values particularly for different geometries. The form of gas hold-up distribution throughout the vessel will be affected by the impeller design, dimensions and location. Only if the system is an homogeneous dispersion could it be expected to conform to a simple correlation of the form of equation (21). Small scale units, in which many of the reported studies have been made, do not possess the fluid characteristics necessary for producing a homogeneous gas-liquid dispersion. Extrapolation to different scales of operation using corrclations of the type described by equation (21) is made with great risk. It has been shown, using air-water systems [Fuchs *et al.*, 1971], that surface aeration contributes 50% of the measured k_La at 10^{-2} m^3 scale but only 1.5% at 50 m^3 scale. This would suggest that scale-up from large scale fermenters to even larger units might be more predictable than information from laboratory units would suggest. The high fluid flow characteristic achieved by large impellers will tend to produce a more homogeneous dispersion than achieved in the small laboratory vessel. Thus results from different geometries of plant scale fermenters should be more readily unified than those from different scales. Plant scale results have been reported [Richards, 1961] in which the correlation included the stirrer speed (N) separately from its implied involvement in the power term:

$$k_La = K\left(\frac{P_g}{V}\right)^{0.4} (V_s)^{0.5} (N)^{0.5} \tag{22}$$

A more fundamental approach has been to obtain correlations for bubble sizes, expressed as the Sauter mean diameter (d_{SM}) and the gas hold-up ratio (H) as functions of the many controlling variables. These two variables can then be combined to give the specific interfacial area 'a':

$$a = \frac{6H}{d_{SM}} \tag{23}$$

Several studies of this kind have been made [Calderbank, 1958; Yoshida and Miura, 1963; Westerterp *et al.*, 1963; Miller, 1974; Robinson and Wilke, 1974; Hassan and Robinson, 1977]. However, these investigations can only be carried out in ideal air-water, or air-electrolyte solutions and the results are only applicable to such systems.

The presence of bacteria in the culture medium does not greatly change the rheological properties of the fluid [Banks, 1977]. Soluble starch in the medium may result in an increase in viscosity above that of water but as the microorganisms grow, they tend to degrade and utilize such compounds. The application of aeration correlations of the form of equation (21) are thus probably justified in the bacterial fermentations. However, in fermentations which contain quantities of filamentous mould mycelium (see Figure 3), the aeration rate is greatly reduced. An empirical approach used with data obtained from a penicillin fermentation [Ryu and Humphrey, 1972] included the effect of mycelium on the absorption coefficient in terms of the apparent viscosity of the suspension (μ_a):

$$k_La = K(\frac{P_g}{V})^y (V_s)^z (\mu_a)^{-0.86} \qquad (24)$$

The rheological properties of fungal culture broths have not attracted a great deal of attention in the past. It was established [Deindoerfer and West, 1960; Richards, 1961] that most mycelial suspensions could be described as pseudoplastic power law fluids. Impeller power requirements can be correlated in terms of the apparent viscosity associated with the condition of shear in the vicinity of the impeller [Calderbank and Moo-Young, 1961b]. However, an understanding of the rheological effects on the oxygen mass transfer is still a subject of research and discussion [Blanch and Bhavaraju, 1976; Banks, 1977; Charles, 1978; Taguchi, 1971]. Procedures for quantifying the morphological characteristics of the microorganisms and relating these to the rheological properties of the suspensions are being explored [Roels *et al.*, 1974].

Clearly some further information is required of absorption coefficient values measured for fungal fermentations in plant-scale equipment for which power input data [Brown, 1977] and broth rheological properties [Bongenaar *et al.*, 1973] have been also noted. The design engineer would then be more able than he is at present to determine the appropriate system required to produce the desired k_La from which x_m was calculated above.

Process Design for Cellulase Production

On the basis of the discussions of enzyme biosynthesis and fermenter design it is possible to consider a design strategy suitable for the batch production of a polysaccharase in a given fermenter. The process to be considered here is the production of filter paper activity (FPA) cellulase.

Selection of the microorganism The fungus *Trichoderma viride* produces the desired type of cellulase system. Workers at the United States Army Natick Laboratories isolated a suitable strain which was designated QM6a [Mandels and Reese, 1957]. Mutation work was carried out and has produced a number of progressively improved strains such as QM9123 and QM9414 [Mandels, 1975]. Selection of such strains is discussed by Montenecourt *et al.* [Chapter 14]. The use of strain QM9123 is described in the following design procedure.

Control mechanisms Cellulase biosynthesis by *Trichoderma viride* is subject to induction by sophorose [Mandels *et al.*, 1962]. In addition it was accepted that the enzyme production would be metabolically repressed. The medium must thus contain sophorose, which can be supplied as a trace contaminant in a commercial glucose called Cerelose [Brown, 1976]. De-repression must be achieved by the suitable choice of carbohydrates, or temperature and pH profiles or controlled feed rates of simple sugars. In this process the use of a suitable uncontrolled pH profile will be the method of metabolic repression control. After available glucose has been used to produce the bulk of the biomass the organism will use stored materials during the stationary phase. The rate at which the stored materials are used must be slowed down in order to remove the metabolic repression. This can be done quite conveniently by developing the suitable pH profile. Some laboratory experiments may be required to determine the exact starting pH and relative quantities of ammonium sulphate and buffering agents, such as potassium dihydrogen phosphate, that will be needed to achieve the desired result.

Maximum cell concentration The design calculation is to be carried out for a 10^{-2} m^3 working volume laboratory fermenter, previously described [Brown and Halsted, 1975], with an air rate of 1 volume/volume culture fluid/minute operating at 400 rev/minute. From Figure 3 [Brown and Zainudeen, 1978] it can be seen that for a cell concentration greater than 5 kg/m^3 the absorption coefficient k_La = 130 h^{-1} (for H' = 1 liter atm/mMO_2). The fermentation process is to be carried out at 28°C and an average specific growth rate of μ = 0.08 h^{-1} can be assumed [Brown and Halsted, 1975]. A value for the specific oxygen uptake rate at μ = 0.08 h^{-1} can be obtained of Q = 1.70 mMO_2/g/h [Brown and Zainudeen, 1977]. If a minimum allowable oxygen tension of 40% saturation is assumed then equation (17) gives x_m = 10.2 kg/m^3. The values of the terms used in equation (17) are listed in Table 1.

Medium composition In order to produce approximately 10 kg cells/m^3 it is necessary to supply sufficient carbon in the form of glucose. It was found [Brown and Halsted, 1975] that for this temperature of 28°C the yield coefficient was 0.40 kg cells/kg glucose. Thus the required glucose concentration = 10/0.4 = 25 kg/m^3.

The second important commodity in the medium is the nitrogen source which must be in relative excess compared with the glucose. A very approximate value of 8.0 can be assumed for the carbon-to-nitrogen ratio of a balanced

TABLE 1

Values of terms in equation (17)

Term	Values assumed	Value in SI units
p_i	0.210 atm.	2.12×10^4 N m^{-2}
T_m	0.084 atm.	8.40×10^3 N m^{-2}
Q	1.70 mmol $g^{-1}h^{-1}$	4.72×10^{-7} kmol $kg^{-1}s^{-1}$
H'	0.028 litre atm. mg^{-1}	2.80×10^6 N m kg^{-1}
M	32	32
k_La	130 h^{-1}	3.6×10^{-2} s^{-1}
V	10 litres	10^{-2} m^3
P	1 atm.	1.01×10^5 N m^{-2}
G_m	26.8 gmol h^{-1}	7.44×10^{-6} kmol s^{-1}
x_m	10.2 g $litre^{-1}$	10.2 kg m^{-3}

medium [MacLennan *et al.*, 1973]. It was assumed here that a ratio of 4.5 would give adequate excess nitrogen. Thus the concentration of nitrogen in the medium is given by $(25 \times 0.4)/4.5 = 2.20$ kg/m^3. This quantity of nitrogen cannot be supplied simply as ammonium sulphate as the pH would go too low as the ammonia nitrogen is consumed. Thus some of the nitrogen is supplied as urea and the exact ratio is determined by experiment together with the initial starting pH, to give the optimum pH profile. The final medium composition (Table 2) includes magnesium sulphate, calcium chloride and a trace element mix in quantities suggested by similar media used for cellulase production by *Trichoderma viride* [Mandels and Weber, 1969].

Results of process design This medium was used in the appropriate fermenter ($k_La = 3.6 \times 10^{-2}$ s^{-1}) to grow *Trichoderma viride* QM9123 and produce cellulase (Figure 1). Cells grew to a final concentration reached about 12 kg/m^3. However, the inoculum quantity of about 3 kg/m^3 must be deducted from this to compare the final yield of cells. The oxygen tension fell smoothly at first and then at about 35h, when the pH had reached 3.5, the fall became slower and a final T_m occurred at about 58% saturation. Thus the final minimum oxygen tension was higher than the 40% saturation used in the calculation. This is probably due to the transient nature of the uncontrolled pH profile which at the low pH of 3.0 had slowed down the respiration below the estimated value of $Q = 4.72 \times 10^{-7}$ kmol/kg/s assumed above. The rise in oxygen tension identifies the point at which the glucose had all been consumed. By this time the

TABLE 2

Medium composition

Composition	Composition
Cerelose (commercial glucose containing sophorose)	25 kg m^{-3}
$(NH_4)_2SO_4$	7 kg m^{-3}
Urea	1.5 kg m^{-3}
KH_2PO_4	10 kg m^{-3}
$MgCl_2.7H_2O$	1.5 kg m^{-3}
$CaCl_2.2H_2O$	2 kg m^{-3}
Trace element mix	5 x 10^{-3} m^3m^{-3}
Starting pH	6.3

pH of the culture had reached 2.8 and remained at this value for approximately 10 hours. As stored materials were used up and the pH was so low that metabolism was very slow, de-repression occurred at about 48 hours. By 52 h very favourable conditions were present and a rapid rise in cellulase activity occurred. At the end of the batch run, the pH value started to rise as autolysis occurred and by 70 hours had reached a value at which enzyme loss was minimized.

Conclusions

The relevance of a decade of developments in microbial genetics, biochemistry and engineering to the commercial production of extra-cellular polysaccharases has been reviewed. Most of these enzymes are produced by either a constitutive strain or as a result of induction, and enzyme formation is repressed by high rates of metabolic activity. Various methods of genetic and metabolic control are now available to the fermentation technologist for use as process improvement procedures. Future progress can be anticipated in the selection of high yielding constitutive mutants [see Montenecourt *et al.*, Chapter 14]. The problem of metabolic repression may be solved by genetic procedures or handled through controlled nutrient feeds and temperature and pH profiles. This progress that has been made in the understanding of the genetic and biochemical control mechanisms involved in the biosynthesis of enzymes is also greatly influencing the formulation of kinetic models for such processes.

Fermenter and process design for enzyme production has hitherto been mainly of an empirical nature. However, some general principles are now becoming established. Complex interactions exist between the aeration capacity of the fermenter, the concentration and morphology of the cells, the scale at which the process is operated and suitable engineering design procedures for producing the aeration capacity at the chosen scale. A single answer may not be possible.

The recent progress which has been made in the biotechnology of enzyme production is illustrated in a simplified process design procedure for the manufacture of cellulase. The various stages that might be followed from the initial selection of a suitable microorganism to the determination of a satisfactory medium composition for a given fermenter are described. The results of a test run show a close agreement with the anticipated performance of the system. With the assistance of computer calculations more sophisticated versions of such a procedure will be made to yield effective solutions to problems of design and optimization.

Nomenclature

a	Rate constant in equation (5)	units kg^{-1} s^{-1}
a'	Rate constant in equation (4)	kg kg^{-1}
b	Rate constant in equation (5)	units kg^{-1}
b'	Rate constant in equation (4)	kg kg^{-1} s^{-1}
C	Liquid phase oxygen concentration	kmol m^{-3}
C*	Interface liquid phase oxygen concentration	kmol m^{-3}
d_{sm}	Sauter mean bubble diameter	m
[E]	Enzyme concentration	units m^{-3}
G_M	Molar gas rate	kmol s^{-1}
H	Hold-up ratio	m^3 m^{-3}
H'	Henry's Law constant	Nm kg^{-1}
[I]	Inducer or co-repressor concentration	kg m^{-3}
k,k'	Rate constants for decay of mRNA	s^{-1}
K	Constant in equation (21)	--
k_1,k_2	Equilibrium constants	--
k_3	Rate constant in equation (9)	s^{-1}
k_4	Rate constant for decay of enzyme	s^{-1}
k_5	Proportionality constant in equation (10)	--
k_La	Absorption coefficient	s^{-1}

k_p	Product formation rate constant	units kg^{-1}
M	Molecular weight of oxygen = 32	--
n	Stoichiometric constant in equation (7)	--
N	Rate of transfer of oxygen Stirrer speed	kmol m^{-3} s^{-1} rev s^{-1}
[O]	Concentration of operator gene	kg m^{-3}
$[O_t]$	Total concentration of operator gene	kg m^{-3}
P	Total pressure Product concentration	N m^{-2} kg m^{-3}
P_g	Gassed power input	W
p_i, p_o	Inlet and outlet gas oxygen concentration	N m^{-2}
PROD	Productivity	kg s^{-1} m^{-3}
Q	Fraction of available operator genes Specific oxygen uptake rate	-- kmol kg^{-1} s^{-1}
r	Quantity of mRNA per unit mass of cells	kg kg^{-1}
[r]	Concentration of mRNA	kg m^{-3}
[R]	Repressor molecule concentration	kg m^{-3}
$[R_t]$	Total concentration of repressor molecule	kg m^{-3}
t	Time	S
T	Dissolved oxygen tension	N m^{-2}
t_m	Maturation time	s
T_m	Minimum dissolved oxygen tension	N m^{-2}
V	Culture volume	m^3
V_s	Superficial gas velocity	m s^{-1}
W	Production rate	kg s^{-1}
x	Cell concentration	kg m^{-3}
x_m	Maximum cell concentration	kg m^{-3}
x_p	Concentration of cells mature enough to make product	kg m^{-3}
y, z	Indices in equation (21)	--
α	Rate constant in equation (1)	kg m^{-3}
β	Rate constant in equation (1)	m^{-3} s^{-1}
ε	Specific rate of enzyme formation	units kg^{-1} s^{-1}

μ	Specific growth rate	s^{-1}
μ_a	Apparent viscosity	$Ns\ m^{-2}$
ρ	Fluid density	$kg\ m^{-3}$
θ	Arbitrary time interval	s
	Total batch time	s

References

Andreotti, R.E., Mandels, M. and Roche, C. (1978). Effect of some fermentation variables on growth and cellulase production by *Trichoderma viride* QM9414. In *Proceedings of Symposium on Bioconversion of Cellulosic Substances into Energy, Chemicals and Microbial Protein*, pp.249-267. Edited by T.K. Ghose. New Delhi and Zurich: Indian Institute of Technology and Swiss Federal Institute of Technology.

Banks, G.T. (1977). Aeration of mould and streptomycete culture fluids. In *Topics in Enzyme and Fermentation Biotechnology* 1, pp.72-110. Edited by A. Wiseman. Chichester: Ellis Horwood.

Banks, G.T., Binns, F. and Cutcliffe, R.L. (1967). Recent developments in the production and industrial applications of amylolytic enzymes derived from filamentous fungi. *Progress in Industrial Microbiology* **6**, 95-139.

Barrett, F.F. (1975). Enzyme uses in the milling and baking industries. In *Enzymes in Food Processing*, 2nd edition, pp.301-330. Edited by G. Reed. New York: Academic Press.

Blain, J.A. (1975). Industrial enzyme production. In *The Filamentous Fungi*, 1, pp.193-211. Edited by J.E. Smith and D.R. Berry. London: Arnold.

Blanch, H.W. and Bhavaraju, S.M. (1976). Non-Newtonian fermentation broths: Rheology and mass transfer. *Biotechnology and Bioengineering* **18**, 745-790.

Blanch, H.W. and Rogers, P.L. (1971). Production of Gramicidin S in batch and continuous culture. *Biotechnology and Bioengineering* **13**, 843-864.

Bongenaar, J.J.T.M., Kossen, N.W.F., Metz, B. and Meijboom, F.W. (1973). A method of characterizing the rheological properties of viscous fermentation broths. *Biotechnology and Bioengineering* **15**, 201-206.

Both, G.W., McInnes, J.L., Hanlon, J.E., May, B.K. and Elliott, W.H. (1972). Evidence for an accumulation of messenger RNA specific for extracellular protease and its relevance to the mechanism of enzyme secretion in bacteria. *Journal of Molecular Biology* **67**, 199-217.

Brown, D.E. (1976). Cellulase production by *Trichoderma viride*. *Biotechnology and Bioengineering Symposium* **6**, 75-77.

Brown, D.E. (1977). The measurement of fermenter power input. *Chemistry and Industry*, 20 Aug., 684-688.

Brown, D.E. and Fitzpatrick, S.W. (1976). Food from waste paper. In *Food and Waste*, pp.139-155. Edited by G.G. Birch, K.J. Parker and J.T. Worgan. London: Applied Science.

Brown, D.E. and Halsted, D.J. (1975). The effect of acid pH on the growth kinetics of *Trichoderma viride*. *Biotechnology and Bioengineering* **17**, 1199-1210.

Brown, D.E., Halsted, D.J. and Howard, P. (1975). Studies on the biosynthesis of cellulase by *Trichoderma viride* QM9123. In *Symposium on Enzymatic Hydrolysis of Cellulose*, pp.137-153. Edited by M. Bailey, T-M. Enari and M. Linko. Helsinki: Finnish National Fund for Research and Development.

Brown, E.D. and Vass, R.C. (1973). Maturity and product formation in cultures of microorganisms. *Biotechnology and Bioengineering* **15**, 321-330.

Brown, D.E. and Zainudeen, M.A. (1977). Growth kinetics and cellulase biosynthesis in the continuous culture of *Trichoderma viride*. *Biotechnology and Bioengineering* **19**, 941-958.

Brown, D.E. and Zainudeen, M.A. (1978). Effect of inoculum size on the aeration pattern of batch cultures of a fungal microorganism. *Biotechnology and Bioengineering* **20**, in press.

Burbidge, E. and Collier, B. (1968). Production of bacterial amylase. *Process Biochemistry* **3**, 53-64.

Calam, C.T. (1969). Automatic in screeing. In *Fermentation Advances*, pp.31-41. Edited by D. Perlman. New York: Academic Press.

Calderbank, P.H. (1958). Physical rate processes in industrial fermentation. Part 1: The interfacial area in gas-liquid contacting with mechanical agitation. *Transactions of the Institute of Chemical Engineers (London)* **36**, 443-463.

Calderbank, P.H. and Moo-Young, M. (1961a). The continuous phase heat and mass transfer properties of dispersions. *Chemical Engineering Science* **16**, 39-54.

Calderbank, P.H. and Moo-Young, M. (1961b). The power characteristics of agitators for the mixing of Newtonian and non-Newtonian fluids. *Transactions of the Institute of Chemical Engineers (London)* **39**, 337-347.

Charles, M. (1978). Technical aspects of the rheological properties of microbial cultures. *Advances in Biochemical Engineering* **8**, 1-62.

Coleman, G. (1967). Studies on the regulation of extracellular enzyme formation by *Bacillus subtilis*. *Journal of General Microbiology* **49**, 421-431.

Constantinides, A., Spencer, J.L. and Gaden, E.L. (1970). Optimization of batch fermentation processes. II. Optimum temperature profiles for batch penicillin fermentations. *Biotechnology and Bioengineering* **12**, 1081-1098.

Cooper, C.M., Fernstrom, G.A. and Miller, S.A. (1944). Performance of agitated gas-liquid contactors. *Industrial and Engineering Chemistry* **36**, 504-509.

Deindoerfer, F.H. and Gaden, E.L. (1955). Effects of liquid physical properties of oxygen transfer in penicillin fermentation. *Applied Microbiology* **3**, 253-257.

Deindoerfer, F.H. and West, J.M. (1960). Rheological examination of some fermentation broths. *Journal of Biochemical and Microbiological Technology and Engineering* **2**, 165-175.

Demain, A.L. (1971). Overproduction of microbial metabolites and enzymes due to alteration of regulation. *Advances in Biochemical Engineering* **1**, 113-142.

Demain, A.L. (1972). Theoretical and applied aspects of enzyme regulation and biosynthesis in microbial cells. *Biotechnology and Bioengineering Symposium* No. **3**, 21-32.

Demain, A.L. (1973). The marriage of genetics and industrial microbiology after a long engagement, a bright future. In *Genetics of Industrial Microorganisms*, Vol. 1, *Bacteria*, pp.19-31. Edited by Z. Vanek, Z. Hostalek and J. Cudlin. Amsterdam: Elsevier.

Emmer, M., De Crombrugghe, B., Pastan, I. and Perlman, R. (1970). Cyclic AMP receptor protein of *Escherichia coli*: Its role in the synthesis of inducible enzymes. *Proceedings of the National Academy of Sciences U.S.A.* **66**, 480-487.

Enari, T-M. and Markkanen, P. (1977). Production of cellulolytic enzymes by fungi. *Advances in Biochemical Engineering* **5**, 1-24.

Eriksson, K.E. (1978). Enzyme mechanisms involved in fungal degradation of wood components. In *Proceedings of Symposium on Bioconversion of Cellulosic Substances into Energy, Chemicals and Microbial Protein*, pp.195-201. Edited by T.K. Ghose. New Delhi and Zurich: Indian Institute of Technology and Swiss Federal Institute of Technology.

Fogarty, M., Griffin, P.J. and Joyce, A.M. (1974). Enzymes of the *Bacillus* species. *Process Biochemistry* **9**, (6) 11-18 and (7) 27-35.

Forss, K. and Fremer, K.E. (1975). A structural model for coniferous lignin and its applicability to the description of the acid bisulfite cook. In *Symposium on Enzymatic Hydrolysis of Cellulose*, pp.41-63. Edited by M. Bailey, T-M. Enari and M. Linko. Helsinki: Finnish National Fund for Research and Development.

Fuchs, R., Ryu, D.D.Y. and Hymphrey, A.E. (1971). Effect of surface aeration on scale-up procedures for fermentation processes. *Industrial and Engineering Chemical Process Design and Development* **10**, 190-196.

Gould, B.J. (1975). Enzyme data. In *Handbook of Enzyme Biotechnology*, pp.128-162. Edited by A. Wiseman. Chichester: Ellis Horwood.

Hassan, I.T.M. and Robinson, C.W. (1977). Stirred-tank mechanical power requirement and gas hold-up in aerated aqueous phases. *Association of Incorporated Chemical Engineers Journal* **23**, 48-56.

Herzenberg, L.A. (1959). Studies on the induction of β-galactosidase in a cryptic strain of *Escherichia coli*. *Biochimica et Biophysica Acta* **31**, 525-538.

Hernandez, E. and Pirt, S.J. (1975). Kinetics of utilisation of a highly polymerised carbon source (starch) in a chemostat culture of *Klebsiella aerogenes*: Pullulanase and α-amylase activities. *Journal of Applied Chemistry and Biotechnology* **25**, 297-304.

Hockenhull, D.J.D. and Mackenzie, R.M. (1968). Preset nutrient feeds for penicillin fermentation on defined media. *Chemistry and Industry*, 11 May, 607-610.

Hulme, M.A. and Stranks, D.W. (1971). Regulation of cellulase production by *Myrothecium verrucaria* grown on non-cellulosic substrates. *Journal of General Microbiology* **69**, 145-155.

Imanaka, T. and Aiba, S. (1977). A kinetic model of catabolite repression in the dual control mechanism in microorganisms. *Biotechnology and Bioengineering* **19**, 757-764.

Kinoshita, S., Okada, H. and Terui, G. (1967). Kinetic studies on enzyme production by microbes (11). Process kinetics of α-amylase production by *Bacillus subtilis*. *Journal of Fermentation Technology* **45**, 504-510.

Kulp, K. (1975). Carbohydrases. In *Enzymes in Food Processing*, 2nd edition, pp.54-122. Edited by G. Reed. New York: Academic Press.

Luedeking, R. and Piret, E.L. (1959). A kinetic study of the lactic acid fermentation. Batch process at controlled pH. *Journal of Biochemical and Microbiological Technology and Engineering* **1**, 393-412.

MacAllister, R.V., Wardrip, E.K. and Schnyder, B.J. (1975). Modified starches, corn syrups containing glucose and maltose, corn syrups containing glucose and fructose, and crystalline dextrose. In *Enzymes in Food Processing*, 2nd edition, pp.331-359. Edited by G. Reed. New York: Academic Press.

MacLennan, D.G., Gow, J.S. and Stringer, D.A. (1973). Methanol-bacterium processes for SCP. *Process Biochemistry* **8** (6), 22-24.

Makman, R.S. and Sutherland, E.W. (1965). Adenosine 3',5'-phosphate in *Escherichia coli*. *Journal of Biological Chemistry* **240**, 1309-1314.

Mandels, M. (1975). Microbial sources of cellulase. *Biotechnology and Bioengineering Symposium* No. **5**, 81-105.

Mandels, M., Parrish, F.W. and Reese, E.T. (1962). Sophorose as an inducer of cellulase in *Trichoderma viride*. *Journal of Bacteriology* **83**, 400-408.

Mandels, M. and Reese, E.T. (1957). Induction of cellulase in *Trichoderma viride*. *Journal of Bacteriology* **73**, 269-278.

Mandels, M. and Sternberg, D. (1976). Recent advances in cellulase technology. *Journal of Fermentation Technology* **54**, 267-286.

Mandels, M. and Weber, J. (1969). The production of cellulases. *Advances in Chemistry Series* No. **95**, 391-414.

Marconi, W. (1974). Enzymes in the chemical and pharmaceutical industry. In *Industrial Aspects of Biochemistry*, Vol. 30, pp.139-186. Edited by B. Spencer. Amsterdam: North-Holland.

Melling, J. and Phillips, B.W. (1975). Practical aspects of large-scale enzyme purification. In *Handbook of Enzyme Biotechnology*, pp.181-202. Edited by A. Wiseman. Chichester: Ellis Horwood.

Meers, J.L. (1972). The regulation of α-amylase production in *Bacillus licheniformis*. *Antonie van Leeuwenhoeck Journal of Serology and Microbiology* **38**, 585-590.

Meyrath, J. and Volavsek, G. (1975). Production of microbial enzymes. In *Enzymes in Food Processing*, 2nd edition, pp.255-300. Edited by G. Reed. New York: Academic Press.

Miller, D.N. (1974). Scale-up of agitated vessels: Gas-liquid mass transfer. *American Institute of Chemical Engineers Journal* **20**, 445-453.

Miura, Y. (1976). Transfer of oxygen and scale-up in submerged aerobic fermentation. *Advances in Biochemical Engineering* **4**, 3-40.

Nisizawa, T., Suzuki, H., Nakayama, M. and Nisizawa, K. (1971). Inductive formation of cellulase by sophorose in *Trichoderma viride*. *Journal of Biochemistry* **70**, 375-385.

Norkrans, B. (1967). Cellulose and cellulolysis. *Advances in Applied Microbiology* **9**, 91-130.

Nyiri, L. (1971). The preparation of enzymes by fermentation. *International Chemical Engineering* **11**, 447-458.

Okazaki, M. and Terui, G. (1967). Kinetic studies on enzyme production by microbes (III) Process kinetics of glucamylase production by *Aspergillus niger*. *Journal of Fermentation Technology* **45**, 1147-1154.

Pardee, A.B. (1969). Enzyme production by bacteria. In *Fermentation Advances*, pp.3-14. Edited by D. Perlman. New York: Academic Press.

Pastan, I. and Perlman, R. (1970). Cyclic adenosine monophosphate in bacteria. *Science, New York* **169**, 339-344.

Pirt, S.J. (1975). *Principles of Microbe and Cell Cultivation.* Oxford: Blackwell Scientific Publications.

Pollard, R. and Shearer, C.J. (1977). The application of chemical engineering concepts to the design of fermenters. *The Chemical Engineer* Feb., 106-110.

Reese, E.T., Siu, R.G.H. and Levinson, H.S. (1950). The biological degradation of soluble cellulose derivatives and its relationship to the mechanism of cellulose hydrolysis. *Journal of Bacteriology* **59**, 485-497.

Richards, J.W. (1961). Studies in aeration and agitation. *Progress in Industrial Microbiology* **3**, 141-172.

Robinson, C.W. and Wilke, C.R. (1974). Simultaneous measurement of interfacial area and mass transfer coefficients for a well-mixed gas dispersion in aqueous electrolyte solutions. *American Institute of Chemical Engineers Journal* **20**, 285-294.

Roels, J.A., Van den Berg, J. and Voncken, R.M. (1974). The rheology of mycelial broths. *Biotechnology and Bioengineering* **16**, 181-208.

Ryu, D.D.Y. and Humphrey, A.D. (1972). A reassessment of oxygen transfer rates in antibiotic fermentations. *Journal of Fermentation Technology* **50**, 424-431.

Schugerl, K., Lucke, J., Lehmann, J. and Wagner, F. (1978). Application of tower bioreactors in cell mass production. *Advances in Biochemical Engineering* **8**, 63-131.

Shinmyo, A., Okazaki, M. and Terui, G. (1969). Kinetics of derepression and synthesis of hydrolase. In *Fermentation Advances*, pp.337-367. Edited by D. Perlman. New York: Academic Press.

Shore, D.T. and Royston, M.G. (1968). The chemical engineering of the continuous brewing process. *The Chemical Engineer* May, CE99-CE109.

Shurz, J. (1978). How to make native lignocellulosic materials accessible to chemical and microbial attack. In *Proceedings of Symposium on Bioconversion of Cellulosic Substances into Energy, Chemicals and Microbial Protein*, pp.37-58. Edited by T.K. Ghose. New Delhi and Zurich: Indian Institute of Technology and Swiss Federal Institute of Technology.

Sideman, S., Hortacsu, O. and Fulton, J.W. (1966). Mass transfer in gas-liquid contacting systems. *Industrial and Engineering Chemistry* **58**, (7), 32-47.

Smith, R.W. and Dean, A.C. (1972). β-Galactosidase synthesis in *Klebsiella aerogenes* growing in continuous culture. *Journal of General Microbiology* **72**, 37-47.

Taguchi, H. (1971). The nature of fermentation fluids. *Advances in Biochemical Engineering* **1**, 1-30.

Terui, G. (1972). Application of molecular biology and physiology to the process kinetics of enzyme production. *Biotechnology and Bioengineering Symposium* No. **3**, 33-35.

Terui, G. Okazaki, M. and Kinoshita, S. (1967). Kinetic studies on enzyme production by microbes. (1) On kinetic models. *Journal of Fermentation Technology* **45**, 497-503.

Toda, K. (1976). Dual control of invertase biosynthesis in chemostat culture. *Biotechnology and Bioengineering* **18**, 1117-1124.

Underkofler, L.A. (1966). Manufacture and uses of industrial microbial enzymes. *Chemical Engineering Progress Symposium Series* **62**, (69), 11-20.

Van Dedem, G. and Moo-Young, M. (1973). Cell growth and extracellular enzyme synthesis in fermentations. *Biotechnology and Bioengineering* **15**, 419-439.

Westerterp, K.R., Van Dierendonck, L.L. and De Kraa, J.A. (1963). Interfacial areas in agitated gas-liquid contactors. *Chemical Engineering Science* **18**, 157-176.

Wood, T.M. and McCrae, S.I. (1978). The mechanism of cellulase action with particular reference to the C_1 component. In *Proceedings of Symposium on Bioconversion of Cellulosic Substances into Energy, Chemicals and Microbial Protein*, pp.111-141. Edited by T.K. Ghose. New Delhi and Zurich: Indian Institute of Technology and Swiss Federal Institute.

Yoshida, F. and Miura, Y. (1963). Gas absorption in agitated gas-liquid contactors. *Industrial and Engineering Chemical Process Design and Development* **2**, 263-268.

References Added in Proof

Hope, G.C. and Dean, A.C.R. (1974). Pullulanase synthesis in *Klebsiella (Aerobacter) aerogenes* strains growing in continuous culture. *Biochemical Journal* **144**, 403-411.

Hope, G.C. and Dean, A.C.R. (1975). Pullulanase synthesis in *Klebsiella aerogenes* growing at high biomass levels in maltose-limited chemostat culture. *Journal of Applied Chemistry and Biotechnology* **25**, 549-553.

Chapter 14

MECHANISMS CONTROLLING THE SYNTHESIS OF THE *TRICHODERMA REESEI* CELLULASE SYSTEM

B.S. MONTENECOURT, D.H.J. SCHAMHART and D.E. EVELEIGH

Department of Biochemistry and Microbiology, Cook College, Rutgers - The State University of New Jersey, New Jersey, USA

Introduction

The energy crisis of the 1970's has focussed attention on the need to develop alternate forms of renewable energy. Biomass is one potential source. In the United States it is estimated that about 6% of its current energy needs of 80 quads (10^{15} Btu) can be met by the end of the century, by developing the "Fuels from Biomass" concept through the production of ethanol, methanol and methane from biomass [Del Gobbo, 1978]. Ethanol has been proposed as one alternate liquid transportation fuel, to be used initially as a supplement to petrol to give "gasahol" (Figure 1). The large scale production of ethanol through fermentation of waste molasses for this purpose is currently practiced in Brazil [Hammond, 1978]. In the United States, wood (cellulose) is considered the most practical potential feedstock and would be obtained principally from forests (energy farms) and also from agricultural and urban cellulosic wastes [Del Gobbo, 1978]. Saccharification of cellulose is a prerequisite in the overall process. We favour enzymatic saccharification (cellulolysis), its principal advantage being high conversion efficiency without the production of undesirable side products. Unfortunately the cost of production of microbial cellulase is high. Our objectives are to reduce this cost by developing a range of regulatory cellulolytic mutants. These mutants will be used for the commercial production of low cost cellulase and for the clarification of the mechanisms controlling enzyme synthesis.

Cellulase is a complex of enzymes that act together synergistically in the hydrolysis of crystalline cellulose (Figure 2). The currently favoured hypothesis of the mode of action is that the substrate is initially hydrolysed by endo-glucanases to yield oligomeric intermediates [see

Fig. 1 Gasahol (5-10% ethanol blend in petrol) has been produced even with the aid of moonshiners in the United States recently [Bernton, 1978]. It has been used in Nebraska in a demonstration of its feasibility for several years [Scheller, 1976, 1977]. (With permission of Tom Chalkley and the Environmental Action Magazine.)

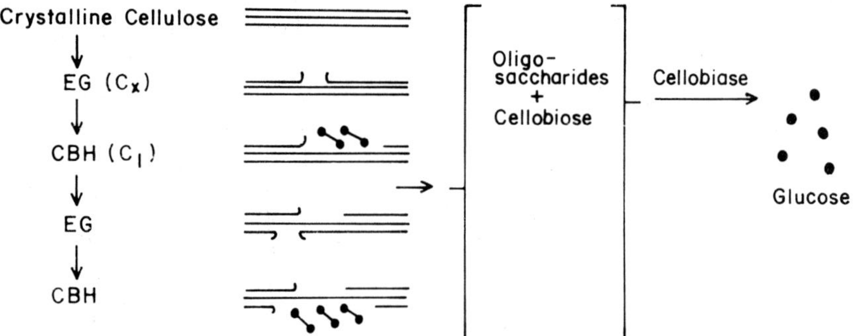

Fig. 2 Enzymatic hydrolysis of cellulose. Three major enzymes act synergistically. Initially an endo-glucanase (C_X) cleaves the crystalline chains to yield reducing termini. These sites allow hydrolysis by cello-biohydrolase (CBH), which directly produce cellobiose. The oligosaccharides continue to be hydrolysed by both enzyme species, the cellobiose product being converted to glucose by cellobiase [see Wood and McCrae, 1978].

Eriksson, Chapter 12]. These are immediately acted on by exosplitting glucanases (glucohydrolase or cellobiohydrolase) which produce respectively glucose or cellobiose from the non-reducing termini. Both types of glucanase continue to hydrolyse the residual oligomers and finally cellobiase cleaves both short chain oligomers and cellobiose to yield glucose [for review see Wood and McCrae, 1978]. A minor exoglucohydrolase component is also present in the enzyme complex from commercial *Trichoderma* cellulase preparations [G. Tsao, personal communication]. The most effective cellulases have both exo- and endo-splitting components and only these 'dual preparations' are able to produce high saccharification conversions of crystalline cellulose. The simultaneous occurrence of both enzymes appears to be relatively restricted. Good yields have only been obtained from a few fungal genera including *Fusarium*, *Penicillium*, *Phanaerochaete* (syn. *Sporotrichum*) and *Trichoderma* [Wood and McCrae, 1978].

We selected *Trichoderma reesei* [Simmons, 1977] for our programme because its cellulase complex appears to give the highest saccharification yields and the enzymes have been fairly well defined [Fägestam *et al.*, 1978; Mandels, 1975]. It has two exocellobiohydrolase components, several distinct endo-glucanases, a cellobiase and an aryl-β-glucosidase. Detailed analysis of the individual enzyme components from *Trichoderma koningi* [Wood and McCrae, 1978] and also of commercial preparations of *Trichoderma viride* [Berghem *et al.*, 1976; Emert *et al.*, 1974; Gong *et al.*, 1977; Okada, 1976; Shoemaker and Brown, 1978; Tomita *et al.*, 1974] have been made. The multiplicity of isozymes of the individual enzymes may be derived in part through proteolytic modification of the initially synthesized components [Nakayama, 1975]. If genetic analysis is to be accomplished with this system, it is hoped that each isozyme is not the product of a separate gene.

Each enzyme is inducible and is sensitive both to catabolite repression and to end-product inhibition (Table 1). Thus, the mechanisms controlling the synthesis and action of the total cellulase complex are intricate. This complexity allows several approaches to unravelling the control systems and realizing high yielding strains. Seven major approaches have been considered (Table 2) including selection of:

i. Catabolite repression resistant mutants.
ii. Antimetabolite resistant mutants [Toda, 1976].
iii. Mutants whose characteristics allow their practical application in the fermenter.
iv. High yielding revertant strains from cellulase negative mutants.
v. End-product resistant mutants.
vi. Secretory mutants [Suzuki *et al.*, 1974].
vii. Constitutive mutants.

Our initial emphasis has been on catabolite repression

resistant (i), antimetabolite resistant (ii), "fermenter" (iii) and cellulase negative revertant (iv) mutants but our current emphasis is now on the latter three mutant classes (v, vi, vii). Case histories of the development of a catabolite repression resistant strain of *Trichoderma reesei* RUT-C30 and also of β-glucosidase mutants D series and I series are given as examples of these methodologies.

TABLE 1

Biochemical mechanisms controlling cellulase synthesis and activity in Trichoderma reesei

Enzyme	End product inhibitor	Catabolite repressor	Inducer
cellobiohydrolase	cellobiose	glucose or	
endoglucanase	cellobiose	metabolites	cellulose
		of glucose	sophorose
cellobiase	glucose		

The genealogy of the *Trichoderma reesei* mutants is outlined in Figure 3. Strains of *Trichoderma reesei* were selected by screening on agar plates. As this mould rapidly forms large diffuse colonies, paramorphic agents (oxgall 1.5% and Phosfon D 500 μg/ml) were incorporated into the culture media. This allowed development of up to 100 small discrete colonies per petri dish. Subsequent testing was in shaker flasks and in fermenters [Montenecourt and Eveleigh, 1977a, b]. Mutagens included ultraviolet irradiation and *N*-methyl-*N*-nitrosoguanidine.

Catabolite Repression Resistant Mutants

Catabolite repression resistant mutants were screened for as they may include hyper-enzyme producing strains [Demain, 1972]. As the enzymes of the *Trichoderma* cellulase complex are repressed by several catabolites (e.g. glucose, glycerol), cellulase catabolite repression resistant mutants should be detectable by observing their cellulolytic activity when grown on cellulosic substrates in the presence of the catabolites. Such mutants have been obtained by screening for mutants capable of producing "clearing zones" around the colony when grown on cellulose-glycerol agar. The selection procedure also included the sequential use of increased substrate concentrations (RUT-M7, 1% cellulose; RUT-NG-14, 2% cellulose; S series, 2.5% cellulose) and also variation in the selective substrate used (RUT-NG-14, cellulose; RUT-C30 cellobiose). The hypercellulolytic activity of these strains was demonstrated in shake flasks and in the fermenter [Montenecourt and Eveleigh, 1977b]. An analogous screening procedure speci-

TABLE 2

Screening procedures for selecting regulatory and high yielding cellulase mutants

Category	Selection criteria	Rationale for high enzyme yield
i. Catabolite repression resistant mutants	Cellulolytic activity in the presence of a catabolite repressor	Loss of control can result in hyperenzyme production
ii. Antimetabolite resistant mutants	Growth on cellulose media in presence of an analogue (2-deoxyglucose)	Wild strains are killed by the antimetabolite. Constitutive and catabolite repression resistant mutants circumvent the toxicity by producing glucose from the cellulose substrates
iii. Fermenter mutants	Selection with regard to fermenter operating parameters	Optimal conditions for economic operation of the fermenter (rapid growth rate, efficient utilization of complex culture medium components, thermotolerant strains, non-foaming strains)
iv. Cellulase positive revertants from cellulase negative mutants	Cellulolytic activity	Derangement of control of enzyme synthesis can lead to hyperenzyme production
v. End-product resistant mutants	Enzyme activity demonstrated in the presence of high concentrations of glucose or cellobiose	A useful characteristic for obtaining a high degree of saccharification
vi. Secretory mutants	Pleiotropic mutants obtained through resistance to membrane active antibiotics	Derangement of control of secretion can result in high yields of extracellular enzyme
vii. Constitutive mutants	Cellulolytic activity shown in a cellulose agar overlay of non-induced colonies	For application when using crude, inexpensive culture media, while maintaining high enzyme yields

fically for β-glucosidase mutants employs the use of aesculin as the substrate, ferric ammonium citrate as a visualizing agent and similar repressors [Montenecourt and Eveleigh, 1978]. β-Glucosidase hydrolyses aesculin to release aesculetin and glucose. As aesculetin forms a black complex with ferric ammonium salts, any mutants that are β-glucosidase, catabolite repression resistant can be readily visualized as colonies with black zones surrounding them. We have used this methodology to confirm the activities of β-glucosidase mutants obtained by other procedures (see below).

Antimetabolite Resistant Mutants

2-Deoxyglucose is a toxic antimetabolite of glucose and growth in its presence results in the death of *Trichoderma reesei*. Resistance to this antimetabolite can arise through several mechanisms, including that of catabolite repression resistance in certain instances. The latter type of resistant organism when grown on a 2-deoxyglucose/cellulose medium, hydrolyses the cellulose to yield glucose and thus circumvents the use of the antimetabolite. In contrast, wild strains can only utilize the toxic 2-deoxyglucose and thus are killed. Visualization of catabolite repression resistant mutants is simply of large colonies growing on such plates. In that resistance to the analog can have arisen through other mechanisms, for example inability to take up 2-deoxyglucose, the antimetabolite resistant mutants are further screened to determine which are catabolite repression resistant strains. For β-glucosidase mutants, production of the enzyme in the presence of glucose using the aesculin screening methodology is a confirmatory screening assay (see i. above). This methodology has resulted in obtaining mutant RUT-C30 (Figure 3). Comparative yields for RUT-C30 are presented in Figure 4.

Fermenter Mutants

Mutants that are amenable to efficient and rapid growth in the fermenter are a prerequisite for enzyme production [see Brown, Chapter 13]. RUT-C30 (Figure 3) developed through the antimetabolite resistance methodology, has been shown to be an extremely rapidly growing strain that is capable of producing cellulase when grown on crude culture media containing corn steep liquor [Montenecourt and Eveleigh, 1978]. Productivities of 100 filter paper units (FPU) per liter per hour have been achieved with this mutant. The culture finally yields 15 FPU/ml, equivalent to about 2% protein (Figure 4).

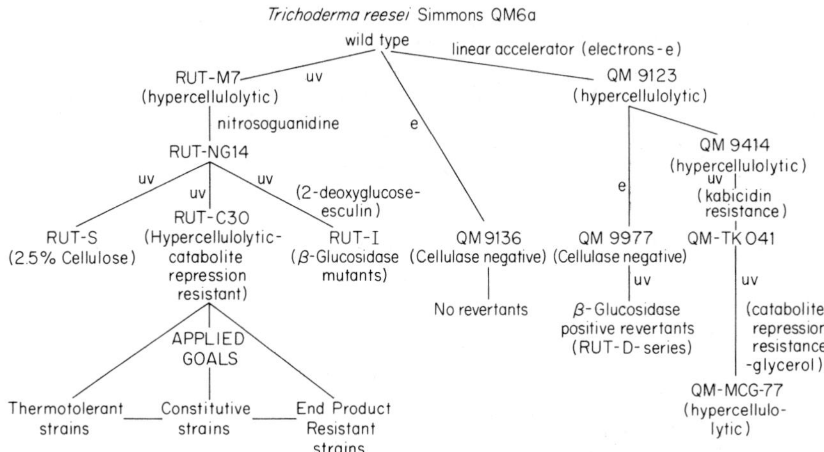

Fig. 3 Genealogy of *Trichoderma reesei* cellulolytic strains. QM, U.S. Army (Natick), Massachusetts; RUT, Rutgers, New Jersey; UV, ultraviolet irradiation; e, high voltage electrons.

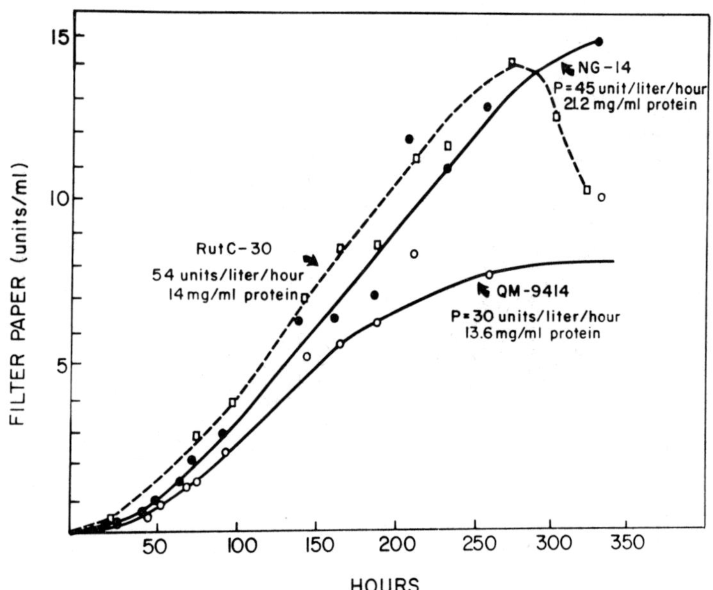

Fig. 4 Yields of cellulase from mutants QM9414, RUT-NG-14 and RUT-C30 when grown under controlled fermenter conditions [M. Mandels, unpublished data]. FPU = Filter paper unit [Mandels, 1975].

Mutants with Perturbed Control of Cellulase Synthesis Derived from Cellulase-negative Strains

Cellulase-negative mutants can be selected using cellulose agar media, as those colonies which fail to produce "hydrolytic" clearing zones around the colony. Back-mutation can potentially yield hypercellulolytic mutants due to derangement of the control of enzyme synthesis. Our initial work has not resulted in any such high yielding mutant strains to date. However, two distinct mutant types have been obtained by this methodology: (i) negative mutants which have yielded β-glucosidase positive strains, e.g. QM9977; (ii) negative mutants which have failed to yield cellulase positive back-mutants, e.g. QM9136 (Figure 3).

From the characteristics of group (i), it is inferred that β-glucosidase is under a separate and distinct control from the other enzymes of the cellulase complex. The mutants in group (ii) can most easily be interpreted to be deletion mutants, but this explanation is perhaps naive, in that they were readily obtained, while deletion mutants are not frequently obtained in analogous studies with other eukaryotic enzyme systems.

Discussion

These initial studies on fungal cellulase have successfully exploited the rationale that catabolite repression mutants will include hyperproducing enzyme strains. The most useful strain to date, RUT-C30, was selected from about 800,000 strains using a combination of these methodologies. RUT-C30 grows rapidly, is catabolite repression resistant, utilizes inexpensive culture media (corn steep liquor), and is quite remarkable in that it produces up to 2% extracellular protein, the majority of which is cellulase (unpublished observations). The U.S. Army Natick laboratory has also obtained high yielding cellulase mutants using a similar approach in conjunction with screening for "secretory" mutants (Table 2, vi; Figure 3). For instance strain QM-MCG 77 was derived through screening for polyene antibiotic resistant strains (QM-TK 041) and subsequently for catabolite repression resistance in the presence of glycerol (Figure 3) [Gallo, 1978]. RUT-C30 shows somewhat higher productivities in the fermenter than QM-MCG 77 [M. Mandels, personal communication]. Furthermore, as one step of the selection of RUT-C30 was carried out in the presence of 5% glucose, it is not too surprising to note that its β-glucosidase is partially resistant to end-product inhibition by glucose. The two mutant series (RUT and QM) (Figure 3) demonstrate how relatively simple screening procedures can allow selection of high yielding cellulase strains. It should be possible to develop even more effective strains through both further screening and genetic manipulations. Screening should focus on constitutive and end-product resistant strains, while genetic applications should include somatic hybridization by protoplast fusion or the parasexual cycle.

These studies have given some initial insight into the *Trichoderma* cellulase system. First, in the production of RUT-C30 all three classes of enzyme (cellobiohydrolase, endo-glucanase and cellobiase) are concomitantly released from catabolite repression. This implies an overriding system controlling the coordinate synthesis of the whole cellulase complex. A similar conclusion can be inferred from the one step production of cellulase negative mutants QM9136 and QM9977. Secondly, there appears a further control system governing the synthesis of each individual enzyme. Thus the production of the "β-glucosidase positive revertants from cellulase negative" strains implies that the synthesis of β-glucosidase is under individual control and supports the view of discrete control of synthesis. Similar results have been obtained by Nevalainen and Palva [1978]. Differentially enhanced levels of β-glucosidase and endo-glucanase in mutants of *Trichoderma reesei* have also been reported [Markkanen *et al*., 1978]. Furthermore, the RUT-NG 14 cellulase produced in the fermenter appears, on the basis of antigenic analysis, to possess greatly enhanced levels of cellobiohydrolase in comparison to that of the mutant QM9414 strain [G. Pettersson, unpublished data]. Thus, it appears that each component of the cellulase complex is under discrete control to some degree. In that cellulolysis operates through a synergistic action of the several cellulase components, it should be possible to define the ratios necessary for their optimal action and then obtain hyper-cellulolytic strains with these characteristics through use of the mutants derived from both the screening and genetic methodologies.

Acknowledgements

This study was supported by the U.S. Department of Energy, Contract Number EG-78-S-02-4591 A 000, the New Jersey Agricultural Experiment Station, and ZWO (D.H.J.S.).

References

Berghem, L.E.R., Pettersson, L.G. and Axio-Fredriksson, U.B. (1976). The mechanism of enzymatic cellulose degradation. Purification and some properties of two different 1,4-β-glucan glucanohydrolases from *Trichoderma viride*. *European Journal of Biochemistry* **61**, 621-630.

Bernton, H. (1978). Paving the way for alcohol fuels. *Environmental Action* **10**, 4-7.

Del Gobbo, N. (1978). Fuels from biomass systems - Program overview. In *Second Annual Fuels from Biomass Symposium*, pp.7-23. Edited by W. Schuster. Troy, New York: U.S. Department of Energy.

Demain, A.L. (1972). Theoretical and applied aspects of enzyme regulation and biosynthesis in microbial cells. *Biotechnology and Bioengineering Symposium* No. **3**, 21-32.

Emert, G.H., Gum, E.K. Jr., Lang, J.A., Liu, T.H. and Brown, R.D. Jr. (1974). Cellulases. *Advances in Chemistry Series* **136**, 74-100.

Fägestam, L., Håkasson, U., Pettersson, G. and Andersson, L. (1978). Purification of three different cellulolytic enzymes from *Trichoderma viride* QM9414 on a large scale. In *Bioconversion of Cellulosic Substances into Energy, Chemicals and Microbial Protein*, pp.165-178. Edited by T.K. Ghose. Faridabad, Haryana, India: Thompson Press (India) Ltd.

Gallo, B.J. (1978). MCG 77, A regulatory mutant of cellulase synthesis. *Abstracts of Third International Symposium on the Genetics of Industrial Microorganisms*. Madison, Wisconsin.

Gong, C-S., Ladisch, M.R. and Tsao, G.T. (1977). Cellobiose from *Trichoderma viride*: Purification, properties, kinetics and mechanism. *Biotechnology and Bioengineering* 19, 959-981.

Hammond, A.L. (1978). Energy: Elements of a Latin American strategy. *Science, New York* 200, 752-753.

Mandels, M. (1975). Microbial sources of cellulase. In *Cellulose as a Chemical and Energy Resource*, pp.81-105. Edited by C.R. Wilkie. *Biotechnology and Bioengineering Symposium* **5**, 81-105.

Markkanen, P., Bailey, M. and Enari, T-M. (1978). Production of cellulolytic enzymes. In *Proceedings of the Symposium on Bioconversion in Food Technology*, pp.111-114. Helsinki, Finland.

Montenecourt, B.S. and Eveleigh, D.E. (1977a). Semi-quantitative plate assay for determination of cellulase production by *Trichoderma viride*. *Applied and Environmental Microbiology* **33**, 178-183.

Montenecourt, B.S. and Eveleigh, D.E. (1977b). Preparation of mutants of *Trichoderma reesei* with enhanced cellulase production. *Applied and Environmental Microbiology* **33**, 777-782.

Montenecourt, B.S. and Eveleigh, D.E. (1978). Hypercellulolytic mutants and their role in saccharification. In *Second Annual Fuels from Biomass Symposium*, pp.613-625. Edited by W. Schuster. Troy, New York: U.S. Department of Energy.

Nakayama, M. (1975). The possible proteolytic modification of cellulase components in the aqueous solutions of a crude enzyme preparation from *Trichoderma viride*. *Memoirs Osaka Kyoiku University* **24**, 127-143.

Nevalainen, K.M.H. and Palva, E.T. (1978). Production of extracellular enzymes in mutants isolated from *Trichoderma viride* unable to hydrolyze cellulose. *Applied and Environmental Microbiology* **35**, 11-16.

Okada, G. (1976). Enzymatic studies on a cellulase system of *Trichoderma viride*. IV. Purification and properties of a less-random type cellulase. *Journal of Biochemistry (Tokyo)* **80**, 913-922.

Scheller, W.A. (1976). *Nebraska 2 million mile gasahol road test program*. Sixth Progress Report. Lincoln, Nebraska: University of Nebraska.

Scheller, W.A. (1977). The use of ethanol-gasoline mixtures for automotive fuel. *Symposium on Clean Fuels from Biomass and Wastes*. Orlando, Florida.

Shoemaker, S.P. and Brown, R.D. Jr. (1978). Characterization of endo-

1,4-β-D-glucanases purified from *Trichoderma viride*. *Biochimica et Biophysica Acta* **523**, 147-161.

Simmons, E.G. (1977). Classification of some cellulase producing *Trichoderma* species. *Abstracts of Second International Mycological Congress Tampa, Florida, U.S.A.*, p.618. Edited by H.E. Aigelow and E.G. Simmons.

Suzuki, M., Kuno, M., Maejma, K. and Nakao, Y. (1974). Production of alkaline protease from n-paraffins by mutants of *Fusarium* sp. *Agricultural and Biological Chemistry* **38**, 135-139.

Toda, K. (1976). Invertase biosynthesis by *Saccharomyces carlsbergensis* in batch and continuous culture. *Biotechnology and Bioengineering* **18**, 1103-1115.

Toda, K. (1976). Dual control of invertase biosynthesis in chemostat cultures. *Biotechnology and Bioengineering* **18**, 1117-1124.

Tomita, Y., Suzuki, H. and Nisizawa, K. (1974). Further purification and properties of "avicelase", a cellulase component of less-random type from *Trichoderma viride*. *Journal of Fermentation Technology* **52**, 233-246.

Wood, T.M. and McCrae, S.I. (1978). The mechanism of cellulase action with particular reference to the C_1 component. In *Bioconversion of Cellulosic Substances into Energy, Chemicals and Microbial Protein*, pp.111-141. Edited by T.K. Ghose. Faridabad, Haryana, India: Thompson Press (India) Ltd.

Chapter 15

THE APPLICATION OF POLYSACCHARIDE DEGRADING ENZYMES IN THE STARCH INDUSTRY

B.E. NORMAN

Novo Industri A/S, Bagsvaerd, Denmark

Introduction

Starch is a polysaccharide consisting of two components: a linear glucose polymer, amylose, which contains α 1,4 glucosidic links, and a branched polymer, amylopectin, in which linear chains of α 1,4 glucose residues are interlinked by α 1,6 glucosidic linkages [Manners, 1974].

Starch is the major source of carbohydrate in our diet, but apart from it being an important food in its own right, it can be hydrolysed readily to produce syrups or solids containing glucose, maltose and other oligosaccharides. The degree of hydrolysis can be controlled so that products with the desired physical properties may be obtained. In certain applications viscosity might be important, in other applications it might be osmotic pressure, sweetness or resistance to crystallization [Palmer, 1975].

The hydrolysis of starch may be carried out using either acid or enzymes as catalysts. The enzymatic hydrolysis of starch has been practised on an industrial scale for many years and is gradually replacing the traditional acid catalysed processes [Underkofler *et al.*, 1965; Barfoed, 1976]. In 1976 over 4 million tons of syrup and solid glucose were produced in the United States alone. Approximately one quarter of this amount was produced by acid hydrolysis, whereas the remainder was produced using industrial enzymes.

Enzymatic hydrolysis has several advantages to offer. It is more specific and fewer by-products are formed, so that yields are higher. The conditions under which hydrolysis takes place are also milder so that subsequent refining stages to remove ash and colour are minimized [Underkofler *et al.*, 1965].

For the purpose of this article, the starch degrading enzymes will be classified into three main groups:

1) endo-amylases
2) exo-amylases

3) debranching enzymes.
Each group will be looked at separately and examples of enzymes and their applications will be given.

Endo-amylases

Endo-amylases are generally alpha-amylases which cleave α 1,4 glucosidic bonds in amylose, amylopectin and related polysaccharides. The products of hydrolysis, which are oligosaccharides of varying chain lengths, have the α-configuration at the C_1 of the reducing glucose unit. As the name suggests, endo-amylases hydrolyse the bonds located in the inner regions of the substrate. This results in a rapid decrease in the viscosity of starch solutions and a decrease in the iodine staining power of amylose. Endo-amylases are also able to by-pass α 1,6 branch points in amylopectin [Robyt and Whelan, 1968a]. We can divide the endo-amylases of industrial importance into two groups: thermostable alpha-amylases which are used mainly for high-temperature liquefaction, and thermolabile alpha-amylases which are used for saccharification.

Thermostable alpha-amylases

There are two distinct types of thermostable alpha-amylases which are commercially available and which are used in large quantities in the starch processing industry. Amylases derived from *Bacillus amyloliquefaciens (Bacillus subtilis* var. *amyloliquefaciens)* have been in use for many years but a major breakthrough was made in 1973 when a more heat stable alpha-amylase isolated from *Bacillus licheniformis* was introduced commercially [Madsen *et al.*, 1973]. Apart from its greater heat stability the *Bacillus licheniformis* alpha-amylase possesses a number of properties which distinguish it from the *Bacillus amyloliquefaciens* enzyme [Saito, 1973].

The effect of temperature on the activity of these two enzymes is illustrated in Figure 1. The analytical method used for this experiment is a modification of the so-called SKB analysis [Sandstedt *et al.*, 1939]. A substrate consisting of 0.46% w/v soluble starch in phosphate buffer at pH 5.7 is incubated with the enzyme at the appropriate temperature. The time taken to reach a standardized colour end-point after the addition of iodine solution is used as a measure of activity. Dilutions of the enzyme are made so that the reaction time lies within the range 7 to 20 minutes and preferably 10 to 12 minutes. The activity is expressed as NOVO α-amylase units (NU) and 1 unit is defined as the amount of enzyme which will hydrolyse 5.26 mg starch per hour under standard conditions.

Under the conditions of the analysis the optimum temperature of *Bacillus licheniformis* amylase is 92°C whereas for the *Bacillus amyloliquefaciens* amylase the optimum is

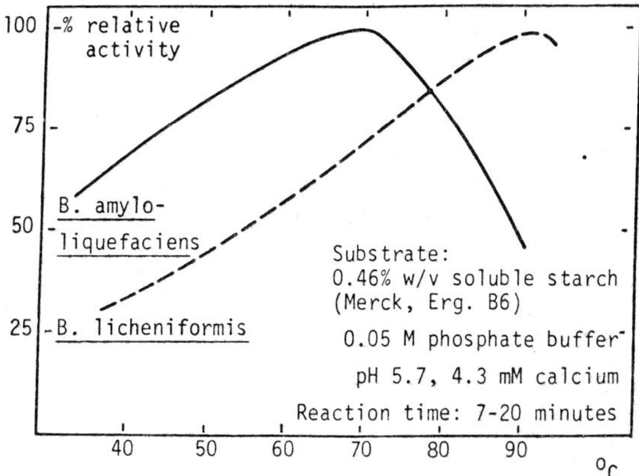

Fig. 1 The influence of temperature on α-amylase activity.

only 70°C. Under industrial application conditions substrate concentrations of 30-40% starch are normally used. This has a considerable effect on enzyme stability and enables the *Bacillus licheniformis* amylase to be used at temperatures of up to 110°C for short reaction periods. The maximum operating temperature for the *Bacillus amyloliquefaciens* amylase under optimum conditions is 85-90°C.

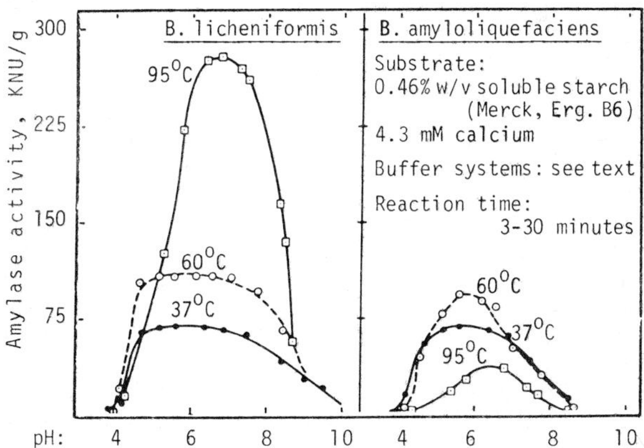

Fig. 2 The influence of pH and temperature on α-amylase activity. Buffer systems: 37°C and 60°C: 0.05 M acetate (pH 4.0-5.6), 0.05 M tris-maleate (pH 6.0-10.0); 95°C: 0.05 M acetate (pH 4.0-6.0), 0.05 M tris-maleate (pH 6.0-7.5), 0.05 M glycine/NaOH (pH 7.6-9.0).

The effect of pH on enzyme activity at different temperatures is illustrated in Figure 2. In these experiments we have plotted the change in amylase activity for the two enzymes with pH at three different temperatures.

At 37°C the activity for both enzymes is approximately 60 kilo NOVO units (KNU) per gram at the optimum pH, which is 5.7. At 60°C the activity of the *Bacillus licheniformis* amylase is approximately 15% higher and the enzyme is active over a much wider pH range. At 95°C the activity of the *Bacillus licheniformis* amylase has increased considerably, but the *Bacillus amyloliquefaciens* amylase has lost most of its activity because it is not sufficiently stable under these conditions.

In Figure 3 the effect of calcium on enzyme stability can be seen. In these experiments a 0.1% enzyme solution was prepared in deionized water in the absence of substrate. Different amounts of calcium chloride were added and the temperature raised to 70°C. Samples were withdrawn periodically and the percentage residual activity determined. An addition of 3.4 p.p.m. calcium stabilized the *Bacillus licheniformis* amylase completely under these conditions. The enzyme preparation itself contributed approximately 1 p.p.m. calcium, so that we can conclude that 5 p.p.m. is sufficient for stabilization at 70°C.

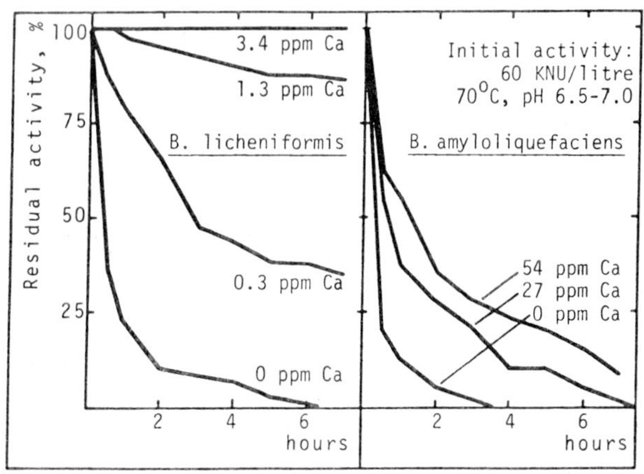

Fig. 3 The influence of added calcium on α-amylase stability in the absence of substrate. Residual activity measured according to the standard NOVO method at 37°C.

The *Bacillus amyloliquefaciens* amylase is less stable under these conditions even when considerably greater amounts of calcium are present. The calcium saturation level for this enzyme is about 150 p.p.m. The level of calcium required for full stability by the *Bacillus licheniformis* amylase is low at 70°C, but increases with increasing temperature. At 105°C an increase in stability can be seen at levels up to about 50 p.p.m. When the enzyme is used under industrial conditions, no further addition of calcium is necessary if the starch slurry to be treated contains at least 50 p.p.m. free calcium.

Action on starch Thermostable α-amylases are used for liquefaction and the reaction is normally terminated before significant starch hydrolysis has taken place, that is when the average degree of polymerization is about 10. However, if hydrolysis is allowed to proceed interesting differences in the action pattern of the two enzymes can be seen.

Figure 4 shows a series of chromatograms which illustrate the action of *Bacillus licheniformis* and *Bacillus amyloliquefaciens* amylases on gelatinized starch. The experimental conditions used for this work were as follows. A suspension of gelatinized starch was made by slurrying 220 g of soluble starch (Merck Art 1263) in 250 ml cold tap water (15° German hardness) and adding this to 750 ml of boiling tap water. The temperature of the slurry was raised to 100°C and held for 10 minutes to complete the gelatinization. The starch suspension was then cooled to 85°C (for the *Bacillus licheniformis* amylase) or 70°C (for the *Bacillus amyloliquefaciens* amylase) and the pH adjusted to 6.5. An amount of enzyme corresponding to 24,000 NOVO alpha-amylase units was added to the suspension and samples withdrawn periodically. After sampling, the pH was adjusted to 3.0 and the temperature raised to 100°C to destroy the enzyme activity. The samples were filtered and purified before being submitted to gel permeation chromatography analysis. This system employs a polyacrylamide gel with a low exclusion limit (BioGel P2, minus 400 mesh) and enables excellent separation of oligosaccharides with a degree of polymerization from 1-13 [Wight, 1976].

A striking difference in the carbohydrate spectrum produced by the action of the two enzymes on starch can be seen. The *Bacillus licheniformis* amylase produces mainly maltose, maltotriose and maltopentaose. The maltohexaose formed initially is almost completely hydrolysed. The *Bacillus amyloliquefaciens* amylase produces mainly maltohexaose and unlike the *Bacillus licheniformis* amylase it does not appear to be able to hydrolyse this further.

As mentioned previously the thermostable endo-amylases are used primarily for liquefaction and the hydrolysis is therefore terminated when the average degree of polymerization is approximately 10. The differences in the carbohydrate spectra described above are probably without significance for this application.

To summarize, the alpha-amylase produced by *Bacillus licheniformis* is a more robust enzyme than that produced by *Bacillus amyloliquefaciens*. It can tolerate higher operating temperatures, it is less dependent on calcium ions for stability and it is active over a wider pH range.

Use of Bacillus licheniformis amylase for high temperature liquefaction The most important application of thermo-

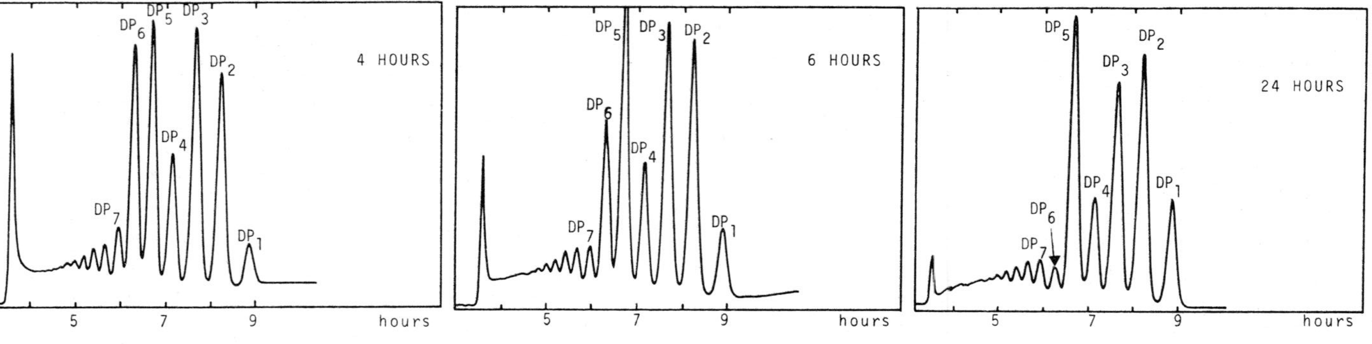

Fig. 4a Carbohydrate spectrum of starch hydrolysed by *Bacillus licheniformis* α-amylase after 4, 6 and 24 h (see text for experimental conditions).

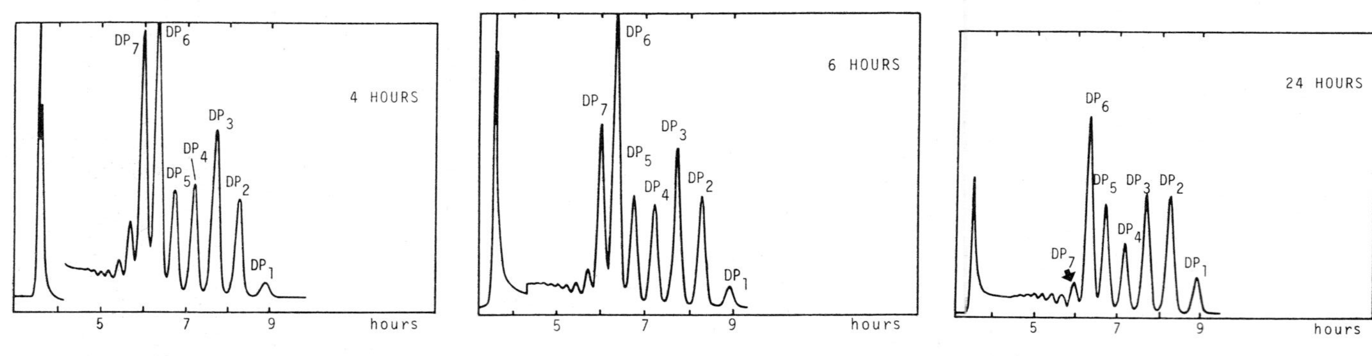

Fig. 4b Carbohydrate spectrum of starch hydrolysed by *Bacillus amyloliquefaciens* α-amylase after 4, 6 and 24 h (see text for experimental conditions).

stable endo-amylases in the starch processing industry is in the so-called liquefaction process. Liquefaction is the term used to describe the dispersion of starch molecules into aqueous solution, followed by partial hydrolysis.

In its native state, starch consists of microscopic granules, each possessing a highly complex internal structure. At ambient temperatures these granules are insoluble in water, but if an aqueous starch suspension is heated to above 60°C the granules swell and eventually disrupt, dispersing the starch molecules into solution. The adhering protein separates and coagulates during this process. The temperature required for complete dispersion or gelatinization depends on the source of starch but 105 to 110°C is sufficient for most starches.

Normally the dry substance content of the starch suspension used for liquefaction is greater than 30% and because of this the viscosity is extremely high after gelatinization. This necessitates the use of a thinning agent, which apart from reducing viscosity also partially hydrolyses the starch so that starch precipitation or "retrogradation" is prevented during subsequent cooling.

The traditional thinning agent used in the starch industry is acid. In the so-called acid liquefaction process (Figure 5) the pH of a 30-40% dry solids starch slurry is adjusted to about 2.0 with a strong acid. Hydrochloric acid is normally used, but in the Far East it is the practice to use oxalic acid.

The starch slurry is heated to 140-150°C either by direct steam injection or indirect heating in a converter for approximately 5 minutes, after which it is flash-cooled to atmospheric pressure and neutralized. This treatment results in complete starch gelatinization and a final product which is easy to filter. However, the non-specific catalytic action of the acid can result in the formation of undesirable by-products such as 5-hydroxymethyl 2-furfuraldehyde and anhydro-glucose compounds [Greenshields and MacGillivray, 1972; Birch and Shallenberger, 1973]. Moreover, the colour and ash content are high so that purification costs become significant.

If a thermostable endo-amylase is used as a catalyst, the processing conditions are milder so that by-product formation is not a problem and furthermore refining costs are lower. The *Bacillus licheniformis* amylase is normally preferred because apart from its greater heat stability it is less dependent on calcium ions for stability. If the starch suspension to be treated contains about 50 p.p.m. Ca^{++}, no further addition of calcium salts is necessary. Two liquefaction processes based on the use of *Bacillus licheniformis* amylase are shown in Figure 5. These will now be described separately.

Single-stage enzyme liquefaction In 1973, Novo Industri A/S, Copenhagen developed and patented a simple enzyme liquefaction process which took advantage of the extreme thermostability of the *Bacillus licheniformis* amylase [Slott and Madsen, 1975]. An outline of this process can be seen in Figure 5 and the layout of our pilot-plant in which the process has been evaluated is shown in Figure 6.

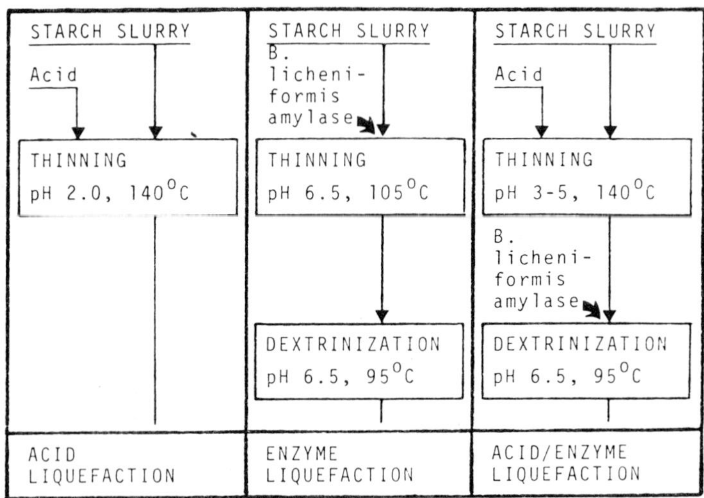

Fig. 5 Liquefaction processes.

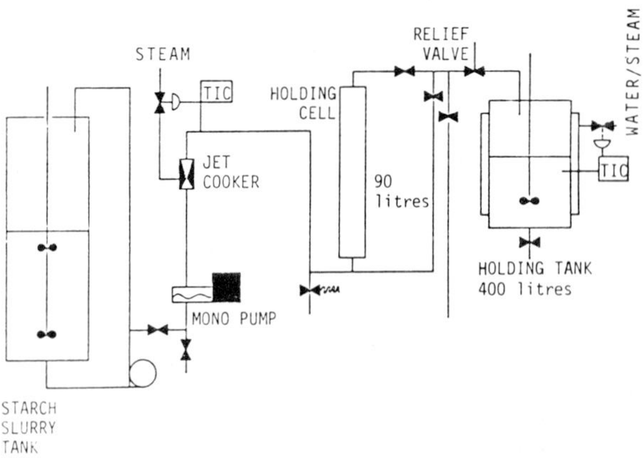

Fig. 6 Pilot plant equipment for liquefaction process.

The process is as follows. A starch slurry containing 30-40% dry solids is prepared in the feed tank. The pH is adjusted to 6.0-6.5 with sodium hydroxide and calcium salts may be added if the level of free calcium ions is below about 50 p.p.m. In our pilot-plant we add the liquefying enzyme to the feed tank, but in a large scale industrial process the enzyme would be metered directly into the stream emerging from the feed tank. The slurry is then pumped continuously through a 'jet-cooker' where the temperature is raised to 105°C by the direct injection of live steam. Tremendous shearing forces are exerted on the slurry as it is pumped through the jet-cooker so in addition to the viscosity reduction action of the enzyme some mechanical thinning also takes place. Peak viscosities are therefore avoided. The slurry is maintained at this high temperature in the pressurized holding cell for about five minutes, after which it is discharged via a spring-loaded release valve into a reaction tank where enzyme action is allowed to continue for about 2 hours at 95°C. On an industrial scale this reaction tank would be replaced by a series of stirred tank reactors to enable the process to be run fully continuously. After this treatment the liquefied starch will have a dextrose equivalent (DE) of 10-20, depending on the amount of enzyme used. (Dextrose equivalent is defined as reducing sugars expressed as dextrose and calculated as a percentage of dry substance.)

A number of factors are important for the smooth running of this process and the operating parameters have to be chosen very carefully. The temperature of the slurry during the five minute 'cooking' stage must be accurately controlled. At temperatures much above 105°C, the enzyme will be rapidly inactivated and below 105°C there will be a risk that the starch is not fully gelatinized. This will give rise to filtration problems during subsequent processing. A range of 103 to 107°C is therefore recommended. A pH operating range of 6.0 to 6.5 is recommended. At lower pH values the enzyme is less stable and at higher pH values there will be a risk of colour and by-product formation. A high dry substance has a stabilizing effect on the enzyme but a low dry substance improves gelatinization and subsequent filtration. A level of 35% dry solids is therefore recommended, based on trials in our pilot-plant. The calcium ion requirement for stabilization is very low and as the starch and processing water normally contain calcium salts, further addition is not required. However, if the Ca^{2+} level in the slurry is low, calcium chloride may be added to bring the level up to about 50 p.p.m.

The most important advantage of this process, apart from its simplicity, is that the energy consumption is relatively low because the maximum operating temperature is only 105°C compared with 140-150°C normally used. However, as has been pointed out previously, careful control of temperature is required. If it is too low the starch will not be completely gelatinized and filtration problems will occur. If the temperature runs too high a significant amount of enzyme acitivity will be destroyed.

Acid/enzyme liquefaction Another process which takes advantage of the thermostability of *Bacillus licheniformis* amylase is the so-called acid/enzyme process. Here the enzyme is added after the starch has been cooked and cooled to 100-95°C (Figure 5).

A starch slurry containing 30 to 40% dry solids is cooked at a high temperature for about 5 minutes. A jet-cooker is used so that sufficient mechanical thinning, due to shearing, takes place. The pH may be in the range 2-5, but if it is too low, by-product formation will be significant and if it is too high there will be no thinning effect from the 'acid' and there will also be increased colour formation. After cooking, the slurry is flash-cooled to about 100°C and the pH adjusted to 6.0-6.5 before the addition of enzyme. Using this type of process the enzyme consumption can be slightly reduced and the filtration properties are also improved because better fat/protein separation is achieved. However, there is a dramatic increase in steam consumption and hence fuel costs due to the high temperature cooking.

To conclude, liquefaction is the first and perhaps most important step in starch processing. Its purpose is to provide a partially hydrolysed starch suspension of relatively low viscosity which is free from by-products, stable to retrogradation and suitable for further processing (saccharification).

Thermolabile alpha-amylases

Thermolabile alpha-amylases are not sufficiently heat stable to be used during the liquefaction process, but they may find a use as saccharifying enzymes. The most widely used enzymes in this group are the maltogenic enzymes.

Maltogenic alpha-amylase An example of a maltogenic alpha-amylase which is used industrially is the fungal alpha-amylase obtained from *Aspergillus oryzae*. This enzyme catalyses the hydrolysis of starch to maltose and maltotriose and is used mainly to produce syrups rich in maltose [Allen and Spradlin, 1974; Barfoed, 1976].

The carbohydrate spectrum produced by the action of a typical maltogenic fungal alpha-amylase on starch is illustrated in Figure 7. A suspension of gelatinized starch was prepared according to the procedure described above. The temperature was lowered to 50°C and the pH adjusted to 5.0. An amount of enzyme corresponding to 32,000 NOVO alpha-amylase units was added to this suspension and samples taken periodically. After sampling the pH was adjusted to 3.0 and the temperature raised to 100°C to inactivate the enzyme. The samples were filtered and purified before being submitted to gel-chromatography analysis. Initially, large amounts of maltotriose are formed, but as hydrolysis proceeds, the level of maltose increases and reaches a maximum at about 60%. The properties of this enzyme will now be described.

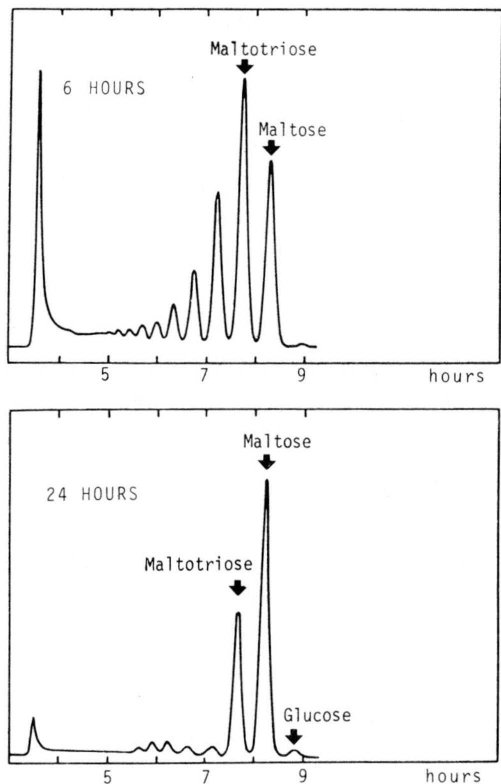

Fig. 7 Carbohydrate spectrum of starch hydrolysed by the maltogenic α-amylase from *Aspergillus oryzae* (see text for experimental conditions).

Figure 8 illustrates the activity of fungal alpha-amylase at different temperatures. The method of analysis used is identical to that described previously for the thermostable α-amylases, except that the pH of the substrate is adjusted to 4.7 with an acetate buffer. Under the conditions of the analysis the optimum temperature is 55°C which is considerably lower than for the thermostable amylases from *Bacillus amyloliquefaciens* and *Bacillus licheniformis*. When the enzyme is used under industrial conditions, the substrate concentration will be in the order of 30 to 40%. This will have a stabilizing effect on the enzyme, enabling saccharification to be carried out at 50-55°C for 48 hours.

Figure 9 illustrates the effect of pH on enzyme activity

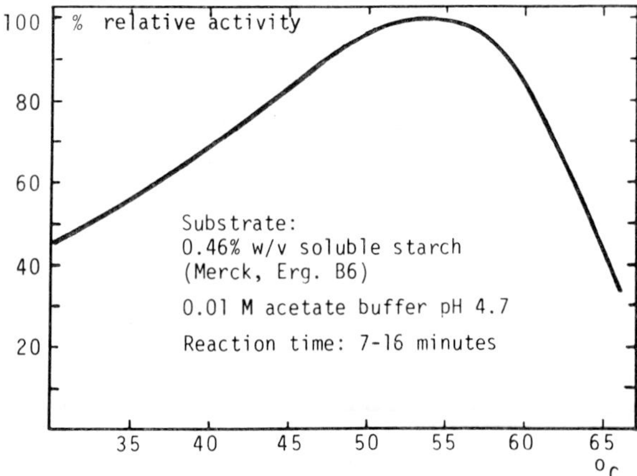

Fig. 8 Activity of *Aspergillus oryzae* α-amylase at different temperatures.

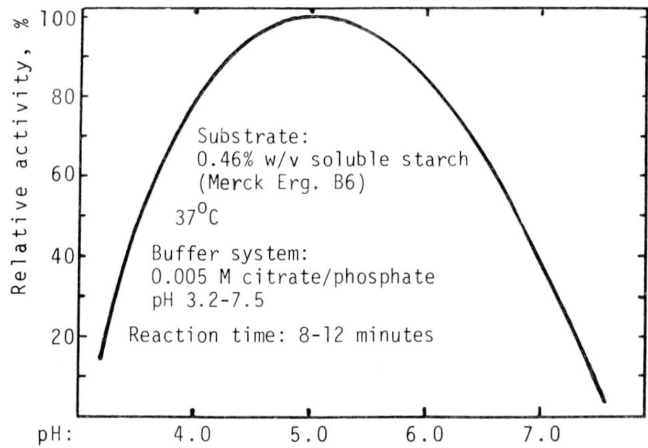

Fig. 9 Activity of *Aspergillus oryzae* α-amylase at different pH values.

at 37°C. Under the conditions of analysis the optimum pH is 4.7. This contrasts markedly with pH/activity curves of the thermostable amylases shown previously where the activity range is broader and the optimum closer to neutral.

Another interesting difference is that the fungal alpha-amylase from *Aspergillus oryzae* is able to hydrolyse Schardinger β-dextrin (cycloheptaose), but this is not a property which has any industrial importance.

Production of maltose containing syrups Syrups with a high maltose content are valuable for many applications in the food industry as they show a decreasing tendency to crystallize and are relatively non-hygroscopic. They are therefore very useful for the manufacture of hard candy and for frozen dessert formulations where they control crystal formation. High maltose syrups are also used in the baking industry [Palmer, 1975; Henry, 1976].

Two stages are involved in the production of high maltose syrups from starch (Figure 10). The starch is first liquefied, and as the glucose level has to be kept as low as possible, a high temperature enzyme liquefaction process is preferred. The liquefied starch is then treated with fungal α-amylase until the required amount of maltose is formed.

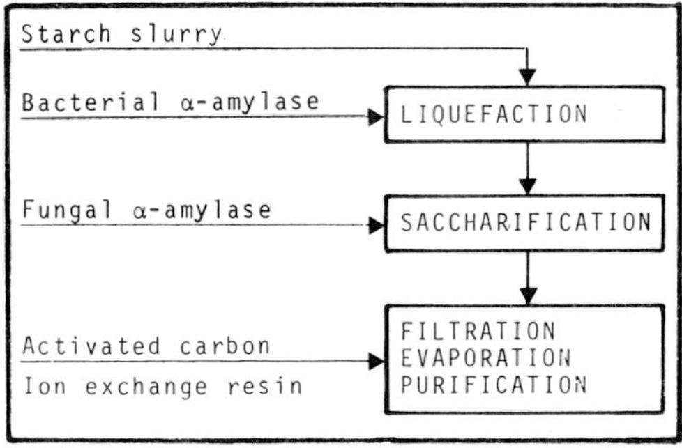

Fig. 10 Production of high-maltose syrup from starch.

The saccharification procedure is as follows. The pH of the liquefied starch slurry, which will have a DE of about 10 - 20, is adjusted to 5.0 and the temperature lowered to 50°C. The substrate concentration should be in the range 35-45% DS. The enzyme is added and the reaction allowed to proceed for 24-48 hours. Figure 11 illustrates the change in carbohydrate composition during saccharification and Figure 12 illustrates the effect of enzyme dosage on maltose and glucose formation. When the saccharification reaction is completed, the crude syrup is filtered, decolourized with activated carbon and ion-exchanged to remove ash. High maltose syrups normally contain 40 - 50% maltose. In recent years there has been an increasing interest in 'extra' high maltose syrups containing more than 80% maltose. In order to achieve such a high level it is necessary to use a maltogenic exo-amylase and a debranching enzyme in combination.

Another application of fungal alpha-amylase is in the production of so-called 'High-Conversion' (60-70 DE) syrups [Underkofler *et al.*, 1965]. This involves a combined use of fungal α-amylase and glucoamylase and the application will be described later.

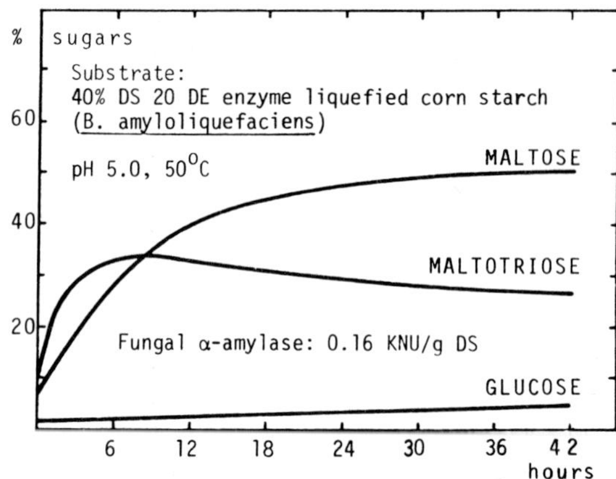

Fig. 11 High-maltose syrup - change in carbohydrate composition during saccharification.

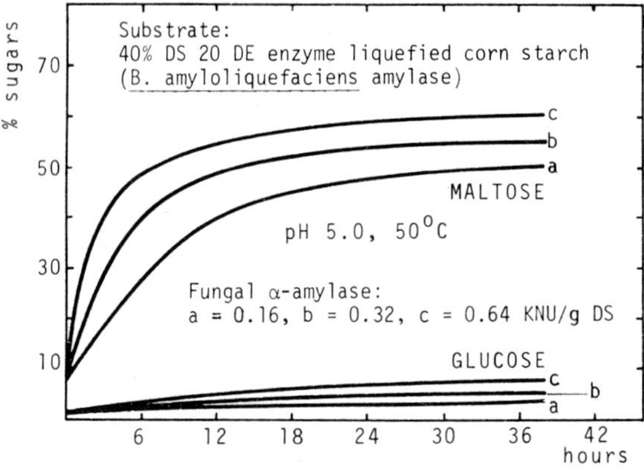

Fig. 12 High-maltose syrup - effect of enzyme dosage on carbohydrate composition.

Other saccharifying alpha-amylases During a routine screening programme for amylase producing *Bacillus* strains, a strain designated BL 458 was isolated which produced a thermolabile saccharifying α-amylase. This enzyme was found to have a pH optimum of 5.5 and a temperature optimum of 55-60°C. Its stability at temperatures above 55°C was rather low. It was also found that maltose strongly activated the enzyme and this property distinguished it from other α-amylases tested.

Another interesting property of BL 458 amylase was its action on starch. Figure 13 shows a series of gel filtration chromatograms which illustrate this. A suspension of

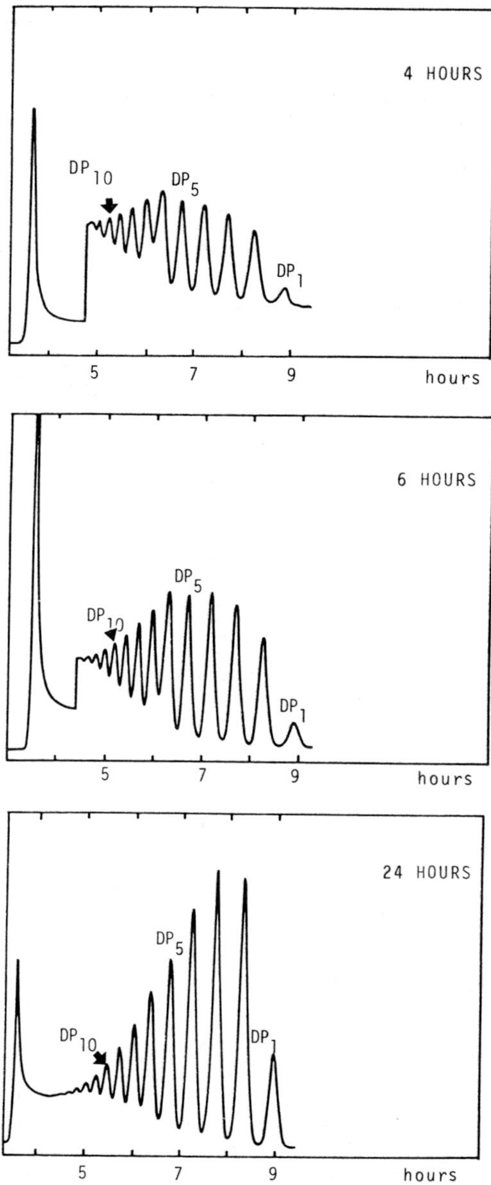

Fig. 13 The action of BL 458 saccharifying α-amylase on starch (see text for experimental conditions).

gelatinized starch was prepared according to the procedure described above. The temperature was lowered to 55°C and the pH adjusted to 6.0. An amount of enzyme corresponding to 24,000 NOVO alpha-amylase units was added to this suspension and samples taken periodically. After sampling the pH was adjusted to 3.0 and the temperature raised to 100°C to inactivate the enzyme. The samples were filtered and purified before being submitted to gel-chromatography analysis.

The carbohydrate spectrum produced is quite different from those seen with *Bacillus licheniformis*, *Bacillus amyloliquefaciens* and *Aspergillus oryzae* amylases. The pattern resembles very closely that produced by the action of acid on starch, apart from the lower glucose level (Figure 14).

In 1976 almost 1 million tons of so-called regular acid converted syrup were produced in the USA alone. This type of syrup could be produced enzymatically using an alpha-amylase similar to BL 458 so there is at least one potential market where it could be applied [Henry, 1976].

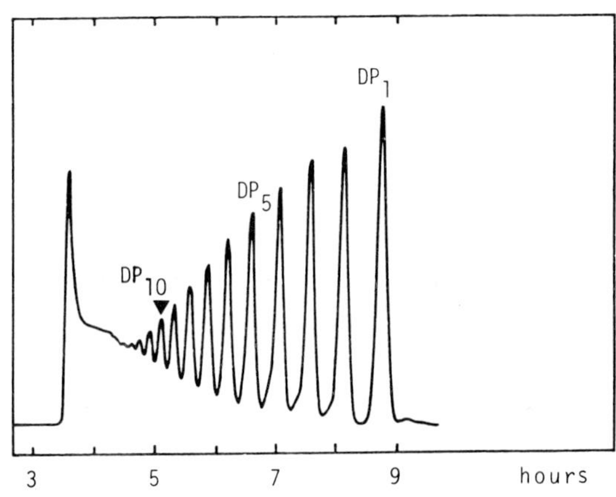

Fig. 14 Gel chromatogram of a 34 DE acid hydrolysed corn starch syrup.

Exo-amylases

Exo-amylases will cleave α 1,4 glucosidic bonds in amylose, amylopectin and related polysaccharides. The glucogenic exo-amylases are also able to cleave α 1,6 glucosidic bonds in isomaltose, panose or branched oligosaccharides, though at a slower rate. On the other hand, maltogenic exo-amylases such as cereal beta-amylase are not able to by-pass α 1,6 glucosidic branch points [Banks and Greenwood, 1975].

Exo-amylases catalyse the hydrolysis of alpha-glucosidic bonds by successively removing low molecular weight products such as glucose or maltose from the non-reducing chain end in a stepwise manner. In contrast to the action of endo-amylases, this results in a slow decrease in the viscosity and iodine staining power of starch. The products of hydrolysis generally have the β-configuration at the C_1 of the reducing glucose unit [Robyt and Whelan, 1968a].

Apart from glucogenic and maltogenic exo-amylases other enzymes are known which release maltotetraose or maltohexaose by an exo-attack mechanism. These will be briefly described at the end of this section.

Glucogenic exo-amylases (glucoamylases)

Glucoamylases (E.C. 3.2.1.3. 1,4-α-D-glucan glucohydrolase) are produced by many different microorganisms, but commercially available products originate from either *Aspergillus niger* or *Rhizopus* sp. [Underkofler *et al.*, 1965].

Glucoamylases have a low degree of specificity and are able to hydrolyse α 1,3 and α 1,6 linkages in malto-oligosaccharides, though at a much slower rate than α 1,4 linkages (see Table 1). The affinity of the enzyme towards different substrates depends on chain length and the rate of hydrolysis increases linearly up to DP 4 (maltotetraose) [Pazur *et al.*, 1973]. In this way glucoamylases can be distinguished from α glucosidases (E.C. 3.4.1.20) which hydrolyse α 1,4 glucosidic links more rapidly in low molecular weight oligosaccharides. Another significant difference is that the glucose residues which are released by the action of α glucosidase have the alpha-configuration rather than the β-configuration produced by glucoamylase [Manners, 1974].

TABLE 1

Relative rate of hydrolysis of disaccharides by purified glucoamylase from Aspergillus niger [Pazur and Kleppe, 1962]

Disaccharide	α-Linkage	Relative rate of hydrolysis
Maltose	1,4	100
Nigerose	1,3	6.6
Isomaltose	1,6	3.6

Aspergillus niger glucoamylase contains at least two components [Freedberg *et al.*, 1975] which are normally referred to as glucoamylase I and glucoamylase II. It is reported that glucoamylase I has more debranching activity than glucoamylase II [Gasdorf *et al.*, 1975], but we have not been able to demonstrate any significant difference in the activity of these two components towards starch which has been partially hydrolysed with α-amylase. Some of the properties of glucoamylase from *Aspergillus niger* will now be described.

At ambient temperatures the enzyme is stable over a very wide pH range (2-10), but if the temperature is raised to 60°C, the stability range is narrowed down to pH 3.0-7.0 (Figure 15). In these experiments the enzyme was incubated at the appropriate pH for 1 hour in 35% w/w glucose. The pH was then adjusted to 4.5 and the residual activity measured using 4-nitrophenyl-α-D-glucopyranoside as substrate. The pH activity range is considerably narrower, the optimum being about 4.0 at 55°C in a 30% w/v maltose substrate with a 30 minutes reaction time (Figure 16). The activity declines very rapidly above pH 5 and at pH 7.0 almost no activity remains, although the enzyme is fairly stable. Figure 17 illustrates the effect of temperature on activity. Under the conditions of the analysis the maximum activity is seen at 75°C, but as the enzyme is unstable at high temperatures (Figure 18), the limit for practical applications is about 60°C.

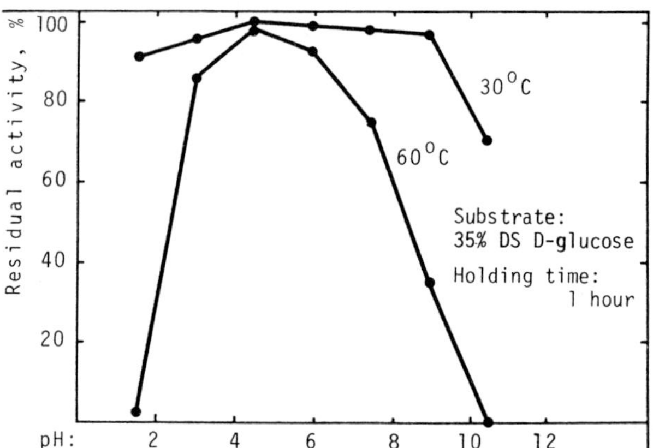

Fig. 15 The influence of pH on glucoamylase stability.

Reversion Glucoamylase is able to polymerize glucose by a reaction which is formally the reverse of hydrolysis. This reversion or condensation reaction is accompanied by the elimination of water [Underkofler *et al.*, 1965; Hehre *et al.*, 1969].

$$\text{glucose} + \text{glucose} \underset{\text{Hydrolysis}}{\overset{\text{Reversion}}{\rightleftharpoons}} \text{isomaltose} + \text{water}$$

The main products of the reversion reaction are maltose and isomaltose, though on prolonged incubation at high substrate concentrations other disaccharides can be iden-

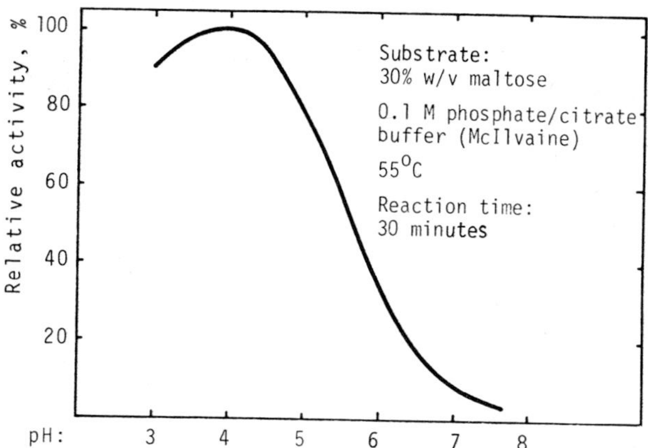

Fig. 16 The influence of pH on glucoamylase activity.

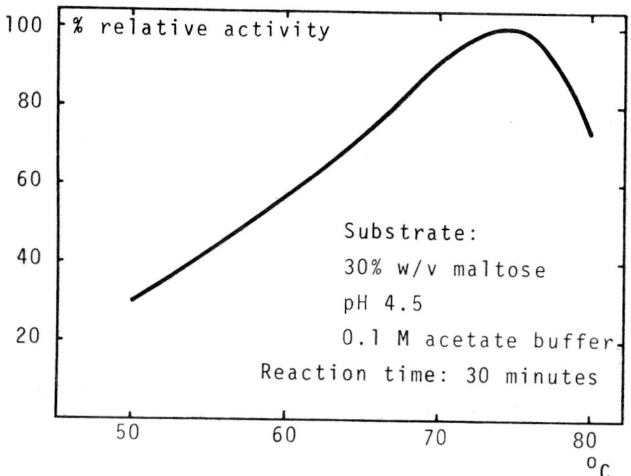

Fig. 17 The influence of temperature on glucoamylase activity.

tified in addition to higher polymers. The synthesis of maltose and other oligosaccharides proceeds at a much slower rate. In order to reach equilibrium it is necessary to use very high enzyme levels. The amount of reversion sugars formed is also very dependent on the substrate concentration. At 30% dry substance the equilibrium mixture will contain 85% glucose and 15% oligosaccharides of which about 11% is isomaltose. At 10% dry substance equilibrium will be established at 95% glucose. When glucoamylase is used in practice for saccharification the reaction is terminated long before equilibrium is reached.

Polymerization can also be catalysed by another mechanism namely glucose transfer. Unlike reversion this reaction is catalysed by a separate enzyme, transglucosylase (E.C. 2.4.1.24 1,4-α-D-glucan 6-α-glucosyltransferase).

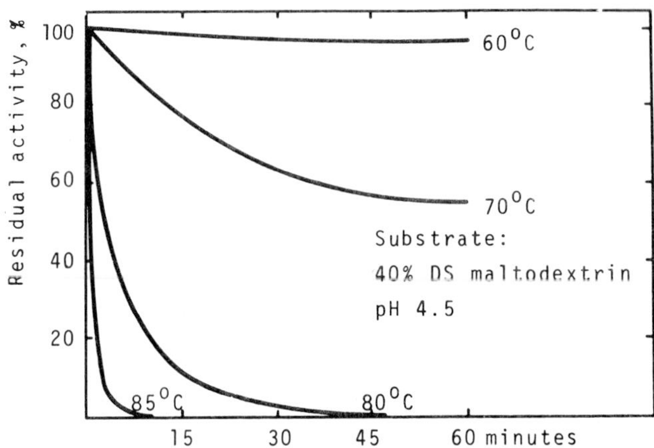

Fig. 18 The influence of temperature on glucoamylase stability.

Transglucosylase Transglucosylase is an enzyme impurity which is often present in crude glucoamylase preparations [Maher, 1968]. It catalyses the formation of non-fermentable glucose oligomers by transferring a glucosyl moiety from an α 1,4 position to an α 1,6 position [Pazur and Ando, 1961].

maltose + maltose ⟶ panose + glucose
maltose + glucose ⟶ isomaltose + glucose

Relatively large amounts of panose and isomaltose are built up during saccharification so that the final yield of glucose will be considerably lower if transglucosylase is present. The following experiment illustrates the problem more clearly.

A crude glucoamylase preparation was isolated from a strain of *Aspergillus* species. The enzyme was fractionated into two major components on a DEAE-cellulose chromatography column. A transglucosylase fraction was eluted first and this was followed by a second fraction which contained alpha-amylase and glucoamylase. These two fractions plus a sample of the crude glucoamylase were evaluated for their action on partially hydrolysed starch (DE 10 maltodextrin) (Table 2). The glucoamylase fraction produced about 95.6% glucose. The transglucosylase fraction had almost no action on the substrate whereas the crude starting material produced a syrup of inferior quality with a low glucose content and a high level of isomaltose and panose.

Many methods have been described for the removal or inactivation of transglucosylase in commercial glucoamylase

TABLE 2

The effect of transglucosylase activity on saccharification

Fraction	Reaction time (hours)	% sugars (HPLC) DP_1	DP_2	DP_3	DP_{4+}
Glucoamylase	24	88.1	2.6	0.5	8.8
+	48	94.0	2.7	0.6	2.6
α-amylase	72	95.6	2.8	0.3	1.3
Transglucosylase	24	3.2	2.2	-	94.5
	48	3.6	2.6	-	93.8
	72	5.3	4.3	-	90.5
Crude enzyme	24	48.6	13.1	6.5	31.9
	48	65.8	11.9	5.7	16.7
	72	75.4	10.9	4.3	9.5

Substrate: 30% DS DE 10 maltodextrin, 60°C, pH 4.5. Enzyme addition: 0.23 AG units/g dry substance (1 AG unit is defined as the amount of enzyme which will hydrolyse 1 μmol maltose per minute at 25°C, pH 4.3).

products. The list includes absorption on bentonite clay, precipitation by heteropolyacids such as phosphorvanadic or silicotungstic acid, inactivation with silver or mercuric ions and inactivation by heat treatment under acidic conditions [Maher, 1968; White and Dworschack, 1973]. At Novo we have chosen another method. By selection an *Aspergillus* mutant has been isolated which does not produce transglucosylase.

Production of high-glucose syrups The most important application of glucoamylases is in the production of high-glucose syrups, that is syrups containing 90 - 97% D-glucose. These syrups could be used directly, for example for fermentation, but transport and storage would have to take place under heated conditions to prevent crystallization and solidification.

High glucose syrups are normally used either for the production of crystalline D-glucose (dextrose) [Kingma, 1969; Knight, 1969] or as a starting material for the production of high fructose syrups [Zittan *et al.*, 1975]. In 1976, half a million tons of dextrose and about one million tons of high fructose syrup were produced in the U.S.A. For both applications the highest possible glucose levels are required. In the case of dextrose this will improve the crystallization yield and in the case of high fructose syrups, this will reduce the amount of glucose isomerase required to give a certain fructose level

[Oestergaard and Knudsen, 1976].

An outline of the process for the production of glucose from starch is given in Figure 19. An aqueous starch slurry (30 - 40% DS) is liquefied and partially hydrolysed by one of the previously described processes. After liquefaction, the pH of the slurry is adjusted to 4.5 and the temperature lowered to 60°C. Glucoamylase is now added and the hydrolysis reaction allowed to continue for 48 - 96 hours in stirred tank reactors until the maximum level of glucose is reached. The syrup is then filtered to remove protein and fat and purified with activated carbon to remove colour and solubilized protein. Carbon purification is normally followed by an ion-exchange stage to remove ash.

The aim of the saccharification process is to produce a syrup with the maximum amount of D-glucose. A number of factors affect the amount of glucose that can be produced and these have to be considered carefully when optimizing the process.

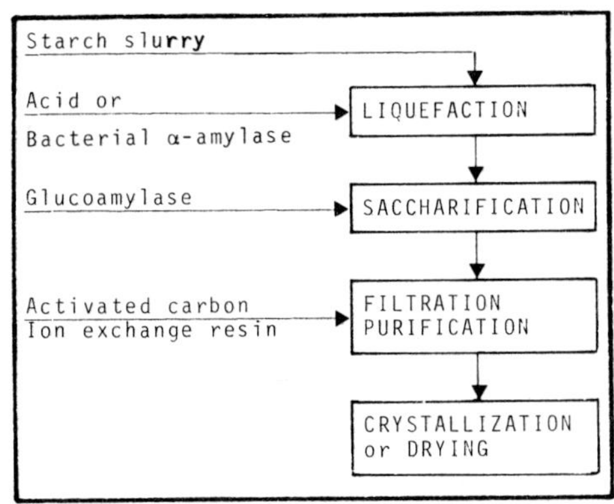

Fig. 19 Production of crystalline D-glucose from starch.

The maximum yield of glucose obtained during saccharification is very dependent on the liquefaction process [Underkofler *et al.*, 1965]. Two distinct processes, namely acid and enzyme liquefaction have been described previously. During acid liquefaction a number of by-products are formed because of the severe conditions under which the process takes place (140-150°C, pH 1.5 - 2.5). The formation of laevoglucosan (1,6-anhydro-β-D-glucopyranose) [Birch and Schallenberger, 1973] and 5-hydroxymethyl 2-furfuraldehyde have been reported [Greenshields and Mac Gillivray, 1972]. We have also detected the presence of oligosaccharides with a chain length of 5 to 6 glucose units which are extremely difficult to hydrolyse with glucoamylase. The amount of isomaltose present is also higher. The maximum level of glucose obtainable after acid liquefaction followed by saccharification with glucoamylase at 30% DS is 92%. Under similar conditions using enzyme liquefaction the maximum level would be 96 - 96.5%.

By-product formation may also take place during enzyme liquefaction if process conditions are not properly controlled. We have observed the presence of up to 3% maltulose (4-*O*-α-D-glucanopyranosyl-D-fructose) in some high glucose syrups prepared using enzyme liquefaction. The origin of maltulose is not altogether clear and it is normally formed by an alkaline-catalysed isomerization of maltose. In our laboratories we have prepared maltulose from maltose using sodium aluminate as catalyst. Almost complete isomerization can be achieved in 3 hours at 50°C at pH 11.5. However, during enzyme liquefaction these conditions are not encountered. Moreover, maltulose cannot be detected in liquefied starch and it appears only on subsequent saccharification. As maltulose is not detected in glucose syrups which have been prepared by acid liquefaction we assume that during enzyme liquefaction maltulose 'precursors' are formed which on saccharification yield maltulose, thus:

$$\text{.. glucose-glucose-glucose} \xrightarrow[\text{heat}]{OH^-} \text{.. glucose-glucose-fructose} \xrightarrow{\text{glucoamylase}} \text{glucose + maltulose}$$

Maltulose formation can be avoided by carrying out liquefaction at a low pH (6.0-6.3) and keeping holding times as short as possible.

The degree of hydrolysis after liquefaction is also important and a DE of about 10 should be aimed for. If the degree of hydrolysis is lower there will be a risk of starch retrogradation after cooling to 60°C and this will lower the yield and affect filtration. Higher degrees of conversion could result in by-product formation if they are the result of prolonged holding times during liquefaction. Also there is a possibility that alpha-limit dextrins will be produced which are less readily hydrolysed by glucoamylase.

The quality of the saccharifying enzyme will also influence the yield of glucose. Certain contaminating activities such as transglucosylase and β-glucosidase should be absent as these will result in lower yields due to the build up of 1,6 α and β-linked glucose oligomers. On the other hand, trace amounts of α-amylase, which are normally present in commercial glucoamylases, are important for efficient conversion to glucose [Marshall, 1975].

It has been reported that glucoamylases from *Rhizopus* sp. contain less 1,6 α glucosidase activity per unit of 1,4 α glucosidase than those from *Aspergillus niger* [Kawamura *et al.*, 1970]. However, we have not been able to demonstrate any significant difference in the composition of the high-glucose syrups produced with these two enzymes (see Table 3).

The saccharification temperature employed should be sufficiently high to prevent microbial contamination during the long reaction times (48-96 hours). 60°C is normally used for glucoamylases from *Aspergillus niger*. Although it would be desirable to operate at a higher temperature, it is not possible because the enzyme would be rapidly inactivated. Glucoamylases from *Rhizopus* species are even less heat stable and a maximum temperature of 55°C has to be used during saccharification (see Table 4).

TABLE 3

Comparative saccharification with glucoamylases from Aspergillus niger and Rhizopus species

Glucoamylase source	Reaction time (hours)	% sugars (HPLC) DP_1	DP_2	DP_3	DP_{4+}	% isomaltose (GLC)
Aspergillus niger	24	91.8	2.8	0.5	4.9	0.4
	48	94.4	3.2	0.5	1.9	0.8
60°C	72	95.5	3.2	0.3	0.9	1.0
pH 4.5	96	95.2	3.7	0.5	0.7	1.3
Rhizopus species	24	91.1	3.2	0.8	4.8	0.4
	48	95.2	2.6	0.6	1.6	0.7
55°C	72	95.5	3.0	0.5	1.0	0.8
pH 5.0	96	95.5	3.4	0.5	0.7	1.5

Substrate: 33% DS DE 10 maltodextrin. Enzyme addition: 0.225 AG units/g DS.

TABLE 4

Residual glucoamylase activity during saccharification

Reaction time, hours		24	48	72	96
Glucoamylase source, pH and temperature		% residual activity			
Aspergillus niger					
pH 4.5	55°C	92	74	64	58
	60°C	57	42	31	24
Rhizopus species					
pH 5.0	55°C	35	33	27	21
	60°C	29	14	5	1

Substrate: 33% DS DE 10 maltodextrin. Enzyme addition: 0.225 AG units/g DS.

Glucoamylases from *Aspergillus niger* are stable over a relatively wide pH range (3.0 - 8.0 at 60°C), but their activity is very pH dependent. At 55°C the activity at pH 6.0 is only 40% of the activity at pH 4.0. In practice, saccharification is normally carried out at pH 4.0-4.5 because in addition to being the optimum range, it reduces colour formation and hence purification costs.

The maximum obtainable glucose level is dependent on the substrate concentration used during saccharification. As the dry substance increases, the glucose level decreases (Figure 20) due to the formation of glucose oligomers (mainly isomaltose). From Figure 20 it can be seen that by decreasing the substrate concentration from 30% to 20% an increase of 1.5% in the glucose level can be obtained. However, this increase has to be balanced against the evaporation costs. Substrate concentrations between 27% and 33% are normally used industrially.

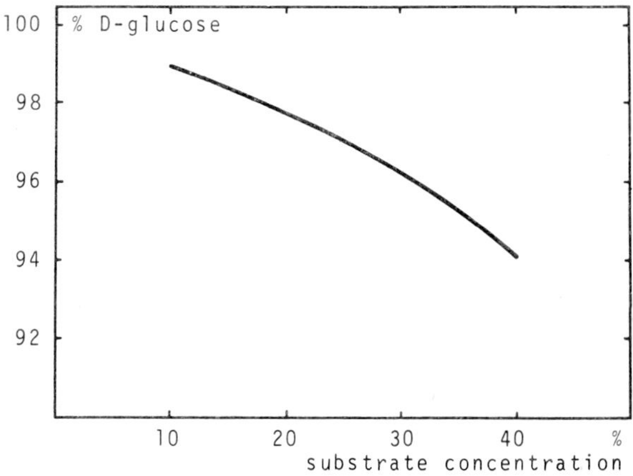

Fig. 20 The influence of substrate concentration on final D-glucose level [Nakamura and Suzuki, 1977].

Figure 21 illustrates the effect of enzyme dosage on reaction time. At the highest enzyme concentration it can be seen that the glucose level reaches a maximum and then declines. This is due to an increased formation of isomaltose and other 'reversion' sugars. The saccharification reaction should be stopped when the maximum glucose level has been obtained. An enzyme dosage of 0.2 to 0.25 AG units per gram dry substance will give a reaction time of 72 to 48 hours.

Production of high-conversion syrups Another important industrial application of glucoamylases is in the production of so-called high-DE or high-conversion syrups [Underkofler *et al.*, 1965; Palmer, 1975; Henry, 1976].

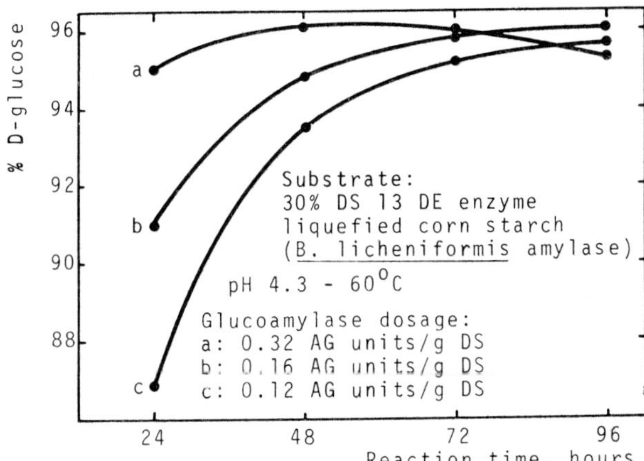

Fig. 21 The influence of enzyme dosage and reaction time on D-glucose level.

These syrups contain high levels of glucose and maltose and have a DE in the range of 60-70. Their composition is as follows (based on total carbohydrate): D-glucose 35-43%; maltose 30-37%; maltotriose 8-13%.

They are widely used in the soft-candy, brewing, baking, soft-drinks and canning industries. In 1976 over 1 million tons of high-conversion syrups were produced in the U.S.A. alone. The most important requirement of these syrups is that they should have a high DE and still be stable enough to resist crystallization at temperatures down to 4°C at 80-83% dry substance. In some applications it is the sweetness of the product that is important whereas in other applications it is the fermentability or hygroscopicity.

High-conversion syrups are produced by treating a 'regular' acid converted syrup (38-42 DE) with a mixture of a maltogenic enzyme and glucoamylase. The composition of the syrup can be 'tailored' by adjusting the ratio of these two enzymes. This is best illustrated by examining the curve shown in Figure 22. In these experiments we have kept the amount of fungal alpha-amylase constant and varied the amount of glucoamylase.

The procedure for producing these syrups is as follows. A starch slurry is liquefied and partially hydrolysed with acid to a DE of about 40. The temperature is then lowered to 55°C and the pH adjusted to 5.0 before the addition of the saccharifying enzymes. If we require a product with 40% glucose and 30-35% maltose, it will be necessary to use 0.023 glucoamylase units (AG units) and 0.12 fungal alpha-amylase units (kilo NU units) per gram dry substance.

Hydrolysis is allowed to proceed for about 40 hours or until the required composition is reached. The reaction

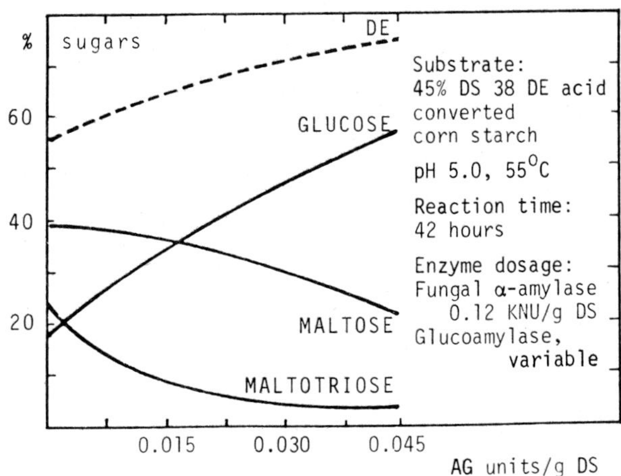

Fig. 22 The effect of glucoamylase dosage on syrup composition.

must then be stopped by heating the syrup to about 85°C to inactivate the enzymes. If the reaction is not properly terminated, hydrolysis will continue and excessive amounts of glucose will be formed. The final product is then filtered, decolourized with activated carbon and evaporated.

The maximum allowable glucose level in these syrups is about 43%. Above that level the glucose will crystallize out. If a higher fermentability is required, the maltose level in the syrup can be increased by using a substrate with a lower degree of conversion and increasing the level of the maltogenic enzyme (Figure 23).

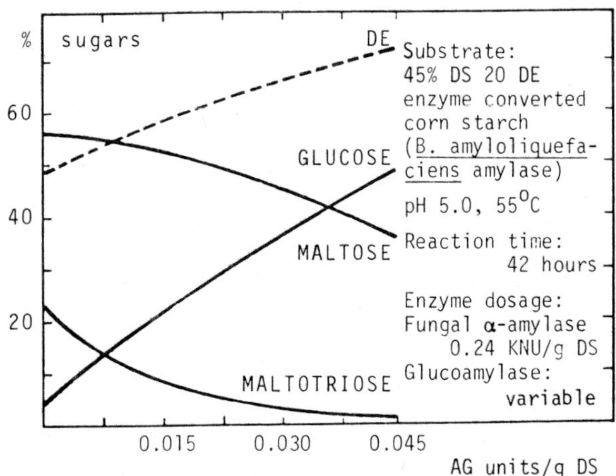

Fig. 23 The effect of glucoamylase dosage on syrup composition.

Maltogenic exo-amylases (β-amylases)

β-Amylases (EC 3.2.1.2. 1,4-α-D-glucan maltohydrolase) of plant origin have been known for many years. They are found in cereals such as barley, wheat, rye, oats, sorghum, and in soya beans and sweet potato [Robyt and Whelan, 1968c]. These enzymes contain sulphydryl groups and are therefore very sensitive to heavy metal ions and oxidizing agents. It was thought that β-amylases were confined exclusively to the plant kingdom, but several microbial sources are now known.

Higashihara and Okada [1976] have recently described the isolation and preparation of a β-amylase from *Bacillus megaterium*. The optimum pH for activity lies between 6.5 and 7.0, and the enzyme is stable over the range 5.0-8.0. *Bacillus megaterium* β-amylase is thermolabile and all activity is destroyed if the enzyme is held for 2 hours at 65°C. At temperatures below 55°C it is relatively stable.

In common with the plant beta-amylases the *Bacillus megaterium* enzyme is sensitive to sulphydryl reagents. It is also strongly inhibited by p-chloromercuribenzoate (PCMB), but the activity can be restored by treatment with excess cysteine. The action pattern of the enzyme on malto-oligosaccharides is also similar to that of plant beta-amylases, maltose being removed in a stepwise manner from the non-reducing end of the molecule.

Another maltogenic β-amylase isolated from *Bacillus cereus* var. *mycoides* has been reported by Takasaki [1976]. The microorganism is of interest because it also elaborates a debranching enzyme simultaneously. The properties of the beta-amylase are very similar to those of the *Bacillus megaterium* enzyme. The pH optimum is 7.0 and the temperature optimum 50°C. The enzyme is inhibited by heavy metal ions such as Hg^{+}, Cu^{++}, Ag^{+} as well as PCMB.

Early work by Robyt and French [1964] suggested that a maltogenic β-amylase isolated from *Bacillus polymyxa* was an endo-acting enzyme in that it could by-pass 1,6 α linkages in amylopectin, and did not require a non-reducing chain end for action. However, more recent work by Marshall strongly indicates that the enzyme is a true exo-amylase. He has compared purified *Bacillus polymyxa* beta-amylase with sweet potato β-amylase and has shown that they have an almost identical action on potato amylopectin, periodate oxidized amylose and Schardinger beta-dextrin [Marshall, 1974]. Hydrolysis of amylopectin stops at about 60% conversion, the partially oxidized amylose is only hydrolysed to a very small extent and the Schardinger dextrin is not hydrolysed.

There are, however, a number of important differences suggesting that *Bacillus polymyxa* beta-amylase is not a sulphydryl enzyme, in contrast to both the cereal beta-amylases and the other microbial beta-amylases mentioned above [Marshall, 1975].

A β-amylase from *Bacillus circulans* has also been described [Napier, 1977] which resembles the *Bacillus polymyxa* enzyme in that it is not significantly inhibited by PCMB, indicating that sulphydryl groups are not involved in the active site.

Application of β-amylases β-Amylases may be used to produce maltose containing syrups similar to those described under the section dealing with maltogenic α-amylases. The distribution of oligosaccharides in the two products will be different because of the difference in action pattern of the two types of enzyme. The syrup produced by the maltogenic α-amylase will contain a high level of maltotriose, whereas the syrup produced by the β-amylase will contain a large amount of β-limit dextrin and less maltotriose.

Other exo-amylases

Several other exo-amylases have been described in the literature. These include a maltotetraose-producing amylase from *Pseudomonas stutzeri* and a maltohexaose-producing amylase from *Klebsiella aerogenes* [Kainuma *et al.*, 1975]. (Although the species *Klebsiella aerogenes* is not recognized in the most recent edition of Bergey's Manual (Buchanan and Gibbons, 1974), this name will be used throughout when referring to the species previously called *Aerobacter aerogenes*.) The action of these enzymes is an exo-attack in that hydrolysis proceeds from the non-reducing end of the polysaccharide molecule resulting in a cleavage of the fourth or sixth glucosidic bond.

These exo-enzymes have not gained any commercial importance because there is at present no market for maltotetraose or maltohexaose syrups. One outlet could be in the production of well defined, non-sweet readily digestible carbohydrates for hospitalized patients.

Debranching Enzymes

The majority of starches which are of industrial importance contain approximately 80% amylopectin (Table 5). When starch is subjected to enzymatic hydrolysis by α-amylase, the amylopectin fraction is only partially degraded. The branch points containing 1,6 α-glucosidic linkages are resistant to attack and their presence also imposes a certain degree of resistance on neighbouring 1,4 α linkages. Thus the prolonged action of α-amylase on amylopectin results in the formation of so-called α-limit dextrins which are not susceptible to further hydrolysis [Robyt and Whelan, 1968b].

Similarly, when amylopectin is treated with β-amylase, hydrolysis stops as a 1,6 α branch point is approached, resulting in the formation of β-limit dextrins [Robyt and Whelan, 1968c].

TABLE 5

The composition of various starches
[Williams, 1968]

Source of starch	Amylopectin content %	Amylose content %
Maize	79	21
Potato	82	18
Rice	86	14
Tapioca	81	19
Wheat	81	19
Amylomaize	48	52
Waxy maize	100	0

Glucoamylases on the other hand are able to hydrolyse 1,6-alpha-glucosidic links, but the reaction proceeds relatively slowly [Pazur and Ando, 1960].

In these cases the use of a debranching enzyme would have obvious advantages. A number of different enzymes which are able to hydrolyse the branch points in amylopectin have been described in the literature [Allen and Dawson, 1975]. The one which has perhaps been most widely studied and which is being used on industrial scale is pullulanase (pullulan 6-glucanohydrolase EC 3.2.1.41).

Pullulanase

Pullulanase is a 1,6 α glucosidase which specifically attacks the linear α glucan pullulan. The structure of pullulan, which is produced by the fungus *Aureobasidium pullulans*, is shown in Figure 24. It is a linear polymer consisting of maltotriosyl units connected by 1,6 α bonds. Pullulanase hydrolyses the 1,6 α links in pullulan at random producing hexa- and nona-oligosaccharides initially and maltotriose finally. Pullulanase will also hydrolyse the 1,6 α-glucosidic links in amylopectin and limit dextrins provided that there are at least two 1,4 α-glucosidic links on either side of the 1,6 α links. The smallest molecule meeting this requirement is the linear tetrasaccharide

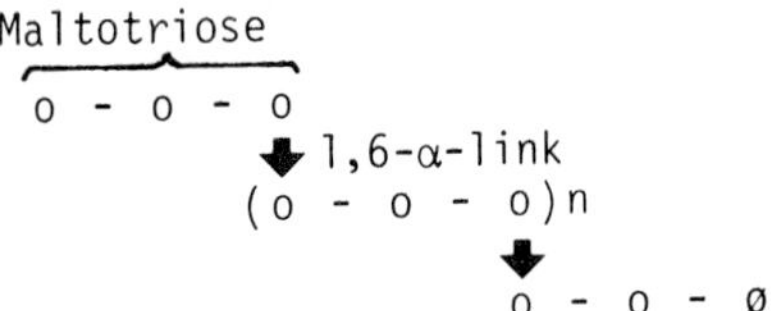

Fig. 24 Structure of pullulan.

6^2-α-maltosylmaltose, which is hydrolysed to maltose [Abdullah and French, 1970; Marshall, 1975].

Pullulanase is produced by a number of microorganisms including *Klebsiella* species, *Bacillus cereus*, *Streptococcus mitis*, *Escherichia intermedia* [Wöber, 1976], but the enzyme which has been studied most thoroughly is that from *Klebsiella aerogenes* [Ohba and Ueda, 1975].

Some of the properties of *Klebsiella aerogenes* pullulanase will now be described.

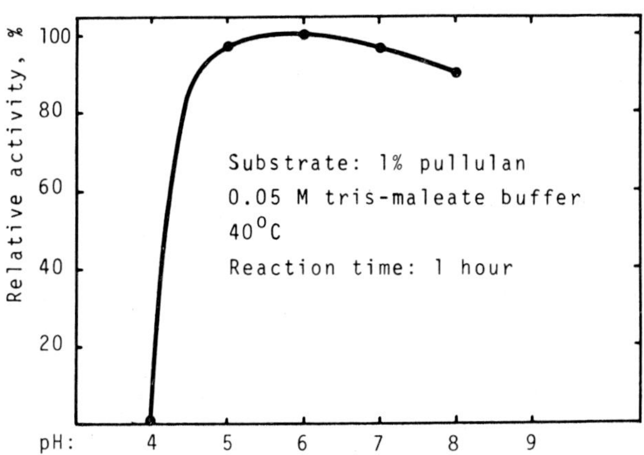

Fig. 25 The influence of pH on pullulanase activity.

The optimum pH of pullulanase lies in the region 5.0-7.0 (Figure 25). The activity has been measured by a modification of the CPC method [British Patent, 1969]. One unit is defined as the amount of enzyme required to produce 180 micrograms of reducing sugar calculated as D-glucose. A 1% pullulan solution in a 0.05 M tris-maleate buffer at the appropriate pH was incubated with the enzyme for 1 hour at 40°C. The reaction was terminated by the addition of 1M HCl and the reducing sugars formed were measured by the alkaline ferricyanide method.

The effect of temperature on activity is shown in Figure 26. A 1% pullulan solution in a 0.014 M phosphate buffer at pH 6.0 was incubated for 3 hours at the appropriate temperature. The reducing sugars formed were measured by the alkaline ferricyanide method. Under these conditions the optimum temperature was 50°C. At temperatures above 50°C the decline in activity was very rapid.

Calcium ions have been shown to have an activating effect on pullulanase. However, at levels above 10^{-2} M inhibition is observed. No other metal ions appear to have this effect whereas copper, iron, aluminium and mercury have a strong inhibitory effect.

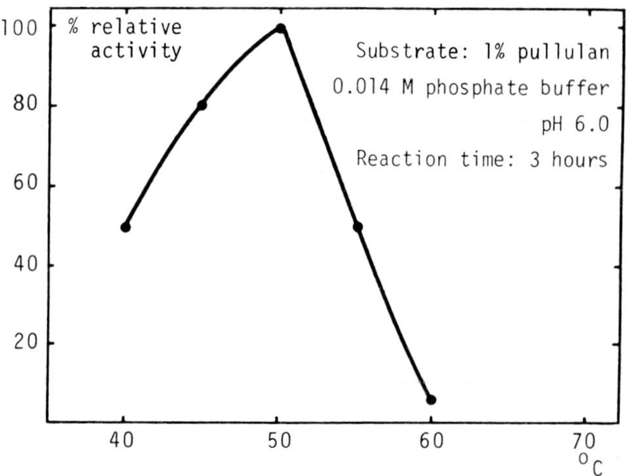

Fig. 26 The influence of temperature on pullulanase activity.

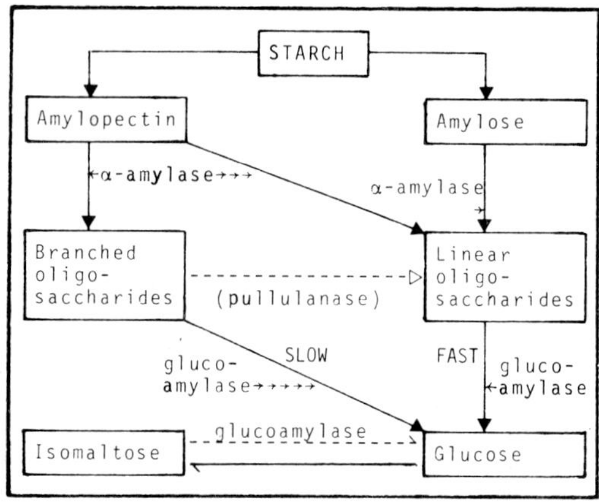

Fig. 27 Enzymatic conversion of starch to glucose.

Uses of pullulanase In dextrose production if pullulanase and glucoamylase are used simultaneously to saccharify a substrate of partially hydrolysed starch, a slight increase in glucose yield can be obtained. α-Limit dextrins, which are formed as a result of the action of α-amylase on amylopectin and which are only slowly hydrolysed by glucoamylase, are readily hydrolysed in the presence of pullulanase (Figure 27). Less glucoamylase activity is required per gram of starch, therefore less isomaltose formation, due to reversion, takes place and the maximum glucose levels achieved are higher.

In a series of experiments carried out in our laboratories the data shown in Table 6 were obtained. In the standard saccharification the pH throughout the reaction was 4.3. In the control and the pullulanase saccharification the initial pH was 6.0, but it was allowed to fall during the course of the reaction. Although the same amount of glucoamylase has been added initially, the activity in the two samples at pH 6.0 is less, therefore less isomaltose formation takes place than in the standard.

TABLE 6

Preparation of high-glucose syrups using glucoamylase and pullulanase

Experiment		Standard	Control	+ pullulanase
Glucoamylase, AG units/g DS		0.225	0.225	0.225
Pullulanase, CPC units/g DS		0	0	0.64
pH	initial	4.5	6.0	6.0
	final	4.3	5.3	5.2
% sugars:				
Glucose		95.9	91.6	97.0
Maltose and isomaltose		2.9	2.7	1.9
Higher sugars		1.1	5.7	1.1

Substrate: 30.6% DS DE 12.6 hydrolysed corn starch (*Bacillus licheniformis* amylase). Temperature: 55°C. Reaction time: 72 hours.

In maltose syrup production the treatment of starch and partially hydrolysed starch (maltodextrin) with β-amylase alone results in the formation of limit dextrins which are resistant to further attack. The maximum amount of maltose formed under these conditions is of the order of 60%. If the reaction is carried out in the presence of pullulanase, β-limit dextrins are not formed and the yield of maltose is higher. The total fermentability (glucose, maltose and maltotriose) is also considerably higher (Table 7).

If pullulanase is used in combination with a maltogenic fungal α-amylase the maltose level remains unaltered, but the maltotriose level is higher and therefore the total fermentability of the syrup also increases.

Other debranching enzymes

Another type of debranching enzyme which has received almost as much attention as pullulanase is isoamylase (amylopectin 6-glucanohydrolase, EC 3.2.1.9) [Yokobayashi *et al.*, 1969; Harada *et al.*, 1972].

According to Yokobayashi and co-workers [Yokobayashi *et al.*, 1973], the isoamylase isolated from *Pseudomonas*

TABLE 7

Use of pullulanase for high maltose syrup production

Sugars (determined by G.L.C.)	Pullulanase, CPC units/g DS 0	1	10
Glucose, %	0.7	0.6	2.0
Maltose, %	59.4	67.6	66.5
Maltotriose, %	15.3	28.5	32.7
Total fermentable sugars (G_1+G_2+G_3), %	75.4	96.7	101.2

Substrate: 31.5% DS DE 12.0 hydrolysed corn starch (*Bacillus licheniformis* amylase). Temperature: 50°C; Reaction time: 46 hours. 0.013 β-amylase units/g DS (barley-β-amylase).

TABLE 8

Comparison between pullulanase and isoamylase

	Pullulanase (*Klebsiella aerogenes*)	Isoamylase (*Pseudomonas* species)
pH optimum	5.5	3.5
Temperature optimum	50°C	45°C
Substrate specificity		
Pullulan	complete debranching	little or no action
Glycogen (native)	little action	complete debranching
Smallest substrate	6^2-α-maltosyl maltose (tetrasaccharide)	6^3-α-maltotriosyl maltotetraose (heptasaccharide)

SB-15 differs from pullulanase in a number of ways (Table 8). Isoamylase has a higher affinity for amylopectin and glycogen, but its activity towards pullulan is very low. It is able to completely debranch native glycogen whereas pullulanase can only attack glycogen if it has first been partially degraded. Another significant difference is that the smallest substrate for pullulanase is a linear tetrasaccharide 6^2-α-maltosyl maltose. The smallest substrate hydrolysed by isoamylase is a branched heptasaccharide 6^3-α-maltotriosyl maltotetraose.

Isoamylases are not used industrially to any great extent, and their most important application is in structural studies.

Conclusion

The number of polysaccharide degrading enzymes available to the starch industry is relatively large so that a variety of special syrups can be produced from starch. Unfortunately the range of properties such as pH and temperature optima is also large so that during processing changes in pH and temperature have to be made and controlled carefully.

To take an extreme example - the production of high fructose syrup from starch. First the pH of the starch slurry has to be adjusted upwards to about 6.0-6.5 for liquefaction. It is then lowered to 4.0-4.5 for saccharification and finally raised to about 8.0 for isomerization. The operating temperature varies from 105°C during liquefaction to 60°C during saccharification.

When screening for new microbial sources of 'polysaccharases' emphasis is normally placed on thermostability and this of course is important for avoiding infection problems during processing. However, it is also important to be able to carry out processing at constant pH, preferably in the range 4.0-5.0 to prevent by-product formation. The introduction of a thermostable acidophilic alpha-amylase and a moderately thermostable (60-65°C) acidophilic debranching enzyme, compatible with glucoamylase, would be an important contribution to the existing product range.

References

Abdullah, M. and French, D. (1970). Substrate specificity of pullulanase. *Archives of Biochemistry and Biophysics* **137**, 483-493.

Allen, W.G. and Dawson, H.G. (1975). Technology and uses of debranching enzymes. *Food Technology* May 1975, 70-80.

Allen, W.G. and Spradlin, J.E. (1974). Amylases and their properties. *The Brewers Digest* July 1974, 48-56, 65.

Banks, W. and Greenwood, C.T. (1975). *Starch and its Components.* Edinburgh University Press.

Barfoed, H.C. (1976). Enzymes in starch processing. *Cereal Foods World* **21**, 588-604.

Birch, G.G. and Schallenberger, R.S. (1973). Configuration, conformation, and the properties of food sugars. In *Molecular Structure and Function of Food Carbohydrates*, pp.9-20. Edited by G.G. Birch and L.F. Green. London: Applied Science.

British Patent (1969). 1,144,950. *Improvements in or Relating to the Conversion of Starch.*

Buchanan, R.E. and Gibbons, N.E. (1974). *Bergey's Manual of Determinative Bacteriology.* 8th ed. Baltimore, Md: Williams and Wilkins.

Freedberg, I.M., Leven, Y., Kay, C.M., McCubbin, W.D. and Katchalski-Katzir, E. (1975). Purification and characterization of *Aspergillus niger* exo-1,4-glucosidase. *Biochimica et Biophysica Acta* **391**, 361-381.

Gasdorf, H.J., Atthasampunna, P., Dan, V., Hensley, D.E. and Smiley, K.L. (1975). Patterns of action of glucoamylase isozymes from

Aspergillus species on glycogen. *Carbohydrate Research* **42**, 147-156.

Greenshields, R.N. and MacGillivray, A.W. (1972). Caramel - part 1. The browning reactions. *Process Biochemistry* 7, 11-13, 16.

Harada, T., Misaki, A., Akai, H., Yokobayashi, K. and Sugimoto, K. (1972). Characterization of *Pseudomonas* isoamylase by its actions on amylopectin and glycogen: comparison with *Aerobacter* pullulanase. *Biochimica et Biophysica Acta* **268**, 497-505.

Hehre, E.J., Okada, G. and Genghof, D.S. (1969). Configurational specificity: unappreciated key to understanding enzymic reversions and de novo glycosidic bond synthesis. *Archives of Biochemistry and Biophysics* **135**, 75-89.

Henry, R.E. (1976). Versatility of corn syrups expands with wider research. *Candy and Snack Industry* May 1976, 61-64.

Higashihara, M. and Okada, S. (1974). Studies on β-amylase of *Bacillus megaterium* strain No. 32. *Agricultural and Biological Chemistry* **38**, 1023-1029.

Kainuma, K., Wako, K., Kobayashi, S., Nogami, A. and Suzuki, S. (1975). Purification and some properties of a novel maltohexaose-producing exo-amylase from *Aerobacter aerogenes*. *Biochimica et Biophysica Acta* **410**, 333-346.

Kawamura, S., Wantanabe, T. and Matsuda, K. (1970). Hydrolysis of glucobioses by *Aspergillus niger* glucoamylase. *Tohoku Journal of Agricultural Research* **21**, 170-175.

Kingma, W.G. (1969). Crystalline dextrose manufacture. *Process Biochemistry* **4**, 19-21.

Knight, J.W. (1969). Conversion products of starch. *The Starch Industry*, pp.102-115. Oxford: Pergamon Press.

Madsen, G.B., Norman, B.E. and Slott, S. (1973). A new, heat stable bacterial amylase and its use in high temperature liquefaction. *Die Stärke* **25**, 304-308.

Maher, G.G. (1968). Inactivation of transglucosidase in enzyme preparations from *Aspergillus niger*. *Die Stärke* **20**, 228-232.

Manners, D.J. (1974). The structure and metabolism of starch. In *Essays in Biochemistry* Vol. 10, pp.37-91. Edited by P.N. Campbell and F. Dickins. London: Academic Press.

Marshall, J.J. (1974). Characterization of *Bacillus polymyxa* amylase as an exo-acting (1→4)-α-D-glucan maltohydrolase. *FEBS Letters* **46**, 1-4.

Marshall, J.J. (1975). Starch-degrading enzymes, old and new. *Die Stärke* **27**, 377-383.

Nakamura, M. and Suzuki, S. (1977). In *Denpun Kagaku Handbook*, p.445. Edited by J. Nikuni. Tokyo: Asakura Shoten.

Napier, E.J. (1977). Biochemical Process. British Patent 4,011,136.

Oestergaard, J. and Knudsen, S.L. (1976). Uses of Sweetzyme in industrial continuous isomerization. *Die Stärke* **28**, 350-356.

Ohba, R. and Ueda, S. (1975). Some properties of crystalline extra- and intra-cellular pullulanase from *Aerobacter aerogenes*. *Agricultural and Biological Chemistry* **39**, 967-972.

Palmer, T.J. (1975). Glucose syrups in food and drink. *Process Biochemistry* **10**, 19-20.

Pazur, J.H. and Ando, T. (1960). The hydrolysis of glucosyl oligosaccharides with α-D-1,4 and α-D-1,6 bonds by fungal amyloglucosidase. *Journal of Biological Chemistry* **235**, 297-302.

Pazur, J.H. and Ando, T. (1961). The isolation and mode of action of a fungal transglucosylase. *Archives of Biochemistry and Biophysics* **93**, 43-49.

Pazur, J.H. and Kleppe, K. (1962). The hydrolysis of α-D-glucosides by amyloglucosidase from *Aspergillus niger*. *Journal of Biological Chemistry* **237**, 1002-1006.

Pazur, J.H., Dropkin, D.J. and Hetzler, C.E. (1973). Glucan gluco-hydrolases: action mechanisms and enzyme-carbohydrate complexes. In *Carbohydrates in Solution*, pp.374-385. Edited by H.S. Isbell. Washington D.C.: American Chemical Society.

Robyt, J.F. and French, D. (1964). Purification and action pattern of an amylase from *Bacillus polymyxa*. *Archives of Biochemistry and Biophysics* **104**, 338-345.

Robyt, J.F. and Whelan, W.J. (1968a). General. In *Starch and its Derivatives*, pp.423-429. Edited by J.A. Radley. 4th edition. London: Chapman and Hall.

Robyt, J.F. and Whelan, W.J. (1968b). The α-amylases. In *Starch and its Derivatives*, pp. 430-476. Edited by J.A. Radley. 4th edition. London: Chapman and Hall.

Robyt, J.F. and Whelan, W.J. (1968c). The β-amylases. In *Starch and its Derivatives*, pp.477-497. Edited by J.A. Radley. 4th edition. London: Chapman and Hall.

Saito, N. (1973). A thermophilic extracellular α-amylase from *Bacillus licheniformis*. *Archives of Biochemistry and Biophysics* **155**, 290-298.

Sandstedt, R.M., Kneen, E. and Blish, M.J. (1939). A standardized Wohlgemuth procedure for alpha-amylase activity. *Cereal Chemistry* **16**, 712-723.

Slott, S. and Madsen, G.B. (1975). Starch liquefaction with α-amylases from *Bacillus licheniformis*. U.S. Patent 3,912,590.

Takasaki, Y. (1976). Productions and utilizations of β-amylase and pullulanase from *Bacillus cereus* var. *mycoides*. *Agricultural and Biological Chemistry* **40**, 1515-1522, 1523-1530.

Underkofler, L.A., Denault, L.J. and Hou, E.F. (1965). Enzymes in the starch industry. *Die Stärke* **17**, 179-184.

White, W.H. and Dworschack, R.G. (1973). Enzyme treatment. U.S. Patent 3,725,202.

Wight, A.W. (1976). Low pressure gelchromatographic separation of oligosaccharides. *Die Stärke* **28**, 311-315.

Williams, J.M. (1968). The chemical evidence for the structure of starch. In *Starch and its Derivatives*, p.102. Edited by J.A. Radley. 4th edition. London: Chapman and Hall.

Wöber, G. (1976). Pullulanase is a characteristic of many *Klebsiella* species and functions in the degradation of starch. *European Journal of Applied Microbiology* **3**, 71-80.

Yokobayashi, K., Misaki, A. and Harada, T. (1969). Specificity of *Pseudomonas* isoamylase. *Agricultural and Biological Chemistry* **33**, 625-627.

Yokobayashi, K., Akai, H., Sugimoto, T., Hirao, M., Sugimoto, K. and Harada, T. (1973). Comparison of the kinetic parameters of *Pseudomonas* isoamylase and *Aerobacter* pullulanase. *Biochimica et Biophysica Acta* **293**, 197-202.

Zittan, L., Poulsen, P.B. and Hemmingsen, S.H. (1975). Sweetzyme - a new immobilized glucose isomerase. *Die Stärke* **27**, 236-241.

Chapter 16

POLYSACCHARIDE DEGRADATION IN THE RUMEN

P.N. HOBSON

Rowett Research Institute, Aberdeen, UK

The Ruminant

Herbivorous animals, including even insects, depend on microbial activity in some part of their gut for the conversion of the polysaccharides of plants to substances which the animal can metabolize. With the exception of starches, which occur to any extent only in seeds and tubers and which are not normally a component of the herbivorous diet, plant polysaccharides are attacked little, if at all, by the gastric and intestinal secretions. Microorganisms in the gut can provide a variety of enzymes which will hydrolyse these polysaccharides.

Since microbial growth on herbivorous dietary constituents is comparatively slow, the rate of passage of the food must be slowed down in some portion of the gut by increasing the size of this portion relative to that in predominantly carnivorous animals or by providing a 'fermentation vessel' not found in the carnivorous alimentary tract. In herbivores such as the horse and rabbit the caecum and colon are enlarged and are the sites of action of 'digestive' microorganisms. (The commensal microflora, which inhabits the gut of carnivores plays little part in digestion of food.) The omnivorous pig, rat and other animals also have some digestive microbial activity in the large gut. However, in these animals the microbial activity comes after the gastric and small intestinal digestive and absorptive sites and so not all the products of microbial activity are available to the animal, although coprophagy may make more available.

In other herbivorous animals, though, the site of microbial activity is before the true stomach and the small intestines and so all products of microbial activity are potentially available for digestion and absorption in the gastro-intestinal tract. This is brought about by development of a compartmented stomach system. In some animals the microbial compartment and the true stomach may be in virtually the same organ, the two compartments differing in the secretory and absorptive linings. In others the microbial compartment may be physically separated from the gastric compartment and this is so to a very marked degree in the ruminant (Figure 1)

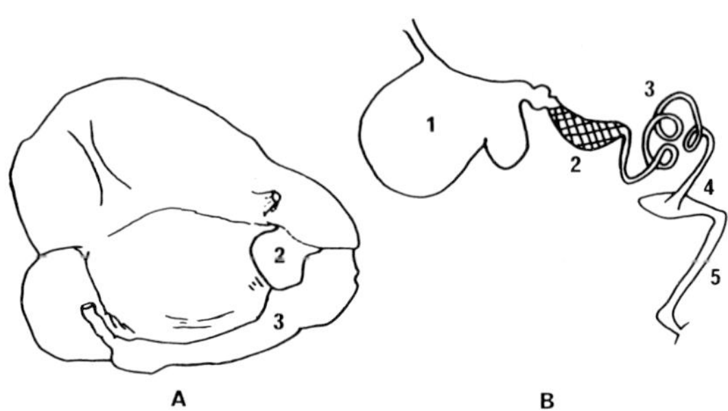

Fig. 1 The stomach system of the ruminant as in the animal (A): (1) rumen-reticulum; (2) omasum; (3) abomasum. The digestive system of the ruminant shown diagrammatically (B): (1) rumen system; (2) abomasum; (3) small intestine; (4) caecum; (5) large intestine.

The importance to man of the ruminant digestive system with its microorganisms lies in the fact that it is found in the commonly domesticated animals, cattle of various kinds, sheep, camels, llamas, and in many of those animals which are semi-domesticated or which man kills for food, e.g. deer and reindeer. Thus much of man's meat, as well as milk, skins and hides, bone and horn implements and ornaments is in effect the product of microbial activity.

The main microbial activity in the ruminant digestive tract takes place in the rumen-reticulum compartment (often referred to just as the 'rumen', as the compartments are virtually one) as fresh food goes directly to the rumen and is immediately subject to microbial action. Some 80% of the plant cellulose (depending on the plant and its maturity) is digested in the rumen, and virtually all the more easily degraded polysaccharides. Some of the residual cellulose and other food material may be digested in the caecum and the acidic fermentation products can be utilized by the animal, but the main cellular components are lost in the faeces.

The rumen is a complex continuous culture; complex not only in its content of both bacteria and protozoa of various activities, but also in its dynamics. There is a flow of 'basal medium' through the system provided to some extent by fresh drinking water but usually in greater amount by water recirculated in the body and entering the rumen as saliva. This saliva contains no enzymes but is a mineral salts solution buffered by phosphate, and by bicarbonate which in the rumen equilibrates with the carbon dioxide of the gas phase. With the ammonia and the volatile fatty acids which are the products of microbial action this provides a buffer solution for microbial growth of pH normally about 6.5. The saliva also contains urea, and urea can enter the rumen through the rumen epithelium. This adds a small amount (in a sheep perhaps some 10% per day) to the food nitrogen available to the bacteria. The saliva also contains some mucus which may have some surface-active properties in the homogenisation of the food. The turnover time of this liquid fraction of the rumen contents varies from about 10 h to 30 h.

Although the saliva provides liquid, minerals and some nitrogen for the microorganisms it provides no energy sources. These and the bulk of the nitrogen, together with some trace elements such as cobalt, sulphur and possibly some vitamins and other growth factors, come from the food of the animal. With the naturally grazing or browsing ruminant the food is leaves and stems torn off in quite large pieces and swallowed with little chewing. These large, fibrous pieces are gradually reduced in size by microbial action aided by regurgitation and chewing ('rumination') until they are small enough in size to leave the rumen: the size of particle which can pass out of the rumen varies with size of animal but is only a few millimetres in dimensions. Large fibrous particles may thus remain in the rumen for some days. It is only with farm feeds of chopped and milled legumes, grasses, hays and cereal grains that small particulate substrates enter the rumen and some of these may have a retention time similar to that of the liquid portion of the rumen contents. This latter, in turn, may be influenced by the salts content of the feed and its particle size.

The food, and so the carbohydrate, nitrogenous sources and other substrates for microbial growth may be available continuously to the animal, as in a grazing, or farm animal fed *ad libitum* (although even here periods of feeding are followed by periods of digestion and rumination) or it may be available only once or twice a day as with some penned farm animals. Different components of the diet may be given at different times, as with hay-concentrates, or grazing with addition of concentrates to the diet of dairy animals.

This variation in the rate of ingestion, and possibly in the type of food ingested, by the animal over a day, added to the fact that although the food may have entered the rumen its constituents may not be immediately available to the microorganisms, induces complex situations with respect to the detailed dynamics of the breakdown of polysaccharides in the rumen. Some aspects of this will now be considered.

The Polysaccharides Available to the Rumen Microorganisms

The natural food of most ruminants consists of leaves and stems of grasses, shrubs and trees. Reindeer consume large amounts of lichens while eating grass, especially in summer [Hobson *et al.*, 1976]. While different species may eat the same plants, they may take different parts of it. For example sheep in Scottish mountains tended to select the soft tips of heather while red deer ate more of the whole plant, including woody stem [Hobson *et al.*, 1976].

Since all these kinds of vegetation provide energy for the animals it follows that the rumen microorganisms can degrade a variety of different types of polysaccharides. Much of the analysis of feedstuffs is done by methods which describe the percentages of 'cellulose', 'hemicellulose', 'starch', etc., but these are in fact heterogeneous fractions defined by the particular method of analysis (solubility or insolubility in detergents, acids or alkalies) and not chemical entities. Nevertheless such analyses have their place, as the percentage of some fraction in various lots of hay, for instance, may be related to the overall digestibility of that hay and its efficiency in promoting animal growth. However, the use of different methods of analysis, or methods which do not define the same material in different plant species, together with differing methods of determining the amounts of material digested (analysis of feed and faeces, incubation of the material in a container in the rumen, incubation *in vitro* with rumen contents whole or fractionated) have probably been the cause of some of the discrepancies in results; e.g. whether cellulose or hemicellulose is degraded faster.

An example of the variety of polysaccharides available to the rumen microorganisms is provided by the detailed analyses of grasses [Waite and Gorrod, 1959]. The analysis varies with the species of grass and the age of the plant, but a young cocksfoot grass contained (% of dry matter): fats, 4.2; waxes, 0.6; organic acids, 6.4; water-soluble carbohydrates, 6.6; phenolic compounds, 1.0; pectin complex, 4.4; crude protein, 15.8; lignin, 6.0; hemicellulose, 14.9; hemicellulose aldobiuronic anhydrides, 2.4; cellulose, 21.8; acetyl, 0.8; ash, 12.2; unidentified and lost, 2.6. The water-soluble carbohydrate contained (% of dried original grass): hexose, 1.4; sucrose, 3.2; fructosan, 1.2; galactan, 0.8. The hemicellulose contained: galactan, 0.7; glucosan, 2.0; araban, 2.2; xylan, 8,2; aldobiuronic xylose, 1.8.

Animals being given concentrate feeds will be getting starch. The starch may be from barley, maize, wheat, oats or potatoes or some other plant. Each starch has a different granule structure and amylose and amylopectin content. The starches are fed as cereal grains or broken tubers and so are accompanied by cellulose and other polysaccharides.

The reindeer ingest lichen polysaccharides. The Scottish reindeer rumens contained a soluble polysaccharide, presumed to come from the lichen, which contained mainly galactose, with traces of glucose, xylose, and ribose (uronic acids and amino sugars were not tested for) [Hobson *et al.*, 1976]. Other lichen polysaccharides are glucans, similar to cellulose, and polyarabitol. Animals whose diet is being supplemented with fish or meat meals will be ingesting some tissue polysaccharides. In addition, bacterial polysaccharides, complex or starch or dextran, will be available from bacterial secretion or lysis.

The Rumen Microorganisms

There is space here for only the briefest outline of the rumen microbiota; more detail is given by Hungate [1966], Hobson [1971] and Bryant [1977]. Here, we wish to emphasize that there is a complex microbiota of some 10^9 bacteria ml^{-1} and up to about 10^6 protozoa ml^{-1}. The total protozoal population is usually made up of mixed genera and species of very different sizes and the largest organisms may be present only as a few thousand or even hundreds per millilitre, the small protozoa making up the bulk of the population.

The main metabolic activity of the rumen population is an anaerobic fermentation. However, some oxygen does get into the rumen and there must be some metabolic activity such as is found at low oxygen tensions with facultative bacteria, as oxygen entering the rumen is 'scavenged' to produce an environment of Eh some -200 mv for the growth and activity of the bacteria and protozoa concerned in the main fermentation pathways. These are all strict anaerobes requiring highly-reduced conditions for growth. Less strict, and facultative anaerobes, and even aerobes, can however be cultured in small numbers from rumen contents. Some may be artifacts from the large numbers of bacteria ingested each day by the animal from food, water and the air, and may not actually be growing in the rumen. Others, such as *Streptococcus bovis* and *Veillonella gazogenes* can multiply, and they have a role in the rumen activities although their numbers are usually only small (10^4-10^6 ml^{-1}) compared with those of the major species (10^8-10^9 ml^{-1}).

The overall reactions in the rumen are shown in Figure 2. These reactions are common to all adult rumens, but one or other may be enhanced or suppressed by the feeding regime of the ruminant. At birth and during the time of milk feeding, the rumen is small and plays no part in the digestive processes. As the animal grows and changes from milk to milk plus solids and then a totally solid diet the rumen develops in size and absorptive epithelium and the predominantly lactobacillary flora of the milk-fed animal is replaced by the mixed flora of the adult; the bacterial and protozoal populations developing as conditions become favourable, from constant inoculation directly from other animal's saliva or faeces, or indirectly via food, water or the air.

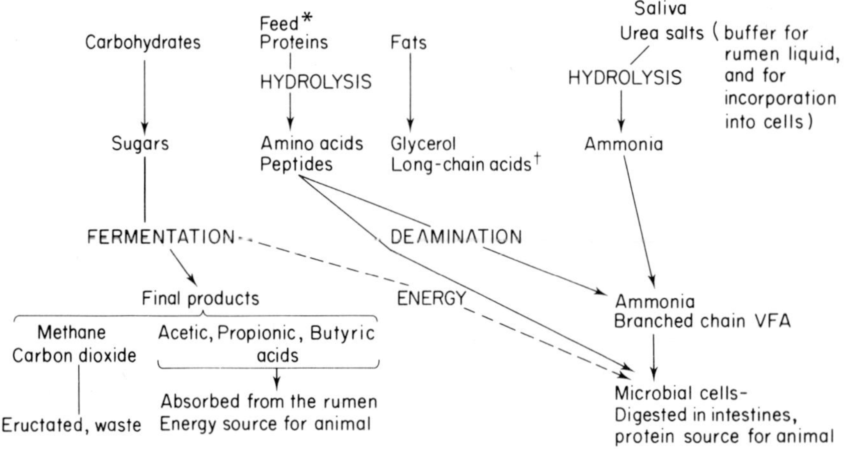

Fig. 2 The main reactions in the rumen. *The feed also contains salts, including trace elements, and sulphur for microbial growth. †Glycerol is fermented. Long-chain fatty acids are hydrogenated if unsaturated. Some of the long-chain fatty acids are incorporated into microbial cells and are available to the animal on digestion of the cells, the rest of the acids pass to the intestines where they are absorbed. VFA, volatile fatty acids.

While utilization of milk by the immature ruminant is the same as that of any other young mammal, utilization of carbohydrate by the adult ruminant is via fermentation to acetic, propionic and butyric acids and the absorption and synthesis from these in the body of glucose and lipid. Nitrogen utilization is via the breakdown of food proteins, the breakdown products along with non-protein nitrogenous compounds being synthesized into microbial cells, the protein of these being digested as the rumen contents pass to the abomasum and intestines.

Carbohydrates are the primary source of the fermentation products required by the ruminant and the fermentations generate the ATP required for the microbial growth which forms the ruminant's protein food.

Since the bacteria and protozoa actually absorb and ferment only mono- and disaccharides, and with a few exceptions carbohydrates in the food are polysaccharides, the primary hydrolysis of polysaccharides is of great importance in the overall rumen reactions and may, in some cases, be the rate-limiting step for the whole system.

The 'few exceptions', apart from the relatively small amounts of sugars contained in herbage, are farm diets where whey, molasses, residues from brewing and distilling and similar materials, or vegetables such as beets, are being fed. In these feeds a large proportion, or all, of the carbohydrate is in the form of sugars, although polysaccharides (vegetable fibres, etc.) may also be present. The presence in all except some defined laboratory feeds of a variety of polysaccharides means that for full utilization of the energy of the feed a similar variety of polysaccharide hydrolases must be present in the rumen contents. While some bacteria produce a number of these enzymes, others, such as some of the starch-fermenting bacteria, are limited in their activities. Cellulose hydrolysis is mainly limited to strains of bacteria from about four or five species of three genera (other cellulolytic bacteria have occasionally been isolated but they seem to be of no general importance). Except for one or two species of starch hydrolysing bacteria, where the hydrolase activity is a specific characteristic, polysaccharide hydrolases may be present or absent in isolates of otherwise similar characteristics; hence the reference to 'strains' of bacteria having cellulolytic activity.

However, whatever the difficulties of classification, there are groups of bacteria in the rumen which are predominantly concerned in hydrolysis of certain polysaccharides, and cultures with differential media can show that these groups become of greater or less dominance in the rumen population as the ruminant's diet is changed [Mackie *et al.*, 1978] say from grass to cereal grain [S.O. Mann, unpublished].

Many of the species of protozoa in the rumen will ingest and ferment starch granules; a few have cellulolytic activity, ingesting and digesting plant fibres [Hungate, 1966].

Various factors affect the polysaccharide hydrolyses. Enzyme activities produced by the bacteria sometimes vary with strain, some strains being designated as 'slow' fermenters some as 'rapid' ones. And in some cases where a complex of enzymes seems to be involved, as in cellulolysis, some bacteria appear to contain the whole complex, some only parts, with consequent differences in their ability to hydrolyse different forms of cellulose.

Factors affecting enzyme activities and actions are discussed in Chapter 11 by Bacon. In this chapter the various factors which can affect bacterial growth and formation of enzymes will be considered.

Factors Involved in Breakdown of Polysaccharides in the Rumen

Colonization of solid feedstuffs

Feed polysaccharides enter the rumen as solids, and in almost every case the solids consist of a number of polysaccharides together with proteins and other constituents of vegetable matter. Once the vegetable material has entered the rumen the action of the warm rumen fluid alone may begin to leach out water-soluble polysaccharides or separate insoluble polysaccharides from each other. An example of the latter is starch granules which will separate from milled grains or chopped potatoes. Such separations are aided by the physical action of mastication which occurs when the solid material is regurgitated for rumination.

However, although soluble polysaccharides become immediately available to bacteria living free in the rumen fluid, the insoluble polysaccharide structures must be colonized by bacteria before they can be degraded.

The colonization of food particle surfaces was noted many years ago by investigators who used light microscopy for observations. Baker [1942], Baker and Harris [1947] and Baker *et al.* [1951] describe bacteria attached to and apparently digesting plant fibres and starch granules with pitting of the plant structure around the bacteria. The early work has been followed by electron microscope studies. Both transmission and scanning electron-microscopy have been used. Examination of thin sections of fixed rumen contents can show bacteria actually 'tunnelling' their way into the plant fibre and can show how some parts of the structure are more easily degraded than others, but it is often difficult or impossible to identify the bacteria seen in thin sections.

Van der Wath and Myburgh [1941] noted coccoid bacteria clustered around starch granules in the rumen, and van der Wath [1948], continuing the studies on starch degradation, later isolated a streptococcus from rumen contents but did not fully characterize it.

At about the same time the author and colleagues were investigating starch breakdown in the rumen and an amylolytic streptococcus was isolated from the rumen and identified as *Streptococcus bovis* [MacPherson, 1953]. This bacterium was shown by serological methods ('capsule swelling' [MacPherson, 1953], fluorescent antibodies [Hobson, Mackay and Mann, 1955; Hobson and Mann, 1957]) to be a true rumen inhabitant and later work has shown its presence in nearly all rumen samples. This bacterium and a *Clostridium butyricum* isolated from the rumen of a maize-fed sheep [Masson, 1951] (but later work has suggested that the bacterium is probably only occasionally of significance in the rumen) were shown to produce an extracellular, constitutive, α-amylase degrading dissolved starch to principally maltose and maltotriose [Hobson and MacPherson, 1952]. Attack by the enzyme, even in high concentration, on starch granules, was slow and the dissolution of the starch granules was shown to originate in either microscopically visible or submicroscopic (shown by a staining method) fissures in the granules. Light ball-milling of potato starch increased the number of 'damaged' granules and the extent of amylase attack. The majority of granules in a normal potato starch sample, which were 'undamaged' according to the tests used, seemed resistant to amylase attack, whereas with smooth pea starch, where extensive granular damage was apparent, over 75% of the granules were attacked by α-amylase [Baker and Hobson, 1952]. This appeared to be very similar to the situation in the rumen where Baker *et al.* [1951] had found potato starch granules to be resistant to bacterial attack when compared with other types of starch.

The amylolytic cocci adhered to starch granules in laboratory culture and it appeared that the proximity of the bacteria to the starch granules might induce very high local concentrations of amylase and that this aided by flaws in the outer surface of the granules could lead to hydrolysis of the granular starch. The experiments of Walker and Hope [1964] suggested a link between the adsorption of amylases, including those from *Streptococcus bovis* and *Clostridium butyricum*, on maize starch granules and their ability to hydrolyse granular starch. More recently McWethy and Hartman [1977] purified an extracellular α-amylase from the rumen *Bacteroides amylophilus* and showed that it had an ability to attack maize and potato starch granules similar to that of *Streptococcus bovis* and *Clostridium butyricum*, but the amylase did not adsorb on to starch granules. This bacterium in pure culture also adheres to maize starch granules [Hamlin and Hungate, 1956].

Akin *et al.* [1974] showed that some parts of grass tissue were apparently degraded without attachment of bacteria, but that other parts were degraded by bacteria attached to the plant cell walls by an electron-dense, extracellular material. Patterson *et al.* [1975] and Latham *et al.* [1978] extended electron microscopy studies to examination of two particular species of cellulolytic bacteria, *Ruminococcus albus* and *Ruminococcus flavefaciens* Latham *et al.* [1978] used both transmission and scanning electron microscopy and showed that *Ruminococcus flavefaciens* adhered to cotton or grass cell walls, but that adhesion varied with different parts of the plant.

Colonization of the cotton (hammer-milled filter paper) was greatest at the fractured ends of fibres, and with the plant cells cut or damaged cell walls were major sites for attachment of the bacteria. This has obvious similarities to the amylase action on damaged portions of starch granules previously mentioned, since heavy growth of the cellulolytic bacteria presupposes a ready availability of fermentable substrates from hydrolysed cellulose.

Although adhesion of *Ruminococcus flavefaciens* seemed to depend to a large extent on damage to the plant cell walls as previously mentioned, there was variability in adhesion with type of plant cell. Adhesion was more rapid to cells of the epidermis and sclerenchyma than to phloem and mesophyll. The bacteria did not readily adhere to the bundle sheath, metaxylem or protoxylem.

Variable adherence to different cell walls seems to be characteristic of the pure cultures and the mixed bacteria of rumen contents which have been studied and parallels earlier studies which showed that rumen bacteria attack different parts of grass or straw in succession [Hanna *et al.*, 1973; Akin *et al.*, 1973; Kawamura *et al.*, 1973]. The electron microscope observations show that lignin in plant tissues inhibits attachment of bacteria, and this is in accord with much previous work, as an inverse relationship between extent of lignification and digestibility, as shown by animal feeding experiments or tests with rumen contents *in vitro*, has long been accepted by workers on ruminant nutrition. Lignin may not be the only factor affecting fibre digestion; plant waxes (suggested by Johnston and Waite [1965] who found that the relationship between lignin and digestibility varied with different grasses) or silica [Smith *et al.*, 1971] may also slow down or prevent digestion.

Hobson and MacPherson [1953] showed by light microscopy that a large proportion of the rumen bacteria were encapsulated and that the largest capsules appeared on the bacteria from a hay-fed sheep. The capsular material of the amylolytic *Streptococcus bovis* was a polysaccharide composed of galactose, rhamnose and uronic acid (probably galacturonic) units with a molecular weight about 90,000 [Hobson and MacPherson, 1953b, 1954; Greenwood, 1954]. Recent investigations using electron microscopy have confirmed the fact that rumen bacteria, in the rumen fluid or grown in pure culture, generally have an outer 'coat' or capsule and have added more details of the coat morphology [Cheng and Costerton, 1975; Costerton *et al.*, 1974].

Adhesion of bacteria to plant cells seems to be by means of a 'capsule' layer of greater or lesser thickness and this layer has been identified by staining as polysaccharide-containing. Latham *et al.* [1978] showed that this 'adhesion' material on *Ruminococcus flavefaciens* contained rhamnose and galactose, and (although uronic acids were not tested for) this obviously suggests a similar polysaccharide to that of the capsule of the starch-adherent *Streptococcus bovis* previously mentioned [Hobson and MacPherson, 1954]. The plant substances (e.g. lignin) which prevent digestion of the plant cells also seem to prevent adherence of the bacteria. These substances are virtually undegradable by the fibre-digesting bacteria, and presumably prevent adhesion of the bacteria by not allowing them to obtain carbohydrate substrate for synthesis of the polysaccharide capsule.

Adherence of bacteria to surfaces is not unique to rumen contents; it occurs on surfaces in all kinds of aqueous habitats, and bacteria adhere to the rumen walls where they digest dead epithelial cells (a recent study with scanning and transmission electron micrographs is that of McGowan *et al.* [1978]).

However, it seems that adherence to the substrate is a prerequisite for degradation of most polysaccharides by the rumen bacteria. Whether the initial adherence is by chance or by chemotaxis is difficult to say; one might suppose that if chemotaxis were involved, a damaged plant cell or starch granule might be a source of slowly-solubilizing carbohydrate which would attract bacteria to these sites first. Recent work has shown that chemotaxis to sugars in solution appears to be a motive force behind movement of the rumen ciliates *Isotricha intestinalis* and *Isotricha prostoma* to plant particles where they attach [Orpin and Letcher, 1978].

There are many factors to be considered in the adhesion of bacteria to surfaces. Electrostatic charges may be one, and studies which could be relevant to rumen organisms have been made with marine bacteria. However, for the purposes of this paper, what is important is that the polysaccharide substrates in the rumen are generally solids and that their hydrolysis is generally brought about by bacteria which adhere to and colonize the solids. The microbial concentration in the rumen is high and colonization of food particles entering the rumen seems to be quite rapid. So probably the chance of initial contact between a particulate substrate and a bacterium which can initiate a colony is not a rate-controlling factor in the overall breakdown of that substrate.

Sequential degradation of substrates

It has already been mentioned that the bacteria appear to digest plant cells in sequence and a factor controlling the rate of degradation of one plant polysaccharide may be the prior rate of removal of another polysaccharide preventing access to the first. Although amylase activity is found in some strains of a number of the rumen bacterial species, the principal starch-hydrolysing species have little, if any, ability to hydrolyse other polysaccharides. This should be no disadvantage as starch occurs in granules which are usually readily liberated by the physical disintegration of the plant material. This is the case when a ground and pelleted barley is being fed. The barley grains are broken by the processing and starch granules are liberated and readily fermented. Cellulolytic activity is low in this case. On the other hand, when whole barley grains are fed rumen fermentation is less rapid and the numbers of cellulolytic bacteria are higher [Mann and Ørskov, 1975]. Presumably liberation and rapid fermentation of starch cannot take place until the cellulose outer layers of the cereal grain have been digested by the cellulolytic bacteria.

Since plant leaves and stems, on the other hand, are complex structures it might be expected that bacteria with a number of polysaccharide hydrolases might be best fitted for attack on these, and in general this is found to be the case. For instance, bacteria isolated on semi-selective xylan media fermented various other polysaccharides, in some cases including cellulose or inulin [Hobson and Purdon, 1961; Dehority, 1966]. The bacteria primarily concerned in cellulose hydrolysis, *Ruminococcus albus*, *Ruminococcus flavefaciens*, *Bacteroides succinogenes* and *Butyrivibrio* spp., often also ferment xylan and pectin. However, the ability to hydrolyse a hemicellulose or pectin may not be detected in the usual fermentation tests, as it has been found that bacteria may have the ability to degrade xylans or pectins to oligosaccharides, and so solubilize them, without the ability to ferment the hydrolytic products [Dehority, 1973; Gradel and Dehority, 1972]. Some of the protozoa appear to be similar in this respect; for instance *Isotricha* although able to hydrolyse pectins does not ferment the hydrolysis products [Abou Akkada and Howard, 1961]. In some cases such activity may enable an organism to expose a fermentable substrate in a plant fibre, as in the case of some cellulolytic bacteria which could degrade xylan to soluble oligosaccharides but could not utilize the products [Dehority, 1973]. In other cases pure culture studies suggest that symbiosis or synergism between two or more species of bacteria may exist in the rumen. For example the presence of non-cellulolytic, but hemicellulolytic, *Bacteroides ruminicola* increased cellulose breakdown in intact forages by various strains of cellulolytic bacteria [Dehority and Scott, 1967], and Coen and Dehority [1970] found that a cellulolytic strain of *Ruminococcus flavefaciens* could degrade the hemicellulose of intact brome grass fibres to oligosaccharides but could not utilize the products. A strain of *Bacteroides ruminicola* could not degrade the hemicellulose but could ferment hydrolytic products. Individually neither could utilize the hemicellulose, but in mixed culture almost complete utilization occurred. Similar effects were shown with pectins [Gradel and Dehority, 1972], and a similar situation seems to exist with starch. While most starch-hydrolysing bacteria can ferment the hydrolytic products, *Succinivibrio dextrinsolvens* seems able to utilize only dextrins and maltose, so it must normally coexist with an organism that can hydrolyse the whole starch of feedstuffs. A different kind of symbiosis was suggested by experiments of Hobson and Fina (unpublished). An amylolytic strain of

Selenomonas ruminantium could not grow alone on granular starch, but could grow together with *Streptococcus bovis* which could degrade the starch granules. The selenomonad seemed to be growing mainly on dissolved starch or starch hydrolysis products provided by the attack of the streptococcus on the starch granules and not on lactic acid formed by the streptococcus.

There is, also, a further symbiosis in that the number of bacteria which can ferment mono- and disaccharides is probably in excess of those which can degrade the polysaccharides of any particular feed, so that many bacteria in the rumen fluid must be 'scavenging' sugars released by the hydrolytic bacteria. Such an action has been demonstrated *in vitro* by Scheifinger and Wolin [1973] who grew the non-cellulolytic, but cellobiose-fermenting *Selenomonas ruminantium* with the cellulolytic *Bacteroides succinogenes*.

There may be other symbiotic or multienzyme reactions not involving hydrolysis of linkages between monosaccharides in the polysaccharides, and these may be responsible for the differences in ability of bacteria to utilize, for instance, the hemicellulose of different grasses or grasses and legumes. The ability to hydrolyse side-chain linkages, for instance acetyl groups in hemicelluloses [Morris and Bacon, 1976] or methyl esters in pectins [Howard, 1961], may be important in exposing the main polysaccharide chain to attack.

These are only some of the factors which must affect the rate at which polysaccharides are digested in the rumen. But even if a bacterium, or bacteria, with the correct enzymic activities is present then various other factors governing bacterial growth and enzyme production come into play, and these will be discussed in the next sections. The bacteria attached to feed particles will be growing as microcolonies on a surface. Even inert surfaces can cause ionic and other effects and here the surface is not inert for the bacteria, so it is difficult to say under exactly what conditions the bacteria are growing [Hobson, 1971]. In the following discussion factors influencing growth of bacteria in culture *in vitro*, usually on dissolved substrates, are mentioned. These factors must influence growth in the rumen but their effects may be modified because of the heterogeneous nature of the rumen contents.

Nutritional factors affecting growth of the bacteria

Like bacteria in other habitats, the rumen bacteria require energy and sources of nitrogen, sulphur, various cations and anions including phosphate, vitamins and other growth factors. It is impossible to more than briefly indicate the nutritional requirements of the rumen bacteria here, more details and references to original papers are given in the reviews already referred to and by Hobson [1979]. What is of relevance here, though, is to point out that while minerals are provided in the animal's food and, with minerals recirculating in the saliva, are in solution in the rumen fluid and so always available to the microorganisms, the other growth requirements are largely the products of microbial action and their availability depends on the growth and activity of particular microorganisms. Availability may also depend on spatial relationships of bacteria, as the probability of a bacterium obtaining a growth factor from another must be greatly increased if the bacteria are in proximity.

The principal nitrogen source for the bacteria is ammonia. This is produced in a two-stage microbial reaction from feed proteins by hydrolysis and deamination, or by hydrolysis of urea entering the rumen in saliva, or is a component of feedstuffs. Actively proteolytic, deaminative, or ureolytic bacteria are a minority of the rumen species. However, with farm feeds or good pasture, ammonia is present in the rumen liquid in higher concentrations than seems necessary purely for maximum bacterial growth rate in a continuous culture system and should be available to all the bacteria, fixed or free. The total amount of microbial growth could be limited by the total amount of nitrogen in the feed, or by the total amount of available carbohydrate.

Overall growth of the bacteria may also be limited by availability of sulphur, as sulphate or S-amino acids, in some feeds, or even by sodium (rumen bacteria seem to be mildly halophilic) or other ions.

The growth of particular species of cellulolytic bacteria could be limited by the availability of the branched-chain volatile fatty acids (products of amino acid deamination) which the bacteria require for synthesis of cellular amino acids from ammonia.

These are only some of the factors involved in the total growth, and so enzymic activity in the rumen. For good activity a good feed is required; poor feed such as winter mountain pasture leads to low numbers of microorganisms and low overall activities [Hobson *et al.*, 1976]. But even if the feed is adequate there must be the correct mixed bacterial population for all the actions and interactions, as has been demonstrated in experiments with gnotobiotic lambs where an adequate, defined population for fibre digestion has not yet been attained [Hobson *et al.*, unpublished; Lysons *et al.*, 1976; Mann and Stewart, 1974].

Continuous culture experiments *in vitro* have demonstrated some of these interactions. Hobson and Summers [1978] demonstrated the ability of a non-ureolytic but ammonia-utilizing *Bifidobacterium* to grow in continuous culture in a urea medium in the presence of very much smaller numbers of either a ureolytic *Peptostreptococcus provotii* or *Streptococcus faecium*, as in the rumen. Hoover and Lipari

[1971] grew mixed continuous cultures of *Bacteroides ruminicola* and the cellulolytic *Ruminococcus flavefaciens* where *Bacteroides ruminicola* hydrolysed casein to provide methionine essential for growth of *Ruminococcus flavefaciens* on the ammonia-N in the medium.

Multiple substrates, removal of fermentation products

Since, with the exception of some amylolytic and acid-fermenting bacteria, most of the rumen bacteria will utilize a number of sugars or acids in addition to products of polysaccharide hydrolysis, possibilities exist of diauxic growth or growth on a simple substrate in preference to slowly-degraded polysaccharide. As an example strains of *Selenomonas ruminantium* var. *lactylitica* will grow on starch or lactate separately and when both substrates are present it will utilize both together. However, with glucose plus lactate, lactate utilization is suppressed. These experiments [Hobson, 1972] and experiments of Hishinuma *et al.* [1968] showed that sugars in high concentration were preferentially used, but that in low concentration, as produced by slow starch hydrolysis, lactate was fermented as well. Since lactic acid is an undesirable end product of carbohydrate fermentations, lowering the rumen pH to inhibit many organisms and being toxic to the animal, inhibition of its removal by high concentrations of sugars, added as such to the diet or from easily hydrolysed, processed polysaccharides, could help to lead to overall inhibition of feed polysaccharide utilization.

The results of van Gylswyck and Labuschagne [1971], who found that the rate of hydrolysis was not growth-rate limiting for ruminococci growing on cellulose but was for *Butyrivibrio*, suggest that *Butyrivibrio* (which have wide fermentative activity) will preferentially grow on simple sugars, if these are present, and not on cellulose.

The end product of cellulose hydrolysis, cellobiose, itself inhibits cellulase formation by ruminococci [Fusee and Leatherwood, 1972]. So, although competition for the products of cellulose hydrolysis by non-cellulolytic bacteria adjacent to ruminococci may tend to limit the growth of the cocci, it could also help to increase cellulolysis and degradation of plant fibres.

In the case of starch hydrolysis, competition by protozoa, which rapidly ingest free starch granules and remove them from bacterial action, can alter the overall rates of feed polysaccharide fermentation in the rumen [Whitelaw *et al.*, 1970].

The removal of the potentially inhibitory fermentation product, lactate, was mentioned above, but the removal of another fermentation product, hydrogen, has been the subject of much recent work. A number of combinations of hydrogen-producing and -utilizing bacteria have been tested, but one more relevant to the present discussion is the growth of *Ruminococcus flavefaciens* on cellulose together with *Methanobacterium ruminantium* [Latham and Wolin, 1977]. As in other cases, utilization of hydrogen (with CO_2) by the *Methanobacterium* altered the *Ruminococcus* fermentation products towards acetic acid. But decrease in formation of possibly inhibitory 'reduced' acids, such as lactate, and the removal of hydrogen itself, a possibly inhibitory product, can also increase bacterial growth and rate of carbohydrate breakdown in the rumen.

Effects of growth conditions and growth rate

The effects described above may apply only in the microhabitat around a feed particle, or may affect the entire rumen population. Similarly, although the effects to be next discussed may apply to the rumen as a whole they may also apply particularly to microhabitats where conditions are unlike the overall ones. The pH of the whole rumen contents can be lowered by the feeding of rapidly-fermented carbohydrate. Not only have the polysaccharide hydrolases a limited pH range for activity (e.g. *Ruminococcus albus* cellulase has a pH range of 6.0-6.8 [Smith *et al.*, 1973]), but the rumen organisms themselves have in general a fairly limited pH range for growth [Hobson, 1972], and Stewart [1977] found that with diluted rumen contents incubation at pH 6.9 gave 10^6 cellulolytic bacteria ml^{-1} whereas at pH 6.0 only 10^3 bacteria ml^{-1} were present.

Enzymes hydrolysing polymers, often solids, must obviously be external to the cell. The amylase of *Bacteroides amylophilus* is about 50% free in the medium and 50% cell-bound [Blackburn, 1965]. However, all the enzyme is available to the substrate. The cellulase of *Ruminococcus albus* seems to be similar [Smith *et al.*, 1973]. Close proximity of cell and solid substrate will allow more enzyme activity to be used and also tend to lessen loss of enzyme activity by denaturation or destruction by proteolytic enzymes from other bacteria. However, just as pH affects growth of the bacteria, the environmental pH affects enzyme production. For example, specific amylase production of *Bacteroides amylophilus* was maximum at pH 6, about in the middle of the pH range for growth of the bacteria [Hobson and Summers, 1967; Henderson *et al.*, 1969].

These workers also found that the *Bacteroides amylophilus* protease production was more uniform than that of the amylase over most of the pH range for bacterial growth. Proteolytic and deaminative activities affect the supply of nitrogen and thus the total bacterial growth. Also the nitrogen sources of the bacteria are generally polymers as well and so are not immediately available as N-sources. Thus, whether the bacteria are growing under C- or N-limited conditions depends more on the rates and extents of hydrolysis and deamination of feed proteins compared with the hydrolyses of feed polysaccharides, rather than on the absolute ratios of C to N in the feeds.

The effect of C- or N-limitation on polysaccharide hydrolysis was shown by Henderson *et al.* [1969] who found that, at the same growth rate, the amylase activity of *Bacteroides amylophilus* decreased by a factor of about 13 on changing from C- to N-limitation. This might be brought about by the bacterium diverting the limited nitrogen to synthesis of more essential cell components than the amylase, but under the continuous culture conditions there was an argument for some degree of inhibition of amylase synthesis by excess maltose present in the N-limited culture. The protease activity decreased only by a factor of about 2.7 under the same change in conditions.

However, whether C- or N-limited growth conditions prevail the synthesis of extracellular, hydrolytic enzymes is affected by the growth rate of the bacteria. The specific amylase and protease activities of *Bacteroides amylophilus* showed a rather complex variation with dilution rate, but both increased rapidly towards the lowest growth rates [Hobson and Summers, 1967; Henderson *et al.*, 1969]. A high amylase activity per cell could possibly be of some ecological significance in helping to compensate for the low cell yields occurring at low growth rates.

Conclusion

In a short chapter such as this it is not possible to examine every subject in detail. What has been attempted is to show some of the various factors which may influence polysaccharide degradation in a microbial habitat which contains a very mixed population of microorganisms and a heterogeneous collection of polysaccharides in many physical states.

Such a situation is very difficult to analyse; the original is complex, but if it is simplified vital parts may be lost. The prepared polysaccharides used in solution as substrates for isolating or counting bacteria with particular hydrolases may be different in chemical structure from those in the intact plant. Hydrolysis of the natural substrate may demand another bacterium or another enzyme capable of removing groupings substituent to the main polysaccharide and which are chemically removed during preparation of the isolated material. Different preparations of solid substrates (a well-known one is cellulose) may vary considerably in physical structure and so in rate and extent of attack by different bacteria, all nominally 'cellulose-fermenting'.

Diauxic growth and other factors mentioned previously may affect different bacteria in different ways. The behaviours of mixed cultures of different species, with nominally the same enzyme activities, may differ considerably.

The actions and interactions described here are not unique to the rumen. Observations suggest that similar considerations apply to the behaviour of all microbial habitats. The pure culture is a laboratory artefact and only more experiments will finally define the life of microorganisms in their natural situation, the mixed culture.

References

Abou Akkada, A.R. and Howard, B.H. (1961). The biochemistry of rumen protozoa. 4. Decomposition of pectic substances. *Biochemical Journal* **78**, 512-517.

Akin, D.E., Amos, H.E., Barton, F.E. and Burdick, D. (1973). Rumen microbial degradation of grass tissue revealed by scanning electron microscopy. *Agronomy Journal* **65**, 825-828.

Akin, D.E., Burdick, D. and Michaels, G.E. (1974). Rumen bacterial interrelationships with plant tissue during degradation revealed by transmission electron microscopy. *Applied Microbiology* **27**, 1149-1156.

Baker, F. (1942). Normal rumen microflora and microfauna of cattle. *Nature, London* **149**, 220.

Baker, F. and Harris, S.T. (1947). Microbial digestion in the rumen (and caecum), with special reference to the decomposition of structural cellulose. *Nutrition Abstracts and Reviews* **17**, 3-12.

Baker, F. and Hobson, P.N. (1952). The selective staining of intact and damaged starch granules by safranin O and niagara blue 4B. *Journal of the Science of Food and Agriculture* **3**, 608-612.

Baker, F., Nasr, H., Morrice, F. and Bruce, J. (1951). Bacterial breakdown of structural starches and starch products in the digestive tract of ruminant and non-ruminant animals. *Journal of Pathology and Bacteriology* **62**, 617-638.

Blackburn, T.H. (1965). Protease Production by *Bacteroides amylophilus*, a Rumen Bacterium. Ph.D. Thesis, University of Aberdeen.

Bryant, M.P. (1977). Microbiology of the rumen. In *Duke's Physiology of Domestic Animals*, 9th ed. Edited by M.J. Stevensen. Ithaca: Cornell University Press.

Cheng, K.J. and Costerton, J.W. (1975). Ultrastructure of cell envelopes of bacteria of the bovine rumen. *Applied Microbiology* **29**, 841-849.

Coen, J.A. and Dehority, B.A. (1970). Degradation and utilisation of hemicellulose from intact forages by pure cultures of rumen bacteria. *Applied Microbiology* **20**, 362-368.

Costerton, J.W., Damgaard, H.N. and Cheng, K.J. (1974). Cell envelope morphology of rumen bacteria. *Journal of Bacteriology* **118**, 1132-1143.

Dehority, B.A. (1966). Characterization of several bovine rumen bacteria isolated with a xylan medium. *Journal of Bacteriology* **91**, 1724-1729.

Dehority, B.A. (1973). Hemicellulose digestion by rumen bacteria. *Federation Proceedings* **32**, 1819-1825.

Dehority, B.A. and Scott, H.W. (1967). Extent of cellulose and hemicellulose digestion in various forages by pure cultures of rumen bacteria. *Journal of Dairy Science* **50**, 1136-1141.

Fusee, M.C. and Leatherwood, J.M. (1972). Regulation of cellulase from *Ruminococcus*. *Canadian Journal of Microbiology* **18**, 347-353.

Gradel, C.M. and Dehority, B.A. (1972). Fermentation of isolated pectin and pectin from intact forages by pure cultures of rumen bacteria. *Applied Microbiology* **23**, 332-340.

Greenwood, C.T. (1954). A physico-chemical examination of the capsular polysaccharide from an amylolytic sheep rumen streptococcus. *Biochemical Journal* **57**, 151-153.

Gylswyck, N.O. van and Labuschagne, J.P.L. (1971). Relative efficiency of pure cultures of different species of cellulolytic rumen bacteria in solubilizing cellulose *in vitro*. *Journal of General Microbiology* **66**, 109-113.

Hamlin, L.J. and Hungate, R.E. (1956). Culture and physiology of a starch-digesting bacterium (*Bacteroides amylophilus* n.sp.) from the bovine rumen. *Journal of Bacteriology* **72**, 548-554.

Hanna, W.W., Monson, W.G. and Burton, G.W. (1973). Histological examination of fresh forage leaves after *in vitro* digestion. *Crop Science* **13**, 98-102.

Henderson, C., Hobson, P.N. and Summers, R. (1969). The production of amylase, protease and lipolytic enzymes by two species of anaerobic rumen bacteria. In *Continuous Cultivation of Microorganisms*, pp.189-204. Edited by I. Málek *et al.* Prague: Academia.

Hishinuma, F., Kanegasaki, S. and Takaheshi, H. (1968). Ruminal fermentation and sugar concentrations. *Agricultural and Biological Chemistry, Japan* **32**, 1325-1330.

Hobson, P.N. (1971). Rumen microorganisms. *Progress in Industrial Microbiology* **9**, 41-77.

Hobson, P.N. (1972). Physiological characteristics of rumen microbes and their relationship to diet and fermentation patterns. *Proceedings of the Nutrition Society* **31**, 135-139.

Hobson, P.N. (1979). The nutrition of rumen bacteria. In *Handbook of Nutrition and Food*. Edited by M. Recheigl. Cleveland: CRC Press.

Hobson, P.N., Mackay, E.S.M. and Mann, S.O. (1955). The use of fluorescent antibody in the identification of rumen bacteria *in situ*. *Research Correspondence* **8**, 30-31.

Hobson, P.N. and MacPherson, M.J. (1952). Amylases of *Clostridium butyricum* and a *Streptococcus* isolated from the rumen of the sheep. *Biochemical Journal* **52**, 671-679.

Hobson, P.N. and MacPherson, M.J. (1953a). Encapsulation in rumen bacterial fractions. *Nature, London* **171**, 129.

Hobson, P.N. and MacPherson, M.J. (1953b). A serologically active polysaccharide from the capsules of amylolytic streptococci. *Proceedings of the Biochemical Society* **53**, xxxviii.

Hobson, P.N. and MacPherson, M.J. (1954). Some serological and chemical studies on materials extracted from an amylolytic streptococcus from the rumen of the sheep. *Biochemical Journal* **57**, 145-151.

Hobson, P.N. and Mann, S.O. (1957). Some studies on the identification of rumen bacteria with fluorescent antibodies. *Journal of General Microbiology* **16**, 463-471.

Hobson, P.N., Mann, S.O. and Summers, R. (1976). Rumen microorganisms in red deer, hill sheep and reindeer in the Scottish Highlands.

Proceedings of the Royal Society of Edinburgh B **75**, 171-180.
Hobson, P.N., Mann, S.O., Summers, R. and Staines, B.W. (1976). Rumen functions in red deer, hill sheep and reindeer in the Scottish Highlands. *Proceedings of the Royal Society of Edinburgh B* **75**, 181-198.
Hobson, P.N. and Purdon, M.R. (1961). Two types of xylan-fermenting bacteria from the sheep rumen. *Journal of Applied Bacteriology* **24**, 188-193.
Hobson, P.N. and Summers, R. (1967). The continuous culture of anaerobic bacteria. *Journal of General Microbiology* **47**, 53-65.
Hobson, P.N. and Summers, R. (1978). Anaerobic bacteria in mixed cultures. Ecology of the rumen and sewage digesters. In *Techniques for the Study of Mixed Populations*. Edited by R. Davies and D. Lovelock. London: Academic Press.
Hoover, W.H. and Lipari, J.J. (1971). Pure and mixed continuous culture of two rumen anaerobes. *Journal of Dairy Science* **54**, 1662-1668.
Howard, B.H. (1961). Fermentation of pectin by rumen bacteria. *Proceedings of the Nutrition Society* **20**, xxiv.
Hungate, R.E. (1966). *The Rumen and its Microbes*. London and New York: Academic Press.
Johnston, M.J. and Waite, R. (1965). Studies in the lignification of grasses. I. Perennial rye-grass (S24) and cocksfoot (S37). *Journal of Agricultural Science* **64**, 211-219.
Kawamura, O., Senshu, T., Horiguchi, M. and Matsumoto, T. (1973). Histochemical studies on the rumen digestion of rice straw cell wall and on the chemical determination of its non-nutritive residue. *Tohoku Journal of Agricultural Research* **24**, 183-189.
Latham, M.J., Brooker, B.E., Pettipher, G.L. and Harris, P.J. (1978). *Ruminococcus flavefaciens* cell coat and adhesion to cotton cellulose and to cell walls in leaves of perennial ryegrass (*Lolium perenne*). *Applied and Environmental Microbiology* **35**, 156-165.
Latham, M.J. and Wolin, M.J. (1977). Fermentation of cellulose by *Ruminococcus flavefaciens* in the presence and absence of *Methanobacterium ruminantium*. *Applied and Environmental Microbiology* **34**, 297-301.
Lysons, R.J., Alexander, T.J.L., Welstead, P.D., Hobson, P.N., Mann, S.O. and Stewart, C.S. (1976). Defined bacterial populations in the rumens of gnotobiotic lambs. *Journal of General Microbiology* **94**, 257-269.
Mackie, R.I., Gilchrist, F.M.C., Roberts, R.M., Hannah, P.E. and Schwartz, H.M. (1978). Microbiological and chemical changes in the rumen during the stepwise adaptation of sheep to high concentrate diets. *Journal of Agricultural Science* **90**, 241-254.
McGowan, R.P., Cheng, K.J., Bailey, C.B.M. and Costerton, J.W. (1978). Adhesion of bacteria to epithelial cell surfaces within the reticulo rumen of cattle. *Applied and Environmental Microbiology* **35**, 149-155.
MacPherson, M.J. (1953). Isolation and identification of amylolytic streptococci from the rumen of the sheep. *Journal of Pathology and Bacteriology* **66**, 95-102.
McWethy, S.J. and Hartman, P.A. (1977). Purification and some proper-

ties of an extracellular alpha-amylase from *Bacteroides amylophilus*. *Journal of Bacteriology* **129**, 1537-1544.

Mann, S.O. and Ørskov, E.R. (1975). The effect of feeding whole or pelleted barley to lambs on their rumen bacterial populations and pH. *Proceedings of the Nutritional Society* **34**, 63A.

Mann, S.O. and Stewart, C.S. (1974). Establishment of a limited rumen flora in gnotobiotic lambs fed on a roughage diet. *Journal of General Microbiology* **84**, 379-382.

Masson, M. (1951). Microscopic studies of the alimentary micro-organisms of the sheep. *British Journal of Nutrition* **4**, viii-ix.

Morris, E.J. and Bacon, J.S.D. (1976). The digestion of acetyl groups and cell-wall polysaccharides of grasses in the rumen. *Proceedings of the Nutritional Society* **35**, 94-95A.

Orpin, C.G. and Letcher, A.J. (1978). Some factors controlling the attachment of the rumen holotrich protozoa *Isotricha intestinalis* and *Isotricha prostoma* in plant particles *in vitro*. *Journal of General Microbiology* **106**, 33-40.

Patterson, H., Irvin, R., Costerton, J.W. and Cheng, K.J. (1975). Ultrastructure and adhesion properties of *Ruminococcus albus*. *Journal of Bacteriology* **122**, 278-287.

Scheifinger, C.C. and Wolin, M.J. (1973). Propionate formation from cellulose and soluble sugars by combined cultures of *Bacteroides succinogenes* and *Selenomonas ruminantium*. *Applied Microbiology* **26**, 789-795.

Smith, G.S., Nelson, A.B. and Boggino, E.J.A. (1971). Digestibility of forages *in vitro* as affected by content of 'silica'. *Journal of Animal Science* **33**, 466-471.

Smith, W.R., Yu, I. and Hungate, R.E. (1973). Factors affecting cellulolysis by *Ruminococcus albus*. *Journal of Bacteriology* **114**, 729-737.

Stewart, C.S. (1977). Factors affecting the cellulolytic activity of rumen contents. *Applied and Environmental Microbiology* **33**, 497-502.

Waite, R. and Gorrod, A.R.N. (1959). The comprehensive analysis of grasses. *Journal of the Science of Food and Agriculture* **10**, 317-326.

Walker, G.J. and Hope, P.M. (1964). Degradation of starch granules by some amylolytic bacteria from the rumen of sheep. *Biochemical Journal* **90**, 398-408.

Wath, J.G. van der (1948). Studies on the alimentary tract of merino sheep in South Africa. 11. Digestion and synthesis of starch by ruminant bacteria. *Onderstepoort Journal of Veterinary Science* **23**, 367-383.

Wath, J.G. van der and Myburgh, S.J. (1941). Studies in the alimentary tract of merino sheep in South Africa. VI. The role of infusoria in ruminal digestion with some remarks on ruminal bacteria. *Onderstepoort Journal of Veterinary Science* **17**, 61-88.

Whitelaw, F.G., Hyldgaard-Jensen, J., Reid, R.S. and Kay, M.G. (1970). Volatile fatty acid production in the rumen of cattle given an all-concentrate diet. *British Journal of Nutrition* **24**, 179-195.

Chapter 17

POLYSACCHARIDE DEGRADATION IN ESTUARIES

N.J. POOLE* and D.J. WILDISH*

Department of Microbiology, University of Aberdeen, Aberdeen, UK

Introduction

The exact pathway by which a polysaccharide is degraded will depend on a large number of both extrinsic and intrinsic factors, including:

1. the chemical and physical nature of the polysaccharide, and any compounds associated with the polysaccharide which interfere with degradation;
2. the physical structure of the estuary, which determines where and for how long the polysaccharide is retained within the estuary;
3. the biological and chemical characteristics of the estuary, such as the concentration of nutrients, oxygen and toxic chemicals, as well as the activity of the microbial and animal communities.

A conceptual framework for investigating polysaccharide degradation is presented in the first part of this chapter. Since the actual process of polysaccharide degradation is complex the discussion is restricted to considering the possible role of microbial/animal relationships and the degradation by anaerobic microbes. The second part considers some of the effects of polysaccharide degradation by outlining the reaction of a shallow estuary which is stressed by the presence of excess cellulose.

The Estuarine Environment

In order to understand polysaccharide degradation within estuaries it is necessary to consider first those factors unique to this environment.

Estuaries are areas, usually semi-enclosed by land, where the freshwater from a river meets the salt water from

---

**For present addresses see List of Contributors.*

the sea. A characteristic feature of an estuary, and a dominant factor controlling the distribution of its fauna and flora, is the gradient of fresh to salt water (oceanic seawater has a salinity of 35°/oo). Distribution of seawater in an estuary is controlled by the discharge rate of freshwater in relation to the topography and volume capacity of the estuary, as well as by the volume of seawater which enters on each tide. Based on salinity distributions and circulation patterns the estuary may be classified as "vertically homogeneous", "salt wedge" - or "partially mixed" [see Poole *et al.*, 1978].

An important factor in studying the degradation of polysaccharides is their transport into and out of estuaries as dissolved, colloidal or particulate forms. Because polysaccharide particles have densities lower than the inorganic sedimentary particles (which usually have a density of 2.6 g cm^{-3}), they may be differentially eroded, transported and deposited. Deposition occurs in specific parts of the estuary, and when currents slacken at specific times of the tidal cycle. The process of flocculation of inorganic particles greater than colloidal size decreases their settling rate. Flocculation occurs as a result of particle collision and aggregation, the floccules having a density, 1.2 - 1.8 g cm^{-3}, which is usually less than that of the original particles, and a surface area which is greater. On the other hand, flocculation of colloidal particles will result in an increase in their rate of settling. Controlling factors for flocculation include the quantity and type of suspended solids, the temperature and the salinity [Dyer, 1972]. The presence of dissolved complex carbohydrates at concentrations above 0.5 mg l^{-1} decreases the settling rates of floccules by up to 40% [Whitehouse *et al.*, 1960]. The process of floccule disaggregation occurs where there is intensive turbulence, or where fresh and salt water mixing occurs, and results in an increased settling rate of particles above colloidal size.

The circulation patterns in many estuaries with a seawater intrusion cause the estuary to become a "particulate matter sink". This is because particles from freshwater rivers enter the estuary in surface water but rapidly settle into the more saline bottom water where residual currents are landwards. Hence these particles, as well as those originating from the sea, are carried landwards. At a point in the estuary where the bottom tidal currents slacken, or where flocculation occurs, an area known as the "turbidity maximum" is found. Here the concentration of suspended solids is highest and cycles of deposition at slack tide, and erosion at half flood and ebb tide, are most marked.

Origin of Polysaccharides in an Estuary

There have been few attempts to identify and quantify all the polysaccharides entering an estuary. This is not surprising considering the wide range of polysaccharides produced by prokaryotic and eukaryotic cells, and the effects of seasonal and other environmental factors on polysaccharide input into an estuary. The polysaccharides within an estuary can be divided into two broad classes; autochthonous and allochthonous.

Autochthonous polysaccharides are produced within the estuary and any surrounding salt-marshes. Examples are:

1. phytoplankton and attached algae producing pectins, cellulose and hemicelluloses;
2. benthic fauna producing chitin;
3. microbes producing chitin and peptidoglycans;
4. flowering plants such as species of *Spartina* and *Zostera* in temperate regions or mangroves and *Thalassia* in tropical regions producing pectins, cellulose and hemicellulose.

Allochthonous polysaccharides are those that are transported, usually by the river or the sea, into the estuary. These polysaccharides often will have undergone some microbial degradation before they reach the estuary. Examples are:

1. litter of both flowering plants and algae;
2. animal remains;
3. microbial cells; for example in some rivers there are extensive growths of *Sphaerotilus natans* and the microbial community known as "sewage fungus" which regularly sloughs off material.

Many of man's activities will add directly or indirectly to the type and quantity of polysaccharides entering an estuary. An example of a direct input is the discharge of pulp or paper mill effluent, which can contain cellulose, starch and hemicelluloses [Poole *et al.*, 1978]. An indirect effect is exemplified by the discharge of domestic sewage which can trigger off an algal bloom or the growth of sewage fungus in the river.

Polysaccharide Degradation

This chapter is concerned with the rates of polysaccharide degradation and its effect on the estuarine environment rather than with the identity of the microbes responsible or the enzymic mechanisms involved, since these topics have been covered by other contributors to this book.

A major factor determining the rate of polysaccharide degradation is the location of that polysaccharide within the estuarine environment. A considerable quantity of the polysaccharides within an estuary will be deposited on the sediments because of the effects of flocculation/deflocculation and of water currents described above. In a study of chitin degradation Hood and Meyers [1977a, b] found that the maximum degradation rate occurred at the water/sediment interface. The literature on chitin degradation illustrates some of the problems confounding any attempt to present a unified view of polysaccharide degradation. Thus, there have been few research investigations concerned with the rates of chitin degradation, though it is one of the major polysaccharides found in the estuarine environment, and the few results which have been published are concerned with dissimilar systems and have been obtained by different methods. For example, Hood and Meyers [1977b] calculated from a study *in situ* using the "litter bag" technique that the chitin degradation rate in a subtropical estuary was 35 to 118 mg g^{-1} day^{-1}. From a study *in vitro*, Seki [1965] calculated that the rate in Japanese waters was in the order of 27 mg g^{-1} day^{-1}, while Liston *et al.* [1965] using a simulated seabed model calculated that the chitin degradation rate in Puget Sound, U.S.A. was 18.9 mg g^{-1} day^{-1}.

In the zones of the sediment or water column where oxygen is freely available the principal agents responsible for polysaccharide degradation are aerobic, heterotrophic microbes. An important influence on the degradative activity of aerobic microbes is believed to be the macrofauna which are to be found in most oxic sediments.

Interactions of Microbes and Macrofauna

The oxic zone of sedimentary deposits in estuaries is inhabited by macrofauna, i.e. animals which do not pass through a 1 mm^2 mesh sieve. Representative trophic life-form types of macrofauna are listed in Table 1.

Macrofauna may influence the magnitude of microbial degradation of polysaccharides within sediments in one of four ways:

1. cropping of the microbial flora associated with polysaccharide particles;
2. the culture of a specialized polysaccharide-degrading gut micro-flora;
3. the production of faecal and pseudofaecal material containing partially degraded polysaccharides;
4. burrowing activities.

1. The macrofauna of soft sediments are generally trophically dependent on the microbes, or at least on the micro-fauna such as ciliates and nematode worms which in turn feed on microbes [Fenchel and Jorgensen, 1977]. Thus macrofaunal production may often be limited by the microbial activity within the sediment. Specific ways in which bacterial grazing by animals increases microbial activity are poorly understood but may include [Hargrave, 1976]:

a) increasing the availability of suitable surfaces;
b) elimination of bacteriostasis during digestion;
c) enrichment by excretion;

TABLE 1

The principal groups of macrofauna inhabiting the oxic zone of the sediment

Group	Characteristics	Examples
Suspension feeders	Extract particles suspended in bottom currents	*Modiolus modiolus*
Surface deposit feeders	Selectively feed from the surface deposits	*Macoma balthica*
Sub-surface deposit feeders	Feed below the surface of the sediment, relatively immobile	*Nucula proxima*
Active deposit-feeding burrowers	Move freely through, and feed on the sediment	*Casco bigelowi*
Sedentary burrowing or tubicolous species	Build U-shaped burrows in the sediment. Feed by ingesting sediment or filtering particles from the water drawn through the burrow	*Corophium volutator* *Arenicola marina*
Conveyor-belt burrowers	Sedentary burrowers feeding on deposits at the deep end of their burrow	*Praxillella praetermissa* *Maldane* sp.
Herbivorous scraper	Live at or near sediment surface, feed directly on seaweeds or algal films	*Strongylocentrotus droebachiensis* *Littorina* sp.
Wrack fauna	Live in or under, and feed on the seaweed and other plant material in the wrack	*Orchestia gammarella*
Predators	Actively capture and eat living prey	*Spio filicornis* *Buccinum undatum*

d) removal of senescent colonies.

Bactivorous animals increase microbial activity in activated sludge, where protozoa are recognized to be important in controlling the efficiency of the process. Fenchel [1977] has shown that the decomposition rate of uniformly labelled plant material, as measured by the release of $^{14}CO_2$, occurs more rapidly when a mixed protozoan fauna is present than when this is excluded. Hargrave [1970] has shown by oxygen uptake studies that macrofauna can increase the true rate of microbial oxygen uptake.

2. The relative importance of a specialized gut microflora which utilizes the more recalcitrant polysaccharides

is uncertain. However, three lines of evidence suggest that for some animals this represents an important symbiotic relationship. The first comes from the wood borers, such as *Chelura* and the gribble, *Limnoria* which inhabit the surface layers of the wood, and from the deeper burrowing bivalve molluscs or shipworms. Shipworms such as *Teredo navalis* reduce the dry weight of wood by 50% during digestion and the bacterium *Cellulomonas* sp. was isolated from their digestive organs and is believed to be the principal source of the hydrolytic enzymes involved in cellulolysis [Cutter and Rosenberg, 1972]. Some free-living molluscs and crustacea, such as the shorehopper, *Orchestia* [Wildish and Poole, 1970], have been shown to possess some cellulolytic capability. In all of these cases, there are no critical experiments to determine whether extracellular cellulolytic enzymes originate from animal or bacterial/fungal sources. Chitinoclastic bacteria able to utilize chitin as the sole carbon source have been found to be associated with a wide range of macrofauna, both in the gut and on animal surfaces [Hood and Meyers, 1973]. In the white shrimp, *Penaeus setiferus*, the enzyme produced by the endosymbiotic gut bacteria is an inducible chitinase whereas the hepatopancreatic cells are claimed to produce a constitutive chitinase system [Hood and Meyers, 1977a]. The third line of evidence comes from the work of Jannasch and Wirsen [1977] who show that the activity of free-living deep sea floor microbes is very low and that much of the heterotrophic microbial activity occurs in the gut of deep sea animals.

3. The production of faecal and pseudofaecal pellets or strings has a number of stimulatory effects on the activity of aerobic heterotrophic microbes. An obvious effect is that such material is a focus for colonization by bacteria. Newell [1965] demonstrated this indirectly by observing a rise in C:N ratios in aged faeces of *Macoma balthica* and *Hydrobia ulvae*. Levinton *et al.* [1978] showed that the rate of microbial activity on *Spartina* detritus, as indicated by an increase in ATP, depended on the intensity of grazing and hence faecal production by the amphipod *Orchestia grillus*.

Because of the increased microbial activity on aged faecal pellets, coprophagy was suggested to be of common occurrence within deposit feeding communities [Rhoads, 1974], but its importance has been questioned [Levinton *et al.*, 1978]. Large quantities of deposited sediment are ingested by deposit feeders and filtered by suspension feeders, and this activity is then important in re-processing sediments. Re-processing involves comminuting and mixing particulate polysaccharides within the sediment as well as the production of extracellular enzymes which hydrolyse the polysaccharides. It has been estimated that five dominant macrofaunal species in a saltmarsh are capable of re-processing 53% of the annual saltmarsh primary production [Kraeuter, 1976]. Other examples of reprocessing rates are given by Rhoads [1974]. Another effect of faecal and pseudo-faecal production is to change the distribution of particle sizes of the deposit, which often originates as a fine silt/clay sediment and becomes one of predominantly sand-sized particles because of faecal production. The larger, lighter faecal pellets are more easily eroded by wave or tidal action. Biogenic resuspension of this kind has been described by Rhoads and Young [1970] and involves an increase in microbial ATP [Erkenbrecker and Stevenson, 1975], presumably as a result of the microbes being exposed to increased concentrations of nutrients.

4. Burrowing activity involving movements of the whole body through unconsolidated sediments by active burrowers (Table 1) assists in mixing the sediment and particulate polysaccharides. Other burrowers construct sedentary burrows or tubes which may be irrigated by water currents. Both types assist in geochemical cycling, by mixing oxygen and sources of carbon, nitrogen and sulphate throughout the surface layers of the sediment. The burrows produced by these animals may also serve to consolidate sediments susceptible to erosion [Mills, 1969].

The Role of Anaerobic Microbes

In estuarine sediments a surface oxic zone usually overlies an extensive anoxic zone (Figure 1). The depth of the oxic zone in the sediment will depend on the amount of oxygen which diffuses or is transported from the overlying water column. The burrowing animals play an important role in oxygenation of the sediment. The redox discontinuity layer (RDL) (Figure 1) in fine sediments which lack macrofauna is usually only a few millimeters from the surface, whereas if animals are present the RDL will be found at a depth of 15 to 30 cm. Some animals, particularly the "conveyor belt" species, will also assist in the transport of nutrients, including polysaccharides, across the RDL and in the mixing of the anoxic and oxic zones [Rhoads, 1974]. One by-product of the activity of burrowing macrofauna will be that of the transport of polysaccharide between active populations of aerobic and anaerobic microbes several times during its degradation.

In the anoxic zones the polysaccharides are degraded by a number of distinct biochemical groups of microbes which form a "food or substrate web" (Figure 2). These food webs arise because of the biochemical nature of the two mechanisms by which anaerobic bacteria can generate energy in the absence of oxygen [Fenchel and Jorgensen, 1977]. During fermentation organic compounds serve as both electron donor and electron acceptor, and the fermentation products are excreted into the surrounding medium, often in amounts equimolar with the substrate being fermented.

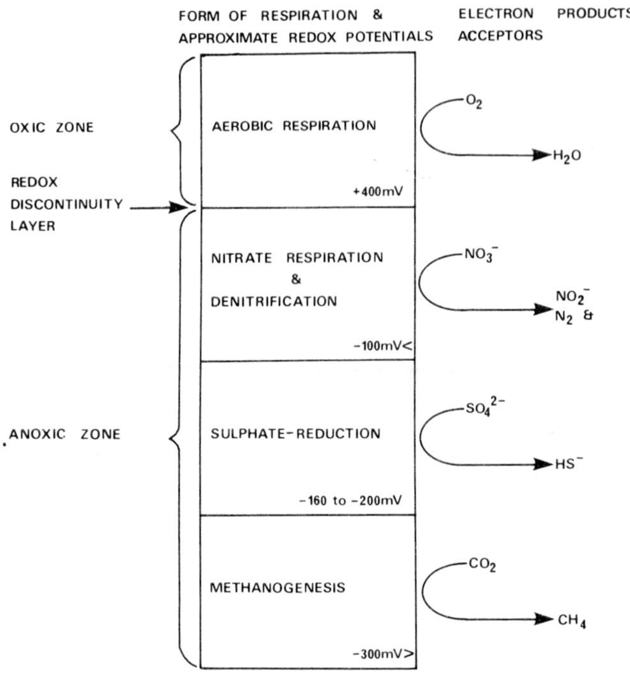

Fig. 1 The relationship between sediment depth, the redox potential, and type of respiration.

In anaerobic respiration an electron acceptor other than oxygen has to be used. The principle electron acceptors are nitrate, nitrite, fumarate, sulphate and CO_2. The bacteria using sulphate or CO_2 as the terminal electron acceptors differ from microorganisms using other compounds for anaerobic respiration as they are obligate anaerobes, and can only use a limited range of compounds as electron donors. These compounds are usually the products of fermentative organisms and hence a "food web" is formed.

Bacterial sulphate reduction in anoxic estuarine sediments is of considerable importance because of the high sulphate concentration in sea water (sulphate can usually be detected at considerable depths in the sediments). Jorgensen [1977a] found that 53% of the mineralization of organic matter in the sediments of a Danish fjord was catalysed by the sulphate-reducing bacteria. Thus it is possible to postulate the relationship shown in Figure 3 between the cellulolytic and non-cellulolytic fermentative bacteria and the sulphate-reducing bacteria [M. Bryder and R. Madden, personal communication]. This is a simplified representation of the process, and gives, for example, no indication of possible product inhibition of the process which might be due to the production of simple organic

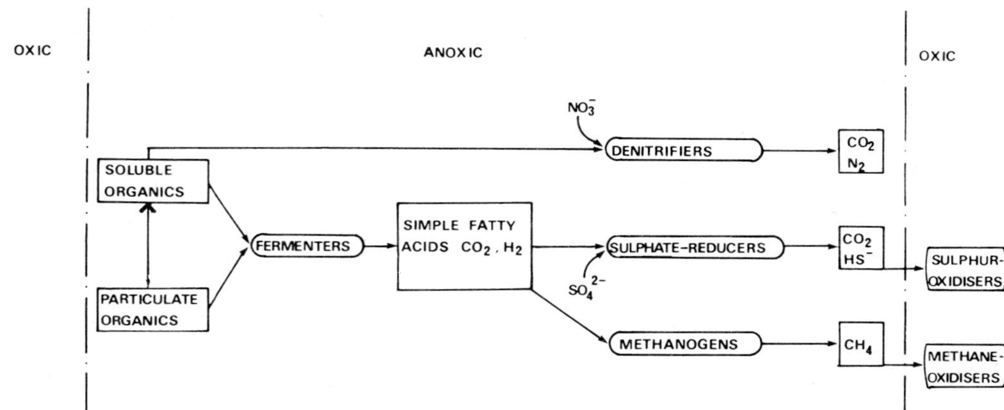

Fig. 2 Diagrammatic representation of the basis of a "food web" within an anoxic microniche.

acids, hydrogen, or sulphide.

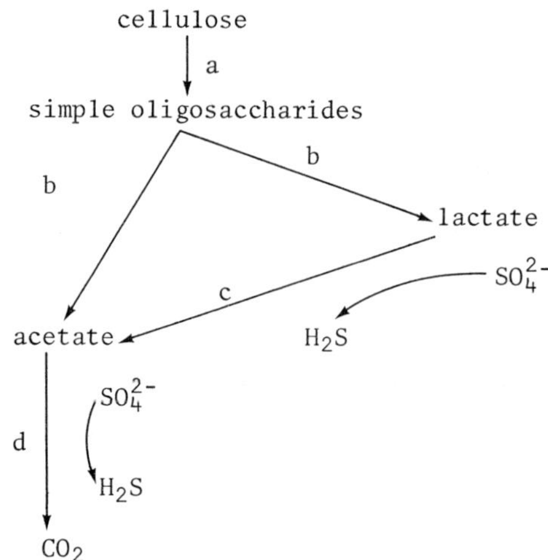

Fig. 3 A proposed nutritional relationship between fermentative and sulphate-reducing bacteria in the degradation of cellulose. a = cellulolytic fermentative bacteria; b = cellulolytic or non-cellulolytic fermentative baceria; c = *Desulfovibrio* spp.; d = *Desulfuromonas acetoxidans*.

By measuring the redox potential of a system it is possible to determine which electron acceptor is being used in anaerobic respiration. Such a succession in both the microbial population and the redox potential is shown as a function of sediment depth in Figure 1. This is an idealized situation since in many sediments all the zones are not distinguishable because of factors such as a low nitrate concentration and the activities of burrowing animals. It must also be realized that Figure 1 is a human or macro-organism's view of the sediment and does not recognize that microbial activity occurs principally within micro-environments. It is therefore to be expected that denitrification or sulphate-reduction may occur in what, on this gross scale, would be called an oxic environment [Jorgensen, 1977b].

We can now consider a specific example of polysaccharide degradation in an estuarine environment and of the pollution arising because of the presence of excess cellulose.

Pollution Resulting from Polysaccharide Degradation

The L'Etang Inlet in New Brunswick, Canada, is a recent example of how the discharge into an estuary of excess polysaccharides can result in the development of pollution.

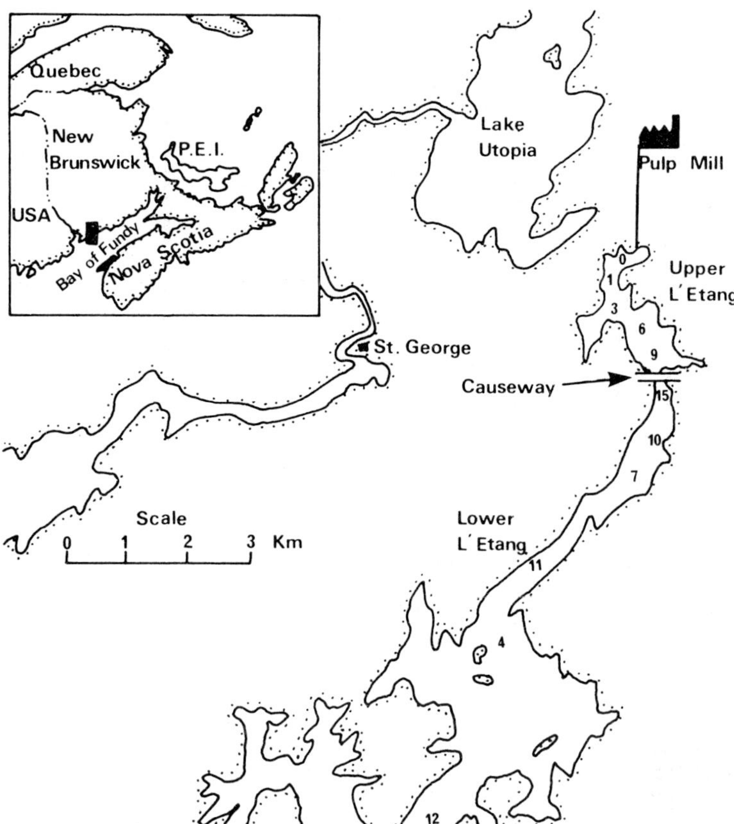

Fig. 4 A map of the L'Etang estuary showing the sampling stations.

In 1967 a causeway (Figure 4) was built across the L'Etang dividing it into a 2 km long upper and a 12 km long lower portion. The causeway contained four 1.6 m diameter culverts which permitted the exchange of water between the upper and lower L'Etang. The construction of the causeway impounded the waters of the upper L'Etang; the tidal amplitude of the upper portion was 0.1 m compared with 8.0 m in the lower L'Etang. In addition the patterns of water circulation [Kristmanson *et al.*, 1977] and oxygen supply were altered (Figure 5).

In 1971 a sulphite pulp mill began to discharge its effluent into the upper L'Etang (Figure 4). The mill was designed to produce 200 tonnes of pulp daily. The effluent produced by this type of mill would be in the order of 62 - 210 m^3 per tonne product.

In 1975 the water column of the upper L'Etang was anoxic and the waters of the lower L'Etang also exhibited an

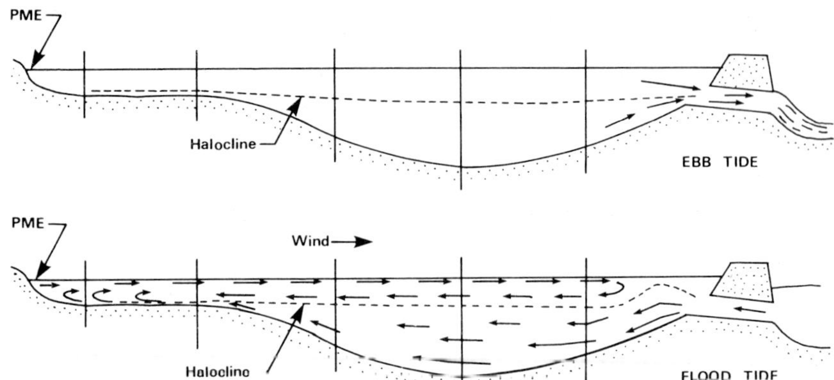

Fig. 5 The movement of water in the upper L'Etang during the ebb and flood tides. PME = pulp mill effluent.

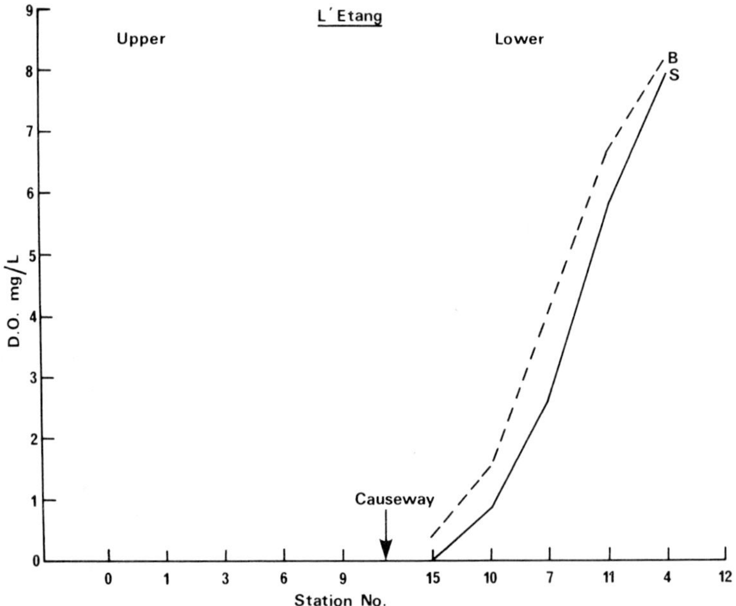

Fig. 6 Dissolved oxygen concentrations in the L'Etang estuary during summer 1975. B = Bottom water; S = surface water.

oxygen deficit (Figure 6). There was also a high sulphide concentration in both the water column and sediments of the upper L'Etang [Poole *et al.*, 1976a]. No macrofauna could be found in the upper L'Etang; as discussed earlier their absence could be expected to affect adversely rates of oxygenation and degradation. In the head regions of the lower L'Etang anoxia-tolerant macrofauna such as *Capitella capitata* and *Nereis diversicolor* were dominant [Wildish *et al.*, 1977].

Sulphite pulp mill effluent is a complex mixture containing three major classes of compounds [Perkins, 1974]: lignosulphonates, sugars and cellulosic pulp fibres, only the first two of which are soluble. The patterns of water circulation in the upper L'Etang meant that the soluble components of the effluent were restricted to the surface layer. The retention time of these soluble compounds in the upper L'Etang was 5 - 10 days, depending on wind direction and force. The lignosulphonates prevent the transmission of light energy to the algae and hence the production of oxygen by photosynthesis [Parker and Sibert, 1973]. The soluble sugars are an excellent source of both carbon and energy to heterotrophic microorganisms. The increased microbial activity in turn placed an increased oxygen demand on the system; the BOD_5 imposed by the effluent was in the order of 13,600 kg day^{-1}.

There was no oxygen in the water below the halocline in the upper L'Etang (Figure 6). This was unexpected as some oxygenated sea water entered this layer on the flood tide, while the oxygen demand exerted by microbial utilization of the sugars was restricted to the waters above the halocline. To explain this deoxygenation of the water column it was proposed that it was due to increased anaerobic microbial activity within the sediments. In the region of 800 kg day^{-1} of suspended fibres were discharged by the pulp mill into the upper L'Etang. These fibres rapidly settled out, and therefore there was a high cellulose concentration in the sediments. As discussed earlier, under anoxic conditions cellulose can be degraded by a mixed population of fermentative and sulphate-reducing bacteria, with sulphide being one of the principal end-products. The seawater entering the upper L'Etang every flood tide would provide the necessary sulphate. Sulphide is readily oxidized, the half life of sulphide in sterile oxygenated sea water being between 10 and 60 minutes [Bella *et al.*, 1972]. The sulphide diffused from the sediments of the upper L' Etang into the overlying water column where its oxidation purged the system of oxygen. The sulphide concentrations in the water column and sediments of the upper L'Etang were extremely high [Poole *et al.*, 1976a, 1977]. The release of reduced compounds from the sediment is reflected in the oxygen demand exerted by that sediment on the overlying water column. In laboratory experiments using model ecosystems the oxygen demand of the sediments of the upper L'Etang was found to be extremely high (13 g O_2 m^{-2} sediment day^{-1}) when compared with the results obtained by other workers for less "polluted" sediments [Poole *et al.*,

1976b, 1977]. The addition of 1% (w/w) of pulp fibre to sediment obtained from an unpolluted estuary (the Digdequash) doubled the oxygen demand of that sediment and led to the appearance of sulphide in the overlying water [Poole *et al.*, 1976b]. In these laboratory experiments it was also shown that there was 24% reduction in the rate of cellulose degradation in those anoxic sediments with a high sulphide concentration [Poole *et al.*, 1977], and it was suggested that this was due to the toxic effects of the sulphide on the bacteria comprising the food web.

These results indicated that in this shallow estuary anaerobic microbial activity in the sediments, which was caused by the presence of excess cellulose, coupled with restricted water circulation, were major factors in the development of the pollution. This hypothesis was used to construct a mathematical model of the oxygen balance of the upper L'Etang. For the purposes of this model the upper L'Etang was divided into a number of boxes (Figure 7) and the volume of water flowing between each box determined [Kristmanson *et al.*, 1977]. The model allowed various remedial actions for the upper L'Etang to be investigated.

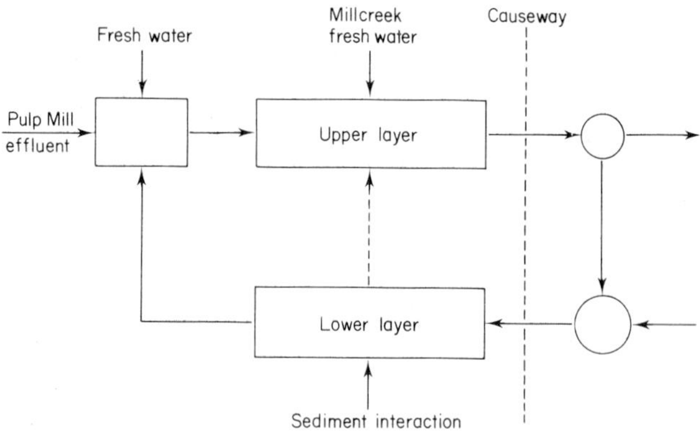

Fig. 7 The basis of the mathematical model of the upper L'Etang. The water column of the upper L'Etang can be divided into a number of well defined areas or boxes. The volume of each box and the volume of water flowing between the boxes can be determined from hydrological data.

The model predicted that because of the restricted water circulation, caused by the construction of the causeway, the sediment and then the water column would have gradually become anoxic. The discharge of pulp mill effluent, especially the particulate cellulose, further stressed the system causing the onset of anoxia to be more rapid and incidentally making remedial action considerably more difficult.

No seasonal study of microbial activity was carried out during the L'Etang research programme, but one has been completed for the Aberdeenshire Don [Parkes, 1978]. The Don estuary has certain similarities to the upper L'Etang. It is shallow and traps a considerable quantity of polysaccharides from paper mills, a sewage treatment plant, and the extensive growths of sewage fungus which occur in the river. It is not as badly polluted as the upper L'Etang and the water is rarely anoxic. During most summers, however, a smell of hydrogen sulphide is associated with the estuary, and in particular with the extensive anoxic mud flats. In this research programme there were two periods when a marked decrease was observed in the numbers of viable bacteria (aerobic and anaerobic, heterotrophic and cellulolytic, in addition to sulphate-reducers) isolated from the anoxic sediments. These decreases occurred during the winter (January to March) and more unexpectedly in the summer (July and August) [Poole *et al.*, 1977; Parkes, 1978]. There was also evidence that the rate of cellulose degradation decreased during the July and August period. It was suggested [Parkes, 1978] that this summer decrease in microbial numbers and activity was caused, in part, by sulphide toxicity. Therefore sulphate-reduction can be considered necessary for the successful operation of the anaerobic food web responsible for cellulose degradation, but only until the sulphide concentration reaches a threshold which inhibits one of the component organisms of that food web.

Conclusions

At present there is little published information on the rates and mechanisms of polysaccharide degradation within the estuarine environment. Polysaccharides are, however, a major source of both carbon and energy in the estuarine ecosystem. Since estuaries are of considerable economic importance for the disposal of industrial effluent, for recreation, and for the part they play in the life cycle of many commercially valuable fish, we suggest that there is need for systematic investigations into polysaccharide degradation. Of particular interest would be the effects of factors such as salinity, season, oxygen and sulphide concentrations. To be meaningful, however, these investigations must include the nutritional and physical relationships which exist between micro- and macro-organisms and the role of this relationship in polysaccharide degradation.

References

Bella, O.A., Ramm, A.E. and Peterson, P.E. (1972). Effects of tidal flats on estuarine water quality. *Journal of the Water Pollution Control Federation* **44**, 541-556.

Cutter, J.M. and Rosenberg, F.A. (1972). The role of cellulolytic bacteria in the digestive processes of the shipworm. II. Requirements for bacterial cellulase in the digestive system of teredine borers. In *Biodeterioration of Materials*, vol. 2, pp.42-51. Edited by A. Harry Walters and E.H. Hueck-Van der Plas. London: Applied Science Publishers.

Dean, R.C. (1974). Cellulose and wood digestion in the marine mollusk *Bankia gouldi* Bartsch. In *Proceedings of the Third International Biodegradation Symposium*, pp.955-965. Edited by J.M. Sharpley and A.M. Kaplan. London: Applied Science Publishers.

Dyer, K.R. (1972). Sedimentation in estuaries. In *The Estuarine Environment*, pp.10-32. Edited by R.S.K. Barnes and J. Green. London: Applied Science Publishers.

Erkenbrecker, C.W. and Stevenson, L.H. (1975). The influence of tidal flux on microbial biomass in salt marsh creeks. *Limnology and Oceanography* **20**, 618-625.

Fenchel, T.M. (1977). The significance of bactivorous protozoa in the microbial community of detrital particles. In *Aquatic Microbial Communities*, 2nd ed. Edited by J. Cairns. New York: Garland Publishing.

Fenchel, T.M. and Jorgensen, B.B. (1977). Detritus food chains of aquatic ecosystems: the role of bacteria. In *Advances in Microbial Ecology*, vol. 1, pp.1-58. Edited by M. Alexander. New York: Plenum Press.

Hargrave, B.T. (1970). The effect of a deposit-feeding amphipod on the metabolism of benthic microflora. *Limnology and Oceanography* **15**, 21-30.

Hargrave, B.T. (1976). The central role of invertebrate faeces in sediment decomposition. In *Role of Terrestrial and Aquatic Organisms in Decomposition Processes*, pp.301-321. Edited by J.M. Anderson and A. MacFadyen. Oxford: Blackwell Scientific Publications.

Hood, M.A. and Meyers, S.P. (1973). The biology of aquatic chitinoclastic bacteria and their chitinolytic activities. *La Mer* **11**, 213-229.

Hood, M.A. and Meyers, S.P. (1977a). Microbiological and chitinoclastic activities associated with *Penaeus setiferus*. *Journal of the Oceanographical Society of Japan* **33**, 235-241.

Hood, M.A. and Meyers, S.P. (1977b). Rates of chitin degradation in an estuarine environment. *Journal of the Oceanographical Society of Japan* **33**, 332-338.

Jannasch, H.W. and Wirsen, C.O. (1977). Deep-sea microorganisms *in situ* response to nutrient enrichment. *Science, New York* **180**, 641.

Jorgensen, B.B. (1977a). The sulfur cycle of a coastal marine sediment (Limfjorden, Denmark). *Limnology and Oceanography* **22**, 814-832.

Jorgensen, B.B. (1977b). Bacterial sulphate-reduction within reduced microniches of oxidized marine sediments. *Marine Biology* **41**, 7-17.

Kraeuter, J.N. (1976). Biodeposition by salt-marsh invertebrates. *Marine Biology* **35**, 215-223.

Kristmanson, D.D., Wildish, D.J. and Poole, N.J. (1977). Mixing of pulp mill effluents in the Upper L'Etang. *Fisheries Research Board of Canada M.S. Report* 1416.

Levinton, J.S., Lopez, G.R., Heidemann Lassen, H. and Rahn, U. (1978). Feedback and structure in deposit-feeding marine benthic communities. In *Biology of Benthic Organisms*, pp.409-416. Edited by B.F. Keegan, P.O. Céidigh and P.J.S. Boaden. Oxford: Pergamon Press.

Liston, J., Wiebe, W.J. and Lighthart, B. (1965). Activities of marine benthic bacteria. *Research in Fisheries, 1964, College of Fisheries, University of Washington* No. 184, 39-41.

Mills, E.L. (1969). The community concept in marine zoology, with comments on continua and instability in some marine communities: a review. *Journal of the Fisheries Research Board of Canada* **26**, 1415-1428.

Newell, R. (1965). The role of detritus in the nutrition of two marine deposit feeders, the prosobranch, *Hydrobia ulvae* and the bivalve *Macoma balthica*. *Proceedings of the Zoological Society of London* **144**, 25-45.

Parker, R.R. and Sibert, J. (1973). Effect of pulpmill effluent on dissolved oxygen in a stratified estuary. 1. Empirical observations. *Water Research* **7**, 503-514.

Parkes, R.J. (1978). The seasonal variation of bacteria within the sediments of a polluted estuary. Ph.D. Thesis. University of Aberdeen.

Perkins, E.J. (1974). *The Biology of Estuaries and Coastal Waters*. London: Academic Press.

Poole, N.J., Wildish, D.J. and Lister, N.A. (1976a). Effects of a neutral-sulphite, pulp effluent on some chemical and biological parameters in the L'Etang Inlet, New Brunswick. L'Etang Inlet survey III. *Fisheries Research Board of Canada* M.S. Report 1404.

Poole, N.J., Wildish, D.J. and Lister, N.A. (1976b). The use of micro-ecosystem models to investigate pollution of the L'Etang Inlet by pulp mill effluent. *Fisheries Research Board of Canada* M.S. Report 1403.

Poole, N.J., Parkes, R.J. and Wildish, D.J. (1977). Reaction of estuarine ecosystems to effluent from pulp and paper industry. *Helgoländer wissenschaftliche Meeresuntersuchungen* **30**, 622-632.

Poole, N.J., Wildish, D.J. and Kristmanson, D.D. (1978). The effects of the pulp and paper industry on the aquatic environment. *CRC Critical Reviews in Environmental Control* **8**, 153-195.

Rhoads, D.C. (1974). Organism-sediment relations on the muddy sea floor. In *Oceanography and Marine Biology*, vol. 12, pp.263-300. Edited by H. Barnes. London: George Allen and Unwin.

Rhoads, D.C. and Young, D.K. (1970). The influence of deposit feeding organisms on sediment stability and community trophic structure. *Journal of Marine Research* **28**, 150-178.

Seki, H. (1965). Microbiological studies on the decomposition of chitin in marine environments. IX. Rough estimation on chitin decomposition in the ocean. *Journal of the Oceanographical Society of Japan* **21**, 253-260.

Whitehouse, U.G., Jeffery, L.M. and Debbrecht, J.D. (1960). Differential settling tendencies of clay minerals in saline waters. *Proceedings of the Seventh Conference on Clays and Clay Minerals*, pp.1-79.

Wildish, D.J. and Poole, N.J. (1970). Cellulase activity in *Orchestia gammarella* (Pallas). *Comparative Biochemistry and Physiology* **33**, 213-716.

Wildish, D.J., Poole, N.J. and Kristmanson, D.D. (1977). Temporal changes of sublittoral macrofauna in L'Etang Inlet caused by sulfite pulp mill pollution. *Fisheries Research Board Canada* Tech. Report No. 718.

Chapter 18

ROLE OF POLYSACCHARIDE-DEGRADING ENZYMES IN MICROBIAL PATHOGENICITY

R.J.W. BYRDE

Long Ashton Research Station, University of Bristol, Bristol, UK

Introduction

Microorganisms are essential to higher animals and plants by breaking down complex molecules and so ensuring the continuation of the carbon cycle. In this process, enzymes degrading polysaccharides play an important part.

Unfortunately for the higher life forms, microorganisms sometimes utilize these same enzymes in pathogenicity. This is particularly true for higher plants, because their cell walls are predominantly polysaccharide in nature. Inhibition of wall synthesis is a remarkably effective way of inhibiting the growth of an organism, e.g. the effects of penicillin on gram-positive bacteria and of polyoxin D on many fungi [Gooday, 1977]. Furthermore, destruction of the wall has a drastic effect on a cell's survival.

Animal cells are predominantly proteinaceous, and so polysaccharide-degrading enzymes are of less importance in pathogenicity towards animals. However, it is now clear that on, or in, the surface of animal cells (and the plasma membranes of plant cells) are glycoproteins and glycolipids that are significant in recognition and attachment processes. Modification of these can affect the pathogenicity of microorganisms towards animals.

Involvement in Pathogenicity Towards Plants

This subject has been comprehensively reviewed [Brown, 1965; Bateman and Millar, 1966; Wood, 1967; Albersheim *et al.*, 1969; Bateman, 1976; Bateman and Basham, 1976].

The ability to secrete these enzymes is not itself sufficient to make an organism pathogenic; the other necessary characteristics are not fully understood, but are believed to involve the ability to suppress, or at least not to elicit, plant defence reactions.

Composition of the higher plant cell wall

The plant cell wall consists predominantly of polysaccharides, with some intrusions of other substances. Its chemical structure is still not fully understood, but the model system proposed by Albersheim and his colleagues [Keegstra *et al.*, 1973] (Figure 1) following exhaustive studies represents a major advance and the most detailed description at present available.

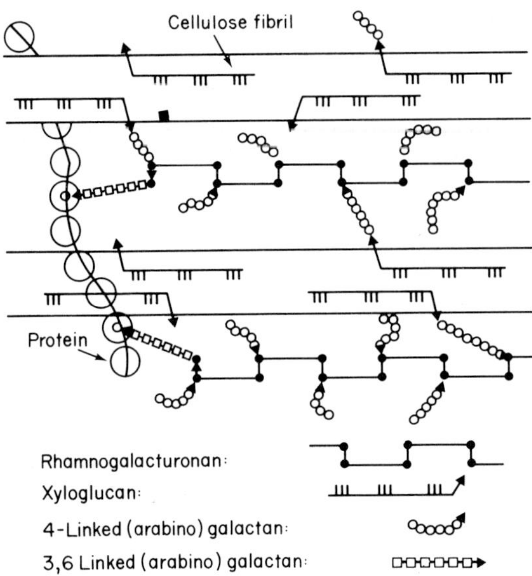

Fig. 1 Proposed model of primary plant cell walls [from Keegstra *et al.*, 1973; as modified by Bateman, 1976].

The main components of the complex structure are cellulose fibrils, consisting of β 1,4 D-glucopyranose units, which confer much of the wall's physical strength. These are embedded in a matrix consisting largely of chains of a rhamnogalacturonan. This comprises mainly α 1,4 D-galacturonan units, with single units of the neutral sugar rhamnose linked 1,2 at intervals which are responsible for the 'kinking' of the chain. The uronic acid carboxyl groups are often methylated, and the uronide units may be acetylated at positions 2 and 3. Associated with the rhamnogalacturonan are chains of neutral sugars (polymers of α L-arabinofuranose linked 1,3 or 1,5 and of β D-galactopyranose, linked 1,4).

The link between the pectic (rhamnogalacturonan) and cellulosic components is believed to be *via* a xyloglucan (consisting of β 1,4 glucopyranose residues with terminal branches of α 1,6 xylopyranose). This is held to the cellulose by hydrogen bonding, and to the rhamnogalacturonan by covalent linkages through the araban and 4-linked galactan.

Also present in the model is a structural protein, rich in hydroxyproline units, that bears tetra-arabinoside groups and is linked to the rhamnogalacturonan *via* a 3,6-linked arabinogalactan on a serine moiety.

Lignin may also be present as secondary thickening in the cell wall, particularly in woody tissues. This is not a polysaccharide, but its presence can often increase the resistance of the wall to degradation [Bateman and Basham, 1976], and its synthesis and inclusion may constitute an important defence mechanism [Ride, 1975].

Another form of wall polysaccharide sometimes found in higher plants, particularly at sites of injury, is callose, which is a β 1,3 glucan. It can be detected by its yellow-green fluorescence in u.v. light after staining with aniline blue [Jensen, 1962]; it has been shown to be present in club root infection [Williams *et al.*, 1973], and several other diseases, including some caused by viruses, cited by Aist [1976a], Bracker and Littlefield [1973] and Pegg [1977].

Enzymes involved in pathogenicity to plants

The enzymes involved in the degradation of plant polysaccharides have been described in detail by Bateman and Basham [1976]. Briefly, the rhamnogalacturonan chain is subject to hydrolytic attack, e.g. by polygalacturonase (PG), or lytic attack, e.g. by polygalacturonate lyase (or trans-eliminase). These enzymes can be sub-divided, firstly on the basis of whether they attack at random ('endo' enzymes) or terminally ('exo' enzymes) and secondly, on their preference for unesterified or esterified carboxyl groups on the galacturonic acid units. Another hydrolytic enzyme (pectinesterase) can remove the methyl ester groups.

Neutral sugars are also degraded: galactanases hydrolyse galactan by random cleavage, and β D-galactopyranosidase (β galactosidase) and α L-arabinofuranosidase hydrolyse β 1,4 galactan and araban respectively by terminal attack.

'Cellulase' activity is now known to result from a complex of enzymes that include endo-β 1,4 glucanase, exo-β 1,4 glucanase, β-glucosidase and cellobiose oxidase [Eriksson, 1977].

Criteria for assessing a role in pathogenicity

The fact that many plant pathogens secrete various polysaccharide-degrading enzymes in culture is not a sufficient reason for ascribing to these enzymes a role in pathogenicity. However, enzyme secretion *in vitro* can nevertheless indicate the types of enzymes that a pathogen is capable of producing and so guide studies relating to infection and disease [Bateman and Basham, 1976]. Many of the enzymes are subject to induction and/or catabolite repression [Cooper, 1977]. Substrates may also be shielded by other polymers *in vivo*. Even the demonstration of degradative enzymes of pathogen origin in diseased tissue is no proof of their involvement in pathogenesis [Hancock, 1967].

The criteria that have been used may be summarized as follows:

Presence in infected tissue and ability to reproduce symptoms It should be demonstrated that the enzyme is present in infected tissues, preferably at an early stage in pathogenesis, and is capable of causing at least some of the disease symptoms when re-introduced into the healthy host. Ideally, this can be done by adapting the criteria for a vivotoxin [Dimond and Waggoner, 1953] as modified by Graniti [1972] for a 'vivo-aggressin'; these are based in turn on Koch's Postulates. Clearly, the enzyme must be purified to homogeneity before it is applied to the host. If the enzyme extracted can be shown by conventional enzymological methods to be identical to that produced in culture, it may be more convenient to obtain the enzyme from culture for treatment of the host.

However, enzyme extraction from infected tissues can be complicated by the presence of inhibitors [Raa *et al.*, 1977; Byrde and Archer, 1977]. Absence of an enzyme from an extract does not prove that it is absent from the infected tissue. Furthermore, a negative result after re-introducing the enzyme into the host does not prove that it is not involved in pathogenesis. In the infection of host plants, enzymes are secreted by the pathogen as it advances through the intercellular tissue. They immediately come into contact with the cell wall matrix, with its own pH status, and presumably not in contact with intracellular inhibitors. This system cannot be reproduced experimentally.

As Bateman and Basham [1976] mention, the case for the involvement of an enzyme is greatly strengthened by the demonstration of the depletion of the appropriate substrate. For example, Cole and Wood [1961] reported changes in pectic polymers following the invasion of fruits by several rot-causing pathogens. In addition, the detection of end-products of enzyme action in infected tissue (but not in healthy tissue) is useful evidence. For example, the demonstration of the presence of free methanol in apple fruits infected by *Monilinia (Sclerotinia) fructigena* confirmed that pectinesterase was not only present (as shown by extraction), but also active [Byrde *et al.*, 1973].

The effect on pathogenesis of known repressors of enzyme

synthesis, and of inhibitors of enzyme action This method is only applicable to those enzymes that can be repressed. It was used by Patil and Dimond [1968], who demonstrated that treatment of *Fusarium oxysporum* f.sp. *lycopersici* with glucose resulted in a decrease in PG, both *in vitro* and *in vivo*; there was an increase in mycelial weight *in vitro* and a decrease in symptom production *in vivo*.

The use of inhibitors may also provide useful information on the involvement of an enzyme in pathogenicity. In some earlier work the inhibitors were relatively non-specific in their action and could be directly toxic to the pathogen, e.g. rufianic acid [Grossmann, 1962]. Inhibitors of greater specificity are now available, and natural PG inhibitors were described by Albersheim and Anderson [1971] that have differential activities against PG activities from different fungi, and which do not inhibit other glycosidases. Derivatives of simple sugars that act as specific inhibitors of glycosidases may also prove valuable.

Differences in enzyme production between compatible and less compatible host-pathogen combinations This approach has been used by several workers as evidence for the involvement of polysaccharases in pathogenicity. Deese and Stahmann [1962a, b] working with *Fusarium* on banana and tomato respectively showed that pectic enzyme activity was lower in the infected resistant host than in the corresponding susceptible plant. Other workers [e.g. Unbehaun and Moore, 1970; Lumsden and Bateman, 1966] have shown that the amount of pectic enzyme extracted coincided with symptom intensity. However, in all such systems an alternative explanation is that the amount of infection is being influenced by some other factor and that the amount of pectic enzymes present is the consequence, and not the cause, of different infection levels. Now that methods are available for estimating total mycelial content [Ride and Drysdale, 1972; Toppan *et al.*, 1976; Sharma *et al.*, 1977] this can be taken into account in assessing the validity of the results.

Correlation of enzyme secretion with virulence in mutant strains of the pathogen Again, correlation is no proof of cause and effect. However, the method has been widely used, with varying results. For example, by use of mutagens Beraha and Garber [1971] obtained a mutant of *Erwinia carotovora* avirulent to celery petioles and subsequently, by further treatment of this, a virulent revertant. Although many attributes were lost in the initial change, only pectinesterase, PG, pectin lyase (trans-eliminase) and cellulase (Cx) were restored with virulence, suggesting that some or all of these enzymes were important or were linked with an important factor not studied. Similarly, Friedman and Ceponis [1959] starting with an isolate of *Pseudomonas marginalis* virulent to lettuce and chicory found that mutants lacking pectic enzymes were avirulent. Such studies should not be restricted to a single enzyme as there are, for example, several enzymes that can degrade a rhamnogalacturonan, any one of which may have a role in pathogenicity.

Comparisons have sometimes been made between the virulence of different species of a genus with regard to enzyme secretion and virulence. So many other factors are likely to be variable that such comparisons are of only limited value; in addition, the cultural conditions used are unlikely to be optimal for each species examined [Brown, 1965]. When small numbers of isolates have been used [Paquin and Colombe, 1962; Singh and Husain, 1964; Lule *et al.*, 1977] the evidence is correspondingly weak. When large numbers of isolates are used, multiple regression analysis can profitably be used to quantify effects [Howell, 1975].

Account must also be taken of the stability of the isolates under test: the experimental design should be such as to minimize genetic changes in the interval between virulence tests and enzyme assays.

The role of multiplicity of enzyme forms

Throughout this section the term 'isoenzyme' is used in its broadest sense, as defined by the IUPAC-IUB Commission on Biochemical Nomenclature [1972]. The role of isoenzymes in biological systems was reviewed by Markert [1975], who stressed that multiple forms have been tailored by evolutionary pressures to fit the fastidious requirements of the cell's metabolic machinery.

The existence of multiple forms of extracellular wall-degrading enzymes of plant pathogens has been recognized for many years; thus Endo [1963] reported the secretion of several forms of PG by *Coniothyrium diplodiella*. Sometimes multiplicity of forms seems to arise as an artifact in purification [Swinburne and Corden, 1969]. The widespread use of such techniques as isoelectric focusing and electrophoresis has led to a greater understanding of the presence, and relevance, of multiple forms.

Hancock [1976] stressed the flexibility that multiple forms of enzymes conferred to a plant pathogen. Several examples of isoenzymes of differing roles in pathogenicity have already been reported. Thus Garibaldi and Bateman [1971] demonstrated that an isolate of *Erwinia chrysanthemi* secreted a polygalacturonate lyase (trans-eliminase) complex that could be resolved into four forms. Three of these (pI values 9.4, 8.4 and 7.9) caused plant tissue to lose coherence; the fourth (pI 4.6) failed to do so although it attacked pectates isolated from several plants and released soluble uronides from isolated walls of bean hypocotyls. On the basis of this paper, Starr and Chatterjee [1972] suggested that different forms may have a role in host specificity with particular isoenzymes being deployed in tissues of different hosts. However, this function did not seem to exist in the pectate lyases of *Erwinia chrysanthemi* re-examined by Pupillo *et al.* [1976].

The brown rot fungus *Monilinia fructigena* produces four forms of extracellular PG in culture [Willetts *et al.*, 1977]. The only form (pI 9.7) found in infected fruits is unable to macerate fruit tissue. However, in extracts of infected cortical tissue of apple stems (Figure 2), another form (pI 5.6) is found; although not yet purified to homogeneity, the purest preparations so far obtained (with only traces of other protein bands) macerate fruit tissue [Fielding and Byrde, 1979]. It is thus tempting to speculate that the existence of multiple forms may have a role in tissue specificity. Analogous multiple enzyme strategies are believed to have an important role in the ability of an organism to adapt to different environments [Moon, 1975; Somero, 1975]. Significantly, of the species or formae of *Monilinia* examined by Willetts *et al.* [1977], the one with the most restricted host and tissue specificity had the simplest pattern of extracellular wall-degrading isoenzymes. Presumably the control of isoenzyme production in a given tissue is based on induction and repression mechanisms [Cooper, 1977].

Where multiple forms of an enzyme are found in an extract of infected tissue, some of these forms may be of host origin [e.g. Pressey and Avants, 1977], and may complement those of pathogen origin; others may be intracellular pathogen isoenzymes [e.g. Byrde and Willetts, 1977] that are in the supernatant only as a result of the extraction process; others may arise from contaminating organisms.

In summary, it is not always sufficient to consider an enzyme, e.g. PG, as a single entity in assigning to it a role in pathogenicity. Where multiple forms exist, the role of each form should be examined separately, using the criteria already outlined.

Damage caused by polysaccharide-degrading enzymes

To the wall Damage to plant cell walls by polysaccharidases has been studied extensively. In general, those enzymes, i.e. PG, pectin lyase and polygalacturonate (pectate) lyase that degrade the rhamnogalacturonan chain are those considered most important in pathogenicity. This is particularly true of 'endo-' enzymes that attack the substrate chain in a random manner and thereby, for a given number of chain cleavages, bring about more drastic changes in the substrate's physical properties.

Time course studies on the sequence of enzyme production by pathogenic fungi *in vitro* have shown that an endo-PG is often the first to be secreted; enzymes attacking neutral sugar polymers are produced later, probably as their respective substrates become exposed [English *et al.*, 1971; Jones *et al.*, 1972; Cooper, 1977]. Lyases are often produced only when the pH is neutral or alkaline [Bateman and Basham, 1976]. Last in the sequence of extracellular enzymes produced are those degrading xylan and cellulose. For this and other reasons the role of cellulases as primary factors in pathogenicity to plants is not well established [Wood, 1967]; however, they clearly affect symptom expression, and Mukhopadhyay and Chakravarty [1976] found that virulence of a xanthomonad causing leaf blight of rice was correlated with cellulolytic, and not pectolytic, activity.

As well as being the first to be produced in a temporal sequence, many endo-PG's and lyases readily attack cell walls *in vivo*. One consequence of this attack is the 'maceration' of plant tissue, which then loses its coherence as a result of the digestion of the middle lamellar region [see Bateman and Basham, 1976].

Less is known of the effects of enzymes degrading neutral sugar polymers, although they probably have a role in pathogenicity [English *et al.*, 1971; Howell, 1975]. Partially purified endo-galactanases have been reported to cause some loss of tissue coherence [Knee and Friend, 1970; Cole and Sturdy, 1973]; but Bauer *et al.* [1977] could not detect any with a pure preparation from *Sclerotinia sclerotiorum*. Arabinosidases normally attack polymers terminally, and would therefore not be expected to cause tissue injury. Such exo-glycosidases may provide the pathogen with sugar monomers that are readily utilized [Howell, 1975], and might be available before cell leakage liberates the host cell contents. However, this does not constitute a direct role in pathogenicity.

To the membrane It has long been recognized that, in addition to their effects on the physical cohesion of tissue, pectic enzymes can bring about its death [Brown, 1915]. Initially this was demonstrated by the inability of treated tissue to be plasmolysed in hypertonic solutions and to take up vital stains such as neutral red [e.g. Tribe, 1955]. The effects are consistent with injury to the plasmalemma resulting in increased permeability; later workers have used the leakage of electrolytes from treated tissue (often by conductivity measurements) as a measure of cell injury. Many workers have noted the correlation between maceration of tissue and cell killing. There is thus the anomaly that enzymes attacking substrates situated in the cell wall can cause injury to the cell membrane which presumably has few, if any, such substrates.

Tribe [1955] attempted to differentiate cell separation from cell killing; he showed that if cells were plasmolysed at the time of treatment they became separated by pectic enzyme treatment, but were not killed. This is indeed the basis of methods used for protoplast preparation.

Fig. 2 Infection by *Monilinia (Sclerotinia) fructigena* of apple fruit (F) and shoot (S), showing fungal conidial pustules (P). In infected fruit, one extracellular form (pI 9.7) of polygalacturonase is present; in the shoot, there is an extra form (pI 5.6) [Fielding and Byrde, 1979].

Many workers have readily confirmed Tribe's observation, and any theory to account for the toxic effects of pectic enzymes must take account of it. In addition, it must explain the fact that all pectic enzymes that cause cell separation simultaneously cause tissue death [Basham and Bateman, 1975a] and that enzymes (or isoenzymes) unable to 'macerate' also do not kill.

The hypothesis most strongly supported by evidence currently available is that damage to the membrane is a direct consequence of injury to the cell wall [Bateman, 1976; Bateman and Basham, 1976; Wood, 1976], possibly as a result of the lack of physical support that the modified wall can give. The only evidence that is difficult to reconcile with the hypothesis is the lack of any lag period sometimes noted between the exposure of the tissue to the enzyme and the onset of electrolyte leakage in unplasmolysed tissue [e.g. Basham and Bateman, 1975b; E.C. Hislop, personal communication].

Binding of polysaccharases to plant tissue

There are several reports of specific binding of PG activities to plant tissue. Raa *et al.* [1977] described the binding of PG from *Cladosporium cucumerinum* but not from *Aspergillus* or *Botrytis* species, to isolated cell walls of its host, cucumber. Mussell and Strand [1977] described differential binding patterns of PG's from a pathogen and a non-pathogen on walls of two cotton cultivars. Fanelli *et al.* [1978] reported that two pure PG's of *Trichoderma koningii* did not bind to potato tuber, carrot root, beet root or French bean seed tissues, and suggested that this might be a factor in the lack of pathogenicity towards plants. Two PG isoenzymes from the strawberry pathogen *Rhizoctonia fragariae*, by contrast, were both absorbed on strawberry tissue, with different affinities, and caused cell separation [Cervone *et al.*, 1978], but they were not absorbed by potato, carrot and beet tissue. Further information is needed to determine whether these are chance phenomena, related to charge affinities, or an important factor in pathogenicity. Albersheim and Anderson-Prouty [1975] have pointed out that PG inhibitors in cell walls are not sufficiently selective to account for race specificity of pathogens.

Involvement of polysaccharases in various types of disease

Space permits only a brief treatment. The clearest involvement is in soft rots [Wood, 1967; Bateman and Basham, 1976], where those caused by *Erwinia* spp. have received particularly detailed study [e.g. Garibaldi and Bateman, 1971; Wood, 1976]. The characteristic cell separation and death can readily be simulated by polygalacturonate lyase enzymes. In 'firm' rots, such as that caused to apple fruit by *Monilinia fructigena*, pectolytic activity is less [Cole and Wood, 1961; Byrde *et al.*, 1973]. In a statistically based survey of 120 mutagen-treated isolates, Howell [1975] accounted for about 35% of the variability in virulence in terms of variation in the secretion *in vitro* of three wall-degrading enzymes. Surprisingly, it was the exo-enzyme α-L-arabinofuranosidase, and not PG, pectinesterase or growth rate *in vitro*, that showed a significant correlation (P = 0.01) with virulence.

For vascular wilt diseases the evidence is often contradictory. Mussell and Strand [1977] summarized the evidence in favour of the involvement of endo-PG in symptom expression in Verticillium wilt of cotton: this includes the presence of the enzyme in infected seedlings and symptom expression following the vascular transfusion of the purified enzyme in dilute salt solution. By contrast, Keen and Erwin [1971], also using an apparently homogeneous PG, applied in water, did not obtain typical symptoms, and

Keen and Long [1972] attributed wilting to a lipopolysaccharide also produced by the pathogen. Furthermore, Howell [1976] showed that mutants of *Verticillium dahliae* deficient in one to four of the pectolytic enzymes likely to be involved (including endo-PG) were able to induce apparently normal symptoms of wilt in stem-inoculated cotton plants. More work is needed to elucidate the factors involved.

It is important not to generalize on the role of these enzymes in vascular wilts [R.M. Cooper, personal communication]. For example, in the silver leaf disease of apple, caused by *Chondrostereum (Stereum) purpureum*, a homogeneous endo-PG has been shown to cause the characteristic silvering [Miyairi *et al.*, 1977]; in Dutch elm disease, caused by *Ceratocystis ulmi*, endo-PG is not correlated with aggressiveness [Appel and MacDonald, 1976].

Rust and mildew fungi are biotrophic pathogens that depend for their existence on the presence of living host cells, so are unlikely to secrete enzymes that cause widespread tissue damage. Lewis [1974], in a consideration of the evolution of parasitism, has stressed that an essential criterion is the control of the synthesis of such enzymes.

On the other hand, there is some evidence from ultrastructural studies of spatially restricted involvement of enzymes in tissue penetration by biotrophs; examples are cited by Bracker and Littlefield [1973] and Aist [1976b]. In electron micrographs, distortion of the wall is generally regarded as evidence of mechanical pressure, and changes in stain density of the wall as an indication of enzymic involvement. However, Aist [1976b] stresses the need for caution in interpreting such visual effects; even the haloes often seen around infection sites may represent a lignification defence reaction rather than an area of enzymic degradation [Ride and Pearce, 1979].

Van Sumere *et al.* [1957] reported the presence of several polysaccharases in extracts of germinating spores of the rust fungus *Puccinia graminis* f. *tritici*; however, in view of the natural origin of these spores, the enzymes may have been formed by contaminating organisms [Wood, 1967]. The role, if any, of such enzymes *in vivo* represents a challenge that should be more readily met now that histochemical [Hislop *et al.*, 1974a], and serological [Hislop *et al.*, 1974b] methods are available for enzyme detection in infected tissues. Possibly such enzymes may be spatially restricted to insoluble forms on the hyphal tip that can aid localized penetration [Stahmann, 1973] and cause no further damage. The main technical problem with most biotrophic pathogens is that they cannot be grown axenically and so it is not possible to produce enzymes to which antibodies can be raised.

Involvement in Pathogenicity Towards Animals

For the reasons already mentioned, microbial enzymes

that degrade polysaccharides have less relevance in pathogenicity to animals. In a recent Society for General Microbiology Symposium on this topic [Smith and Pearce, 1972], there are few index entries for these enzymes. Mims [1976] stated that, although bacteria in general produce a great variety of such enzymes, very few have been shown to be of any clear pathological significance. As frequently occurs also in plant systems, their function is often related to nutrition rather than to a role in the process of infection.

Thus hyaluronidase, at one time regarded as a 'spreading factor' liberated by invasive Streptococci, was thought to liquefy the hyaluronic acid cement of the connective tissue matrix to produce erysipelas. However, it is produced also by the non-invasive *Staphylococcus aureus* and it is no longer regarded as an essential factor in pathogenicity [Mims, 1976].

The microorganisms present in dental plaque [Hardie and Bowden, 1974] utilize ingested carbohydrates, which they oxidize incompletely to release the organic acids responsible for dental caries [Newman and Poole, 1974; Fitzgerald, 1976]. Carbohydrates may also be transformed to intracellular polysaccharides and, in some species, to extracellular polysaccharides. These can be reconverted by glucanases and fructanases to soluble sugars which then serve as acid precursors [Gibbons and Socransky, 1962; Critchley, 1969; Fitzgerald, 1976; Walker and Hare, 1977]. The involvement of the glycanases in caries is thus peripheral.

However, one enzyme - neuraminidase (sialidase) - seems to be involved directly in pathogenicity of bacteria. Its importance stems from the presence of sialic acid units in animal cell surface gangliosides [Cook and Stoddart, 1973]. King and van Heyningen [1973] showed that sialidase-sensitive di- and tri-sialosyl gangliosides do not react with cholera toxin. When treated with neuraminidase, they yield the sialidase-resistant ganglioside GG_nSLC, which binds the toxin. Similar treatment of intestinal mucosal scrapings increased their binding capacity. The nature of the toxin receptor suggests that the neuraminidase of *Vibrio cholerae* may be relevant to the development of the disease. Gascoyne and van Heyningen [1975] subsequently showed that the ability of various tissues to bind cholera toxin was proportional to their content of GG_nSLC. Preincubation of guinea-pig intestinal tissues with sialidase increased toxin binding four-fold. Along similar lines, Staerk *et al.* [1974] showed an enhanced response of canine ileal loops to cholera toxin after a short treatment with neuraminidase, and attributed this to the breakdown of higher gangliosides. Neuraminidase may also be involved in the pathogenicity of *Erysipelothrix* spp. [e.g. Valerianov *et al.*, 1976], *Trichomonas fetus* [Mueller *et al.*, 1974], *Clostridium welchii* in gas gangrene [Fraser and Collee,

1975] and two pathogenic fungi [Mueller, 1975].

Neuraminidase is a constituent of influenza virus, comprising 7 to 15% of the total viral protein [Laver, 1963]. Infection and cytopathic effects can be modified by neuraminidase inhibitors [Scholtissek *et al.*, 1967] or an antibody [Schulman *et al.*, 1968]. However, the anti-neuraminidase modified infection by preventing release of virus from cells rather than as a direct effect on the infection process [Schulman *et al.*, 1968; Schulman, 1970]. A role for neuraminidase in the virulence of Newcastle fowl disease virus, of which it is a structural component [Waterson *et al.*, 1967] is still in doubt [Waterson *et al.*, 1967; Alexander *et al.*, 1970; McNulty *et al.*, 1975].

Conclusions

Against plants, which have much structural polysaccharide, endo-polysaccharases are the enzymes most important in pathogenicity. Against animals, it is an exo-glycosidase (neuraminidase) that is of greatest significance; it modifies important receptor sites on cell surfaces. We should not ignore the possibility that some exo-enzymes have a similar role in plant systems.

Acknowledgements

I thank Prof. D.C. Ellwood and Dr. J.H. Pearce for helpful suggestions on pathogenicity towards animals. Long Ashton Research Station is financed by the Agricultural Research Council.

References

Aist, J.R. (1976a). Papillae and related wound plugs of plant cells. *Annual Review of Phytopathology* **14**, 145-163.

Aist, J.R. (1976b). Cytology of penetration and infection - fungi. In *Physiological Plant Pathology*, pp.197-221. Edited by R. Heitefuss and P.H. Williams. Berlin, Heidelberg and New York: Springer Verlag.

Albersheim, P. and Anderson, A.J. (1971). Proteins from plant cell walls inhibit polygalacturonase secreted by plant pathogens. *Proceedings of the National Academy of Sciences of the United States of America* **68**, 1815-1819.

Albersheim, P. and Anderson-Prouty, A.J. (1975). Carbohydrates, proteins, cell surfaces and the biochemistry of pathogenesis. *Annual Review of Plant Physiology* **26**, 31-52.

Albersheim, P., Jones, T.M. and English, P.D. (1969). Biochemistry of the cell wall in relation to infective processes. *Annual Review of Phytopathology* **7**, 171-194.

Alexander, D.J., Reeve, P. and Allan, W.H. (1970). Characterisation and biological properties of the neuraminidase of strains of Newcastle disease virus which differ in virulence. *Microbios* **6**, 155-165.

Appel, D.N. and MacDonald, W.L. (1976). Endo-polygalacturonase production by selected isolates of *Ceratocystis ulmi*. *Proceedings of the American Phytopathological Society* **3**, 323.

Basham, H.G. and Bateman, D.F. (1975a). Killing of plant cells by pectic enzymes: the lack of direct injurious interaction between pectic enzymes or their soluble reaction products and plant cells. *Phytopathology* **65**, 141-153.

Basham, H.G. and Bateman, D.F. (1975b). Relationship of cell death in plant tissue treated with a homogeneous endo-pectate lyase to cell wall degradation. *Physiological Plant Pathology* **5**, 249-261.

Bateman, D.F. (1976). Plant cell wall hydrolysis by pathogens. In *Biochemical Aspects of Plant-Parasite Relationships*, pp.79-103. Edited by J. Friend and D.R. Threlfall. London, New York and San Francisco: Academic Press.

Bateman, D.F. and Basham, H.G. (1976). Degradation of plant cell walls and membranes by microbial enzymes. In *Physiological Plant Pathology*, pp.316-355. Edited by R. Heitefuss and P.H. Williams. Berlin, Heidelberg and New York: Springer Verlag.

Bateman, D.F. and Millar, R.L. (1966). Pectic enzymes in tissue degradation. *Annual Review of Phytopathology* **5**, 119-146.

Bauer, W.D., Bateman, D.F. and Whalen, C.H. (1977). Purification of an endo-beta-1,4-galactanase produced by *Sclerotinia sclerotiorum*. Effects on isolated plant cell walls and potato tissue. *Phytopathology* **67**, 862-868.

Beraha, L. and Garber, E.D. (1971). Avirulence and extracellular enzymes of *Erwinia carotovora*. *Phytopathologische Zeitschrift* **70**, 335-344.

Bracker, C.E. and Littlefield, L.J. (1973). Structural concepts of host-pathogen interfaces. In *Fungal Pathogenicity and the Plant's Response*, pp.159-318. Edited by R.J.W. Byrde and C.V. Cutting. London and New York: Academic Press.

Brown, W. (1915). Studies in the physiology of parasitism. I. The action of *Botrytis cinerea*. *Annals of Botany* **29**, 313-348.

Brown, W. (1965). Toxins and cell-wall dissolving enzymes in relation to plant disease. *Annual Review of Phytopathology* **3**, 1-18.

Byrde, R.J.W. and Archer, S.A. (1977). Host inhibition or modification of extracellular enzymes of pathogens. In *Cell Wall Biochemistry Related to Specificity in Host-Plant Pathogen Interactions* pp.213-245. Edited by B. Solheim and J. Raa. Tromsø: Universitetsforlaget.

Byrde, R.J.W. and Willetts, H.J. (1977). *The Brown Rot Fungi of Fruit: their Biology and Control*. Oxford: Pergamon Press.

Byrde, R.J.W., Fielding, A.H., Archer, S.A. and Davies, E. (1973). The role of extracellular enzymes in the rotting of fruit tissue by *Sclerotinia fructigena*. In *Fungal Pathogenicity and the Plant's Response*, pp.39-54. Edited by R.J.W. Byrde and C.V. Cutting. London and New York: Academic Press.

Cervone, F., Scala, A. and Scala, F. (1978). Polygalacturonase from *Rhizoctonia fragariae*: further characterization of two isoenzymes and their action towards strawberry tissue. *Physiological Plant Pathology* **12**, 19-26.

Cole, A.L.J. and Sturdy, M.L. (1973). Hemicellulolytic enzymes associated with infection of potato tubers by *Fusarium caeruleum* and

Phytophthora erythroseptica. Abstracts, 2nd International Congress of Plant Pathology, Minneapolis, No. 964.

Cole, M. and Wood, R.K.S. (1961). Types of rot, rate of rotting and analysis of pectic substances in apples rotted by fungi. *Annals of Botany* **25**, 417-434.

Cook, G.M.W. and Stoddart, R.W. (1973). *Surface Carbohydrates of the Eukaryotic Cell.* London and New York: Academic Press.

Cooper, R.M. (1977). Regulation of synthesis of cell wall degrading enzymes of plant pathogens. In *Cell Wall Biochemistry Related to Specificity in Host-Plant Pathogen-Interactions*, pp.163-211. Edited by B. Solheim and J. Raa. Tromsø: Universitetsforlaget.

Critchley, P. (1969). The breakdown of carbohydrate and protein matrix of dental plaque. *Caries Research* **3**, 249-265.

Deese, D.C. and Stahmann, M.A. (1962a). Pectic enzymes and cellulase formation by *Fusarium oxysporum* f. *cubense* on stem tissues from resistant and susceptible banana plants. *Phytopathology* **52**, 247-255.

Deese, D.C. and Stahmann, M.A. (1962b). Pectic enzymes in Fusarium-infected susceptible and resistant tomato plants. *Phytopathology* **52**, 255-260.

Dimond, A.E. and Waggoner, P.E. (1953). On the nature and role of vivotoxins in plant disease. *Phytopathology* **43**, 229-235.

Endo, A. (1963). Studies on pectolytic enzymes of molds. Part VI. The fractionation of pectolytic enzymes of *Coniothyrium diplodiella* (2). *Agricultural and Biological Chemistry* **27**, 751-757.

English, P.D., Jurale, J.B. and Albersheim, P. (1971). Host-pathogen interactions. II. Parameters affecting polysaccharide-degrading enzyme secretion by *Colletotrichum lindemuthianum. Plant Physiology, Lancaster* **47**, 1-6.

Eriksson, K-E. (1977). Degradation of wood cell walls by the rot fungus *Sporotrichum pulverulentum* - enzyme mechanisms. In *Cell Wall Biochemistry Related to Specificity in Host-Plant Pathogen Interactions*, pp.71-83. Edited by B. Solheim and J. Raa. Tromsø: Universitetsforlaget.

Fanelli, C., Cacace, M.G. and Cervone, F. (1978). Purification and properties of two polygalacturonases from *Trichoderma koningii. Journal of General Microbiology* **104**, 305-309.

Fielding, A.H. and Byrde, R.J.W. (1979). Multiple forms of polygalacturonase in fruit and stem tissues infected by *Monilinia fructigena, Monilinia laxa* and *Monilinia laxa* f. *mali*. (in preparation).

Fitzgerald, R.J. (1976). The microbial ecology of plaque in relation to dental caries. In *Proceedings Microbial Aspects of Dental Caries - Special Supplement Microbiology Abstracts*, Vol. III, pp. 849-858. Edited by H.M. Stiles, W.J. Loesche and T.C. O'Brien. Arlington and London: Information Retrieval Ltd.

Fraser, A.G. and Collee, J.G. (1975). Production of neuraminidase by food poisoning strains of *Clostridium welchii (C. perfringens). Journal of Medical Microbiology* **8**, 251-263.

Friedman, B.A. and Ceponis, M.J. (1959). Relationship of loss of pectolytic enzyme synthesis to avirulence of induced mutant strains of a soft rot bacterium. *Phytopathology* **49**, 227.

Garibaldi, A. and Bateman, D.F. (1971). Pectic enzymes produced by *Erwinia chrysanthemi* and their effects on plant tissue.

Physiological Plant Pathology **1**, 25-40.
Gascoyne, N. and van Heyningen, W.E. (1975). Binding of cholera toxin by various tissues. *Infection and Immunity* **12**, 466-469.
Gibbons, R.J. and Socransky, S.S. (1962). Intracellular polysaccharide storage by organisms in dental plaques. *Archives of Oral Biology* **7**, 73-80.
Gooday, G.W. (1977). Biosynthesis of the fungal wall - mechanisms and implications. The first Fleming lecture. *Journal of General Microbiology* **99**, 1-11.
Graniti, A. (1972). The evolution of the toxin concept in plant pathology. In *Phytotoxins in Plant Disease*, pp.1-18. Edited by R.K.S. Wood, A. Ballio and A. Graniti. London and New York: Academic Press.
Grossmann, F. (1962). Therapeutische Wirkung von Pektinase-Hemmstoffen gegen die *Fusarium*-Welke der Tomate. *Naturwissenschaften* **49**, 138-139.
Hancock, J.G. (1967). Hemicellulose degradation in sunflower hypocotyls infected with *Sclerotinia sclerotiorum*. *Phytopathology* 57, 203-206.
Hancock, J.G. (1976). Multiple forms of endo-pectate lyase formed in culture and in infected squash hypocotyls by *Hypomyces solani* f.sp. *cucurbitae*. *Phytopathology* **66**, 40-45.
Hardie, J.M. and Bowden, G.H. (1974). The normal microbial flora of the mouth. In *The Normal Microbial Flora of Man*, pp.47-83. Edited by F.A. Skinner and J.G. Carr. London and New York: Academic Press.
Hislop, E.C., Barnaby, V.M., Shellis, C. and Laborda, F. (1974a). Localization of α-L-arabinofuranosidase and acid phosphatase in mycelium of *Sclerotinia fructigena*. *Journal of General Microbiology* **81**, 79-99.
Hislop, E.C., Shellis, C., Fielding, A.H., Bourne, F.J. and Chidlow, J.W. (1974b). Antisera produced to purified extracellular pectolytic enzymes from *Sclerotinia fructigena*. *Journal of General Microbiology* **83**, 135-143.
Howell, C.R. (1976). Use of enzyme deficient mutants of *Verticillium dahliae* to assess the importance of pectolytic enzymes in symptom expression of Verticillium wilt of cotton. *Physiological Plant Pathology* **9**, 279-283.
Howell, H.E. (1975). Correlation of virulence with secretion *in vitro* of three wall-degrading enzymes in isolates of *Sclerotinia fructigena* obtained after mutagen treatment. *Journal of General Microbiology* **90**, 32-40.
IUPAC-IUB Commission on Biochemical Nomenclature (1972). The nomenclature of multiple forms of enzymes. *Biochemical Journal* **126**, 769-771.
Jensen, W.A. (1962). *Botanical Histochemistry*. San Francisco and London: Freeman.
Jones, T.M., Anderson, A.J. and Albersheim, P. (1972). Host-pathogen interactions. IV. Studies on the polysaccharide-degrading enzymes secreted by *Fusarium oxysporum* f.sp. *lycopersici*. *Physiological Plant Pathology* **2**, 153-166.
Keegstra, K., Talmadge, K.W., Bauer, W.D. and Albersheim, P. (1973). The structure of plant cell walls. III. A model of the walls of

suspension-cultured sycamore cells based on interconnections of the macromolecular components. *Plant Physiology, Lancaster* **51**, 188-197.

Keen, N.T. and Erwin, D.C. (1971). Endo-polygalacturonase: evidence against involvement in Verticillium wilt of cotton. *Phytopathology* **61**, 198-203.

Keen, N.T. and Long, M. (1972). Isolation of a protein-lipopolysaccharide complex from *Verticillium albo-atrum*. *Physiological Plant Pathology* **3**, 307-315.

King, C.A. and van Heyningen, W.E. (1973). Deactivation of cholera toxin by a sialidase-resistant monosialosylganglioside. *Journal of Infectious Diseases* **127**, 639-647.

Knee, M. and Friend, J. (1970). Some properties of the galactanase secreted by *Phytophthora infestans* (Mont.) de Bary. *Journal of General Microbiology* **60**, 23-30.

Laver, W.G. (1963). The structure of influenza viruses. 3. Disruption of the virus particle and separation of neuraminidase activity. *Virology* **20**, 251-262.

Lewis, D.H. (1974). Microorganisms and plants: the evolution of parasitism and mutualism. In *Evolution in the Microbial World*, pp.367-392. Edited by M.J. Carlile and J.J. Skehel. *Symposia of the Society for General Microbiology* 24. Cambridge: Cambridge University Press.

Lule, J., Reme-Delga, A-M. and Barthe, J-P. (1977). Extracellular polygalacturonase activity and surface products *in vitro* in two strains of *Colletotrichum lagenarium* differing in aggressiveness. *Compte rendu hebdomadaire des séances de l'Academie des sciences, Series D* **284**, 29-32.

Lumsden, R.D. and Bateman, D.F. (1966). Pectic enzymes detected in culture filtrates of *Thielaviopsis basicola* and in extracts of Thielaviopsis-infected bean root tissue. *Phytopathology* **56**, 585.

McNulty, M.S., Gowans, E.J., Houston, M.J. and Fraser, G. (1975). Neuraminidase content of strains of Newcastle disease virus which differ in virulence. *Journal of General Virology* **27**, 399-402.

Markert, C.L. (1975). Biology of isozymes. In *Isozymes. I. Molecular Structure*, pp.1-9. Edited by C.L. Markert. New York and London: Academic Press.

Mims, C.A. (1976). *The Pathogenesis of Infectious Diseases*, p.70. London and New York: Academic Press.

Miyairi, K., Fujita, K., Okuno, T. and Sawai, K. (1977). A toxic protein causative of silver leaf disease symptoms on apple trees. *Agricultural and Biological Chemistry* **41**, 1897-1902.

Moon, T.W. (1975). Temperature adaptation, isozymic function and the maintenance of heterogeneity. In *Isozymes. II. Physiological Function*, pp.207-220. Edited by C.L. Markert. New York and London: Academic Press.

Mueller, H.E. (1975). Occurrence of neuraminidase in *Sporothrix schenckii* and *Ceratocystis stenocereas* and its role in ecology and pathomechanism of these fungi. *Zentralblatt für Bakteriologie, Parasitenkunde, Infektionskrankheiten und Hygiene, Abteil 1* **232**, 365-372.

Mueller, H.E., Nicolai, H. and Zilliken, F. (1974). Neuraminidase bei *Trichomonas fetus* und ihre mögliche pathogenetische Bedeutung. *Pathologia et Microbiologia* **41**, 323-333.

Mukhopadhyay, D.K. and Chakravarty, D.K. (1976). Studies on different isolates of *Xanthomonas oryzae*, the incitant of bacterial leaf blight of rice. *Zeitschrift für Pflanzenkrankheiten und Pflanzenschutz* **83**, 674-678.

Mussell, H. and Strand, L.L. (1977). Pectic enzyme involvement in pathogenesis and possible relevance to tolerance and specificity. In *Cell Wall Biochemistry Related to Specificity in Host-Plant Pathogen Interactions*, pp.31-70. Edited by B. Solheim and J. Raa. Tromsø: Universitetsforlaget.

Newman, H.N. and Poole, D.F.G. (1974). Structural and ecological aspects of dental plaque. In *The Normal Microbial Flora of Man*, pp.111-134. Edited by F.A. Skinner and J.G. Carr. London and New York: Academic Press.

Paquin, R. and Colombe, L.J. (1962). Pectic enzyme synthesis in relation to virulence in *Fusarium oxysporum* f. *lycopersici* (Sacc.) Snyder and Hansen. *Canadian Journal of Botany* **40**, 533-541.

Patil, S.S. and Dimond, A.E. (1968). Repression of polygalacturonase synthesis in *Fusarium oxysporum* f.sp. *lycopersici* by sugars and its effects on symptom reduction in infected tomato plants. *Phytopathology* **58**, 676-682.

Pegg, G.F. (1977). Glucanohydrolases of higher plants: a possible defence mechanism against parasitic fungi. In *Cell Wall Biochemistry Related to Specificity in Host-Plant Pathogen Interactions*, pp.305-345. Edited by B. Solheim and J. Raa. Tromsø: Universitetsforlaget.

Pressey, R. and Avants, J.K. (1977). Occurrence and properties of polygalacturonase in *Avena sativa* and other plants. *Plant Physiology, Bethesda* **60**, 548-553.

Pupillo, P., Mazzucchi, U. and Pierini, G. (1976). Pectic lyase isozymes produced by *Erwinia chrysanthemi* Burkh. *et al.* in polypectate broth or in *Dieffenbachia* leaves. *Physiological Plant Pathology* **9**, 113-120.

Raa, J., Robertsen, B., Solheim, B. and Tronsmo, A. (1977). Cell surface biochemistry related to specificity of pathogenesis and virulence of micro-organisms. In *Cell Wall Biochemistry Related to Specificity in Host-Plant Pathogen Interactions*, pp.11-30. Edited by B. Solheim and J. Raa. Tromsø: Universitetsforlaget.

Ride, J.P. (1975). Lignification in wounded wheat leaves in response to fungi and its possible role in resistance. *Physiological Plant Pathology* **5**, 125-134.

Ride, J.P. and Drysdale, R.B. (1972). A rapid method for the chemical estimation of filamentous fungi in plant tissue. *Physiological Plant Pathology* **2**, 7-15

Ride, J.P. and Pearce, R.B. (1979). Physiological Plant Pathology, in press.

Scholtissek, C., Becht, H. and Drzeniek, R. (1967). Biochemical studies on the cytopathic effect of influenza viruses. *Journal of General Virology* **1**, 219-225.

Schulman, J.L. (1970). Effects of immunity on transmission of influenza: experimental studies. *Progress in Medical Virology* **12**, 128-160.

Schulman, J.L., Khakpour, M. and Kilbourne, E.D. (1968). Protective effect of specific immunity to viral neuraminidase on influenza virus infection of mice. *Journal of Virology* 2, 778-786.

Sharma, P.D., Fisher, P.J. and Webster, J. (1977). Critique of the chitin assay technique for estimation of fungal biomass. *Transactions of the British Mycological Society* **69**, 479-483.

Singh, G.P. and Husain, A. (1964). Relation of hydrolytic enzyme activity with virulence of strains of *Colletotrichum falcatum. Phytopathology* **54**, 1100-1101.

Smith, H. and Pearce, J.H. (editors) (1972). *Microbial Pathogenicity in Man and Animals.* Cambridge: University Press.

Somero, G.N. (1975). The roles of isozymes in adaptation to varying temperatures. In *Isozymes. II. Physiological Function*, pp.221-234. Edited by C.L. Markert. New York and London: Academic Press.

Staerk, J., Ronneberger, H.J. and Wiegandt, H. (1974). Neuraminidase. Virulence factor in *Vibrio cholerae* infection. *Behring Institut Mitteilungen* **55**, 145-146. (*Chemical Abstracts* **82**, 122714z, 1975).

Stahmann, M.A. (1973). Contribution to discussion. In *Fungal Pathogenicity and the Plant's Response*, pp.156-157. Edited by R.J.W. Byrde and C.V. Cutting. London and New York: Academic Press.

Starr, M.P. and Chatterjee, A.K. (1972). The genus *Erwinia:* enterobacteria pathogenic to plants and animals. *Annual Review of Microbiology* **26**, 389-426.

Swinburne, T.R. and Corden, M.E. (1969). A comparison of the polygalacturonases produced *in vivo* and *in vitro* by *Penicillium expansum* Thom. *Journal of General Microbiology* **55**, 75-87.

Toppan, A., Esquerré-Tugayé, M.T. and Touzé, A. (1976). An improved approach for the accurate determination of fungal pathogens in diseased plants. *Physiological Plant Pathology* **9**, 241-251.

Tribe, H.T. (1955). Studies in the physiology of parasitism. XIX. On the killing of plant cells by enzymes from *Botrytis cinerea* and *Bacterium aroideae. Annals of Botany* **19**, 351-368.

Unbehaun, L.M. and Moore, L.D. (1970). Pectic enzymes associated with black root rot of tobacco. *Phytopathology* **60**, 304-308.

Valerianov, Ts., Nikolov, P. and Abrashev, I. (1976). On the correlation in decreasing the virulence and the neuraminidase activity in lysogenic variants of *Erysipelothrix rhusiopathiae. Doklady Bulgarska Akademiya na Naukite* **29**, 1813-1814. (*Chemical Abstracts* **86**, 167809h, 1977).

Van Sumere, C.F., Van Sumere-de Preter, C. and Ledingham, G.A. (1957). Cell wall-splitting enzymes of *Puccinia graminis* var. *tritici. Canadian Journal of Microbiology* **3**, 761-770.

Walker, G.J. and Hare, M.D. (1977). Metabolism of the polysaccharides of human dental plaque. Part 2. Purification and properties of *Cladosporium resinae* 1-3-α-D-glucanase EC 3.2.1.59 and the enzymic hydrolysis of glucans synthesized by extracellular D-glucosyl transferases of oral streptococci. *Carbohydrate Research* **58**, 415-432.

Waterson, A.P., Pennington, T.H. and Allan, W.H. (1967). Virulence in Newcastle disease virus: a preliminary study. *British Medical Bulletin* **23**, 138-143.

Willetts, H.J., Byrde, R.J.W., Fielding, A.H. and Wong, A-L. (1977). The taxonomy of the brown rot fungi (*Monilinia* spp.) related to their extracellular cell wall-degrading enzymes. *Journal of*

General Microbiology **103**, 77-83.

Williams, P.H., Aist, J.R. and Bhattacharya, P.K. (1973). Host-parasite relations in cabbage clubroot. In *Fungal Pathogenicity and the Plant's Response*, pp.141-158. Edited by R.J.W. Byrde and C.V. Cutting. London and New York: Academic Press.

Wood, R.K.S. (1967). *Physiological Plant Pathology*. Oxford: Blackwell.

Wood, R.K.S. (1976). Killing of protoplasts. In *Biochemical Aspects of Plant-Parasite Relationships*, pp.105-116. Edited by J. Friend and D.R. Threlfall. London, New York and San Francisco: Academic Press.

Chapter 19

A SURVEY OF POLYSACCHARASE PRODUCTION: A SEARCH FOR PHYLOGENETIC IMPLICATIONS

G.W. GOODAY

Department of Microbiology, University of Aberdeen, Aberdeen, UK

Introduction

Polysaccharides are the most abundant organic materials on the earth. They represent enormous resources of carbon and nitrogen in all ecosystems. Thus it is not surprising to find organisms feeding on them - on the contrary, it is surprising that the ability to utilize directly many common polysaccharides is restricted to only a few types of organism. These organisms consequently play central roles in the recycling of carbon and nitrogen in environments rich in polysaccharides, and have been exploited as symbiotic partners by animals as diverse as the gardening ambrosia beetles and ruminants.

This chapter will describe the organisms that degrade polysaccharides, with the aim of encouraging some thoughts on their comparative biochemistry and on the evolution of their polysaccharide-degrading systems. Only the genuine polysaccharases will be considered; those enzymes hydrolysing the glycosidic linkages of macromolecules, and of these, only some of those used by the organisms as extracellular digesters of potential food sources. Thus enzymes that are very important in their own right in polysaccharide degradation, such as lyases, phosphorylases and glycosidases are not discussed, neither are the polysaccharases involved in morphogenesis of the cells, such as autolysins [see Berkeley, Chapter 9; Rogers, Chapter 10], nor those regulating the utilization of intracellular reserve polysaccharides.

Cellulases

The simple description by the International Commission for Enzyme Nomenclature of the activity of 'cellulase', EC 3.2.1.4, as 'endohydrolysis of 1,4 β-glucosidic links in cellulose' belies the large amount of effort that has gone into unravelling the mechanisms of the microbial systems capable of hydrolysing cellulose. Cellulose is a widespread source of carbon, as it is a major component of the cell walls of all angiosperms, gymnosperms, ferns, liverworts, mosses, some algae and a few fungi. Thus possession of a means of breaking down cellulose is advantageous to pathogenic microbes so that they can enter the cells of these organisms, and so that they can make full use of the available food [see Byrde, Chapter 18]; to saprophytic microbes which use the cellulosic material in dead plant tissue as a carbon source [see Poole, Chapter 17]; and to the symbiotic microbes in the digestive systems of animals feeding on plants or plant remains such as termites and ruminants [see Hobson, Chapter 16].

However, cellulose exists in a highly ordered structure, and there is no one enzyme that can degrade completely native cellulose to its constituent cellobiose or glucose units. Rather, cellulase complexes are produced [see Bacon, Chapter 11; Eriksson, Chapter 12], with the C_1 enzymes attacking the crystalline structure but causing little degradation, and the C_X enzymes rapidly degrading the resultant accessible substrate in an endo- and exo-fashion [Wood, 1975]. Any one cellulolytic microbe will produce several (hence the 'X' of C_X) of these enzymes, which can be separated by molecular exclusion chromatography and subsequently purified to homogeneity.

Such separable enzyme activities from any one organism characteristically show different properties such as kinetic parameters for action on model substrates, products of action on model substrates, thermostability, pH and temperature optima, and amount of associated carbohydrate (ranging from nil to about 50%) [Wood, 1975]. Examples are cellulases A, B and C from the bacterium *Pseudomonas fluorescens* var. *cellulosa* [Yamane *et al.*, 1970] and cellulases II, III and IV from the fungus *Trichoderma viride* [Okada *et al.*, 1968].

It is generally considered that the C_1 and the different C_X activities act synergistically to degrade the macromolecular crystalline cellulose but Goksøyr *et al.* [1975] question whether all microbes have a C_1-type cellulase, and suggest that this is confined to the Ascomycotina, their related imperfect fungi, and the white-rot Basidiomycotina.

Unfortunately it is not yet possible to answer the question: are the cellulases of any one organism or of the different organisms phylogenetically related and do they share a common ancestral protein? There are no available amino acid sequences for cellulases - all that we have are amino acid compositions for a handful of fungal and bacterial enzymes. For example, the amino acid compositions for two cellulases each from the bacterium *Sporocytophaga myxococcoides* and the fungus *Trichoderma viride* and five for the fungus *Sporotrichum pulverulentum* have been reported [Osmundsvag and Goksøyr, 1975; King and Vessal,

1969; Eriksson and Pettersson, 1975]. Goksøyr *et al.* [1975] have compared the amino acid compositions of fourteen cellulases, from five fungi and two bacteria, by a procedure of calculating the sum of the squares of the differences between the mole % values for each amino acid in two proteins. They conclude that these cellulases can be classified into two groups, those from the fungi and *Sporocytophaga myxococcoides*, and those from *Pseudomonas fluorescens*. Within each group, the similarity in composition was close enough to suggest a relationship between these enzymes. The closest relationships were between the enzymes I and II of *Sporocytophaga myxococcoides*, and also between the enzymes II, III and IV of *Trichoderma viride*. Thus in at least some cases the different cellulases of one organism might have a common ancestor.

As is true for most of the polysaccharases described here, the ability to degrade cellulose is a property that occurs sporadically among microbes. Early comprehensive reviews include Thaysen and Bunker [1927], Norman and Fuller [1942], Imschenezki [1959], Gascoigne and Gascoigne [1960].

The most strongly cellulolytic bacteria are among the gliding bacteria found in soil. Thus *Cytophaga hutchinsonii* and *Sporocytophaga myxococcoides* require cellulose, cellobiose or glucose as carbon source [Stanier. 1942a; Leadbetter, 1974]. These organisms characteristically digest cellulose fibres by directly investing them; creeping around them and into any interstices. Other less strongly cellulolytic species of *Cytophaga* are discussed by Christensen [1977], and another species in the same family with some cellulolytic ability is the hot spring bacterium, *Herpetosiphon geysericola* [Lewin, 1970]. Among the myxobacteria, cellulose digestion is confined to species of the Polyangiaceae, most notably *Polyangium cellulosum* [Peterson, 1974].

The family Pseudomonadaceae contains species that have been used for detailed studies of cellulase production, such as the plant pathogenic isolate termed *Pseudomonas fluorescens* var. *cellulosa* [Yamane *et al.*, 1970], and *Cellvibrio gilvus* (= *Pseudomonas* sp.?) [King and Vessal, 1969]. This latter species grows well on cellulose and its oligosaccharides, but poorly on glucose. Other Gram-negative cellulolytic bacteria include isolates of the phytopathogen *Erwinia carotovora* [Tseng, 1974], and of *Flavobacterium capsulatum* [Weeks, 1974].

Cellulose forms a significant part of the diet of ruminants [see Hobson, Chapter 16], so that it is not surprising to find anaerobic cellulolytic bacteria in the rumen: *Ruminococcus flavefaciens*, *Ruminococcus albus*, *Butyrivibrio fibrisolvens*, *Eubacterium cellulosolvens* (= *Cillobacterium cellulosolvens*), although not all strains of these species are cellulolytic. Van Gylswyk and Labuschagne [1971] investigated the relative efficiency of these bacteria in solubilizing cellulose *in vitro*. They found that strains of *Eubacterium cellulosolvens* gradually but irreversibly lost their cellulolytic ability when grown on cellobiose-containing medium; *Butyrivibrio fibrisolvens* grew better on cellobiose than on cellulose; while the *Ruminococcus* strains grew as well on cellulose as on cellobiose. Thus for *Ruminococcus flavefaciens* and *Ruminococcus albus* cellulose degradation is not a rate-limiting step for growth, and these species can be regarded as the "true" cellulolytic inhabitants of the rumen. Other anaerobic cellulose digesters include *Bacteroides succinogenes*, *Clostridium cellobioparum*, and the actinomycete *Micromonospora ruminantium*, all isolated from the rumen [Hungate, 1950; Malwszynska and Janota-Bassalik, 1974], and *Clostridium thermocellum*, a thermophilic species from soil which only ferments cellulose, cellobiose and a few other carbohydrates [Weimer and Zeikus, 1977].

Some strains of *Bacillus* species, such as an isolate of *Bacillus polymyxa* [Fogarty and Griffin, 1973] have been reported to be weakly cellulolytic.

All strains of the coryneform bacteria of the genus *Cellulomonas* are cellulolytic. Only one species, *Cellulomonas flavigens*, is recognized by Keddie [1974]. This is a soil organism. Isolates grow poorly on 'ordinary media', but Stewart and Leatherwood [1976] illustrate the colonies surrounded by clear zones obtained after growth for two to three days on a cellulose-containing agar medium.

Many actinomycetes, occurring in soils, composts, manures and fodders, can degrade cellulose [Lacey, 1973]. Thus Stutzenberger [1972] describes the production of cellulases by the thermophilic *Thermomonospora curvata* isolated from compost, but their activity was much lower than fungal cellulases, as was that of cellulases from a mesophilic *Streptomyces* species [Mandels, 1975].

Cellulases are produced more widely by fungi than by bacteria. This production has been studied in great detail because of the economic interest from two sides, the recycling of waste cellulose, and the biodegradation of cellulosic materials. Many general saprophytes produce cellulases, as do wood-rotting fungi, and plant pathogens [see Byrde, Chapter 18]. Such fungi are chiefly of the Ascomycotina and the related imperfect fungi, and Basidiomycotina. An extensive survey of many isolates of fungi [Reese *et al.*, 1950] shows how widespread cellulolysis is among the saprophytic fungi. Four species of imperfect fungi are particularly rich sources of cellulase enzyme systems: *Trichoderma viride*, *Trichoderma koningii*, *Penicillium funiculosum* and *Fusarium solani* [Wood, 1975]. Reese [1969] gives quantitative results of a survey for cellulase production, showing appreciable activities from Basidiomycotina such as *Polyporus cinnabarinus* and *Lenzites trabea*. However, particularly amongst the imperfect fungi, there seems to be little pattern to the occurrence of cellulolysis. For example, Fergus [1969] found that of the genus *Humicola*, *Humicola insolens* and *Humicola grisea* were strong cellulose degraders, whereas *Humicola lanaginosa* and *Humicola stellata* had no cellulolytic ability.

There are few authentic reports of cellulolytic activity by members of the Oomycotina. Thus Park [1975], in describing a cellulolytic *Pythium* species isolated from a river, questions the validity of methods used by earlier workers, and Thomas and Mullins [1969], describing the production of extracellular cellulase by *Achlya ambisexualis*, suggest that it is not a 'nutritional enzyme', but a 'morphogenetic enzyme' acting on the organism's own cellulosic cell wall. However, McIntyre and Hankin [1978] report that 54 out of 55 isolates of plant pathogenic *Phytophthora* species produced appreciable cellulase activity. This observation, coupled with Park's study of *Pythium* Cl, indicates that it would be worthwhile to reinvestigate these fungi.

Members of the Zygomycotina also show very little propensity for cellulose degradation, with the C_X activity of a strain of the thermophilic *Mucor pusillus* [Somkuti *et al.*, 1969] being one of the few examples. However, Fergus [1969] was unable to detect cellulose digestion by this species.

Glucanases other than Cellulase

There are at least a dozen classes of polysaccharases responsible for the endo-hydrolysis of the plethora of glucans produced by living organisms. Production of these glucanases is widespread among microbes and Littlejohn [1973] and Brown [1973] tabulate the occurrence and activities of microbial amylases, dextranases and β-glucanases. Reviews include Bull and Chesters [1966] on laminarinase, Lee and Whelan [1971] on glycogen and starch debranching enzymes and Takagi *et al.* [1971] on amylases.

That subtle differences can occur in the patterns of glucanase production in closely related organisms is shown by the work of Wöber [1973] and Norrman and Wöber [1975] on glucan utilization by three species of *Pseudomonas*. *Pseudomonas saccharophila* produced an α-amylase and a pullulanase (de-branching the α 1,6 links) and completely degraded amylopectin and glycogen *in vivo*; *Pseudomonas stutzeri* produced a specific amylase (giving maltotetraose as product from starch as opposed to the maltose from α-amylase) and a pullulanase but only partially degraded amylopectin and glycogen *in vivo*; while *Pseudomonas amyloderamosa* produced only a debranching isoamylase, only partially utilizing the branched α-glucans at pH 6.0, and not growing at all at pH 6.5 (probably due to low activity of the isoamylase at this pH). After the respective actions of these four different endo-glucanases, all three organisms assimilated the resultant maltodextrins via a permease, a glucanotransferase, and a phosphorylase.

Also significantly different are the properties of purified endo-β 1,3-D-glucanases from the zygomycetes *Rhizopus arrhizus* QM 1032 [Clark *et al.*, 1978] and *Rhizopus chinensis* R-69 [Yamamoto and Nagasaki, 1975]. Clark *et al.* present their results of molecular weight determinations for the enzyme of 28,800 (by equilibrium sedimentation), 29,800 (by sodium dodecyl sulphate gel-electrophoresis), and 42,700 (by gel permeation chromatography - the larger value probably being the result of a shape factor), and contrast these estimates with those of 10,200 and > 100,000 for the same enzyme by previous authors. They also present its amino acid composition, and its carbohydrate composition, suggesting that it is a glycoprotein, with five mannose and one galactose linked to the protein by an N-acetylglucosamine residue. The enzyme has preferential hydrolytic activity on stretches of β 1,3-linked D-glucosyl residues, although it hydrolyses β 1,4 linkages if these are flanked by β 1,3-linked glucose units, but it cannot attack β 1,6 linkages. It has significant transglycosylase activity. Clark *et al.* suggest that 'the active site of the *Rhizopus* enzyme is analogous to that of lysozyme in that it contains a number of binding sites designed to fit the glucosyl residues in the β 1,3 D-glucose flexible helix and that these binding sites extend on both sides of the location of the catalytic mechanism'.

Hemicellulases

The hemicelluloses are a heterogeneous group of polysaccharides associated with the cellulose of the plant cell wall. They are usually heteroglycans, with the hexose constituents D-galactose and D-mannose and the pentose components D-xylose and L-arabinose. The occurrence and biochemistry of the hemicellulases: D-galactanases, D-mannanases, β 1,3- and β 1,4-D-xylanases, and L-arabinanases - have been reviewed by Dekker and Richards [1976].

There are relatively few reports of these enzymes among the bacteria, but perhaps this is for want of looking. Among the gliding bacteria, the potent cellulolytic organism *Sporocytophaga myxococcoides* also produces mannanases and xylanases [Clermont *et al.*, 1970]; various *Bacillus* species produce galactanases, mannanases and xylanases [Emi and Yamamoto, 1972; Inaoka and Soda, 1956; Fogarty and Ward, 1973]; amongst anaerobic bacteria, arabinanase activity has been characterized from the plant pathogen *Clostridium felsineum* [Kaji *et al.*, 1963], and galactanases, xylanases and arabinanases from the cellulolytic rumen bacteria *Butyrivibrio fibrisolvens*, *Bacteroides succinogenes*, *Ruminococcus albus* and *Ruminococcus flavefaciens* [Clarke *et al.*, 1969; Dehority, 1967]; and amongst the actinomycetes, xylanases have been described from *Streptomyces* species [Sakai *et al.*, 1975]. Fujisawa and Murakami [1971] isolated sixty four strains of β 1,3 xylan-degrading bacteria from seawater, seaweeds and marine sediments. Of these, thirty six also had the ability to degrade β 1,4 xylan.

From the compilations of Dekker and Richards [1976] it is clear that hemicellulases are widespread amongst the fungi, all classes having been described from plant pathogenic and saprophytic isolates of oomycetes, zygomycetes, ascomycetes, basidiomycetes and the imperfect fungi. Bailey and Gaillard [1965] describe their production by rumen protozoa, such as *Epidinium ecaudatum*, *Entodinium* sp. and *Eremoplastron bovis*.

Chitinases

Chitin is the major structural polysaccharide of most invertebrates and fungi. It is produced in very large quantities on land, but also in enormous quantities in the sea where for example the planktonic crustaceans continually make and cast off their exoskeletons. Idealized chitin is a homopolymer of β 1,4-linked *N*-acetylglucosamine units [see Berkeley, Chapter 9], and so can act as a nitrogen source as well as a carbon source for the organisms able to utilize it. Chitin is analogous to cellulose, in that it can be thought of as made of linear assemblages of β 1,4-linked substituted glucose units. Like cellulose, it forms microfibrils by adjacent chains crystallizing in alignment, being strongly consolidated by interchain hydrogen bonds. However, although it must be said that chitinase activity has not been investigated in anywhere near the detail of cellulase activity, there seems to be little suggestion of an activity equivalent to that of C_1 cellulase, apart from the CH_1 described by Monreal and Reese [1969]. There are activities equivalent to those of C_X cellulases. Thus chitinolytic culture filtrates from the slime mould *Physarum polycephelum* can be resolved into several components showing a range of endo- and exo-activities [Stirling, 1979]. There is an overlap in activities between chitinase (EC 3.2.1.14) and lysozyme (EC 3.2.1.17), in that many lysozymes have some activity against chitin [Jollès *et al.*, 1974]. There is clearly a continuum of specificities between enzymes that solely recognize oligomers of β 1,4-linked *N*-acetylglucosamine and those that solely recognize the alternating linkages of *N*-acetylglucosamine and *N*-acetylmuramic acid and hydrolyse at one or other glycosidic bond [see Rogers, Chapter 10].

There are characteristic chitin-degrading microfloras in environments rich in chitin [see Berkeley, Chapter 9]. Such environments include: fragments of exoskeleton of marine zooplankton, whether suspended in the sea [Kaneko and Colwell, 1973], in sediments [Timmis *et al.*, 1974], in fish guts [Goodrich and Morita, 1977], or still attached to the living animal [Hood and Meyers, 1977]; remains of fungi and insects in soil [Okafor, 1966]; rotting toadstools [Johnson, 1932].

Among the gliding bacteria, *Cytophaga johnsonae* is strongly chitinolytic. This is the only cytophaga which degrades chitin [Christensen, 1977], as surprisingly none of the marine species have this ability despite the widespread occurrence of chitin in the sea. *Cytophaga johnsonae* is a ubiquitous soil organism [Stanier, 1947; Veldkamp, 1955]. Sundarraj and Bhat [1972] divided *Cytophaga* strains into two groups, those that only hydrolysed chitin in close contact with them, and those that released cell-free chitinase activity. An isolate of the first type, C35, grew much slower than one of the second type, C31, in an unshaken liquid medium with chitin as carbon source, and the cells adhered to the chitin particles. Shaking the cultures slowed the growth of C31, and prevented growth of C35.

Studies of chitin decomposition in the sea, particularly in sediments and in fish guts, have yielded many isolates of Gram-negative chitinolytic bacteria. Thus of nearly three hundred chitinolytic isolates reported by Chan [1970], chiefly from sediments in Puget Sound, the great majority were assigned to the genus *Vibrio*, with some to *Pseudomonas* and a few to *Aeromonas*; Kaneko and Colwell [1973] report *Vibrio parahaemolyticus* as the major organism associated with chitin and copepods in Chesapeake Bay; Aribisala and Gooday [1978] report *Vibrio alginolyticus* (= biotype II, *Vibrio parahaemolyticus*) as the major organism in the gut and on the surface of Nigerian shrimps, together with *Pseudomonas* species in lesser abundance; Hood and Meyers [1977] report *Beneckea* (= *Vibrio* species?) from white shrimps in Louisiana; Spencer [1961] reports that 30 out of 33 isolates ascribed to the genus *Photobacterium* were chitinolytic, and Hendrie *et al*. [1970], revising the taxonomy of luminous bacteria, state that chitin is degraded by *Photobacterium phosphoreum*, *Photobacterium mandapamensis* and *Lucibacterium harveyi*; and Campbell and Williams [1951] describe chitinolytic bacteria, including *Flavobacterium indolthecium*, from marine mud.

Chitin degradation has been reported for some species of the Enterobacteriaceae. *Serratia marcescens* and *Enterobacter liquefaciens* (which Ewing *et al*. [1973] reclassify as a colourless species of *Serratia*) were by far the most strongly chitinolytic organisms of 30 bacteria and 70 fungi tested by Monreal and Reese [1969], releasing into the medium 68 and 74 units of enzyme activity respectively compared to 9.6 units and 1.9 units produced by the next two most active bacteria, a *Streptomyces* species and *Cellvibrio vulgaris* (= *Pseudomonas* species?) respectively. Production of the chitinase was induced by 1% chitin in the medium but completely repressed by the addition of 0.5% glucose to the chitin medium. No other polysaccharase activities could be detected in the culture filtrates. *Serratia marcescens* is a widespread organism, occurring in soil, water, food, in clinical specimens, and Monreal and Reese [1969] quote it as being able to grow vigorously on insects, when presumably its chitinolytic ability could serve a nutritional role. In another extensive survey for extracellular chitinolysis by bacteria, Clarke and Tracey [1956] report that among the Enterobacteriaceae, strains of *Klebsiella* species produced chitinase activity (together with cellulolytic activity against carboxymethylcellulose), whereas those of *Escherichia coli* and *Salmonella* species did not. However, their assay for chitinase was sensitive; Molise and Drake [1973] being unable to detect any signs of zones of clearing in chitin-agar plates "under any condition even after several months of incubation" by any members of the family Enterobacteriaceae other than *Serratia* species, all strains of which were chitinolytic. Molise and Drake suggest that chitinolysis could be a useful character with which to differentiate serratiae from other members of the Enterobacteriaceae.

Amongst Gram-positive bacteria, chitinolysis has been reported in: *Micrococcus colpogenes* from marine mud [Campbell and Williams, 1951]; isolates of *Bacillus* species, such as a *Bacillus cereus* strain from Nigerian shrimps [Aribisala and Gooday, 1978]; and the 15 isolates of *Clostridium* species described by Timmis *et al.* [1974] from anaerobic marine mud.

Actinomycetes can commonly use chitin as a source of carbon and nitrogen, so much so that chitin agar is the most widely used selective medium for the enumeration and isolation of these organisms [Küster, 1973]. Veldkamp [1955] and Okafor [1966] describe chitin-decomposing actinomycetes as being very common in soils that they examined, in particular species attributed to the genera *Streptomyces*, *Micromonospora*, *Nocardia* and *Actinoplanes*. Jeuniaux [1963] and Skujins *et al.* [1970] describe the purification and properties of chitinase from *Streptomyces* species, and estimate the molecular weight of the major activity to be about 30,000.

Extracellular chitinolysis (as opposed to 'morphogenetic' autolysis in the chitinous fungi) seems fairly widespread amongst fungi, but only in a few cases is it possible to ascribe a major nutritional role to this. Thus chitinolytic arthropod pathogens are found among the cellulosic Oomycotina [Unestam, 1966], in the zygomycete *Conidiobolus villosus* [Monreal and Reese, 1969], in the imperfect fungus *Beauvaria bassiana* [Leopold and Samsinakova, 1970], and in the ascomycete genus *Cordyceps* [Huber, 1958], which has mycoparasitic species as well as insect pathogenic members. Chitinolytic chytrids such as *Chytriomyces* species are found on chitinous residues in soil and fresh water [Reisert and Fuller, 1962], and one species, *Karlingia astereocysta*, was found by Murray and Lovett [1966] to have a nutritional requirement for chitin that could only be relieved by *N*-acetylglucosamine; it is an 'obligate chitinophile'. Amongst the Zygomycotina, Veldkamp [1955] describes the isolation of chitinolytic strains of *Mortierella* from soil, and Monreal and Reese [1969] found *Mucor subtilissimum* to be moderately chitinolytic. Both Veldkamp [1955] and Monreal and Reese [1969] report strongly chitinolytic isolates of the imperfect fungus *Aspergillus fumigatus*, a species found in soil and as a pathogen in lungs of man and animals. *Trichoderma viride*, a species well-renowned for its potent cellulolytic capability, also produces strong chitinolytic activities [Vries and Wessels, 1973].

The Myxomycetes, 'true slime moulds', appear to be potentially rich sources of extracellular lytic enzymes, and *Physarum polycephalum* produced a complex of extracellular chitinases [Stirling, 1979]. Soil amoebae, *Hartmanella* and *Schizopyrenus* species, can produce chitinases (together with cellulases active against carboxymethylcellulose) [Tracey, 1955].

Chitosanases

Chitosan, poly-β 1,4 glucosamine, is a major component of the hyphal walls of the zygomycete fungi. It occurs in these walls together with chitin as microfibrils, and is apparently formed by the deacetylation of nascent chitin. These fungi are very common, particularly in soil, as genera such as *Mucor, Rhizopus* and *Mortierella*. By screening organisms for their ability to lyse cell wall preparations of *Rhizopus* species, Monaghan *et al.* [1973] showed that chitosanase activity is widespread in microbes. Fenton *et al.* [1978] describe chitosan-degrading organisms as being very common in soil. They tested chitosan-degrading organisms for the ability also to degrade chitin and cellulose, and found that some could utilize only chitosan, some chitosan and chitin, some chitosan and cellulose, and some all three.

Some chitosanases have been investigated in more detail. None have any chitinase or lysozyme activities. Those from *Penicillium islandicum* and *Streptomyces* No. 6. have a specific endo-hydrolytic activity towards chitosan [Monaghan *et al.*, 1973; Price and Storck, 1975; Fenton *et al.*, 1978]. However, purified chitosanases from *Myxobacter* AL-1 [Hedges and Wolfe, 1974] and two unidentified bacteria [Fenton *et al.*, 1978] also have cellulolytic activity against carboxymethylcellulose. Thus the enzyme from *Myxobacter* sp. had Km values of 0.3 and 1.68 g l^{-1} and V_{max} values of 0.75 x 10^{-9} and 2.2 x 10^{-9} mol min^{-1} against chitosan and carboxymethylcellulose respectively. The enzyme was a glycoprotein, with about 1.3% carbohydrate. This dual lytic activity is not a property of cellulases *per se*, as purified cellulase preparations from

Pseudomonas fluorescens var. *cellulosa* and *Trichoderma viride* had no activity against chitosan. However, the existence of this enzyme from *Myxobacter* sp. suggests that chitosanases might evolve from cellulases or vice versa.

The taxonomic spectrum for the production is similar to those for chitinases and cellulases. Thus amongst the gliding bacteria it is produced by a *Myxobacter* sp. [Hedges and Wolfe, 1974] and by *Sporocytophaga myxococcoides* [Monaghan *et al.*, 1973]; amongst the Bacillaceae by *Bacillus* R-4 [Tsujisaka *et al.*, 1975] and *Bacillus* S6E [Monaghan *et al.*, 1973]; amongst the Actinomycetes and related organisms by *Arthrobacter* sp. GMJ-1 and several *Streptomyces* species [Monaghan *et al.*, 1973]; and in the fungi by zygomycetes such as *Rhizopus rhizopodiformis*, by imperfect fungi such as *Aspergillus ruber*, *Penicillium islandicum* and *Beauvaria bassiana*, by ascomycetes such as *Chaetomium* sp., and by basidiomycetes such as the very rich source of glucanases, 'Basidiomycete sp. QM 806' [Monaghan, 1975].

Bacteriolytic Polysaccharases

Two classes of enzyme hydrolyse the polysaccharide backbone of peptidoglycan: endo-β-*N*-acetylmuramidases (= lysozymes), and endo-β-*N*-acetylglucosaminidases [Wadström, 1970; Rogers, Chapter 10]. As stated above, the preparations of these two enzymes, and of chitinase, show a continuum of activities. This account is restricted to those enzymes whose role is probably to act extracellularly to digest bacterial walls.

Nearly all of the fruiting myxobacteria are bacteriolytic. Thus Sudo and Dworkin [1972] describe extracellular endo-β-*N*-acetylglucosaminidase activities from *Myxococcus xanthus*. Such enzymes clearly have a nutritional role, but are probably very similar to the endo-*N*-acetylglucosaminidases whose major role may well be autolytic, such as that from *Clostridium perfringens* which Martin and Kemper [1970] describe as active on Gram-negative peptidoglycans but not on *Micrococcus lysodeikticus*; and that from *Staphylococcus aureus* active on both [Wadström, 1970]. Leonenko *et al.* [1974] term the latter enzyme a chitinase, and characterize it as a lipoprotein with 80% lipid content. Bacteriolysis is a widespread property among Actinomycetes [Stolp and Starr, 1965], but the enzymes have not been investigated in detail. The lysis of dead or living bacteria does not seem to be a mode of nutrition favoured by filamentous fungi or yeasts, but the slime moulds, Myxomycetes and Acrasiales, as well as soil amoebae, avidly engulf and digest bacteria. Thus a powerful bacteriolytic β-*N*-acetylhexosaminidase has been described from *Dictyostelium discoideum* [Every and Ashworth, 1973], as have a bacteriolytic glycanase from the amoeba *Hartmanella globae* [Upadhyay and Difulco, 1972], and a lysozyme from *Acanthamoeba castellanii* [Drozanski, 1972].

A remarkable group of lysozymes are those coded for by the genomes of bacteriophages, such as that produced by *Escherichia coli* infected by T4 [Matthews and Remington, 1974], and the phage-induced lysozyme in *Salmonella typhimurium* [Rao and Burma, 1971] [see review by Tsugita, 1971].

Agarases and Alginases

Algae, particularly the seaweeds, produce large quantities of a range of characteristic polysaccharides [Percival and McDowell, 1967]. Most work on the microbial degradation of these has concentrated on the alginates, produced by the brown algae and widely used as gels in foods and other industries, and the agars, produced by species of the red algae and widely used as gels for microbial growth media, for chromatography and electrophoresis, and in foods.

From different seaweeds, there are a range of structures in agaroses and agaropectins, the constituents of agar, but the structure of agarose is chiefly alternating α 1,4-linked 3,6-anhydro-L-galactose (or L-galactose) and β 1,3-linked D-galactose (or 6-O-methyl-D-galactose) [Percival and McDowell, 1967]. Thus agarases are of two types, those that cleave the α-link and those that cleave the β-link. Related red-algal polysaccharides are porphyran and carrageenan, which are sulphated to different degrees [Percival and McDowell, 1967]. Alginic acid consists of polyuronide chains of β 1,4-linked D-mannuronic acid and L-guluronic acid.

As these polysaccharides are marine products, it is not surprising that most microbes degrading them are isolated from marine sources. They are also found away from the sea, for example in sewage [Hofsten and Malmquist, 1975] and soil [Sampietro and Sampietro, 1971] but the ability to produce agarases and alginases is a rare attribute in non-marine organisms. Agarolytic marine cytophagas are regularly isolated [Stanier, 1941] and Christensen [1977] lists nine of fifteen species as having the ability. Of these nine, the six that have been tested are also alginolytic. Of the non-agarolytic species in Christensen's list, two have had agarolytic varieties described by Veldkamp [1961]. Several of the species are facultative anaerobes. Turvey and Christison [1967] isolated a *Cytophaga* species from seawater that degraded porphyran, and an extracellular agarase from this species has been characterized [Duckworth and Turvey, 1969; Young *et al.*, 1971]. Meulen *et al.* [1974] isolated a new agarolytic species, *Cytophaga flevensis*, from freshwater. This species can also utilize inulin and pectin, but not chitin, starch, alginate or cellulose. The extracellular enzyme has been further characterized and shown to be a β-agarase, the enzyme specific for the β 1,3 linkage [Meulen and Harder, 1975].

Marine agarolytic and alginolytic Gram-negative rods have been isolated and attributed to the genera *Pseudomonas* and *Vibrio* [see, for example Stanier, 1941; Girard *et al.*, 1968] but their taxonomic status remains to be established. Yaphe [1962] has isolated two strains, designated *Pseudomonas atlantica* and *Pseudomonas carrageenovora*, that have been the subjects of detailed work on the enzymic degradation of seaweed polysaccharides. *Pseudomonas atlantica* degrades agar and alginate, producing an agarase complex of at least three enzymes [Young *et al.*, 1971; Day and Yaphe, 1975]. *Pseudomonas carrageenovora* degrades alginates and carrageenan, but not agar. It produces kappa- and lambda-carrageenases, to hydrolyse both forms of carrageenan [Weigl and Yaphe, 1966; Johnson and McCandless, 1973]. However, these bacteria may degrade alginates by lyases, not by hydrolases [F.B. Williamson, personal communication]. Both strains can also degrade fucoidan (a sulphated fucose polymer produced by brown algae) when grown on it as sole carbon source [Yaphe and Morgan, 1959].

Yaphe [1963] criticized the contemporary practice of classifying such Gram-negative bacteria by their utilization of a particular polysaccharide, with the consequent erection of genera such as *Agarbacterium* and *Alginomonas*. Thus of 29 marine agarolytic isolates, Girard *et al.* [1968] found that 27 also used alginate, but only 13 used starch as a substrate. The species *Agarbacterium uliginosum* is now described as *Flavobacterium uliginosum*, and may be related to the cytophagas [Weeks, 1974].

Isolates of *Bacillus* and *Clostridium* have been described that are agarolytic or alginolytic; Vattuone and Sampietro [1973] describe the agarase of a *Bacillus* species, and Hunger and Claus [1978] re-describe the 'lost' species *Bacillus agar-exedens*; Billy [1965] isolated an alginolytic species of *Clostridium* from black mud, where it was associated with *Desulfovibrio desulfuricans*.

Finally, some agarolytic actinomycete strains have been described. Stanier [1942b] reports that the ability to hydrolyse agar was an unstable character in his isolates of a strain that he designated of the *Actinomyces coelicolor* species group. Only about 5% of colonies were agarolytic, whether or not they had been sub-cultured from an agarolytic colony.

Pectinases

Pectin is the methyl-esterified α 1,4-linked D-galacturonan that is a major part of the matrix material of the cell walls of higher plants. Microbial enzymes degrade this material in a number of ways [Rexova-Benkova and Markovic, 1976]: by pectin esterases converting the methyl galacturonate residues to galacturonate residues, that is pectin to pectate; by endo- and exo-lyases acting on the resulting pectate; and by endo- and exo-hydrolases acting on the pectate. This account will only consider the endo-polygalacturonases, or 'endogalacturonanases'.

These enzymes are very important in plant pathogenicity [see Byrde, Chapter 18], and are produced by most plant pathogens, as well as saprophytes growing on plant material. However, the hydrolases seem not to be produced as widely as the lyases by bacteria. Thus species of *Cytophaga*, *Xanthomonas*, *Aeromonas*, *Flavobacterium* and *Streptomyces* have been reported as having lyase but not hydrolase activity [Fogarty and Ward, 1974].

Amongst the Gram-negative bacteria, endopolygalacturonase activities have been described from the plant pathogenic isolates of *Pseudomonas marginalis* (= *Pseudomonas fluorescens* biotype II) [Nasuno and Starr, 1966a] and *Erwinia carotovora* [Nasuno and Starr, 1966b]; and amongst the Gram-positive bacteria from an alkalophilic soil isolate *Bacillus* No. P-4-N [Horikoshi, 1972], and isolates of *Clostridium felsineum* [Lund and Brocklehurst, 1978]. These latter isolates had lyase but not pectin esterase activities, whereas isolates of potato-rotting *Clostridium* species and *Clostridium aurantibutyricum* tested at the same time produced lyase and esterase but no hydrolase activities. Amongst the fungi, hydrolase activity is more widely distributed than lyase activity, being found in species of oomycotina, zygomycotina, ascomycotina, basidiomycotina and the imperfect fungi [Fogarty and Ward, 1974]. Several of the endopolygalacturonases have been purified from fungi, and as with the fungal cellulases they show a range of activities and molecular properties.

Conclusions

These polysaccharases have some similarities: they tend to be fairly small proteins, typically with molecular weights less than 5×10^5; they tend to be single polypeptide chains, and not multiple sub-unit enzymes; many are glycoproteins. They tend to be produced as families of enzymes by particular organisms, and it is found that several enzyme activities are required for the complete hydrolysis of ordered substrates such as cellulose and heteropolysaccharides such as the hemicelluloses, and for the degradation of polysaccharides as they occur in nature: in a heterogeneous state - in association with and perhaps covalently linked to other molecules, and secondarily modified by chemical substituents - for example the pectin/hemicellulose/cellulose mixtures in the higher plant cell wall. In general, they are specific for a particular sugar and a particular conformation and linkage, exceptions being for example the enzyme from *Myxobacter* AL-1 that hydrolyses β 1,4-linked glucose and β 1,4-linked glucosamine in cellulose and chitosan [Hedges and Wolfe, 1974] and the glucanase from *Rhizopus arrhizus* that hydrolyses β 1,4 linkages as well as the preferred β 1,3 linkages [Clark *et al.*, 1978].

In some cases, the ability to produce a particular polysaccharase is a very stable character in culture, for example cellulases of *Cellulomonas* isolates [Keddie, 1974]; in others, this ability is irreversibly 'lost' in repeated sub-culture, for example cellulases of isolates of *Eubacterium cellulosolvens* [Van Glyswyk and Labuschagne, 1971].

The taxonomic spectra of organisms capable of hydrolysing extracellular polysaccharides are similar, with genera such as *Cytophaga*, *Bacillus* and *Clostridium*, and *Streptomyces* having the lion's share of the activities among the prokaryotes. In these cases it is tempting to speculate that families of polysaccharases with a range of activities arose from ancestral ones with limited activities. Thus the active site, hydrolysing a glycosidic linkage, would take on a specificity for a different polysaccharide by alterations in its binding sites. Further speculation will have to await knowledge of protein sequences and tertiary structures of these enzymes.

Lysozyme is the only polysaccharase for which we have estimates of a rate of evolution. This enzyme has been fully characterized from bacteriophages, from the egg whites of birds, and from mammals. There are sufficient sequences known for lysozyme c, found in mammals and some birds notably the hen, to show that it is subject to very rapid evolution, with a value of 2.5 for its unit evolutionary period (the time in millions of years for a one per cent difference in amino acid sequence to arise between two lineages) [Wilson *et al.*, 1977]. Values for intracellular enzymes are typically much larger, for example 55 for glutamate dehydrogenase. Wilson *et al.* discuss why its rate of evolution is so fast, and suggest that for vertebrates it might be a dispensible protein, so that functional constraints would have little action in conservation of its structure. This might also explain the very large differences in lysozyme concentration found between different species, for example with human tear glands producing very much more than those of rabbit. Indeed, in mammals, lysozyme c has given rise to a homologous non-enzymic protein with a quite different function - the lactalbumin that regulates lactose synthesis in mammary gland. Lysozyme g is present in the egg white of goose and some other birds [Prager *et al.*, 1974]. It has similar enzymic activity to lysozyme c, but has no sequences in common - the two proteins do not share a common ancestor - they are analogous not homologous. The lysozyme of bacteriophage T4 is yet another analogous protein, which despite an early suggestion to the contrary [Dunnill, 1967] has no amino acid sequence similarities to hen egg white lysozyme c, but has similar enzymic activity

[Matthews and Remington, 1974]. Lysozymes from some invertebrates have kinetic properties closest to those of lysozyme c, while those from plants have properties closer to those of lysozyme g and T4 lysozyme [Jollès *et al.*, 1974]. Unfortunately, amino acid sequences for these or for bacterial lysozymes are not yet available to allow wider phylogenetic relationships to be assessed.

It is clear however, that the enzymic activity of lysozyme has arisen completely separately at least three times and is thus polyphyletic. This is in contrast to the monophyletic evolution of some intracellular proteins such as cytochrome c. There are some remarkable interrelationships between serine proteinases of very different organisms [Hartley, 1974]. Thus *Streptomyces griseus* has a trypsin activity with a 43% sequence homology with that of cow, as well as elastase- and subtilisin-type enzymes that can be related to those of mammals. Hartley states 'the obvious conclusion is that each of these structures is a consequence of great evolutionary conservatism', but cautions that the closeness of structure in the trypsins leads to the suspicion that some form of gene transfer might have occurred between very unrelated organisms.

Certainly the examples of very sporadic occurrence of polysaccharases in some microbes suggest the likelihood of their original acquisition via plasmids. Thus among the many *Pseudomonas* species, only a handful utilize starch, and these by different polysaccharases [Norrman and Wöber, 1975], and only a few isolates of a very few species can utilize cellulose, or chitin, or agar. Again, protein sequences and tertiary structures of these enzymes would be of great interest.

References

Aribisala, O.A. and Gooday, G.W. (1978). Chitinolytic bacteria from the Nigerian shrimp *Panaeus duorarum*. *Proceedings of the Society for General Microbiology* **5**, 106.

Bailey, R.W. and Gaillard, B.D.E. (1965). Carbohydrases of the rumen ciliate *Epidinium ecaudatum* (Crawley). Hydrolysis of plant hemicellulose fractions and β-linked glucose polymers. *Biochemical Journal* **95**, 758-766.

Billy, C. (1965). Etude d'une bactérie alginolytique anaérobie: *Clostridium alginolyticum* n. sp. *Annales de l'Institut Pasteur (Paris)* **109**, 147-151.

Brown, R.G. (1973). Dextranases and β-glucanases. In *Handbook of Microbiology*, Vol. 3, pp.631-636. Edited by A.L. Laskin and H.A. Lechavalier. Cleveland: CRC Press.

Bull, A.T. and Chesters, C.G.C. (1966). Biochemistry of laminarin and the nature of laminarinase. *Advances in Enzymology* **28**, 325-364.

Campbell, L.C. and Williams, O.B. (1951). A study of chitin-decomposing micro-organisms of marine origin. *Journal of General Microbiology* **5**, 894-905.

Chan, J.G. (1970). The occurrence, taxonomy and activity of chitinoclastic bacteria from sediment, water and fauna of Puget Sound. PH.D. Thesis. University of Washington, U.S.A.

Christensen, P.J. (1977). The history, biology, and taxonomy of the *Cytophaga* group. *Canadian Journal of Microbiology* **23**, 1599-1653.

Clark, D.R., Johnson, J., Chung, K.H. and Kirkwood, S. (1978). Purification, characterization, and action-pattern studies on the endo-(1-3)-β-D-glucanase from *Rhizopus arrhizus* QM 1032. *Carbohydrate Research* **61**, 457-477.

Clarke, R.T.J., Bailey, R.W. and Gaillard, B.D.E. (1969). Growth of rumen bacteria on plant cell wall polysaccharides. *Journal of General Microbiology* **56**, 79-86.

Clarke, P.A. and Tracey, M.V. (1956). The occurrence of chitinase in some bacteria. *Journal of General Microbiology* **14**, 188-196.

Clermont, S., Charpentier, M. and Percheron, F. (1970). Polysaccharidases der *Sporocytophaga myxococcoides*, β-mannanase, cellulase et xylanase. *Bulletin de la Societe de Chimie Biologique* **52**, 1481-1495.

Day, D.F. and Yaphe, W. (1975). Enzymatic hydrolysis of agar: purification and characterisation of neoagarobiose hydrolase and pNP-α-galactoside hydrolase. *Canadian Journal of Microbiology* **21**, 1512-1518.

Dehority, B.A. (1967). Rate of isolated hemicellulose degradation and utilization by pure cultures of rumen bacteria. *Applied Microbiology* **15**, 987-993.

Dekker, R.F.H. and Richards, G.N. (1976). Hemicellulases: their occurrence, purification, properties, and mode of action. *Advances in Carbohydrate Chemistry and Biochemistry* **32**, 277-352.

Drozanski, W. (1972). Enzymatic mode of action of the bacteriolytic enzyme from *Acanthamoeba castellanii*. *Acta Microbiologica Polonica Series A* **4**, 53-62.

Duckworth, M. and Turvey, J.R. (1969). An extracellular agarase from a *Cytophaga* species. *Biochemical Journal* **113**, 139-142.

Dunnill, P. (1967). Sequence similarities between hen egg white and T4 phage lysozymes. *Nature, London* **215**, 621-622.

Emi, S. and Yamamoto, T. (1972). Purification and properties of several galactanases of *Bacillus subtilis* var. *amylosaccharitícus*. *Agricultural and Biological Chemistry (Japan)* **36**, 1945-1954.

Eriksson, K.E. and Pettersson, B. (1975). Extracellular enzyme system utilized by the fungus *Sporotrichum pulverulentum (Chrysosporium lignorum)* for the breakdown of cellulose. 1. Separation, purification and physico-chemical characterization of five endo-β-1,4-glucanases. *European Journal of Biochemistry* **51**, 193-206.

Every, D. and Ashworth, J.M. (1973). The purification and properties of extracellular glycosidases of the cellular slime mould *Dictyostelium discoideum*. *Biochemical Journal* **133**, 37-47.

Ewing, W.H., Davis, B.R., Fife, M.A. and Lessel, E.F. (1973). Biochemical characterisation of *Serratia liquefaciens* (Grimes and Hennerty) Bascomb *et al.* (formerly *Enterobacter liquefaciens*) and *Serratia rubidaea* (Stapp) *comb. nov.* and designation of type and neotype strains. *International Journal of Systematic Bacteriology* **23**, 217-225.

Fenton, D., Davis, B., Rotgers, C. and Eveleigh, D.E. (1978). Enzymatic hydrolysis of chitosan. In *Proceedings of the First*

International Conference on Chitin and Chitosan, pp.525-541. Edited by R.A.A. Muzzarelli and F.R. Pariser, MIT Sea Grant Programme Report MITSG 78-7.

Fergus, C.L. (1969). The cellulolytic activity of thermophilic fungi and actinomycetes. *Mycologia* **61**, 120-129.

Fogarty, W. and Griffin, P.J. (1973). Some preliminary observations on the production and properties of cellulolytic enzyme elaborated by *Bacillus polymyxa*. *Biochemical Society Transactions* **1**, 1297-1298.

Fogarty, W.M. and Ward, O.P. (1973). A preliminary study on the production, purification and properties of a xylan-degrading enzyme from a *Bacillus* sp. isolated from water-stored sitka spruce (*Picea sitchensis*). *Biochemical Society Transactions* **1**, 260-262.

Fogarty, W.M. and Ward, O.P. (1974). Pectinases and pectic polysaccharides. *Progress in Industrial Microbiology* **13**, 59-119.

Fujisawa, H. and Murakami, M. (1971). Studies on xylan-decomposing bacteria in the marine environment - V. Grading xylan-decomposing activity of β-1,3-xylan-decomposing bacteria by the determination of the P.S. value. *Bulletin of the Japanese Society of Scientific Fisheries* **37**, 119-123.

Gascoigne, J.A. and Gascoigne, M.M. (1960). *Biological Degradation of Cellulose*. London: Butterworths.

Girard, A.E., Buck, J.D. and Cosenza, B.J. (1968). A nutritional study of some agarolytic marine bacteria. *Canadian Journal of Microbiology* **14**, 1193-1198.

Goksøyr, J., Eidsa, G., Eriksen, J. and Osmundsvag, K. (1975). A comparison of cellulases from different microorganisms. In *Symposium on Enzymatic Hydrolysis of Cellulose*, pp.217-230. Edited by M. Bailey, T-M. Enari and M. Linko. Helsinki: Finnish National Fund for Research and Development.

Goodrich, T.D. and Morita, R.Y. (1977). Bacterial chitinase in the stomachs of marine fishes from Yaquina Bay, Oregon U.S.A. *Marine Biology* **41**, 355-360.

Hartley, B.S. (1974). Enzyme families. In *Evolution in the Microbial World*, pp.151-162. Edited by M.J. Carlile and J.J. Skehel. *Symposia of the Society for General Microbiology* No. 24. Cambridge: Cambridge University Press.

Hedges, A. and Wolfe, R.S. (1974). Extracellular enzyme from myxobacter AL-1 that exhibits both β-1,4-glucanase and chitosanase activities. *Journal of Bacteriology* **120**, 844-853.

Hendrie, M.S., Hodgkiss, W. and Shewan, J.M. (1970). The identification, taxonomy and classification of luminous bacteria. *Journal of General Microbiology* **64**, 151-169.

Hofsten, B. and Malmquist, M. (1975). Degradation of agar by a Gram-negative bacterium. *Journal of General Microbiology* **87**, 150-158.

Hood, M.A. and Meyers, S.P. (1977). Microbiological and chitinoclastic activities associated with *Panaeus setiferus*. *Journal of the Oceanographical Society of Japan* **33**, 235-241.

Horikoshi, K. (1972). Production of alkaline enzymes by alkalophilic micro-organisms. Part III. Alkaline pectinase of *Bacillus* No. P-4-N. *Agricultural and Biological Chemistry* **36**, 285-293.

Huber, J. (1958). Untersuchungen zur Physiologie insektentötender Pilze. *Archiv für Mikrobiologie* **29**, 257-276.

Hungate, R.E. (1950). The anaerobic mesophilic cellulolytic bacteria. *Bacteriological Reviews* **14**, 1-49.

Hunger, W. and Claus, D. (1978). Reisolation and growth conditions of *Bacillus agar-exedens*. *Antonie van Leeuwenhoek Journal of Microbiology and Serology* **44**, 105-113.

Imschenezki, A.A. (1959). *Mikrobiologie der Cellulose*. Berlin: Akademie Verlag.

Inaoka, M. and Soda, H. (1956). Crystalline xylanase. *Nature, London* **178**, 202-203.

Jeuniaux, C. (1963). *Chitine et Chitinolyse*. Paris: Masson et Cie.

Johnson, D.E. (1932). Some observations on chitin-destroying bacteria. *Journal of Bacteriology* **24**, 335-340.

Johnson, K.H. and McCandless (1973). Enzymic hydrolysis of the potassium chloride soluble fraction of carrageenan: properties of "λ-carrageenases" from *Pseudomonas carrageenovora*. *Canadian Journal of Microbiology* **19**, 779-788.

Jollès, P., Bernier, I., Berthou, J., Charlemagne, P., Faure, A., Hermann, J., Jollès, J., Perin, J.P. and Saint-Blancard, J. (1974). From lysozyme to chitinases: structural, kinetic, and crystallographic studies. In *Lysozyme*, pp.31-54. Edited by E.F. Osserman, R.E. Confield and S. Begchoke. New York and London: Academic Press.

Kaji, A., Anabuki, Y., Taki, H., Oyama, Y. and Okada, T. (1963). Studies on the enzymes acting on arabans. Action and separation of arabanase produced by *Clostridium felsineum (Eubacteriales)*. *Kagawa Daigaka Nogakubu Gabuzyuta Hokuku* **15**, 40-44.

Kaneko, T. and Colwell, R.R. (1973). Ecology of *Vibrio parahaemolyticus* in Chesapeake Bay. *Journal of Bacteriology* **113**, 24-32.

Keddie, R.M. (1974). Cellulomonas. In *Bergey's Manual of Determinative Bacteriology*, pp.629-631. Edited by R.E. Buchanan and N.E. Gibbons. Baltimore: Williams and Wilkins.

King, K.W. and Vessal, M.I. (1969). Enzymes of the cellulose complex. *Advances in Chemistry Series* **95**, 7-25.

Küster, D. (1973). Techniques for the isolation and identification of aerobic Actinomycetales: summary of the discussion session. In *Actinomycetales: Characteristics and Practical Importance*, pp.335-339. Edited by G. Sykes and F.A. Skinner. London: Academic Press.

Lacey, J. (1973). Actinomycetes in soils, composts and fodders. In *Actinomycetales: Characteristics and Practical Importance*, pp.231-251. Edited by G. Sykes and F.A. Skinner. London: Academic Press.

Leadbetter, E.R. (1974). Cytophagaceae. In *Bergey's Manual of Determinative Bacteriology*, pp.99-112. Edited by R.E. Buchanan and N.E. Gibbons. Baltimore: Williams and Wilkins.

Lee, E.Y.C. and Whelan, W.J. (1971). Glycogen and starch debranching enzymes. In *The Enzymes*, Vol. 5, pp.191-234. Edited by P.D. Boyer. New York: Academic Press.

Leonenko, V.A., Kravchenko, N.A., Cherkasov, I.V., Afanaseva, T.I. and Ermoleva, Z.V. (1974). Isolation of endo-β-(1-4)-N-acetylglucosaminidase from culture liquid of *S. aureus* by method of affine chromatography and its properties. *Biochemistry U.S.S.R.* **39**, 912-916.

Leopold, J. and Samsinakova, A. (1970). Quantitative estimation of chitinase and several other enzymes in the fungus *Beauveria bassiana*. *Journal of Invertebrate Pathology* **15**, 34-42.

Lewin, R.A. (1970). New *Herpetosiphon* species (Flexibacteriales). *Canadian Journal of Microbiology* **16**, 517-520.

Littlejohn, J.H. (1973). Amylases of microbial origin. In *Handbook of Microbiology*, Vol. 3, pp.651-658. Edited by A.L. Laskin and H.A. Lechavalier. Cleveland: CRC Press.

Lund, B.M. and Brocklehurst, T.F. (1978). Pectic enzymes of pigmented strains of *Clostridium*. *Journal of General Microbiology* **104**, 59-66.

McIntyre, J.L. and Hankin, L. (1978). An examination of enzyme production by *Phytophthora* spp. on solid and liquid media. *Canadian Journal of Microbiology* **24**, 75-78.

Malwszynska, G.M. and Janota-Bassalik, J. (1974). A cellulolytic rumen bacterium *Micromonospora ruminatum* sp. nov. *Journal of General Microbiology* **82**, 57-65.

Mandels, M. (1975). Microbial sources of cellulases. In *Cellulose as a Chemical and Energy Resource*, pp.81-105. Edited by C.R. Wilke. New York: John Wiley.

Martin, H.H. and Kemper, S. (1970). Endo-*N*-acetyl-glucosaminidase from *Clostridium perfringens*, lytic for cell wall murein of Gram-negative bacteria. *Journal of Bacteriology* **102**, 347-350.

Matthews, B.W. and Remington, S.J. (1974). The three dimensional structure of lysozyme from bacteriophage T4. *Proceedings of the National Academy of Sciences of the United States of America* **71**, 4178-7182.

Meulen, H.J. van der and Harder, W. (1975). Production and characterization of the agarase of *Cytophaga flevensis*. *Antonie van Leeuwenhoek Journal of Microbiology and Serology* **41**, 431-447.

Meulen, H.J. van der, Harder, W. and Veldkamp, H. (1974). Isolation and characterization of *Cytophaga flevensis* sp. nov., a new agarolytic flexibacterium. *Antonie van Leeuwenhoek Journal of Microbiology and Serology* **40**, 329-346.

Molise, E.M. and Drake, C.H. (1973). Chitinolysis by Serratiae including *Serratia liquefaciens (Enterobacter liquefaciens)*. *International Journal of Systematic Bacteriology* **23**, 278-280.

Monaghan, R.L. (1975). *The Discovery, Distribution and Utilization of Chitosanase*. Ph.D. Thesis, Rutgers University, U.S.A.

Monaghan, R.L., Eveleigh, D.E., Tewari, R.P. and Reese, E.T. (1973). Chitosanase, a novel enzyme. *Nature New Biology* **245**, 78-80.

Monreal, J. and Reese, E.T. (1969). The chitinase of *Serratia marcescens*. *Canadian Journal of Microbiology* **15**, 689-696.

Murray, C.L. and Lovett, J.S. (1966). Nutritional requirement of the chytrid, *Karlingia astereocysta*, an obligate chitinophile. *American Journal of Botany* **53**, 469-476.

Nasuno, S. and Starr, M.P. (1966a). Pectic enzymes of *Pseudomonas marginalis* (soft-rot pathogen isolated from *Cichorium intybus* and *Lobelia erinus*). *Phytopathology* **56**, 1414-1415.

Nasuno, S. and Starr, M.P. (1966b). Polygalacturonase of *Erwinia carotovora*. *Journal of Biological Chemistry* **241**, 5298-5306.

Norman, A.G. and Fuller, W.H. (1942). Cellulose decomposition by microorganisms. *Advances in Enzymology* **2**, 239-264.

Norrman, J. and Wöber, G. (1975). Comparative biochemistry of α-

glucan utilization in *Pseudomonas amyloderamsoa* and *Pseudomonas saccharophila*. Physiological significance of variations in the pathways. *Archives of Microbiology* **102**, 253-260.
Okada, G., Nisizawa, K. and Suzuki, H. (1968). Cellulase components from *Trichoderma viride*. *Journal of Biochemistry* **63**, 591-607.
Okafor, N. (1966). Ecology of micro-organisms on chitin buried in soil. *Journal of General Microbiology* **44**, 311-327.
Osmundsvag, K. and Goksøyr, J. (1975). Cellulases from *Sporocytophaga myxococcoides*. Purification and properties. *European Journal of Biochemistry* **57**, 405-409.
Park, D. (1975). A cellulolytic pythiaceous fungus. *Transactions of the British Mycological Society* **65**, 249-257.
Percival, E. and McDowell, R.H. (1967). *Chemistry and Enzymology of Marine Algal Polysaccharides*. London: Academic Press.
Peterson, J.E. (1974). Polyangium cellulosum. In *Bergey's Manual of Determinative Bacteriology*, pp.94-95. Edited by R.E. Buchanan and N.E. Gibbons. Baltimore: Williams and Wilkins.
Prager, E.M., Wilson, A.C. and Arnheim, N. (1974). Widespread distribution of lysozyme g in egg white of birds. *Journal of Biological Chemistry* **249**, 7295-7297.
Price, J.A. and Storck, R. (1975). Production, purification, and characterization of an extracellular chitosanase from *Streptomyces*. *Journal of Bacteriology* **124**, 1574-1585.
Rao, G.R.K. and Burma, D.P. (1971). Purification and properties of phage P22-induced lysozyme. *Journal of Biological Chemistry* **246**, 6474-6479.
Reese, E.T. (1969). Estimation of exo-β-1- 4-glucanase in crude cellulase solutions. *Advances in Chemistry Series* **95**, 26-33.
Reese, E.T., Levinson, H.S., Downing, M.H. and White, W.L. (1950). Quartermaster culture collection. *Farlowis* **4**, 45-86.
Reisert, P.S. and Fuller, M.S. (1962). Decomposition of chitin by *Chytridiomyces* species. *Mycologia* **54**, 647-657.
Rexova-Benkova, L. and Markovic, O. (1976). Pectin Enzymes. *Advances in Carbohydrate Chemistry and Biochemistry* **33**, 323-385.
Sakai, Y., Okawa, K. and Kamiyama, Y. (1975). A method of determining penta-, tetra-, tri- and disaccharides from xylan by ion-exchange chromatography. *Agricultural and Biological Chemistry (Japan)* **31**, 545-546.
Sampietro, A.R. and Sampietro, M.A.V. de (1971). Characterisation of the agarolytic system of *Agarbacterium pastinator*. *Biochemica et Biophysica Acta* **244**, 65-76.
Skujins, J., Pukite, A. and McLaren, A.D. (1970). Chitinase of *Streptomyces* sp.: purification and properties. *Enzymologia* **39**, 353-369.
Somkuti, G.A., Babel, F.J. and Somkuti, A.C. (1969). Cellulolysis by *Mucor pusillus*. *Applied Microbiology* **17**, 888-892.
Spencer, R. (1961). Chitinoclastic activity in the luminous bacteria. *Nature, London* **190**, 938.
Stanier, R.Y. (1941). Studies on marine agar-digesting bacteria. *Journal of Bacteriology* **42**, 527-559.
Stanier, R.Y. (1942a). The *Cytophaga* group: a contribution to the biology of myxobacteria. *Bacteriological Reviews* **6**, 143-196.

Stanier, R.Y. (1942b). Agar-decomposing strains of the *Actinomyces coelicolor* species-group. *Journal of Bacteriology* **44**, 555-569.
Stanier, R.Y. (1947). Studies on nonfruiting myxobacteria. 1. *Cytophaga johnsonae*, n.sp. a chitin-decomposing myxobacterium. *Journal of Bacteriology* **53**, 297-315.
Stewart, B.J. and Leatherwood, J.M. (1976). Derepressed synthesis of cellulase by *Cellulomonas*. *Journal of Bacteriology* **128**, 609-615.
Stirling, J. (1979). Chitinase. In *Fungal Wall and Hyphal Growth*. Edited by J.H. Burnett and A.P.J. Trinci. Cambridge: Cambridge University Press.
Stolp, H. and Starr, M.P. (1965). Bacteriolysis. *Annual Review of Microbiology* **19**, 79-104.
Stutzenberger, F.J. (1972). Cellulolytic activity of *Thermomonospora curvata:* Optimal assay conditions, partial purification, and product of the cellulase. *Applied Microbiology* **24**, 83-90.
Sudo S. and Dworkin, M. (1972). Bacteriolytic enzymes produced by *Myxococcus xanthus*. *Journal of Bacteriology* **110**, 236-245.
Sundarraj, N. and Bhat, J.V. (1972). Breakdown of chitin by *Cytophaga johnsonii*. *Archiv für Mikrobiologie* **85**, 159-167.
Takagi, T., Toda, H. and Isemura, T. (1971). Bacterial and mold amylases. In *The Enzymes*, Vol. 5, pp.235-290. Edited by P.D. Boyer. New York: Academic Press.
Thaysen, A.C. and Bunker, H.J. (1927). The *Microbiology of Cellulase, Hemicelluloses, Pectin and Gums*. London: Oxford University Press.
Thomas, D.S. and Mullins, J.T. (1969). Cellulase induction and wall extension in the water mold *Achlya ambisexualis*. *Physiologia Plantarum* 22, 347-353.
Timmis, K., Hobbs, G. and Berkeley, R.C.W. (1974). Chitinolytic clostridia isolated from marine mud. *Canadian Journal of Microbiology* **20**, 1284-1285.
Tracey, M.V. (1955). Cellulase and chitinase in soil amoebae. *Nature, London* **175**, 815.
Tseng, T.C. (1974). Multiple cellulases produced by a soft rot bacterium *Erwinia carotovora*. *Botanical Bulletin of Academic Sinaca (Taipei)* **15**, 49-53.
Tsugita, A. (1971). Phage lysozyme and other lytic enzymes. In *The Enzymes*, Vol. 5, pp.343-411. Edited by P.D. Boyer. New York: Academic Press.
Tsujisaka, Y., Tominaga, Y. and Iwai, M. (1975). Purification and some properties of the lytic enzyme from *Bacillus* R-4 which acts on *Rhizopus* cell wall. *Agricultural and Biological Chemistry (Japan)* **39**, 145-152.
Turvey, J.R. and Christison, J. (1967). The hydrolysis of algal galactans by enzymes from a *Cytophaga* species. *Biochemical Journal* **105**, 311-316.
Unestam, T. (1966). Chitinolytic, cellulolytic and pectinolytic activity in vitro of some parasitic and saprophytic oomycetes. *Physiologia Plantarum* **19**, 15-30.
Upadhyay, J.M. and Difulco, T.J. (1972). Isolation, purification and properties of a second bacteriolytic enzyme from *Hartmanella glebae*. *Archives of Biochemistry and Biophysics* **149**, 470-475.
Van Gylswyk, N.O. and Labuschagne, J.P.L. (1971). Relative efficiency of pure cultures of different species of cellulolytic rumen

bacteria in solubilizing cellulose *in vitro*. *Journal of General Microbiology* **66**, 109-113.

Vattuone, M.A. and Sampietro, M.A.R. (1973). Presence de systemes hydrolysant de liaisons β (1-4) de l'agarose dans le genre *Bacillus*. *Comptes Rendus de l'Academie des Sciences* **276D**, 3225-3228.

Veldkamp, H. (1955). A study of the aerobic decomposition of chitin by microorganisms. *Mededelingen van de Landbouwhogeschool te Wageningen* **55**, 127-174.

Veldkamp, H. (1961). A study of two marine agar-decomposing, facultatively anaerobic myxobacteria. *Journal of General Microbiology* **26**, 331-342.

Vries, O.M.H. de and Wessels, J.G.H. (1973). Release of protoplasts of *Schizophyllum commune* by combined action of purified α 1-3 glucanase and chitinase derived from *Trichoderma viride*. *Journal of General Microbiology* **76**, 319-330.

Wadström, T. (1970). Bacteriolytic enzymes from *Staphylococcus aureus* Properties of the endo-β-*N*-acetylglucosaminidase. *Biochemical Journal* **120**, 745-752.

Weeks, O.B. (1974). Flavobacterium. In *Bergey's Manual of Determinative Bacteriology*, pp.357-364. Edited by R.E. Buchanan and N.E. Gibbons. Baltimore: Williams and Wilkins.

Weigl, J. and Yaphe, W. (1966). The enzymic hydrolysis of carrageenan by *Pseudomonas carrageenovora:* purification of a K-carrageenase. *Canadian Journal of Microbiology* **12**, 939-947.

Weimer, P.J. and Zeikus, J.G. (1977). Fermentation of cellulose and cellobiose by *Clostridium thermocellum* in the absence and presence of *Methanobacterium thermoautotrophicum*. *Applied and Environmental Microbiology* **33**, 289-297.

Wilson, A.C., Carlson, S.S. and White, T.J. (1977). Biochemical evolution. *Annual Reviews of Biochemistry* **46**, 573-639.

Wöber, G. (1973). The pathway of maltodextrin metabolism in *Pseudomonas stutzeri*. *Hoppe-Seyler's Zeitschrift für Physiologische Chemie* **354**, 75-82.

Wood, T.M. (1975). Properties and mode of action of cellulase. In *Cellulose as a Chemical and Energy Resource*, pp.111-137. Edited by C.R. Wilke. New York: John Wiley.

Yamamoto, S. and Nagasaki, (1975). Purification, crystallization and properties of an endo β-1,3-glucanase from *Rhizopus chinensis* R-69. *Agricultural and Biological Chemistry* **39**, 2163-2169.

Yamane, K., Suzuki, H. and Nisizawa, K. (1970). Purification and properties of extracellular and cell-bound cellulase components of *Pseudomonas fluorescens* var. *cellulosa*. *Journal of Biochemistry* **67**, 19-35.

Yaphe, W. (1962). Detection of marine alginolytic bacteria. *Nature, London* **196**, 1120-1121.

Yaphe, W. (1963). Proposals on the classification of microorganisms which utilize the polysaccharides of marine algae and a definition for agar. In *Symposium on Marine Microbiology*, pp.588-593. Edited by C.H. Oppenheimer. Springfield: Charles C. Thomas.

Yaphe, W. and Morgan, K. (1959). Enzymic hydrolysis of fucoidin by *Pseudomonas atlantica* and *Pseudomonas carrageenovora*. *Nature, London* **183**, 761-762.

Young, K., Hong, K.C., Duckworth, M. and Yaphe, W. (1971). Enzymic hydrolysis of agar and properties of bacterial agarases. *Proceedings of the Seventh International Seaweed Symposium*, pp. 469-472. University of Tokyo Press.

ORGANISMS INDEX

SUBJECT INDEX